MW01284531

Logic Design of NanoICs

Svetlana N. Yanushkevich
University of Calgary

Vlad P. Shmerko
University of Calgary

and

Sergey Edward Lyshevski
Rochester Institute of Technology

CRC PRESS

Boca Raton London New York Washington, D.C.

Library of Congress Cataloging-in-Publication Data

Yanushkevich, Svetlana N.
Logic design of nonICs / Svetlana N. Yanushkevich, Vlad P. Shmerko, Sergey E. Lyshevski.
p. cm. — (Nano- and microscience, engineering, tehcnology, and medicine series)
Includes bibliographical references and index.
ISBN 0-8493-2766-0 (alk. paper)
1. Molecular electronics. 2. Logic design. 3. Nanotechnology. 4. Integrated circuits. I. Shmerko, Vlad P. II. Lyshevski, Sergey Edward. III. Title. IV. Series.

TK7874.8.Y36 2004
621.381--dc22 2004055360

Direct all inquiries to CRC Press, 2000 N.W. Corporate Blvd., Boca Raton, Florida 33431.

Visit the CRC Press Web site at www.crcpress.com

International Standard Book Number 0-8493-2766-0
Library of Congress Card Number 2004055360
Printed in the United States of America 1 2 3 4 5 6 7 8 9 0
Printed on acid-free paper

Logic Design of NanoICs

Nano- and Microscience, Engineering, Technology, and Medicine Series

---

Series Editor
Sergey Edward Lyshevski

Titles in the Series

MEMS and NEMS: Systems, Devices, and Structures
Sergey Edward Lyshevski

Microelectrofluidic Systems: Modeling and Simulation
Tianhao Zhang, Krishnendu Chakrabarty, and Richard B. Fair

Nano- and Micro-Electromechanical Systems: Fundamentals of Nano- and Microengineering
Sergey Edward Lyshevski

Nanoelectromechanics in Engineering and Biology
Michael Pycraft Hughes

Microdrop Generation
Eric R. Lee

Micro Mechatronics: Modeling, Analysis, and Design with MATLAB®
Victor Giurgiutiu and Sergey Edward Lyshevski

Contents

Preface

This book is aimed to analyze and design nano integrated circuits (nanoICs) that can be fabricated using nanotechnology.

Rationale and audience

Two approaches to the design of computer devices and systems can be recognized in nanoscience and nanotechnology:

The first approach is based on the principle of *assembling* a complex system from simple quantum-effect components. Since quantum effects in such a networked system are *local*, the global logic design inherits traditional paradigms. To design, for example, a massive parallel adder, one must choose a method, develop an algorithm, and design the logic circuit based on the library of cells implementing an elementary logic function, or based on the processing elements constituting computing arrays.

The key point of the second approach is a *global* modeling of computing processes by appropriate physical phenomena. These processes require nontrivial algorithms whose major principle is so-called quantum superposition, or coherent superposition of the two states of an atom. The design paradigm is based on the control of the chosen phenomena and data streams, including a representation of input and output data in the required forms (electrical, optical, or electromagnetic). For example, in the ion trap, laser pulse directed to the ion, cause electrostatic repulsion from an ion to other ions.

This book emphasizes and contributes to the first direction, i.e., design of nanoelectronic devices to document the technological feasibility in synthesis of complex two- and three-dimensional systems. We argue that from a fabrication viewpoint, complex nanoICs can be designed based on multiterminal nanoelectronic devices, and can be created using self-assembly and various synthesis processes. It draws connections to contemporary approaches to computer hardware design based on traditional Boolean algebra. However, it abstracts from the traditional orientation of very large scale integration (VLSI) technology and ultra large scale integration (ULSI) technology at gate-level implementation of switching functions, and characterizes structural requirements of existing and predictable nanoscale devices.

It is difficult to apply fundamental macroscale architectures and computer aided design (CAD) of integrated circuits (ICs) tools to the nanoscale domain. There has been an active pursuit of alternative computing paradigms, which employ intrinsically nanoscale components. Because today's quantum devices are weak and sensitive, they are not suited to conventional logic gate architectures, which require robust devices. The focus of investigation is the developing of design paradigms that can discover novel architectures, perform functional synthesis, as well as carry out modeling, analysis, design, and optimization for nanoICs. Progress is impossible without timely techniques and tools of CAD of nanoICs, carefully synchronized with current technology trends and opportunities.

With the motivation of making the field of logic design in nanodimensions accessible to engineers, the goals of this book are as follows:

To introduce the data structures that satisfy criteria of massive parallel processing, homogeneity, and fault tolerance; this enables an engineer to choose models appropriate to technological possibilities;

To introduce models and data structures for synthesis of circuits in nanodimensions; the focus on hypercube and hypercube-like structures;

To introduce methods for analysis of data structures and models in nanodimensions; the concept of Akers's change and Shannon's information theoretical measures are keys of the approach.

To achieve these goals, the authors deploy selected methods of contemporary logic design, as well as specific methods for design of nanoICs. These are built on a background with which most electrical engineers are already familiar: logic functions and methods of design of discrete devices. We introduce our vision of logic design of nanoICs through:

- ▶ Synthesis of spatial data structures and assembling topological models;
- ▶ Word-level technique, emphasizing linear word-level representation;
- ▶ Multilevel circuit design in spatial dimensions;
- ▶ Concept of change in spatial structures for analysis, i.e., event-driven analysis;
- ▶ Multiple-valued logic for circuit design;
- ▶ Fault tolerance computing in spatial structures;
- ▶ Systolic arrays of nanodevices; cellular arrays and neural-like networks;
- ▶ Information theoretical measures in spatial structures.

The key features that distinguish this book from others include the following:

The central role of topological models (embedding in hypercubes, assembling the topology);

The revised technique of advanced logic design (algebraic, matrix and graph-based) with respect to new possibilities of processing in spatial dimensions;

Innovative ideas and solutions inspired by recent advents in nanotechnologies, such as a stochastic character of calculations in nanodevices. The best of information theoretic facilities and stochastic computing are applied to modeling and simulation of the proposed nanostructures.

This book is oriented toward two groups of readers. The first group involves researchers who are working on design of nanodevices. On the other hand, nanodesign has not yet had a significant impact on mainstream technologies, and most computer and electrical engineers and graduate students will probably never have to deal with them. However, this book will be useful to them. The reason is that a keystone of this book is the selected methods of advanced logic design introduced in unified style directed toward the dynamic challenges of technologies, including today's frontiers. This volume should be especially valuable to those active and innovative engineers, scientists, and students who are interested in logic design of discrete devices in nanodimensions. This is the second group of readers.

This book does not attempt to be theoretically comprehensive in logic design of nanoICs in scope: it is rather a first attempt at creating theory in the area of logic design of spatial computing structures, an introduction to the subject. By combining a focus on basic principles of logic design in spatial dimensions with a description of current technology and future trends, this book aims to inspire the next generation of engineers to continue to develop the theory and practice of logic design for nanoICs.

There are several excellent textbooks on related topics that we have used in our courses on advanced logic design:

- ▶ *Switching Theory for Logic Synthesis*, Kluwer Academic Publishers, 1999, by T. Sasao,
- ▶ *Algorithms and Data Structures in VLSI Design*, Springer, 1998, by C. Meinel, and T. Theobald,
- ▶ *Logic Synthesis and Verification*, Kluwer Academic Publishers, 2002, S. Hassoun, and T. Sasao (Eds), R. K. Brayton (Consulting Ed.),
- ▶ *Spectral Interpretation of Decision Diagrams*, Springer, 2003, by R. S. Stanković and J. T. Astola.

Though the above books are outstanding references to selected areas of logic design, there are certain aspects of design that are special to nanoICs. They are not covered in the mentioned references.

Scope of the book

Perhaps the largest group of readers will consist of people who want to read a full and unambiguous representation of the methods and tools of logic design of nanoICs. For these readers, the most important chapters of this book are Chapter 2 and Chapters 5 through 12, that give its comprehensive description. In order to follow our representation (description), it might be helpful to read the preliminaries given in Chapters 3 and 4.

A large part of this book is aimed at the readers who want to know why computing logic functions is not straightforward in nanospace, and why it must be carefully reviewed in the way we did. For them, we explain the ideas and principles underlying the design of nanodevices (Chapter 3), culminating in our wide trail design strategy. In Chapter 6, 9, 10, we explain our approach to parallel computations organization focused on word-level computational models.

The book is organized into 12 chapters.

Chapter 1, *Introduction*, gently introduces the directions and methodology of logic design in nanodimensions.

Chapter 2, *Nanotechnologies*, lays the technology groundwork of CAD of NanoICs. The key idea is that data structure is a crucial point of nanotechnologies. In general, this chapter justifies the appropriateness of data structures that are introduced in the next chapters.

Chapter 3, *Basics of logic design in nanospace*, includes a brief introduction in graphical representation of switching functions and networks, and focuses on uniform representation of data structures: sum-of-products, Reed-Muller and arithmetic expressions. All known methods are utilized to introduce the technique of representation and manipulation of these data structures, namely, algebraic equation, matrix (spectral) method, flowgraphs of algorithms, decision trees and decision diagrams.

Chapter 4, *Word-level data structures*, is a continuation of the previous chapter. Word-level technique is a less studied area of contemporary logic design, carrying opportunities for massive parallel computing. It provides the motivation to study their properties in detail. Three kinds of word-level expression are discussed: arithmetic, sum-of-products, and Reed-Muller. Chapters 3 and 4 are recommended for engineers and students who want to study state-of-the-art logic design.

Chapter 5, *Nanospace and hypercube-like data structures*, covers 3-D topological structures. It starts with a classic hypercube, and focuses on assembling the hypercube, in contrast to the well studied problem of embedding a guest graph into the hypercube. Some renowned solutions based on hypercube interpretation and manipulation are introduced from the position of assembling the topological structures from primitives. It is shown that the classical hypercube is limited to carrying information about switching functions, and, therefore new topological constructions are required. Next, a hypercube-like data structure called an $\mathcal{N}$-hypercube is introduced. Its topology resembles the classic hypercube. However, a number of features make it suitable for effective representation of switching functions in spatial dimensions with respect to some physical constraints. In particular, dynamic characteristics are discussed that are related to the polarity of variables and the order of variables in decision trees and diagrams. This is a key chapter in overall understanding of the problem of calculation switching functions in spatial dimensions.

Chapter 6, *Nanodimensional multilevel circuits*, presents the technique of a multilevel circuit design based on an $\mathcal{N}$-hypercube. The general flow of the chapter is as follows: a library of 3-D gates is introduced, and various techniques for implementing multilevel computation are applied. Methods for the evaluation of resulting circuits in spatial dimensions are discussed.

Chapter 7, *Linear word-level models of multilevel circuits*, is a development of data structure considered in Chapter 4 but with respect to the condition of linearity. The study of this boundary case of the word-level data structure is motivated by its ability to carry the same information as word-level data does but in a simpler way. For example, a linear decision tree or diagram is directly mapped into linear parallel-pipelining topology. The library of $\mathcal{N}$-hypercube linear models for primitives allows design of arbitrary combinational and sequential circuits in nanodimensions. This chapter is recommended to researchers looking for parallel organization of logic computations in the advent of nanotechnologies.

Chapter 8, *Event-driven analysis of hypercube-like topology*, introduces the technique of analysis in spatial dimensions based on the concept of change. It shows that logic difference is a useful model in some cases for understanding the relations between different data structures and representations of switching functions. For example, logic Taylor expansion produces Reed-Muller coefficients. However, the coefficients of a logic Taylor expansion carry information in a form acceptable for analysis, i.e., in terms of change. In general, the Taylor expansion (logical and arithmetic) allows us to introduce most of the diversification in transformation of switching functions in a simple way and from a unified position. The value of this material for design of nanoICs is justified by the fact that Taylor expansion in spatial dimensions is the general form used to generate various useful forms of switching functions such as Reed-Muller, arithmetic, Walsh, and their modifications. This chapter will be useful for engineers and students interested in advanced logic design.

Chapter 9, *Nanodimensional multivalued circuits*, contributes in spatial dimensions to logic design of devices based on multivalued signals. The formal extension of switching theory is multivalued logic. It is expected that nanotechnologies will provide us with the opportunity to utilize the concept of multivalued signals in nanodevices. This chapter introduces various aspects of 3-D multivalued circuits design following, however, the line of previous chapters by generalizing of results for switching functions. The material of this chapter will be useful for engineers and students who want to take advantage of algebra of multivalued logic for modeling post-binary devices.

Chapter 10, *Parallel computation in nanospace*, extends the concept of logic calculations and data processing on 2-D linear arrays to spatial dimensions. The main motivation for introducing this extension is the simplification of design in spatial dimensions. Bit-level systolic processing algorithms and linear systolic arrays have been chosen as they perfectly translate into locally-interconnected hardware. This chapter addresses the reader's interest in massive parallel computations in nanospace.

Chapter 11, *Fault-tolerant computation*, discusses the problem of computation using non-reliable elements. Nanodevices created hitherto are weak and sensitive, and the issue of reliable computations on non-reliable devices is essential as never before. This chapter will be useful for innovative engineers and students who are interested in logic design of nanoICs.

Chapter 12, *Information measures in nanodimensions*, contributes to the technique of information-theoretical measures in spatial structures. The reason that the Shannon theory of information should be one of the most important measurement characteristics in nanospace is that it reflects the physical nature and restrictions that nanostructures on a molecular level pose to information carriers, due to quantum effects and other features of ultra-small structures.

Style

A textbook style has been chosen for this book. There are about 250 examples throughout the text illustrated by about 250 figures. These examples are the keys to solving the set of about 100 problems at the end of each chapter. Authors' solutions to the problems can be found in the Manual for Instructors.

Some of the problems can be considered as potential further work, both theoretical and applied, to inspire suggestions, algorithms, or hypotheses, which may extend the ideas presented in this book. If you, gentle and ambitious reader, actually work out any of these problems, we would be interested in seeing your results. The problems in each set range from simple applications of procedures developed in the book to challenging solutions of more complex problems.

In each chapter, recommendations for further reading are given.

Svetlana N. Yanushkevich
Vlad P. Shmerko
Sergey E. Lyshevski

Calgary (Canada)
Rochester (U.S.A.)

Acknowledgments

We would like to acknowledge several people for their useful suggestions and discussion:

Prof. D. Bochmann, Technical University of Chemnitz, Germany.

Prof. J. T. Butler, Naval Postgraduate School, Monterey, CA, U.S.A.

Prof. R. Drechsler, Institute of Computer Science, University of Bremen, Germany.

Prof. G. Dueck, Department of Computer Science, University of New Brunswick, Canada.

Prof. P. Mc Kevitt, University of Ulster (Magee), Nothern Ireland, U.K.

Prof. T. Luba, Institute of Telecommunications, Warsaw University of Technology, Poland.

Prof. K. Likharev, Department of Physics, State University of New York, Stony Brook, U.S.A.

Prof. V. Maluygin, Institute of Control Problems, Russian Academy of Sciences, Russia.

Prof. D.M. Miller, Department of Computer Science, University of Victoria, Canada.

Prof. C. Moraga, Department of Computer Science, Dortmund University, Germany.

Prof. J. Muzio, Department of Computer Science, University of Victoria, Canada.

Prof. M. Perkowski, Department of Electrical Engineering, Portland State University, OR, U.S.A.

Prof. S. Rudeanu, Faculty of Mathematics, University of Bucharest, Romania.

Prof. T. Sasao, Center for Microelectronic Systems, Kyushu Institute of Technology, Japan.

Prof. D. Simovici, Department of Computer Science, University of Massachusetts at Boston, U.S.A.

Prof. R. Stanković, Department of Electrical Engineering, University of Nis, Serbia.

Prof. B. Steinbach, Institute of Computer Science, Freiberg University of Mining and Technology, Germany.

Prof. H. Watanabe, University of Tokyo, Japan.

Prof. A. Zakrevskij, National Academy of Sciences, Belarus.

We would like to thank our graduate students, and also *Ian Pollock*, for their valuable suggestions and assistance that were helpful to us in ensuring the coherence of topics and material delivery "synergy."

1

Introduction

The supercomputer, a computing architecture concept that has inspired the creators of Cray T3D and Ncube, is an example of implementation of 3-D design. The components of a supercomputer, as a system, are designed based on classical paradigms of 2-D architecture that becomes 3-D because of:

- *3-D topology* (of interconnects), or
- *3-D data structures* and corresponding algorithms, or
- *3-D communication flow.*

Spatial models are widely used for representation and manipulation of digital data. For example, the hypercube is a classic model for manipulation and minimization of switching, or Boolean functions in low-level logic design. On the highest level, e.g., a communication system, or network of workstations, hypercube-like topologies are deployed as well.

Nanostructures are confined to the 3-D nature of the physical world. Since low-level design deals with molecular/atomic structures, the physical platform has a 3-D structure instead of the 2-D "macro" layout of silicon ICs. This chapter aims at a global perspective on how contemporary logic design techniques fit into nanosystems.

There is a number of particular features to representing logic functions in nanospace:

- The logic functions have to be represented by a data structure in which information about the function satisfies the requirements of the implementation technology;
- An appropriate data structure must be chosen in order to ensure an information flow that complies with the implementation topology;
- This data structure must be effectively embedded into the topological structure.

Therefore, designing the architectures for computing logic functions supposes a resolution to the problem of finding the appropriate data structure and topology while taking into account some implementation aspects.

In this book, data structures that are efficiently embedded into the hypercube-like topology are introduced. The main criteria of the relationships between data structure, topology and technology are discussed in this chapter.

The structure of this chapter is as follows. A brief overview of current trends in 3-D architectures design is given in Section 1.2. The methodology of logic design in spatial dimensions is represented in Section 1.4. In Section 1.3 the structure of computer aided design (CAD) of integrated circuits (ICs) is introduced. Illustration by example of the application of different hypercube-like topologies is the focus of Section 1.5. Recommendations for "Further reading" are given in Section 1.8, and a number of problems in Section 1.7, some of which can be used as topics of students' projects, are formulated as well.

1.1 Progress from micro- to nanoelectronics

Scaling of microelectronics down to nanoelectronics is the inevitable result of technological evolution (Figure 1.1).

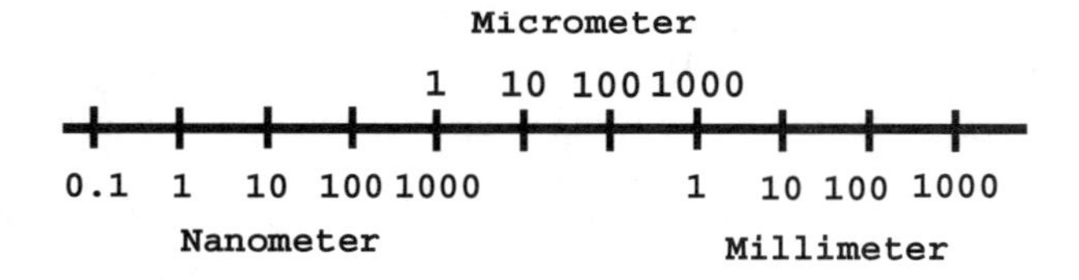

FIGURE 1.1
Revolutionary progress from micro- to nanosize.

The following can be compared against this scale:

- The size of an atom is approximately 10^{-10} m. Atoms are composed of subatomic particles, e.g., proton, neutron and electron. Protons and neutrons form the nucleus with a diameter approximately 10^{-15} m.
- 2-D molecular assembly (1 nm)
- 3-D functional nanoICs topology with doped carbon molecules ($2 \times 2 \times 2$ nm)
- 3-D nanobioICs (10 nm),
- *E.coli* bacteria (2 mm) and *ants* (5 mm) have complex and high-performance integrated nanobiocircuitry,
- 1.5×1.5 cm 478-pin Intel® Pentium® processor with millions of transistors, and Intel 4004 Microprocessor (made in 1971 with 2,250 transistors)

Significant evolutionary progress has been achieved in microelectronics. This progress (miniaturization, optimal design and technology enhancement) is achieved by scaling down microdevices approaching 10 *nm* sizing features for structures. CMOS technology has been enhanced. Currently, nanolithography, advanced etching, enhanced deposition, novel materials and modified processes are used to fabricate ICs. The channel length of metal-oxide-semiconductor field effect transistors (MOSFETs) has been decreased from

- 50 μm in 1960, to
- 1 μm in 1990, and to
- 35 *nm* in 2004.

Example 1.1 *Progress made in Intel processors:*

- *From Intel 4004 (1971, 2,250 transistors), to 286 (1982, 120,000 transistors),*
- *From Pentium (1993, 3,100,000 transistors), to Pentium 4 (2000, 42,000,000 transistors), to Itanium™ 2 Processor (2002), and to Pentium® M Processor (2003) with hundreds of millions of transistors.*

By conservative estimates, it is anticipated that field-effect CMOS transistors scaled approximately to 100 *nm* in total size must meet the 18 *nm* gate-length projection. Fabricating high-yield, high-performance planar (2-D) multilayered ICs with hundreds of millions of transistors on a single 1 cm^2 die is achieved by applying 65 *nm* CMOS technology.

However, there are fundamental technological differences among

- Nanoelectronic devices vs. microelectronic (which can even be nanometers in size) devices,
- Nanoelectronics vs. microelectronics, e.g., nanoICs vs. ICs.

These enormous differences are due to differences in basic physics and other phenomena. The dimensions of nanodevices that have been made and characterized are a hundred times less than newly designed microelectronic devices (including nanoFETs with 10 *nm* gate length). Nanoelectronics sizing leads to 1,000,000 volume reduction in packaging, not to mention revolutionary enhanced functionality due to multiterminal and spatial features.

1.2 Logic design in spatial dimensions

Traditional logic design models and techniques may not satisfy the requirements and properties of nanoscale computing devices:

- While the information flow in today's semiconductor devices is associated with surges of electrons, in new devices this is likely to be associated with states, count of electrons, etc.

- The nature of the signals and processes in ultra-small devices of size compared to wave length is stochastic. These devices are very sensitive to many physical factors (thermal fluctuation, wave coherence, random tunneling, etc.)
- The need for fault-tolerance computation increases as device to device fluctuations become larger at the furthest limits of integration. In nanostructures, it is likely to be achieved through introducing redundant hardware, which is not an acceptable practice for silicon devices (on the transistor level, not the block level where duplication is used), due to power dissipation, clocking and area constraints.
- While voltage state logic can still be acceptable in some types of nanodevices, most of them are supposed to be locally connected arrays of elements (e.g., molecular ones). This means that traditional gate-level, randomly networked circuits, such as AND/OR/NOT (or other gates) and corresponding data structures (directed acyclic graphs of the netlists, symbolic structures) may not work for the purposes of optimization and manipulation of logic functions implemented on nanostructures.

The class of spatial architectures has proven itself to be useful and reasonable, especially in the area of network communication (communication hypercubes) and parallel and distributed algorithms in supercomputers. One can observe a certain relationship to existing parallel algorithms and programming models, some of which are 3-D. However, the application of 3-D techniques to design of nanoICs is not simple.

We draw connections to the contemporary approach to computer hardware design. This direction comprises of two distinct approaches to logic design in nanodimensions:

- Development of a new theory and technique for logic design in spatial dimensions; this direction can be justified, in particular, by nanotechnologies that implement devices on a reversibility principle, and
- Using the advanced logic design techniques and methods of computational geometry in spatial dimensions.

This book focuses on the second approach, inheriting knowledge about distributed parallel processing paradigm. The latter involves (Table 1.1):

- Selected methods of advanced logic design, and
- Appropriate spatial topologies.

Data structure plays a crucial role in logic design of nanoICs. We adopt certain methods of advanced logic design including techniques for representation and manipulation of different data structures (algebraic, matrix, decision trees, etc.). These methods are selected based on the criteria shown in Table 1.1:

- Graph based models suitable for embedding and manipulation;

TABLE 1.1
Logic design in spatial dimensions: which methods of advanced logic design are compatible with nanotechnology requirements.

LOGIC DESIGN	TECHNOLOGY
▶ *Graph based models*	▶ *Interpretation of logic signals*
▶ *Massive parallel computing*	▶ *Interpretation of data structures*
▶ *Testability*	▶ *Locally interconnected architecture*
▶ *Observability*	▶ *Superior density and scalability*
▶ *Fault tolerance computing*	▶ *Unreliable low-gain elements*

▶ Technique for massive parallel computing;
▶ Testability and observability;
▶ Fault tolerance and reliable computing.

Selected methods of computation geometry include elements of graph theory, measures in spatial dimensions, manipulation, transformation and design of topological structures. The topological structures are used in the representation of data structures. These topologies must satisfy a number of requirements:

Scalability. Topologies and algorithms for their construction and manipulation must be scalable in the size of the circuit that can be processed. For example, in 1985, a single run of ICs synthesis system dealt with about 1,000 gate equivalence. In 2004, the gate equivalence in CAD of an IC system has reached 1,000,000 or more in order to support the design of multimillion gate circuits;
Suitability for nanotechnology, i.e., implementation at a minimal cost;
Topological compatibility;
Recursive calculations of size/direction.

The problem of logic design in nanodimensions is a multifaceted problem. For instance, it is impossible to choose a data structure if the topology is not determined. On the other hand, the topology can be chosen if the technology requirements are known. To remove this uncertainty, the following aspects are highlighted in this book:

A technology independent approach that takes into account the main advantages and drawbacks of nanotechnologies (massive parallel computing, event-driven analysis, information flow optimization based on entropy evaluation);
Hypercube-like topology has been proven as efficient for massive parallel computing, and for switching function representations.

Based on these statements, which are reasonable assumptions in problem formulation, this book introduces a systematic approach to logic design of nanoICs.

1.3 Towards computer aided design of nanoICs

CAD of IC systems have to produce circuits which satisfy many design rules of the target technology. These rules include, in particular, limitations on the type of primitives, maximum fanout, and connections.

The quality of the synthesized circuit is measured by a set of parameters: the area, speed, and power of the circuit after physical design.

1.3.1 Contemporary CAD of ICs

The goal of CAD of ICs tools is to automatically transform a description of ICs in the algorithmic or behavioral domains to one in the physical domain, i.e., down to a layout mask for chip production. Traditionally this process is divided into

System level (major units of information processing);
Behavioral level (information flows);
Logic level (the behavior of the circuit is described by switching functions);
Layout level (mapping of logic network to physical layout topology).

Today, design tends to one-pass synthesis from behavioral description down to layout, and the most popular data structure for switching functions is decision diagrams.

A CAD system has to produce correct circuits. This means that circuits are logically equivalent to the source, i.e., initial description in the form of logic equations, networks, etc. The CAD system should produce logically correct results, but because of the complexity of the design process, verifying the correctness of results is a necessary phase of a design. Usually, formal verification techniques deal with different data structures and descriptions.

1.3.2 CAD of nanoICs

In nanotechnology, behavioral, logic, and sometimes, layout levels of design are eventually merged. The efficiency of the algorithms applied in these levels depends largely on the chosen data structure.

An efficient representation of logic functions is of fundamental importance for CAD of ICs and nanoICs design. For example, in deep-submicron technology, which precedes nanotechnology, levels of design are merged, and the decision diagrams are the integrated data structure in this unified design process, called one-pass synthesis.

State-of-the-art and specific new methods are utilized in the design of NanoICs. Due to the spatial nature of nanostructures, each of the facets of CAD employs spatial topological structures. Detailed structure of CAD of nanoICs depends on technology.

Due to the fact that all these approaches originated from traditional logic design "on the plane" (though silicon microelectronic devices are 3-D), they tend to consider 2-D models of the nanoelectronic devices as well. However, on the system level, the nanodevices (perhaps not only electronic but also molecular) might be assembled into large arrays. Parallel and distributed computation on arrays (on macrolevel) has been well-studied; for example, systolic arrays of cells and programmable-logic devices. Most of the distributed architectures are 2-D, however, 3-D ones have been proposed as the most cutting-edge models, for instance, hypercube-configured networks of computers, hypercube supercomputer etc.

A certain type of spatial topology, such as the singular hypercube, representing the truth values of the function, was introduced long ago. However, the hypercube structure serves not solely for representation of logic functions but for computing as well. The most universal data structures, binary decision diagrams (BDDs), are suitable for computations, as models of logic functions, but do not have direct mapping to physical silicon structure except for pass-gate logic. BDD structures have not been associated with hypercube topology as yet.

1.3.3 Topology: 2-D vs. 3-D

On a system level, there exists a topology of ICs:

- *1-D arrays,* e.g., linear cellular arrays, linear systolic processors, pipelines,
- *2-D arrays,* e.g., matrix processors, systolic arrays,
- *3-D arrays,* e.g., of hypercube architecture.

On the physical level, very large scale integration (VLSI), for instance, are 3-D devices, because they have a layered structure, i.e., interconnection between layers while each layer has a 2-D layout.

Therefore, on the way to the top VLSI hierarchy (the most complex VLSI systems), linear and 2-D arrays eventually evolved to multiunit architectures such as 3-D arrays. Their properties can be summarized as follows:

- As stated in the theory of parallel and distributed computing, processing units are packed together and best communicate only with their nearest neighbors.
- In optimal organization, each processing unit will have a diameter comparable to the maximum distance a signal can travel within the time required for some reasonable minimal amount of processing of a signal,

for example to determine how it should be routed or whether it should be processed further.

- Internally to each processing unit, 2-D architecture may be favored, since at that smaller scale, communication times will be short compared with the cost of computation.

However, 3-D architectures need to contain a fair number of mesh elements before the advantages of the mesh organization become significant compared with competing topologies. This does not yet make itself felt much in current multiprocessor designs. Moreover, this structure may suffer from gate and wire delay. These are disadvantages of 3-D architectures.

As technology for logic devices goes down to nanoscale, 3-D array architecture will become important:

- At high speed of nanoscale devices, the distance light can travel per cycle is only around 3 *mm*, which means that a reasonable number of 3-D array elements of that size may be integrated on a single tiny chip and the advantages of the 3-D architecture will begin to become apparent.
- They are desirable for their ideal nature for large scale computing.
- They are simple and regular and relatively straightforward to program.
- There are many 3-D algorithms and designs for existing microscale components that are arranged in 3-D space, which computer designers already have experience with.

1.3.4 Prototyping technologies

Two approaches to mapping designs into nanoscale (electronic and molecular) circuits, that is traditionally called *prototyping technology*, can be distinguished:

- Randomly wired networks of gates (adopted from conventional electronics),
- Locally-interconnected arrays (adopted from massive parallel computation on macrolevel), mostly 3-D by nature.

Conventional electronics emphasized the development of logic families, consisting of gates that are networked. This approach can be adopted for nanoelectronics. While this approach exists, design of large circuits is problematic, due to

- Reliability issues, and
- Interconnection limitations.

The reliability problem is the problem of the design of a reliable machine from unreliable elements formulated by von Neumann.

Several interconnection limitations are already an issue in present deep submicron silicon technology, and will become even worse in nanodimensions.

It is likely that not only would the conventional random wiring demands be extreme but the discrete-physics devices, having poor fan-out and load-driving capabilities, would be unable to drive the inevitable long interconnects.

From this perspective, nanoelectronics technology demands *locally interconnected circuit architectures*. One may say that this restricts application to locally-interconnected computations. However, the rich experience of massive parallel computation (in today's macroscale multiprocessors) show that almost any algorithm can be presented in a form capable of implementation on such structures.

1.4 Methodology

The methodology of logic design in nanodimensions includes methods from several fields:

- Switching theory,
- Multiple-valued logic,
- Graph theory and computing geometry,
- Theory of massive parallel computation,
- Information theory,
- Fault-tolerant computing, and
- Stochastic computing.

Chapters 3 through 12 of this book present these methods in their relevance to logic design in nanodimensions. If, due to the space limit, some aspects are not mentioned or described very briefly, we refer the reader to the sections "Further Reading" at the end of each chapter.

1.4.1 Data structures

The fundamentals of switching theory, or Boolean algebra, cannot be changed while technology and even carriers of logic data are being changed. However, an appropriate choice of data structure is the way to adjust implementation of the logic function to the existing technology. That is why this book will give particular attention to the data structures suitable for logic data processing in nanodimensions.

Graph-based data structures and circuit topology. In design of high parallel systems and communication, a number of hypercube-like topologies have proved themselves to satisfy the above mentioned requirements of design methodology. However, the data structures that carry information about logic functions have a number of specific properties. This means that their representation and manipulation are different from the same methods in high parallel systems and communication.

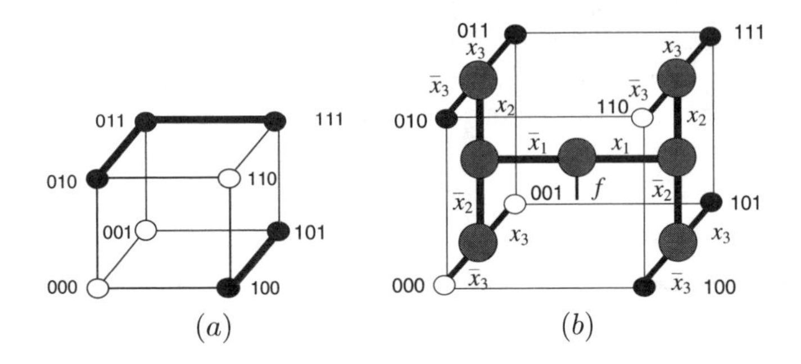

FIGURE 1.2
Representation of a switching function $f = \overline{x}_1 x_2 \vee x_1 \overline{x}_2 \vee x_1 x_2 x_3$ by the classical hypercube (a) and $\mathcal{N}$-hypercube (b) in three dimensions.

In Figure 1.2 two hypercubes are represented: the left is referred to as a *classical* hypercube, the right is referred to as a *hypercube-like* topology called $\mathcal{N}$-*hypercube*. Both hypercubes represent the same switching function

$$f = \overline{x}_1 x_2 \vee x_1 \overline{x}_2 \vee x_1 x_2 x_3$$

but in different ways. To distinguish these topologies, let us analyze the carriers of information.

Hypercube and $\mathcal{N}$-hypercube. The classical hypercube is used for *representation* of switching functions. The carriers of information are the coordinates of links. In the hypercube depicted in Figure 1.2a, three links – 01**x**, **x**11, and 10**x** – connect five nodes. These nodes correspond to 1 of switching function f.

In $\mathcal{N}$-hypercube,information about the function f given in the root node is modified by transformations in the intermediate nodes. In contrast to the classical hypercube, the $\mathcal{N}$-hypercube generates the information by processing in nodes. The $\mathcal{N}$-hypercube is used for *computing* a switching function by *implementation* of this model in hardware.

Naturally, the hypercube-like structures and $\mathcal{N}$-hypercube, in particular, inherit most properties of the classical hypercube. However, there are a number of specific features that make the modified hypercube suitable for representing different data structures for logic functions (decision tree, linear decision diagrams and logical network).

As mentioned above, the components of the methodology of logic design in spatial dimensions consist of selected methods of advanced logic design. In this book, the focus is on appropriate data structures that fit 3-D topology, the $\mathcal{N}$-hypercube. For application of the $\mathcal{N}$-hypercube, the appropriate data structure of logic function and method of embedding this structure in the $\mathcal{N}$-hypercube must be chosen. There are three phases in logic function manipulation aimed at changing the carrier of information from the algebraic form (logic equation) to the hypercube-like structure:

Phase 1 : The logic function (switching or multivalued) is transformed to the appropriate algebraic form (Reed-Muller, arithmetic, word-level, linear word-level, etc. in matrix or algebraic representation).

Phase 2 : The obtained algebraic form is converted to the graphical form (decision tree, decision diagram or logic network).

Phase 3 : The obtained graphical form is embedded into a hypercube-like structure, the $\mathcal{N}$-hypercube.

Schematically, the above is represented as follows

$$\underbrace{\textsf{Logic function}}_{Phase\ 1} \Longleftrightarrow \underbrace{\textsf{Graph}}_{Phase\ 2} \Longleftrightarrow \underbrace{\mathcal{N}\textsf{-hypercube structure}}_{Phase\ 3}$$

Each of the above forms requires specific methods and techniques for manipulation:

- *Algebraic* representations and rules of manipulations with switching and multivalued logic functions.
- *Matrix* representations and rules of manipulations with switching and multivalued logic functions. In some cases, matrix representations provide a better understanding of logic relationships of variables and functions, for example, from the viewpoint of spectral theory.
- *Graph-based* representations are introduced by decision trees, decision diagrams, and logical networks. These representations have their origin in state-of-the-art advanced logic design, and are appropriate data structures for embedding into the $\mathcal{N}$-hypercube.

In the above representations, the term "switching, or logic, function" means switching that take two values, 0 and 1, and multivalued functions which are an extension of switching functions and take k values, $k \in \{0, 1, 2, \ldots, k-1\}$. The information content of multivalued logic functions is higher compared to switching functions (see "Further Reading" Section). There are a number of successful implementations of multivalued functions in advanced VLSI and ultra large scale integration (ULSI) technologies. Nanotechnologies give promising opportunities for the utilization of the advantages of k-valued logic.

Extension of dimensions. The hypercubes depicted in Figure 1.2 are called 3-D hypercubes. They can carry limited information because the number of nodes and links is limited. The unique property of hypercubes is the possibilities for extension. This extension is expressed by notation of *dimension*, i.e., by the 3-D nature of the physical world, we design many-dimensional hypercubes. For example, the 4-D classical hypercube is designed by multiple copies of the 3-D hypercube illustrated in Figure 1.3a. Hypercube-like topology inherits this property from the hypercube: the 4-D $\mathcal{N}$-hypercube is designed by connections of root nodes (Figure 1.3b). This is reasonable architecture from a physical point of view.

On this basis, *assembly* of complex multidimensional structures can be accomplished.

Extension of parallelism by a word-level data structure. The natural parallelism of the $\mathcal{N}$-hypercube can be increased by the appropriate data structure of logic functions, so-called *word-level* representation. In this case each node of the $\mathcal{N}$-hypercube computes a set of logic functions. Hence, the hypercube topology is very flexible in this extension.

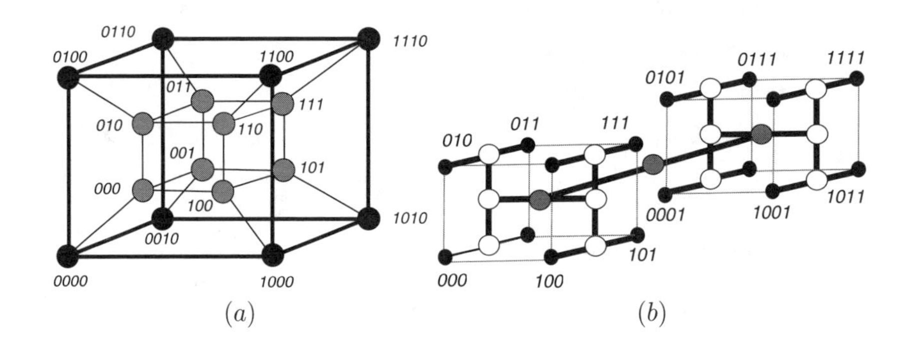

FIGURE 1.3
4-D hypercube (a) and 4-D $\mathcal{N}$-hypercube (b).

1.4.2 Assembling in 3-D

The assembly philosophy of nanoICs design in spatial dimensions differs significantly from the usual ideas of building complex computer systems.

Assembly means the construction of more complex systems from the components provided, in particular, with features identical to the components which began the process.

Self-assembly is the process of construction of a unity from components acting under forces/motives internal or local to the components themselves, and arising through their interaction with the environment. Self-assembling structures create their own representations of the information they receive. That is, assembly according to a distributed plan or goal implicit in the structure and behavior of the parts, not explicitly stated or held by a central controlling entity. Components in self-assembling systems are unable to plan but respond to their surroundings. The stimuli to which a component is capable of responding are dictated by that component's physical composition and properties, for example, minimum energy states. This is quite different from traditional programming methodology, which requires all data to be explicitly specified by the programmer. Self-assembling structures are well suited for problems for which it is either difficult or impossible to define an explicit model, program or rules for obtaining the solution.

A self-assembling system is able to process noisy, distorted, incomplete or imprecise data. This feature of self-assembling systems makes them particular suited for classification, pattern recognition and optimization problems. Self-assembling systems typically have higher level properties that cannot be observed at the level of the elements, and that can be seen as a product of their interactions. In self-assembling, topology plays the crucial role.

Self-organizing. The organism, for example, is a self-organizing system. In the organism, self-organizing is implemented through the local physical and chemical interactions of the individual elements themselves.

Adaptive self-assembling is the ability of structure to learn how to perform assembling (appropriate architecture) aiming to solve certain tasks by being presented with examples.

1.4.3 Massive and parallel computation in nanodimensions

Massive parallel computing of logic functions can be accomplished in nanodimensional structures via word-level representation and also through borrowing some approaches developed in the theory of parallel computing on the "macroscale." Relevant problems are analysis, synthesis, information measures and testability issues.

Hypercube-based massive parallel computing. The hypercube-like topologies inherit high parallelism of computing due to their

- Regular and homogeneous structures,
- Local connectivity, and
- Ability to assemble and extend the structure.

Additional opportunities for parallelization are provided by data structures that are embeddable to 3-D. Logic functions can be represented by word-level decision trees and diagrams that are a powerful resource of parallel computing. In this case, the node implements the processing of a set of functions grouped into a word. Hence, there are two levels of parallelization in the $\mathcal{N}$-hypercube:

- *Natural* parallelism of hypercube-like structures, $\mathcal{N}$-hypercube, in particular. This property can be efficiently exploited. For this, a decision tree that represents a logic function is embedded into the $\mathcal{N}$-hypercube. The node functions are simple – for example, switch only in the case of a tree representation based on Shannon expansion.
- *Extra* parallelism is provided by word-level representation of a logic function. In this case, each node computes logic computations on the bits in the words in parallel.

Cellular arrays. The term *cellular systolic array* refers to networks composed of some regular interconnection of logic cells. These arrays may be either 1-D, 2-D, or theoretically of any higher dimension of three or more. Practical considerations usually constrain them to 1-D, and 2-D cases (Figure 1.4).

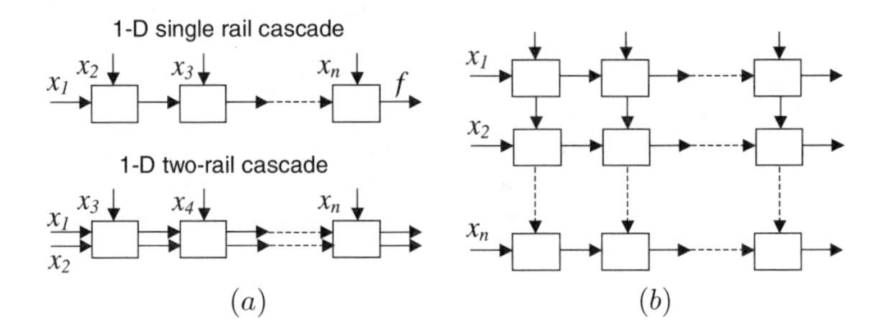

FIGURE 1.4
1-D (a) and 2-D (b) cellular arrays.

Cellular automata are a reasonable model for study of self-assembling and self-reproducing phenomena. For example, with many simplifications, the organisms can be considered as self-assembling, parallel machines whose many and varied components maintain a stable organization when disturbed.

Systolic arrays is another name for parallel-pipelined computing structures. In these structures, data input and output is organized in a sequential or partially parallel way, and the processing is accomplished by parallel computing on the array of the unified processing elements. The topology is usually linear (e.g., for matrix-vector multiplication) or 2-D (for matrix-matrix multiplication). Locality of data input/output and pipelining computing makes this organization of data processing attractive to implementation on nanoarrays. Chapter 10 considers arrays of nanodevices including arrays that deploy systolic principles of data processing.

1.4.4 Fault tolerance computing

Traditional logic circuits are very sensitive to failure: if even a single gate or single wire malfunctions then the computation may be completely wrong. If one is worried about the physical possibility of such failures, then it is desirable to design circuits that are more resilient.

Any computer with nanoscale components will contain a significant number of defects, as well as massive numbers of wires and switches for communication purposes. It therefore makes sense to consider architectural issues and defect tolerance early in the development of a new paradigm.

An incorrect result is defined as a fault. In the presence of faults, a fault-tolerant system reconfigures itself to exclude the faulty elements from the system. A system so reconfigured may or may not change its topology. Ideally, a fault-tolerant system retains the same topology after faults arise.

A hypercube is called *faulty* if it contains any faulty node or link. For hypercubes of large dimensions, the probability of occurrence of faults increases. Since efficient cooperation between the nonfaulty nodes (computing elements) is desirable, a hypercube network is *robust* if the performance does not decrease significantly when its topology changes.

Two main aspects are critical in design of nanodevices:

- *The probabilistic behavior* of nanodevices (electrons, molecules); this means that the valid switching function can be calculated with some probability.
- *The high defect rates* of nanodevices; this means that because many of the fabricated devices have defects, their logic correctness is distorted.

This problem is tackled from two directions:

The first direction: designed architecture must recognize and correct errors, and

The second direction: technological defects must be recognized and removed. The basic concept of these directions is redundancy (additional hardware resources or resources at time of computing).

There are two types of fault tolerance exhibited by a nanosystem: fault tolerance with respect to

- Data that is noisy, distorted or incomplete, which results from the manner in which data is organized and represented in the nanosystem, and
- The physical degradation of the nanosystem itself.

If certain nanodevices or parts of the network are destroyed, it will continue to function properly. When damage becomes extensive, it will only affect the system's performance, as opposed to a complete failure. Self-assembling nanosystems must be capable of this type of fault tolerance because they store information in a distributed (redundant) manner, in contrast to traditional storage of data in a specific memory location in which data can be lost in case of the hardware fault.

The methods of stochastic computing provide another approach to overcoming the problem of design of reliable computers from unreliableelements, nanodevices. For example, a signal is represented by the probability that a logic level be 1 or 0 at a clock pulse. In this way, a random noise is being deliberately introduced into the data. A quantity is represented by a clocked sequence of logic levels generated by a random process. Operations are performed via the completely random data (details are considered in Chapter 11).

1.4.5 Analysis, characterization, and information measures

There are a number of reasons why application of Shannon theory is useful in logic design of nanodevices:

- The information-theoretical approach is a *specific level of abstraction* that is useful in some steps of logic design. It offers approximate estimations of signal streams and often demonstrates impressive results in combination with traditional, exact methods.
- The information-theoretical approach based on Shannon theory is relevant to information measures in nanotechnologies on a physical level (thermodynamic). This coherence of measures can be useful in evaluation of the characteristics of nanodevices, their testing and verification.
- Because nanodevices are very sensitive and processing of information by these devices is described by probabilistic and statistical methods, the information theoretic approach is the most reasonable in fault-tolerant logic design.
- The effectiveness of information-theoretical methods is justified in many heuristic algorithms. For example, it is difficult to distinguish a gate function based on information measures, however it is possible to drastically reduce the search space when exact methods fail.

1.5 Example: hypercube structure of hierarchical FPGA

Topological structures of a field programmable gate array (FPGA), the popular computing device, are a good example to introduce the properties and advantages of hypercube-like topologies. We consider here three topologies (other topologies are discussed in the form of problems at the end of this chapter):

- ▶ The simplest topology (conventional FPGA), characterized by about 50% of area in implementation; this topology is represented by the multi-rooted k-ary decision tree;
- ▶ The hierarchical topology represented by the completed binary decision tree; this tree is embedded into the hypercube-like structure;
- ▶ The hierarchical topology represented by the completed k-ary decision tree; this tree is embedded into the hypercube-like structure too but with other topological properties.

1.5.1 FPGA based on multiinput multioutput switching

A conventional FPGA (Figure 1.5a) consists of an array of logic blocks that can be connected by routing resources. The logic blocks denoted by L contain combinational and sequential circuits which are used to implement logic functions. Logic blocks are grouped into clusters which are recursively grouped together.

The routing resources (switch blocks) denoted by SB consist of wire segments and switch blocks. Switch blocks can be configured to connect wire segments and logic blocks into networks. An example of a switching block is given in Figure 1.5b. With some simplification, the FPGA can be represented by the multirooted k-ary decision tree in 1.5c.

1.5.2 Hierarchical FPGA as hypercube-like structure

Hierarchical FPGA based on single-input two-output switching. Consider a 2×2 cluster of logic blocks L. To connect each logic block to another it is enough to use a single-input two-output switch block. In Figure 1.6a), four copies of the cluster are organized into a "macro" cluster.

A switching block has a more simple structure compared to that of the conventional FPGA (Figure 1.6b). The structure of this FPGA is represented by a binary decision tree of depth 4 in which the root and levels correspond to switching blocks SBs and the 16 terminal nodes correspond to logic blocks (Figure 1.6c, where ■ denotes logic block L and $\circ$ denotes SB).

This tree is embedded in the hypercube-like structure of two dimensions depicted in Figure 1.6d.

Hierarchical FPGA based on single-input four-output switching. Consider 2×2 cluster of logic blocks L connected by single-input 4-output switch blocks SB. In Figure 1.7a four copies of the cluster are organized into a "macro cluster." SB structure is given in Figure 1.7b. The 4-ary decision tree is shown in Figure 1.7c).

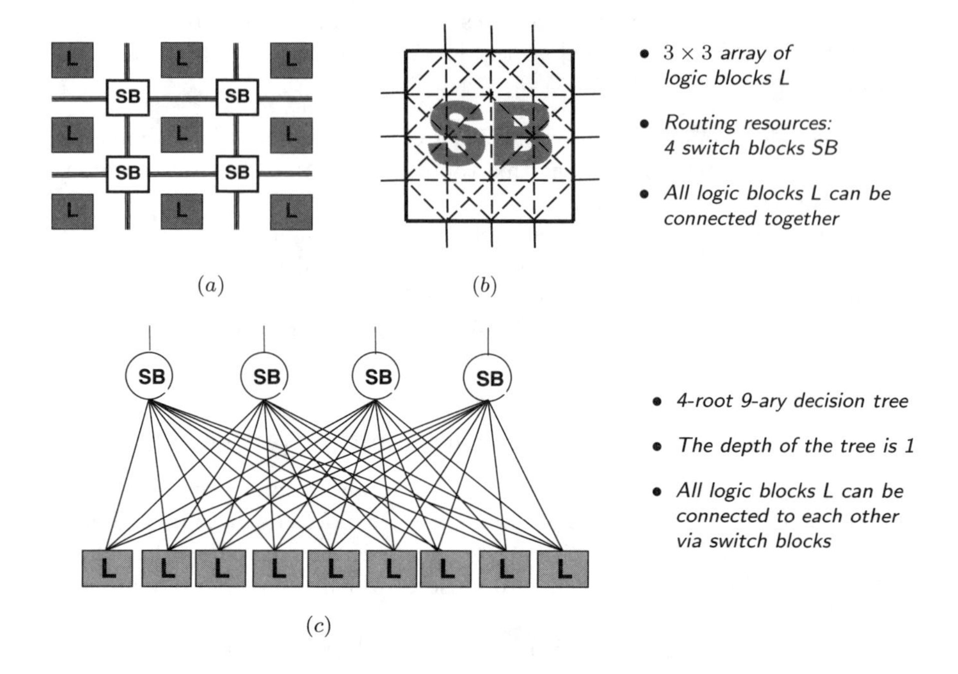

FIGURE 1.5
The topology of conventional FPGA (a), the structure of SB (b), and representation by the 9-ary decision tree (c).

From these three examples related to the popular computing device, FPGA, we can observe that:

- Hypercube and hypercube-like topology can be used for interpretation of a data structure represented by a decision tree;
- Hypercubes consist of clusters that can be patterned to produce arrays; and
- Hypercube topology complies with parallel and distributed architectures.

1.6 Summary

1. The methodology for designing logic circuits in spatial dimensions includes:

(*i*) Advanced methods of logic design (logic optimization, logic differential calculus, decision diagram techniques);
(*ii*) Methods of massive parallel computing;

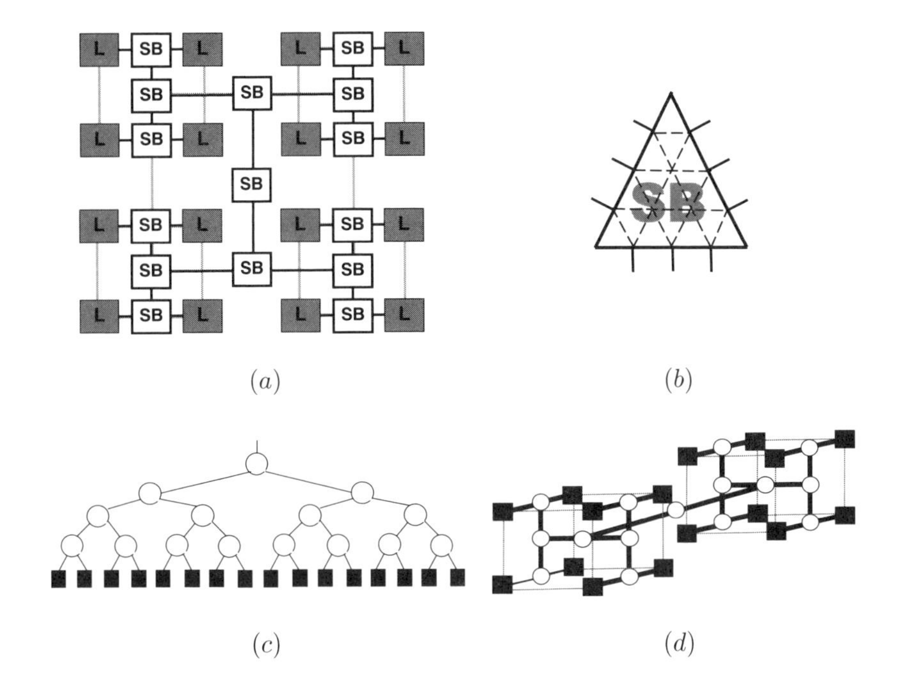

FIGURE 1.6
A hierarchical FPGA (a), the structure of a switching block (b), tree representation (c) and generalization towards 3-D hypercube-like structure (d).

(*iii*) Probabilistic modeling and simulation;
(*iv*) Information-theoretical methods.

2. Graph-based data structures are the "bridge" between logic design and 3-D topology of nanomaterials. Among them, hypercube topology is a useful model of computing in spatial dimensions. This topology can be used at all levels of abstractions from the gate level (nodes implement the simplest logic functions) to macrolevel (nodes implement complex devices).

3. Tree-like and hypercube-like topology is common in parallel and distributed architectures; this fact reflects that the principle of an optimal computing scheme is being preserved while scaling it down to molecular/atomic structure.

4. Hypercube topology can be used for modeling of nanostructures that implement logic functions, in particular, single-electron neuromorphic networks.

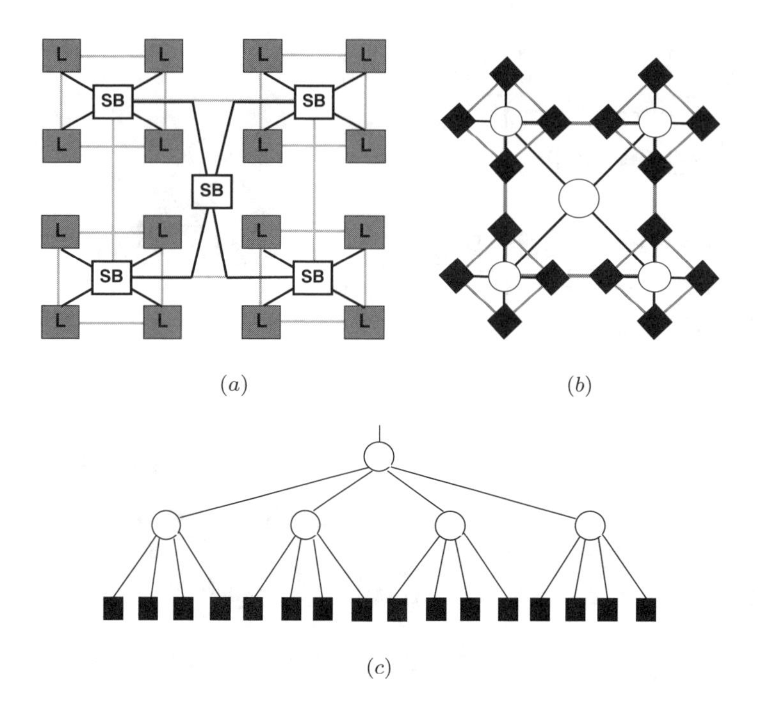

FIGURE 1.7
A hierarchical FPGA topology design: 2-D hypercube-like topology (a), hypercube-like structure (b), and representation by the 4-ary decision tree (c).

1.7 Problems

Problems 1.1 and 1.2 can be solved without any additional information and can be used by both readers and instructors for testing. Detailed information is given in the Manual for solutions. Problem 1.3 needs more time and can be considered as the project.

Problem 1.1 Extend the 2-D hypercube-like topology; construct a complete decision tree. Follow examples given in Figure 1.6 and Figure 1.7.
(a) to 3-D (Figure 1.8a);
(b) to 3-D and 4-D (Figure 1.8b);
(c) to 3-D and 4-D (Figure 1.8c);
(d) to 3-D (Figure 1.8d).

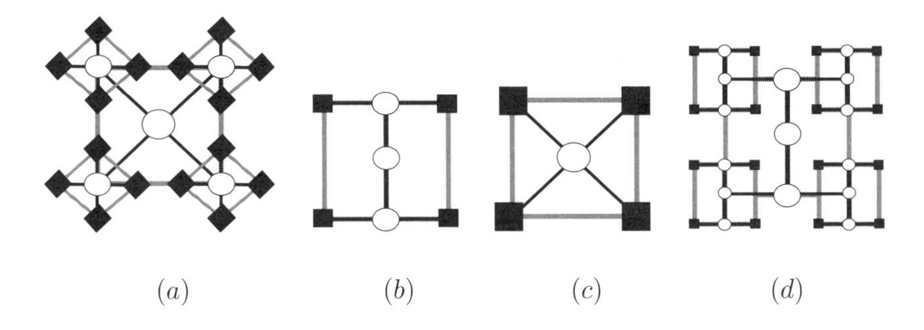

(a) (b) (c) (d)

FIGURE 1.8
Hypercube-like topology for a hierarchical FPGA for Problem 1.1.

Problem 1.2 Define the hypercube-like topology given by a complete decision tree. Follow examples given in Figure 1.6 and Figure 1.7.
(a) to 3-D (Figure 1.9a);
(b) to 3-D and 4-D (Figure 1.9b).

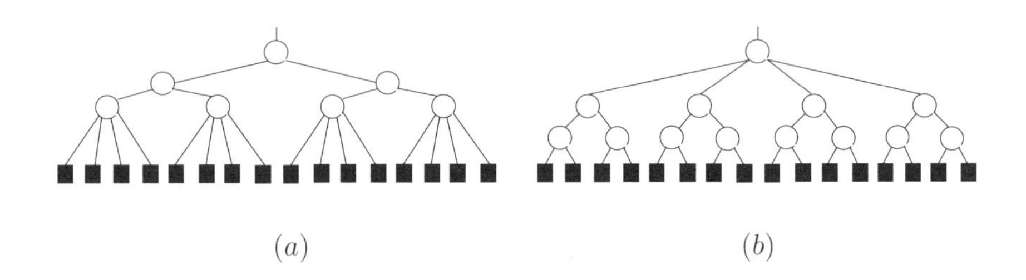

(a) (b)

FIGURE 1.9
A k-ary decision trees for Problem 1.2.

Problem 1.3 The goal of this problem is to generate different FPGA spatial topologies. A study of different FPGA topologies is aimed at reducing interconnections and improving related electrical and technological characteristics.

In Figure 1.7 and Figure 1.6, the design of hierarchical FPGA topology has been interpreted by the hypercube-like structure.

The first topology (Figure 1.7a) corresponds to a binary decision tree. The second topology is based on a 4-ary tree, which 3-D hypercube-like structure is given in Figure 1.8. These hierarchical FPGA topologies have been reported by Lai and Wang, and Shyu et al. [11, 23]. The methods are based on earlier results in parallel supercomputers, in particular, Greenberg and Leiserson [7, 12]. There are a lot of other hypercube-like topologies that can be useful in hierarchical FPGA design, for example, pyramid and fractal structures. Perhaps in this way you can propose novel topologies for FPGA.

1.8 Further reading

Survey on advanced logic design can be found in a special issue *Electronic Design Automation at the Turn of Century* in IEEE Transactions on Computer-Aided Design of Integrated Circuits and Systems, vol.19, no. 12, 2000.

Survey on hypercube architectures. The hypercube structure started to be investigated in the 1970s and provided a good alternative to operations like the fast Fourier transform (FFT). In the 1980s, a few machines based on the hypercube topology had been designed and started to be utilized. This experience stimulated applications of hypercube topology in data communications. However, the cube algebra based on functional hypercube has been applied in logic circuit design in the 1950s. The problem of embedding the complete binary tree in the hypercube has been studied since the hypercube configured processor was proposed. The problem arises from the necessity of the implementation of tree-structured computation. Linial et al. [14], have been shown that a fairly high dimension is needed to embed a general graph with reasonable distortion. Additional references can be found in Chapter 3.

Fault-tolerance computing. Classic von Neumann's work on probabilistic logics and reliable computation upon nonreliable computing elements [27, 28] is the focus of many recent investigations [4, 10, 18].

Cellular arrays are widely studied as fault-tolerance models of computations. Cellular array consists of simple identical cells organized as 1-D, 2-D or 3-D arrays. Interactions between cells are modelled by a small number of transition rules (instructions). These rules act at a local level. Cellular arrays are divided into synchronous (global timing control) and asynchronous (local timing control) arrays. In fault-tolerance cellular arrays, errors of cell are corrected. Note that von Neumann in his classical model of fault-tolerance computing, has used error correction.

Cellular automata have been introduced in [29]. Using cellular automata in quantum computing is discussed in [13]. Fault-tolerance cellular arrays have been studied by Porod [26]. Additional references can be found in Chapters 10 and 11.

Survey on nanotechnologies and nanocircuit design. The book *Ultimate Computing* by Stuart R. Hameroff [8] is useful in understanding the evolution of information processing systems. Representation of today's problems of nanotechnology is given in *Technology Roadmap for Nano-electronics* [2] and papers [15, 16, 21]. The collection of 16 papers in introduction to molecular electronics [20] discusses problems and trends in molecular electronics: molecular materials, techniques for molecular scale electronics, molecular properties (piezoelectric and pyroelectric effects, molecular magnets, polarization, photochromism, etc.), and molecular architectures. Computer architectures for molecular electronics have been studied in [20, 24, 25, 26].

Various programs of research have been reviewed by Ferry [5]. For more detail see references in Chapter 2.

In [1, 3, 6, 22], the design of circuits and systems is discussed from the point of view of nanotechnology. In a paper by Mange et al. [17] discusses the problems of robust circuit design.

In addition. FPGAs are useful structures in nanosystem design. For example, FPGAs can be used in quantum dot cellular automata. The topologies of FPGA are often related to the hypercube-like structures. The reader can find useful results on FPGA topologies in [7, 9, 11, 12, 23].

1.9 References

[1] Al-Rabady AN. *Reversible Logic Synthesis: From Fundamentals to Quantum Computing*, Springer-Verlag, Heidelberg, 2004.

[2] Compano R. *Technology Roadmap for Nano-electronics*. European Commission IST Programme – Future and Emerging Technologies, 2000.

[3] DeHon A. Array-based architecture for FET-based, nanoscale electronics. *IEEE Transactions on Nanotechnology*, 2(1):23–32, 2003.

[4] Depledge PG. Fault-tolerant computer systems. *IEE Proceedings*, 128:257–263, 1981.

[5] Ferry DK. Silicon single-electron devices and a review of nanodevice research in the USA. *FED Journal*, 10(1):5–25, 1999.

[6] Goser K, Pacha C, Kanstein A, and Rossmann ML. Aspects of systems and circuits for nanoelectronics. *Proceedings of the IEEE*, 85:558–573, April, 1997.

[7] Greenberg RI. The fat-pyramid and universal parallel computation independent of wire delay. *IEEE Transactions on Computers*, 43(12):1358–1364, 1994.

[8] Hameroff SR. *Ultimate Computing: Biomolecular Consciousness and Nanotechnology*. Elsevier, Amsterdam; New York, 1987.

[9] Hauck S. The roles of FPGA's in reprogrammable systems, *Proceedings of the IEEE*, 86(4):615–638, 1998.

[10] Heath JR, Kuekes PJ, Snider GS, and Williams RS. A defect-tolerant computer architecture: opportunities for nanotechnology. *Science*, 280:1716–1718, 1998.

[11] Lai YT, and Wang PT. Hierarchical interconnection structures for field programmable gate arrays. *IEEE Transactions on VLSI Systems*, 5(2):186–196, 1997.

[12] Leiserson CH. Fat-trees: universal networks for hardware-efficient supercomputing. *IEEE Transactions on Computers*, 34(10):892–901, 1985.

[13] Lent CS, Tougaw PD, Porod W, and Bernstein GH. Quantum cellular automata. *Nanotechnology*, 4(1):49–57, 1993.

[14] Linial N, London E, and Rabinovich Y. The geometry of graphs and some of its algorithmic applications. In *Foundations of Computer Science*, pp. 577–591, 1994.

[15] Luryi S, Xu J, and Zaslavsky A., Eds., *Future Trends in Microelectronics: The Nano Millennium.* Wiley-IEEE Press, New York, 2002.

[16] Lyshevski SE. Nanocomputers and Nanoachitectronics. In Goddard W, Brenner D, Lyshevski S, and Iafrate G., Eds., *Handbook of Nanoscience, Engineering and Technology*, vol. 6, pp. 1–39, CRC Press, Boca Raton, FL, 2002.

[17] Mange D, Madon D, Stauffer A, and Tempesti G. Von Neumann revisited: a Turing machine with self-repair and self-reproduction properties. *Robotic Autonomous Systems*, 22(1):35–85, 1997.

[18] Mitra S, Saxena NR and McCluskey EJ. Common-mode failures in redundant VLSI systems: a survey. *IEEE Transactions on Reliability*, 49:285–299, 2000.

[19] Peper F, Lee J, Abo F, Isokawa T, Adachi S, Matsui N, and Mashiko S. Fault-tolerance in nanocomputers: a cellular array approach. *IEEE Transactions on Nanotechnology*, 3(1):187–201, 2004.

[20] Petty MC, Bryce MR, and Bloor D., Eds., *Introduction to Molecular Electronics.* Oxford University Press, Oxford, 1995.

[21] Poole CP. Jr, and Owens FJ. *Introduction to Nanotechnology.* John Wiley & Sons, New York, 2003.

[22] Porod W. Nanoelectronic circuit architecture. In Goddard III WA, Brenner DW, Lyshevski SE, and Lafrate GI., Eds., *Handbook of Nanoscience, Engineering and Technology*, CRC Press, Boca Raton, FL, chap. 5, 2002.

[23] Shyu M, Wu GM, Chang YD, and Chang YW. Generic universal switch blocks. *IEEE Transactions on Computers*, 49(4):348–359, 2000.

[24] Reed MA, and Tour JM. Computing with molecules. *Scientific American*, pp. 86–93, June 2000.

[25] Tour JM. *Molecular Electronics: Commercial Insights, Chemistry, Devices, Architecture and Programming.* World Scientific, Hackensack, NJ, 2003.

[26] Tour JM, Zandt WLV, Husband CP, Husband SM, Wilson LS, Franzon PD, and Nackashi DP. Nanocell logic gates for molecular computing. *IEEE Transactions on Nanotechnology*, 1(2):100–109, 2002.

[27] Von Neumann J. Probabilistic logics and the synthesis of reliable organisms from unreliable components. In Shannon CE, and McCarthy J., Eds., *Automata Studies.* Princeton University Press, Princeton, NJ, 1955.

[28] Von Neumann J. *The Theory of Self-Reproducing Automata.* University of Illinois Press, Urbana, IL, 1966.

[29] Wolfram S. *Theory and Applications of Cellar Automata.* Elsevier, Amsterdam; New York, Science Publishers, BV, 1986.

2

Nanotechnologies

In this chapter, nanotechnology is discussed in its capacity for design of digital circuits. Logic circuit design, at present, is applied solely to microelectronics. The process of transfer of circuitry to nanoelectronics and relevant hybrid technologies (e.g., molecular electronics) has already been started. Other meanings of the term "nanotechnology," as it is understood in biochemistry, biophysics etc., are not the subject of this book.

Tremendous evolutionary progress has been accomplished within the space of only 60 years from invention of the transistor to computers with 2 cm^2 processors that integrate hundreds of millions of transistors. Despite the progress in microelectronics, further developments are needed. This can be accomplished by

- Developing and applying new theoretical fundamentals,
- Uniquely utilizing phenomena and novel physics observed at nanoscale,
- Designing and utilizing novel architectures departing from traditional 2-D planar microelectronics to 3-D nanoICs,
- Applying novel fabrication technologies.

This chapter introduces and covers the fundamentals and practice of nanoelectronics, which is a pioneering development compared to microelectronics.

In microelectronic circuits, an operation on one unit or bit of information, registered by voltage/current, involves billions of electrons. In recent years there have been significant advances in the fabrication and demonstration of molecular wires, diode switches, single-electron transistors, and other nanoelectronic circuitry, whereby a single carrier of charge (or a few carriers or particles) controls the motion of other particles. Not only does this provide a more efficient means of storing and processing information, but it also allows a drastic increase in computing speed compared to the existing complementary metal-oxide semiconductor (CMOS) technology.

The most important features of nanotechnology, as it applies to nanoelectronics, are introduced in Section 2.1. Section 2.2 focuses on nanoelectronic devices that can be understood as more realistic components for logic design of computers based on the design paradigms presented in this book. In Section 2.3, potential libraries of nanogates are discussed. Section 2.4 emphasizes molecular electronics: CMOS- and CMOS-molecular (CMOL) devices, neuromorphic circuits and nanowires. Scaling and operational limits are discussed in Section 2.5. After a brief summary (Section 2.6) and Problems (Section 2.7), recommendations for "Further Reading" are given in Section 2.8.

2.1 Nanotechnologies

Nanoelectronics are a natural consequence of the evolution of microelectronics. It is, however, not simply a matter of scaling electronics down to a smaller size. There are certain constraints, mainly physical, that must be taken into account when considering nanoscaling:

- The physics of the conventional approach may be characterized as being continuum-based whereas "scaled" physics is microscopic (discrete), thus enabling quantum effects which make extreme scaling possible.
- Traditional logic devices and models of computation are characterized by limited information capacity. This is because they involve the irreversible production of some minimum amount of entropy per operation, so that they do not reach maximum computational efficiency as permitted by the second law of thermodynamics. This law states that entropy (uncertainty, or unknown information) cannot be destroyed, and can be viewed as a consequence of the reversibility of microscopic physical dynamics.

There are fundamental and technological differences between nanoelectronic devices versus microelectronic (even possibly nanometers in size) devices, e.g., nanoICs vs. integrated circuits (ICs). Even though field-effect transistors can possibly reach 100 nm dimensions for a complete microdevice, they still cannot be called nanoelectronic devices. Novel physics, integrated with design methods and nanotechnology, leads to far-reaching revolutionary progress. For example, multiterminal $1 \times 1 \times 1$ nm electronic nanodevices (endohedral fullerenes, doped fullerenes, functional carbon molecules, etc.) are not submicron microelectronic devices.

Microelectronic devices can be fabricated using nanotechnology-enhanced processes and techniques, for example, transistor channels can be formed using carbon nanotubes. Hence, microelectronics can be nanotechnology-enhanced, and one can define this as nanotechnology-enhanced microelectronics. Therefore, microelectronic devices, even if scaled down to the 100 nm or smaller, are unlikely to be classified as nanotechnology-based electronic nanodevices (see Section 2.4.3).

2.2 Nanoelectronic devices

Quantum effects are present, in general, in

- Single-electron devices,
- Single-flux quantum devices, and
- Resonant-tunneling devices.

Among these, the concept of single-electronics is less dependent on the implementation technology than the other two. It can be implemented in many ways, e.g., on superconductors, at a molecular level, and by hybrid approaches.

2.2.1 Single-electronics

Single-electron charging effects are exploited to implement logic functions in two different ways:

- These effects are confined to the interior of the transistor, so that the logic 0/1 is presented with low/high voltage that is not quantisized.
- The electrons are confined to one or a few islands, so that bits of information correspond to presence or absence of electrons in the specific islands. This is in harmony with the concept of logic switching and is achieved via electrostatic coupling (spin) of the islands.

The implementation of quantum-effect devices of both groups deploys single electrons. The first class inherits some inert knowledge from semiconductor circuits. The second class uses truly new concepts of implementation of logic.

The basic concept of single-electronics is the *Coulomb blockade* observed in a single-electron device including a conducting island. An additional electron can be injected from outside by tunneling it through an energy barrier created by a thin insulating layer. The resulting electric field is inversely proportional to the square of the island size, and is strong enough to repulse adding yet more electrons. This phenomenon is called *energy quantization* and it is expressed in terms of charging energy. This energy E_c is proportional to the electron charge $e \approx 1.6 \times 10^{-19}$ Coulomb and is inversely proportional to the capacitance of the island C:

$$E_c = e^2/C.$$

This is possible when the island size becomes comparable with the wavelength of the electrons inside the island. If the island size is reduced to below 10 nm, the charging energy approaches 100 meV, and some single-electron effects become visible at room temperature. There are several physical problems here:

- Thermally induced random tunneling events make it impossible for single-electron devices to operate at room temperature. The thermal effects are reduced as the device's temperature reaches 4 to 5 *Kelvin.*
- Islands must be smaller than 1 *nm* (called *quantum dots*) in order that the electron's additional energy be as large as a few electron-volts. This is difficult to achieve in nanofabrication technology.
- Transport properties are very sensitive to small variations of quantum dot size and shape.

Single-electron box and trap

Single-electron box (Figure 2.1a) is capable of trapping an electron (see "Further Reading" Section). Once an electron occupies the "island," or "well," it blocks the motion of other electrons travelling along the atomic "wire" (at least 50 atoms long). Thus, the trapped electron repels the mobile electrons. When no trapped electron is present, electrons travel along the atomic wire.

Two major drawbacks of the single-electron box are:

- It is a low-temperature device (down to sub-*Kelvin* scale).
- It cannot be used as memory: the number of electrons in the box is a unique function of the applied voltage.
- The box cannot carry a DC current.

A *single-electron trap* is obtained by replacing the single tunnel junction with an array of several islands separated by tunnel barriers in the single-electron box (Figure 2.1b).

Single-electron transistor

The basic building block of single electron devices and circuits is the tunnel-junction, which forms a single-electron tunneling (SET) transistor originally created experimentally on a metal-insulator system at a low temperature (see "Further Reading" Section). The SET has three terminals (Figure 2.2). Tunnel junctions are implementable on many technologies including ultrasmall metal-oxide-metal junctions, *GaAs/AlGaAs* structure, and the contacts to carbon nanotubes.

At a small source-drain voltage there is no current, since at low enough temperatures the tunneling rate is exponentially low. This suppression of current at low voltages is known as the Coulomb blockade. At a certain threshold voltage the Coulomb blockade is overcome, and at much higher voltages the current exhibits quasi-periodic oscillations similar to the Coulomb staircase in the single-electron box. Thus, the threshold voltage and the source-drain current are periodic functions of the gate voltage. The effect of the gate voltage is equivalent to the injection of charge into the island, that changes the balance of the charges at tunnel barrier capacitances, which determines the Coulomb blockade threshold.

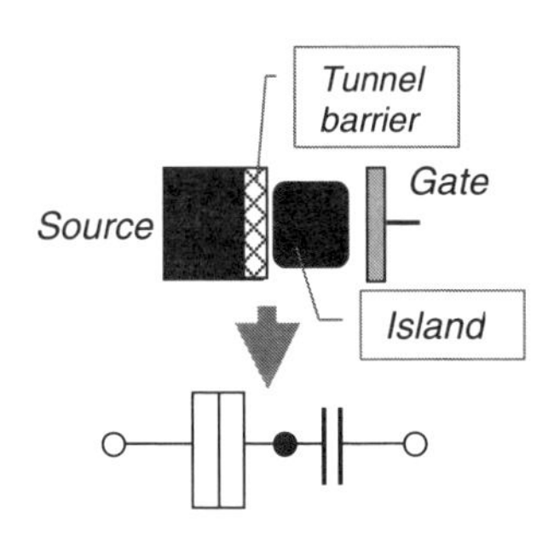

The single-electron box consists of just one small island separated from a larger electrode (electron source) by a tunnel barrier.
An external electric field is applied to the island using a gate separated from the island by a thicker insulator which does not allow noticeable tunneling.
Increasing gate voltage attracts more and more electrons to the island.
The discreteness of electron transfer through low-transparency barriers necessarily makes this increase step-like(the Coulomb staircase).

(a)

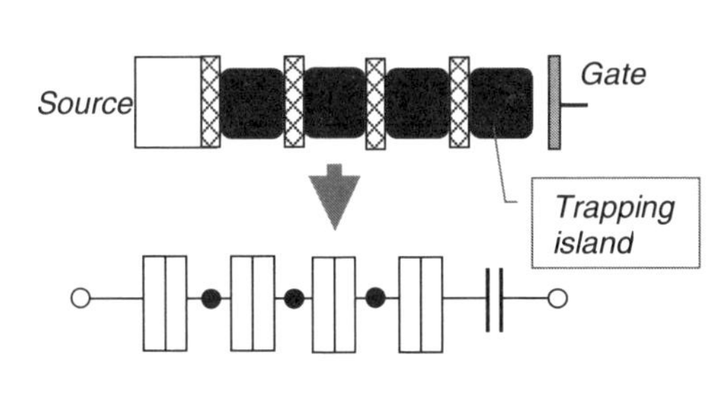

The single-electron trap includes several islands between the source and single-electron box.
The main new feature of this system is its internal memory (bi- or multistability): certain ranges of applied gate voltage cause the system to be in one of two (or more) charged states of its edge island. This is due to electric polarization effects: an electron located in one of the islands of the array extends its field to a certain distance outside.

(b)

FIGURE 2.1
Single-electron box (a) and trap (b).

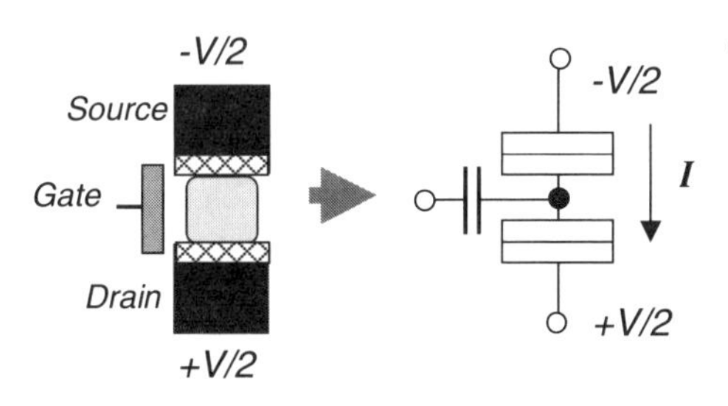

SET consists of a pair of tunnel junctions (barriers) (instead of the usual inversion channel as in MOSFET) separated by a small conducting island with an applied source-drain bias.
The island itself (which represents an isolated conducting region) is capacitively coupled through C_g to a gate bias.

FIGURE 2.2
Single-electron transistor.

Single-electron pump and turnstile

The single-electron pump (Figure 2.3) is a multiterminal device, that has a source and a drain like SET (see "Further Reading" Section). The drive input is fed by a triangular shaped radio-frequency (RF) waveform. The time shift between the triangular pulses applied to neighboring electrodes causes transfer of one electron per period between neighboring islands. This wave carries an electron from the source to the drain. The transmission in the device is bidirectional and does not need source-drain voltage: the direction of the transferred electron is determined by that of the running wave of electric potential.

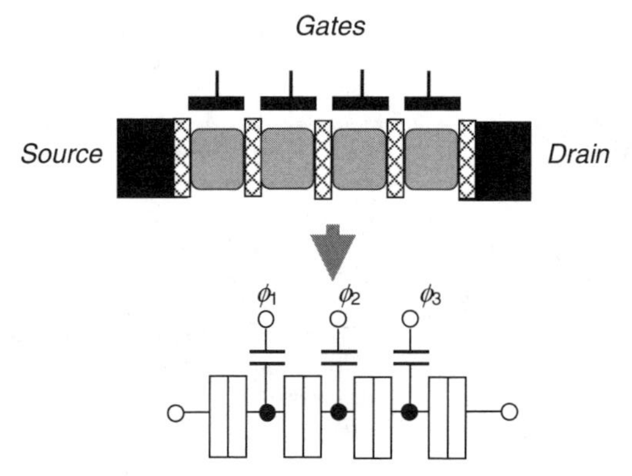

The single-electron pump operates like the single-electron trap: one electron may be pulled into the central island, increasing the gate voltage beyond a certain threshold; then it may be pushed out by decreasing the voltage.

RF waveforms applied to each gate electrode are phase shifted, forming a potential wave gliding along the island array.

FIGURE 2.3
Single-electron pump.

A similar design is a single-electron turnstile (Figure 2.4). The single-electron turnstile has one of the islands in the pump connected to an RF source; other islands are connected to external electrodes to compensate background charges with supply voltages. The voltage applied to the gate is periodical. Each period, one electron is transferred from the source to the drain.

Quantum-effect mesoscopic devices: quantum dot arrays

A hypothesis about the possibility to exploit the wave-mechanical properties of the electron has led to the development of the theory of *quantum effect mesoscopic* devices. These devices have been proposed to implement quantum computing using interference phenomena. However, it has been discovered at the same time that the utilization of these devices is ultimately limited by quantum decoherence.

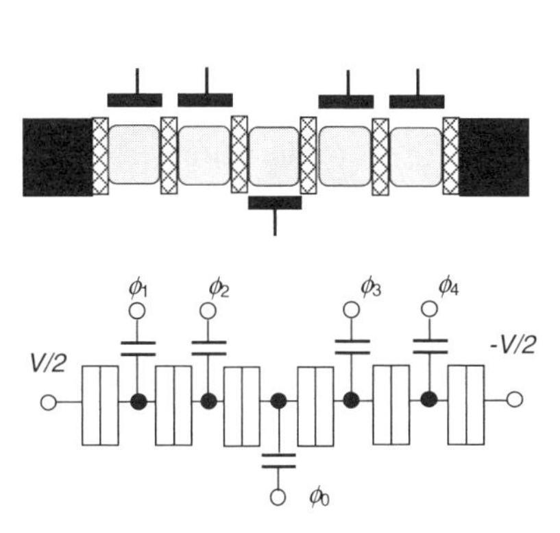

An electron is picked up from the source when the bias voltage ϕ_0 increases and is delivered to the drain when it decreases.

The voltage applied to the gate determines the direction of electron transfer in the first half of the cycle, while the RF potential applied to the central island lowers its electron addition energy.

Attracting more electrons is blocked by the Coulomb repulsion created by the first electron (assuming the amplitude of the RF signal is not too large).

In the second half of the time period, the potential in the middle island increases until the electron escapes into the drain.

FIGURE 2.4
Single-electron turnstile.

This category is associated with the organization of computing based on quantum waveguide behavior in the presence of mutually influenced "quantum dots." This category includes so-called *quantum-dot cellular automata.*

2.2.2 Rapid single flux quantum devices

The other class of similar voltage-state low-temperature devices, is that of rapid single flux quantum (SFQ) devices (see "Further Reading" Section). They exhibit a clock frequency beyond hundreds of gigahertz and extremely low power consumption. A rapid SFQ is a two-terminal device formed by a Josephson junction. It has unusual dynamics due to the macroscopic quantum nature of charge carriers called Cooper pairs (not Fermi particles, single-electron and hole) that have integer spin. These junctions:

- Allow generation of picosecond waveforms and
- Recover weak incoming pulses (this is called "return-to-zero").

The above mentioned properties make the rapid SFQ circuits quite robust, providing isolation between input and output. Digital logic circuits on *dynamic* SFQ devices use transient dynamics for information transfer: during the switching between the neighboring flux states, a short voltage pulse is formed across the junction. This flux change is quantized, as is the pulse area. These pulses are passed to other devices along superconductor Josephson transmission lines. Logical 1 and 0 are represented by the existence or absence of a single flux quantum.

Advantages of SFQ devices include:

- The speed of SFQ devices is very high while energy dissipation is very low.
- The SFQ pulse splitter and merger provide the necessary fan-in and fan-out.
- The pulse merger implements, at the same time, an OR function.
- Latches and flip-flops are also easily implementable in SFQ logic.
- Also, the natural internal memory is present in the quantizing SFQ loops.
- The clocked inverter, AND and EXOR gates are implementable around the quantizing loop.
- The interconnect problem is solved in rapid SFQ circuits by using Josephson transmission lines or passive superconductor lines.

The main drawback of the rapid SFQ devices is their low-temperature operation mode (4 to 5 K). Because rapid SFQ logic uses Josephson transmission lines for distributing clock signals to all gates to reset them, and for wiring gates, it complicates the design of logic circuits.

However, an SFQ circuit can effectively implement binary decision diagrams (BDDs). BDD topology can be directly mapped into a circuit by replacing the BDD node with a SFQ D_2 flip-flop (see "Further Reading" Section on details of this type of flip-flop).

2.2.3 Resonant-tunneling devices

There is one group that must be mentioned as the path on the evolution of nanoelectronics. Today, resonant tunneling transistors are the most established nanoscale devices because they already operate at room-temperature. Moreover, from the viewpoint of circuit applications their fabrication and interfacing with field-effect transistors (FET) and bipolar junction transistors (BJT) has reached an advanced level that allows the investigation of small scale circuits. Resonant tunneling devices are based on electron transport via discrete energy levels in double barrier quantum well structures (see "Further Reading" Section).

Resonant tunneling is applied in the bipolar quantum resonant tunneling transistor and similar devices as well as in gated resonant tunneling diodes.

2.3 Digital nanoscale circuits: gates vs. arrays

We will discuss the design of digital logic of the particular types of quantum-effect devices. The existing approach to design of digital nanoelectronic devices can be divided into two main streams corresponding to classification of single-electron interpretation of logic signals

- Voltage-state logic, and
- Charge-state logic.

2.3.1 Voltage-state logic: library of gates

SET-based devices are based on voltage-state logic, originating from traditional silicon logic.

Logic gates

The simple implementation of voltage-state NOT and EXOR gates is given in Figure 2.5 (see "Further Reading" Section).

In voltage state mode, the input gate voltage controls the source-drain current of the transistor which is used in digital logic circuits, similarly to the usual FETs. This means that the single-electron charging effects are confined to the interior of the transistor, while externally it looks like the usual electronic device switching multi-electron currents, with binary unity/zero presented with high/low DC voltage levels (physically not quantized).

The advantages of voltage state devices on SET include:

- The alternating transconductance of the single-electron transistor makes possible a very simple design of complementary circuits using transistors of just one type.
- Possible compatiblity with CMOS technology if the signals are treated in the same fashion, i.e., as voltage/current.

The following constitute drawbacks of SET based devices:

- They make the exact copying of CMOS circuits impossible; in order to get substantial parameter margins, even the simplest logic gates have to be redesigned.
- Their operation range starts shrinking under the effect of thermal fluctuations as soon as their scale reaches approximately 1 *nm*.
- Neither of the transistors in each complementary pair is closed too well, so static leakage current in these circuits is fairly substantial.
- The corresponding static power consumption is negligible for relatively large devices operating at helium temperatures. However, at prospective room temperature operation this power becomes on the order of 10^{-9} *W/transistor*. This number turn to an unacceptable static power dissipation for high-density circuits.

The maximum temperature may be increased by replacing the usual (single island, double junction) single-electron transistors with short 1-D arrays with distributed gate capacitances. For example, five-junction transistors allow a threefold increase of temperature at the price of bulkier circuits. However, in order to operate at room temperature even with this increase the transistor island size has to be extremely small in *nm*.

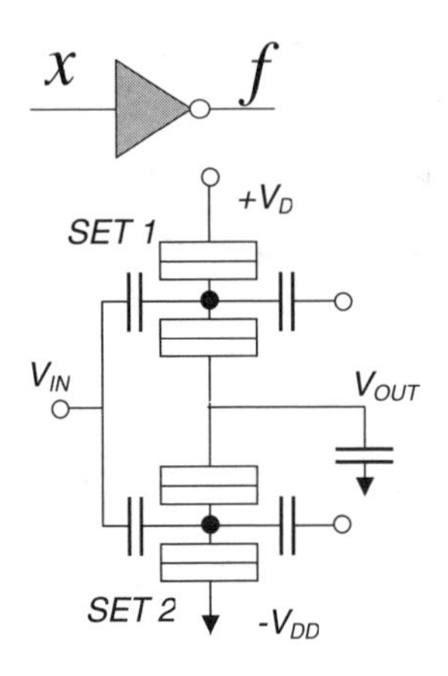

The circuit consists of:

- *Two single-electron transistors, which play the role of the pMOS, and*
- *nMOS transistors in a conventional CMOS inverter.*

In a similar manner to a standard CMOS circuit, the switch will be ON when the gate voltage is high, and conversely when the gate voltage is low, the switch will be OFF.

Unlike nMOS and pMOS devices

- *SET junctions are physically identical.*
- *CMOS inverters, output characteristics are actually periodic functions of the input rather than truly bistable behavior, due to the periodicity of the Coulomb oscillations.*

(*a*)

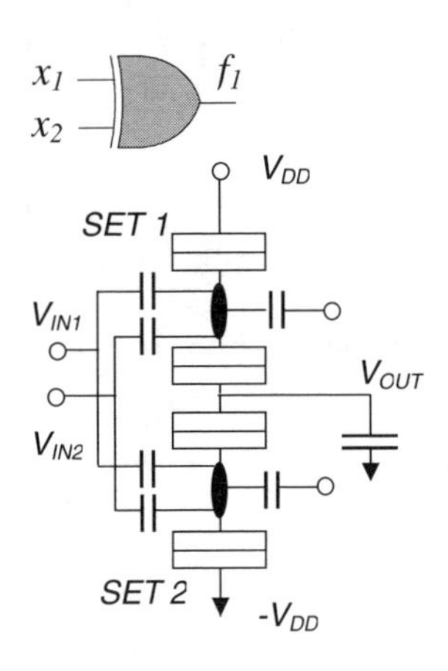

Two multiple-gate SETs are separated by coupled tunneling junctions.
Implementation involves only two transistors, in contrast to CMOS implementation. This effect is achieved due to the Coloumb blockade, which makes the current amplitude consistent with the electron charge.

(*b*)

FIGURE 2.5
NOT SET-based gate (a) and EXOR SET-based gate (b).

Multiterminal devices

The natural property of the multiterminal single-electron device is deployed for implementation of the *majority* operation. A majority gate can be created using a single-electron box (Figure 2.6, see also Section "Further Reading"). Binary logic values 0 and 1 correspond to negative and positive voltage of equal amplitude respectively.

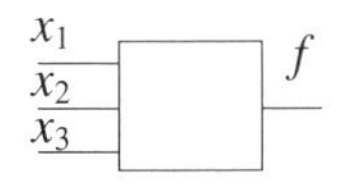

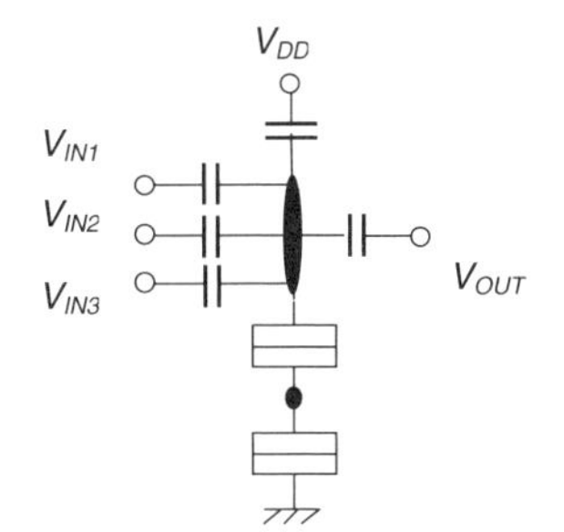

A majority gate consists of:

(*i*) *A double-junction single-electron box,*
(*ii*) *An island between them, and*
(*iii*) *Three input capacitances connected to the other island.*

The input capacitors:

(*i*) *Form the voltage-summing network, and*
(*ii*) *Produce the mean on the island.*

FIGURE 2.6
Single-electron majority gate with three inputs.

2.3.2 Charge state logic

In single-electron logic, the Coulomb blockade is used to suppress quantum fluctuations thereby enabling an elegant representation of bits by single electrons. This makes possible extreme device-level scalability.

Parametron

The most robust single-electron logic circuits suggested as yet are those based on the so-called *single-electron parametron*. (Figure 2.7).

Unfortunately, the parametron has several disadvantages:

- It has very narrow parameter margins, and
- No "wiring" is available in this type of logic – parametron is essentially a "shift register."

The latter means that this device possesses the internal memory, combining the functions of combinational logic and latches – a perfect property for cellular arrays that implement pipelined logic.

Parametron-based logic devices

In parametron-based circuits, an extra electron could be propagated along considerable externally timed shift register type segments of the circuit, while resistively coupled transistors provide splitting of the signal and binary logic operations. The sign of the electric dipole moment of the device (the field that eventually appears in between left and right islands) presents one bit of information. This "wireless" logic circuit can be constructed of parametron-

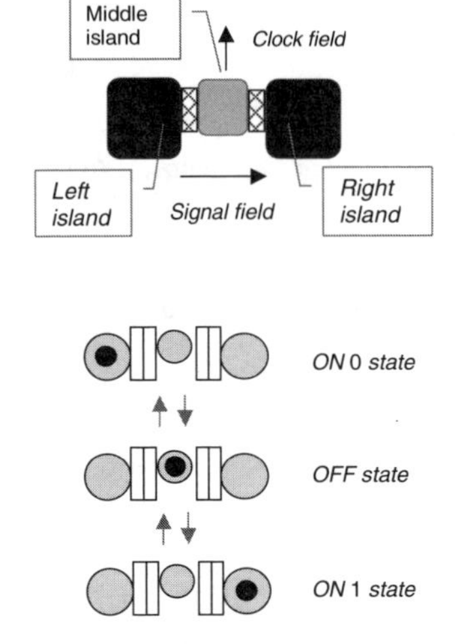

The simplest version of the device uses three small islands separated by two tunnel barriers and galvanically detached from electron sources. The central island is slightly shifted away from the center line. The parametron is biased by a periodic (clocking) vertically oriented electric field that keeps an extra electron in the central island during a part of the clock period:

(i) *If the field reaches a threshold value, the electron is transferred to one of the edge islands.*

(i) *A small additional field applied by a similar neighboring device determines the direction of electron tunneling.*

(i) *Further change of the clock field causes the electron to be trapped in one of the edge islands; the field is turned off, and the device becomes a source of the dipole signal field for the neighboring cells. The sign of this field presents one bit of information.*

FIGURE 2.7
Single-electron parametron.

based logic gates and fan-outs, due to a "fork-shape" geometry that provides NAND and NOR gate implementation considering one direction of signal (charge state) propagation through the chain of cells (parametron-based shift-registers), or fan-out considering opposite direction. The remarkable property of this shift-register is its quasi-reversible character, since the information is preserved by the cells (see "Further Reading" Section for details).

2.3.3 Single-electron memory

In charge state logics, the natural internal memory of logic gates can be used. The gates combine the functions of combinational logic and latches that enables the implementation of deeply pipelined and cellular automata architectures. A single-electron memory represents the ultimate scalability of current semiconductor memory technology, with potential memory storage densities on the order of 10^{11} bit/cm^2.

2.3.4 Switches in single-electron logic

Switches might be a clue to the resolving of the poor input-output gain in the device-centered (based on the library of logic gates) design considered above. Many nanocircuits, such as arrays of nanowires, charge state logic and other devices, are nonuniliteral, i.e., they have no clearly distinguished input and output voltage. In terms of logic design, these circuits act as BDD-based pass-gate logic, or switch (simplest multiplexer) based devices.

Switches and BDD-based models

The pass-gate logic requires representation of a function by a network of switches but gates. The corresponding model is a BDD. It also allows avoiding the interconnection problems, that are present in gate-level implementation, by mapping BDDs directly to the nanowire lattice. This approach is distinguished from netlist-based (network of logic gates), and requires representation of functions as BDD. It is also the closest to implementation on SET or rapid SFQ devices, in which the electron island in the SET can act as a switch. This is the reason this nonconventional (at hardware level) representation of a logic function in terms of BDD is the basic prerequisite behind the hypercube structures considered in the further chapters of this book.

Figure 2.8 considers a demultiplexer-based representation of a logic function of two variables (also called BDD-based structures, as a node of such diagrams is a simple demultiplexer).

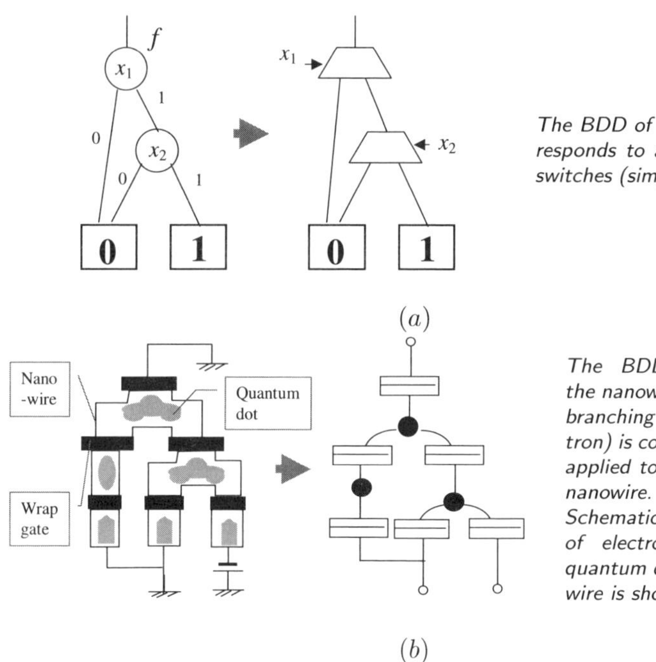

The BDD of the AND gate corresponds to a circuit formed of switches (simplest multiplexers).

The BDD is mapped to the nanowire network, where branching of the signal (electron) is controlled by voltage applied to a wrap-gate on a nanowire.
Schematically, the network of electron junctions and quantum dots formed on the wire is shown on the right.

FIGURE 2.8
A demultiplexer-based single-electron circuit.

2.3.5 Interconnect problem in voltage-state devices

In some circuits consisting of serially connected tunneling junctions, electrons can be transmitted as signals from a node to an adjacent node one by one by the use of clocked control signals, or by applying potential difference across the circuit. However, in these circuits, signals cannot be transmitted over long distances during one clock cycle, or the circuit parameters should be changed depending on the circuit length. In the circuits consisting of all automata, where each cell is electrically isolated, the displacement of electrons in a cell is transmitted as a signal, and, therefore, it is necessary to inject the same number of electrons into all cells in advance.

In some circuits consisting of single electron inverters the number of electrons can be controlled as a signal. In this design, signals cannot be transmitted bidirectionally, since single-electron inverters do not have such a function.

Charge-induced signal transmission (CIST) circuits use single electron junctions that function as two-way transmission passes and branch transmission passes (see "Further Reading" Section). CIST transmits the state of the presence or absence of an electron and a hole as a binary signal. This circuit can also transmit signals over long distances during one clock cycle.

2.3.6 Neuron cell and cellular neural network design using SETs

Once the electronic device can implement transfer of a signal restricted by a certain threshold T, it is called a *neuron.* The neuron's function is:

$$I = f(\sum_{i=1} w_i x_i) = \begin{cases} 1, \sum_{i=1} w_i x_i \geq T, \\ 0, \, otherwise, \end{cases} \tag{2.1}$$

Threshold logic can be easily implemented on single-electron devices. In a multigate SET, the current I_D is a function of input voltages and capacitances:

$$I_D(V_1, V_2, \ldots, V_N) = f(\sum_{i=1}^{N} \frac{C_i V_i}{e}), \tag{2.2}$$

where e is a single-electron charge ($e = 1.6 \cdot 10^{-19}$ *Coulomb*). Equation 2.2 indicates that the device implements a neuron function. However, the threshold understanding is different from the traditional one:

- ▶ In a traditional understanding, the neuron output is 1 (a device is ON) if the output supercedes some threshold value.
- ▶ In multigate SET, the current I_D takes a minimum when $\sum_{i=1}^{N} \frac{C_i V_i}{e}$ is an integer because the Coulomb blockade sets in.

Here, $\frac{C_i V_i}{e}$ corresponds to the number of excess electrons on the i-th input (gate electrode). When $\sum_{i=1}^{N} \frac{C_i V_i}{e}$ is a half-integer, the current flows because the Coulomb blockade is lifted. So, the multigate SET will implement the following threshold function

$$I_D(V_1, V_2, \ldots, V_N) = f(\sum_{i=1}^{N} \frac{C_i V_i}{e}) = \begin{cases} 1, \sum_{i=1}^{N} \frac{C_i V_i}{e} = \frac{2l-1}{2}, \\ 0, \, otherwise, \end{cases}$$

where l is an integer, C_i is a weight, and V_i is an input signal.

The two-input threshold gate and corresponding multigate SET is given in Figure 2.9. This gate can be a cell for a cellular neural network (CNN). In a CNN, a cell's input and output are locally connected in a weighted fashion to a neighborhood of identical cells. This CNN network is nonlinear, since the cells operate in a bistable mode, which is dependent on the weighted sum of the input voltages to the SET in the cell.

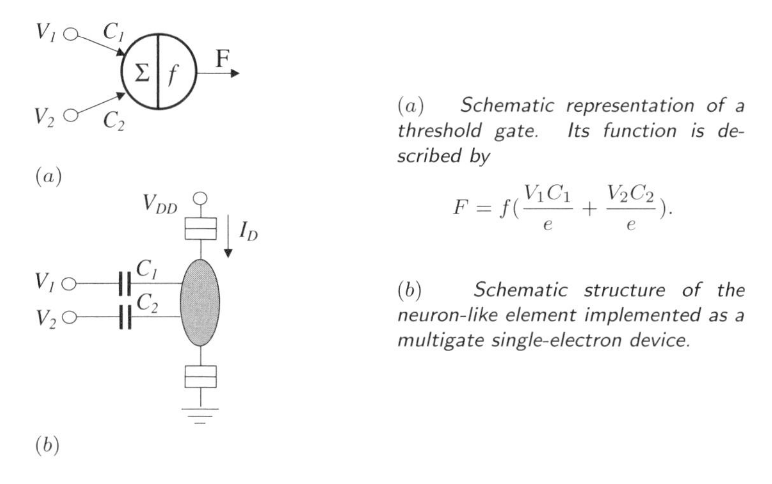

(*a*) Schematic representation of a threshold gate. Its function is described by

$$F = f(\frac{V_1 C_1}{e} + \frac{V_2 C_2}{e}).$$

(*b*) Schematic structure of the neuron-like element implemented as a multigate single-electron device.

FIGURE 2.9
Two-input neuron (a), and two-input multigate SET (b).

2.3.7 Single-electron systolic arrays

Locally-interconnected, synchronous networks can be built of SET devices, which makes possible representation of bits by single electrons and causes extreme device-level scalability. These have the potential for very high performance because both their devices and their architecture are highly scalable, perhaps even to molecular dimensions. They possess the following features:

- Digital representation in the devices is pushed to the ultimate limit and is provided directly by the quantization of electron charge.

- They are locally-interconnected to overcome interconnect bandwidth limitations and to mitigate discrete-physics device problems associated with limited gain, fan-out and impedance matching.
- The cell biasing is regular, which simplifies the design and clock distribution and makes the layout quite similar to a conventional charge-coupled-device.
- In circuits containing multiple electrons/bits, the screening length is short compared to the spacing between electrons. In the most existing Coulomb blockade circuits, only one electron is contained at a time.

The candidates for implementing these devices are:

- Electron-pump devices, and
- Electron-pump-like switches considered earlier.

Single-electron pump switch

In an electron-pump, electron motion occurs both as a result of gate biasing and via interaction with other single electrons in the circuit. In the electron-pump-like switch (Figure 2.10), the same biasing principle is used (see "Further Reading" Section). The switch can be used to design a logic family, in the same manner as the BDD-based circuit considered above.

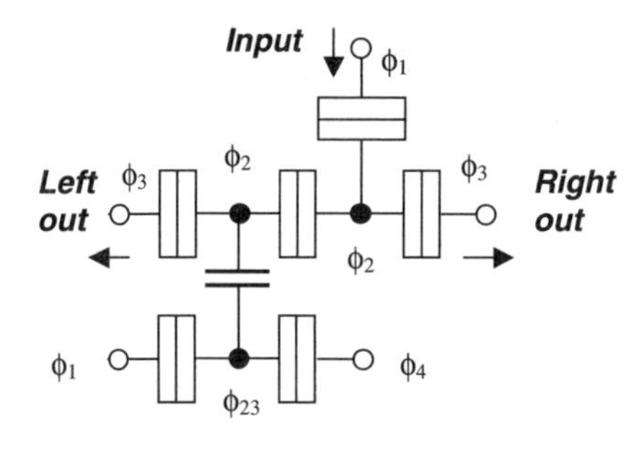

A single-electron pump switch includes

- *Coulomb islands,*
- *Gates biased with phases ϕ, and*
- *Tunnel junctions.*

In operation, an electron will go right or left, in a fully Coulomb-blockaded manner, based solely on whether or not there is an electron at the control node.

The switching direction does not depend on the gate biases, which are occupancy independent.

FIGURE 2.10
Single-electron pump-based switch.

A SET-based systolic processor consists of an array of processing cells, with each cell connected to its nearest neighbors only. It operates in a highly-parallel, pipelined fashion with data and results streaming through in regular, rhythmic patterns. Such processors can implement a variety of important signal processing functions such as convolution, correlation, Fourier transform and other matrix based computations. For instance, in this book matrix transforms are used for representation of switching and multivalued functions in various forms.

These schemes process anywhere from one bit (bit-serial approach) to one word (bit-parallel approach) in parallel. Nanoelectronic implementation seems to be more suitable for bit-parallel methods.

2.3.8 Parallel computation in nanoscale circuits: bit-level vs. word-level models

In Chapters 3, 4, and 10, bit-level and word-level graph based spatial models of computation will be considered in context of organization of massive parallel computation on nanostructures. It will be proven that:

- ▶ The parallel-pipelined architecture of cellular networks such as systolic arrays is well-suited for ultrasmall scale devices.
- ▶ They do not demand long global interconnection – a requirement that nanotechnology can hardly ever provide.
- ▶ They perfectly fit the idea of manipulation with single bits or words of bits represented by single particles, i.e., single-electrons.
- ▶ Locality of cellular network is easily achieved in the devices that are wave-guarded (such as mesoscopic devices).
- ▶ The expected 3-dimensionality of the nanoscale devices is consistent with the idea of cellular design for massive parallel computation.

2.4 Molecular electronics

The principles of logic that utilize charge states, or counting electrons, are not material-sensitive. Therefore, single-electron devices can be implemented using any acceptable basis. This broadens the perspective for single electronics.

2.4.1 CMOS-molecular electronics

The chemically directed self-assembly of molecular single-electron devices on prefabricated nanowires is considered as a promising way toward integrated circuits using, for example, single-electron tunneling.

CMOL

The single-molecule single-electron transistors are characterized as follows:

- ▶ They can operate at room temperature,

- ▶ The molecules can self-assemble on gaps of a few *nm* between prefabricated nanowires,
- ▶ The nanowires may allow us to connect molecular single-electron devices to larger (and hence more sparse) silicon nanoMOSFET devices.

The general idea of such CMOS/nanowire/MOLecular (CMOL) hybrid circuits is to combine the advantages of single-electron devices (very high density) with those of advanced field-effect transistors (high voltage gain and high reproducibility).

This approach uses an inorganic molecule between two electrodes and makes use of quantum tunneling. Such single-electron devices may be used to perform the functions that are natural for them:

- ▶ Compact memory cells or
- ▶ Externally-controlled switches.

They may leave the more difficult functions (e.g., signal amplification with high voltage gain) for CMOS circuits.

Neuromorphic circuits

The architecture can be based on so-called *distributed crossbar arrays* for *neuromorphic networks* based on switches (see "Further Reading" Section). Cells that may be implemented in nanoscale CMOS technology) are embedded into a 2-D array of single-electron latching switches working as single-bit-weight synapses. In a neuromorphic network,

- ▶ Each cell is hard wired to a limited number of other cells, with discrete synaptic weights controlling which of these connections are currently active, and
- ▶ Vice versa, the signal activity of the network determines whether the synapses are open or closed, though the state of any particular synapse is also affected by the underlying randomness of single-electron tunneling.

Such self-adaptive networks can be also trained, using global reinforcement training techniques, to perform more complex information processing. For example, this methodology can be used in minimization of logic functions that are incompletely specified or too large for exact minimization. Moreover, it is expected that a sufficiently large hierarchical system will be able, after a period of initial training, to self-evolve and self-improve on the basis of its interaction with the informational environment.

Molecular technology is considered as the possible technological platform for single-electronics, as molecular devices can operate at room temperature, since the principal drawback of the current single-electron devices is low temperature.

2.4.2 Other structures: nanowires

Nanowires are of particular interest for two reasons:

- Nanowires show quantum-mechanical confinement effects and have the ability to connect to individual molecules,
- They have a very high surface-to-volume ratio, enabling them to be used as sensors.

There is no simple method to fabricate nanowires over large areas in arbitrary material combinations (see "Further Reading" Section). The most interesting are the nanowire patterns of branched "nanotrees." There can be three branches at 60-degree angles, four branches at 90-degree angles, and six branches at 30-degree angles. Using today's technology, nanotrees can be created on 2-D thin film. However, with the progress of technology, 3-D nanotrees, e.g., trees embedded in hypercubes (inherently crystal topology) are becoming a feasible possibility (see "Further Reading" Section).

2.4.3 Nanotechnology-enhanced microelectronics

The carbon nanotube FET (CNTFET) is a typical example of nanotechnology-enhanced microelectronics. The fabrication of CNTFET is straightforward and can be performed utilizing modified CMOS processes. The gold electrodes can be formed on a doped Si with a thermally grown silicon oxide that serves as the gate insulator. Then, carbon nanotubes are dispersed on the wafer. The substrate serves as the gate electrode. A transition metal (titanium or cobalt) is patterned on the wafer as the source/drain contact metal. The CNTFET is equivalent, in terms of performance, with that of conventional MOSFETs. 2-D functional structures have been utilized as crossbar assemblies that can be viewed as quasi 3-D nanoICs.

Another nanotechnology-enhanced type of a circuit is one that combines MOSFET and SET to achieve superior integration packaging (see "Further Reading" Section). Inconsistency in the operation temperature is the main obstacle in this direction; this can be, perhaps, resolved by the progress of technology.

2.5 Scaling and fabrication

Further constraints posed by technology are an important aspect to consider for designing the nanoscale devices.

2.5.1 Scaling limits of electronic devices

The steady downscaling of transistor dimensions over the past two decades (from 10 mm to 0.2 mm) has significantly affected the growth of silicon-based ICs, ensuring high performance (high switching frequency, reliability with lower power consumption and dissipation), density, reliability, etc. Further rescaling to shorter channel length, simultaneously leads to improvements and drawbacks. It is anticipated that room temperature CMOS processes can be scaled down to 18 *nm* channel length, while low-temperature CMOS can potentially extend the scaling limit to 9 *nm*. As the channel length is scaled to the nanometer range (less than 100 *nm*), the electrical barriers in the semiconductor device begin to lose their insulating properties due to thermal injection and quantum tunneling. This results in a rapid rise of standby power, limiting integration level and switching speed. When the MOSFET dimensions are scaled down, both the voltage level and the gate-oxide thickness must also be reduced. Since the electron thermal voltage (kT/q) is constant for room temperature, the ratio between the operating voltage and the thermal voltage decreases. This leads to higher source-to-drain leakage currents due to the thermal diffusion of electrons. The gate oxide has been scaled to a thickness of a few atomic layers (the atomic radius of *Si* is 0.1176 *nm*), where quantum tunneling leads to a significant increase in gate leakage currents degrading the performance.

The major fundamental limiting factors are:

- Electron thermal diffusion,
- Tunneling leakage through the gate oxide, and
- 2-D electrostatic scale length.

Emphasizing the power consideration with the corresponding limits, some well-known physical limitations are reported for MOSFETs.

The threshold voltage V_t is defined as the gate voltage at which significant current begins to flow from the source to the drain. Below the threshold voltage, the current does not decrease to zero instantly. It decreases exponentially with a slope in the log scale inversely proportional to the thermal energy kT. This is due to the fact that the thermally excited electrons at the source of the transistor have the energy to overcome the potential barrier controlled by the gate voltage and flow to the drain. This subthreshold behavior is due to thermodynamics and is independent of power-supply voltage and channel length.

The other consequence of scaling is the gate-oxide tunneling effect. To keep adverse 2-D electrostatic effects on threshold voltage (short-channel effects) under control, gate-oxide thickness is reduced proportionally to the channel length. For CMOS devices with channel lengths of 100 *nm* (0.1 *mm*) or less, a required silicon oxide thickness must be on the order of 3 *nm*. This thickness approaches fundamental limits due to quantum-mechanical tunneling. The gate leakage current increases exponentially as the oxide thickness is reduced through scaling it down. This tunneling current significantly affects the standby power. The limit for the gate-oxide thickness is in the range of 1 *nm*. It should be emphasized that for thin gate oxide, inversion charge and transconductance degrade due to inversion-layer quantization and polysilicon-gate depletion effects. The density of inversion electrons peaks at 1 *nm* below the silicon. This effectively reduces the gate capacitance. Therefore, the inversion charge changes to that of an equivalent oxide 0.4 *nm* thicker than the physical oxide. Hence, the scaling limit of silicon oxide thickness leads to 1.5 *nm* of thickness.

Scaling below 100 *nm* channel length faces several fundamental difficulties. Using properly optimized doping profiles, the silicon depletion width, meanwhile approaching the tunneling limit, it is likely that 20 *nm* channel length with nanometer gate oxides and below 1 *V* voltage levels can be achieved for high-yield fabrication.

Considerable progress has been made in optimizing oxide/nitride and oxynitride dielectrics to reduce boron penetration and dielectric leakage compared to pure silicon. Promising alternative materials have emerged, and are capable of 1 *nm* equivalent oxide thickness. However, even using these novel materials, it seems that the equivalent oxide thickness cannot be reduced to less than 0.6 or 0.8 *nm*. For junctions, the main challenge lies in providing low parasitic series resistance as depths are scaled down in order to reduce short-channel effects. Because contacts dominate parasitic resistance, low-barrier-height contacts and/or very heavily doped junctions are required. While ion implantation and annealing processes can be enhanced to meet junction-depth and series-resistance requirements, low-temperature deposition processes may be needed.

Thus, scaling down semiconductor devices beyond 65 *nm* CMOS technology requires overcoming formidable fundamental and technological limits. In addition to quantum-mechanical tunneling of carriers (through the thin gate oxide, from source to drain, and from drain to the MOSFET body), other issues must be resolved. In particular, control of the density and location of dopant atoms in the channel and source/drain region to ensure high on and off currents, as well as finite subthreshold slope.

Theoretically, for optimal (ideal) ballistic MOSFET, the ultra-thin (2 *nm*) undoped channel length may be 10 *nm* (only hundreds of *Si* atoms) with two gates (below and above the channel). However, this may be the absolute limit. Finally, as was emphasized, this nanotechnology-enhanced microelectronics cannot be viewed as nanoelectronics. The CNTFET, considered above, is a typical example of nanotechnology-enhanced microelectronics.

2.5.2 Operational limits of nanoelectronic devices

The conditions affecting operation. Physical effects in nanoelectronic devices are different from microelectronic ones, but they also have limitations that must be taken into account in logic design, modeling and analysis. The conditions affecting operation of single-electron devices include:

- Temperature,

- Fabrication technology,
- Control.

Temperature. The most important condition for operation of SETs is temperature. For a room-temperature SET with $V \approx 1$ *Volt*, the total current would be below 10 μA. Since the width of such a transistor has to be very small, 1 nm, the available current density may be well above $1,000$ $\mu A/\mu m$, i.e., even higher than that of standard silicon MOSFETs. The latter condition makes the practical implementation of SETs operating at room temperature rather problematic.

The technological processes must be ultraprecise to achieve the required characteristics. Both theory and experiments show that single-electron tunneling effects become visible at the electric potential $E_a \approx 3$ kBT. This means that in order to notice these effects at $T \approx 100$ mK, E_a should be above 25 μeV, corresponding to the island capacitance $C \approx 5 \times 10^{-15} F$ and island size of the order of 1 micron, with tunnel junction area 0.1×0.1 μm^2. Such dimensions can be reached by several methods such as metal evaporation from two angles through a hanging resist mask. However, for reliable operation of most digital single-electron devices, the single-electron addition energy should be at least 100 times larger than 1 kBT. This means that for room temperature operation, this value corresponds to an island size of about 1 nm. The parameters of such transistors are rather unpredictable, so that they can hardly ever be used in nanoICs.

The other option is fabrication of discrete transistors with scanning probes, by nanooxidation of metallic films or manipulation with carbon nanotubes. Unfortunately, the current in these transistors is very low (below 10^{-11} A), though for discrete devices this approach may be promising. Several methods taken from the standard CMOS technology have also been used to fabricate single-electron transistors, mostly by the oxidation of a thin silicon channel until it breaks into one or several tunnel-coupled islands. The disadvantage of this approach is that the parameters of the resulting transistors are difficult to reproduce.

2.6 Summary

In this chapter, we reviewed both technological possibilities for nanoscale fabrication:

- Electronic (semiconductor/superconductor), and
- Molecular.

The first group involves two main groups of nanoelectronic devices:

- Those based on phenomena of quantum interference, and
- Those based on single electron effects.

1. Single-electron devices are frequently considered as the most probable candidates for the replacement of silicon MOSFETs. The reason for this expectation is that the physics of single electron device operation are not sensitive to fabrication material, and hence these devices may be fabricated from a broad variety of materials, in particular, chemically synthesized molecular devices on prefabricated metallic nanowire structures.
2. The challenges on the way to digital application of single-electron devices are:
   - Low voltage gain, and
   - High sensitivity to single charge impurities in the dielectric environment at room temperatures.
3. The most realistic implementation is a few-electron memory cell that can be scaled down to 30 nm^2 cell area and demonstrate 1-ns-scale write/erase time. Moreover, for room temperature operation, they should not be larger than 3 nm, as opposed to 1 nm for logic circuits. Logic circuits are heavily affected by temperature, an increase above few K causing random background charge effects.
4. The hybrid circuit single-electron devices called *CMOL*, together with nanoscale MOSFETs and nanowires, may become the basis for implementation of novel, massively parallel architectures for advanced information processing. Such systems may eventually replace traditional digital processors.
5. The tremendous technological achievements of the past few decades lead us to address some architecture design issues in nanoelectronic circuits and systems in the future:
   - Nanodevices, both nanoelectronic and molecular, are expected to be locally interconnected 3-D structures. They can implement massive parallel computations and non traditional logic because of interconnect bandwidth considerations and because of device limitation in gain, fan-out and impedance matching.
   - Nanoelectronic devices can be classified based on the observed phenomena (waveguide or single electron) into two groups: those relevant to quantum computing (so far theoretical only) and those relevant to traditional von Neumann computing model implementation. In both, however, spatial and geometrical properties are critical, due to the ultra-small size of the proposed particles.
   - Computer aided design (CAD) of nanoelectronic ICs and nanoICs relies on the data structure inherited from classic switching theory as well as that especially for logic circuit design in 3-D. The problem of nanoIC design is formulated in this book as a problem of

creation of appropriate 3-D data structures and algorithms of their manipulation.

2.7 Problems

Problems 1 and 2 can be used as assignments. The rest of the problems can be considered as the projects.

1. *Sketch the schematic of the BDD-based single-electron network representing the BDD given in:*

(a) Figure 2.11a
(b) Figure 2.11b.

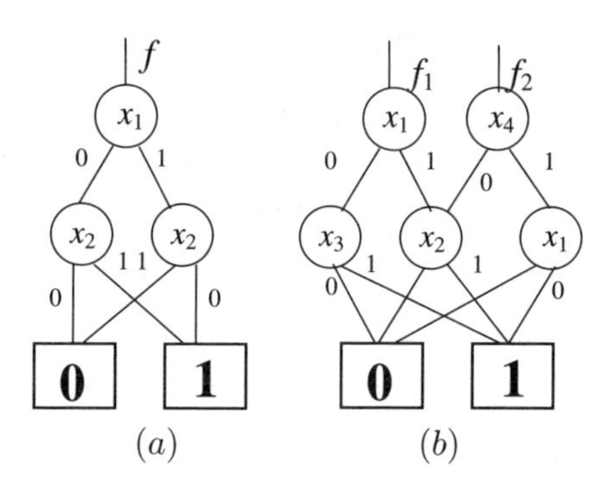

FIGURE 2.11
BDD that represent EXOR function, and shared BDD (Problem 3).

2. *Evaluate the parameters (number of tunnel junctions, quantum dots, and terminals) of the following implementations of the BDD shown in Figure 2.11a:*

(a) on a SET based multigate device
(b) on a BDD based nanowire network
(c) on single-electron pump switches

To prove functionality use PSPICE or SIMON simulation tools.

3. *Propose your design of the voltage state circuits:*

(a) AND gate
(b) OR gate
(c) 3-input EXOR.

4. *Consider implementation of a 7-input majority gate on a 3-input single-electron majority gate (see Figure 2.6).*

2.8 Further reading

Theoretical models of design of nanodevices. Although an advanced deep submicron electronic circuit will be the mainstream technology in at least ten years, novel nanoelectronic device concept is being actively investigated [16, 18, 19, 39, 40, 50].

Quantum-effect devices. Solid-state nanoelectronic devices are generally understood as devices that demonstrate quantum effect and single electron phenomena. The latter have been studied in physics for a long time, without consideration of electronic applications. It has been discovered that the mentioned phenomena at quantum level does not perfectly fit the deterministic models of conventional electronic due to quantum decoherence and other implications [21].

Single-electronics. The key theory of single electronics, *orthodox* theory was developed by Kulik and Shekhter [32]. This theory states that the tunneling of a single electron through a particular tunnel barrier is always a random event, with a certain probability per unit time. For complex systems, the space of all possible charge states may become too large, and the only practical method is to simulate the random dynamics of the system by a Monte Carlo method [63].

Single-electron transistor. The concept of nanocomputing based on Josephson junction technology is discussed in [38, 58, 63]. This is based on the theory of the Bloch-wave oscillations in small Josephson junctions [34] and the phenomenon of Coulomb blockade of tunneling [35]. Coulomb blockade and associated single-electron behavior such as the Coulomb staircase were first observed in metal-oxide tunnel junction systems in the 1980s [1].

A voltage-state single-electron inverter design that is a pair of SET transistors that is similar to CMOS design paradigm, has been proposed in [62]. Likharev provides an exhaustive review of SET-based voltage state logic gates

[38]. For the simulation of single electron tunneling, Monte Carlo techniques are utilized [63]. In Monte Carlo simulation, stochastic tunneling events across all possible junctions are simulated in time using a random number generator to generate the time between tunneling events. Commercial simulators are available such as SIMON 2.0 [63], which provide schematic capture for design and simulation of single electron circuits. Modeling of SET devices on SPICE, SIMON and other tools have demonstrated the expected behavior of separate logic gates.

Poor input/outout transfer makes it difficult to implement large circuits of the above voltage logic devices, unless another principle such as charge state and cellular design are used. However, some resolutions are possible: a dual-rail transmission line based on the principle similar to one for single-electron pump and called charge-induced signal transmission (CIST), has been proposed by Yamamura [65]. It is not necessary to place electrons in advance in each cell of CIST because the circuit has no node that is electrically isolated. Since these properties are suitable not only for a signal transmission pass but also for a logical circuit, the feasibility of realizing basic logic elements with CIST circuits: AND, OR, NOT logic, and so on, has been investigated.

Memory on SET. Memory on charge state logic gates can be implemented on a parametron-based circuit that operates in shift-register fashion. This feature is very similar to that of the ultrafast RSFQ logic [37]). Various proposals include the single-electron flip-flop composed of cross-coupled SETs [30], and the single-electron trap [46]. The most commercialized single electron memory [66] is based on naturally occurring islands formed in single grains in a polysilicon film. This technology has been employed by Hitachi to fabricate a 128 *Mb* memory [67].

Switches and BDD nodes on SET. Mapping a ROBDD into a hexagonal nanowire network controlled by Schottky wrap gates has been introduced in [1, 4]. A T-gate demultiplexing an entry into two branches can be easily implemented using the designated technology called PADOX [1]. Both implementation and modeling aspects have been reported in [1, 2, 4]. Physical prototyping of the gates, including T-gate, based on PADOX technology has been reported.

Multivalued logic on SET. Takahashi et al. proposed to implement a quantized multivalued signal, a multigate circuit [58]. Signal quantization is a property of SET-based devices that is remarkably well-suited for representation of multivalued systems. Other multivalued logic devices using multigate SET have been reported by Inokawa and Takahashi [27]. Recently, the digital circuits combining SET and MOSFET have been reported by Degawa et al. [10]. Oya et al. [48] proposed to implement a majority device using an irreversible single-electron box. This implementation is based on a multigate single-electron device (up to three inputs), that can be connected into a chain to increase the number of inputs.

Neural networks on SET. Threshold logic gate implementation on SET has been reported in [27, 58]. In the multigate single-electron device, the output current is the function (not threshold) of the weighted sum of the input voltages, with weights determined by the capacitance value of each input signal (see Section 2.3.6). This device can serve as a cell in a neural-like network [56, 68].

Another approach for designing analog neural network is based on the concept of CNN. The CNN can be defined as an array of $a \times b \times c$ identical cells arranged in a spatial grid. Each cell mutually interacts with its nearest neighbors. In the most common case, a set of $2(2 \times r + 1)^n) + 1$ parameters, where r is a neighbor radius, completely defined a functional behavior of a CNN in n dimensions. The SET realization of a neuron in such a CNN [19] is based on the circuit proposed by Goossens et al. [22]. It consists of the Tucker SET inverter [62] which provides a bistable behavior and has multiple capacitive inputs to the inverter to form a voltage summing node such that the input voltage is the weighted sum of the external voltages, with weights determined by the capacitance value of each input signal.

Charge-state logic: parametron. A device that implements charge state logic using small conducting islands separated by tunnel barriers, while timing and power are provided by an external RF field, was first explored theoretically in 1987 [36]. It has been shown by Korotkov and Likharev [31] that a special geometry arrangement of such a device enables implementation of logic gates as well as fan-out, and assembling the "wireless" logic circuits.

Quantum effect mesoscopic devices: dots and waveguides. This class of devices are also called *electron waveguides*, and are believed to be solid-state implementation of quantum computing. Two single-mode waveguides, coupled to each other by means of a short potential barrier have been proposed as a realization of a quantum-mechanical bit (qubit) [5]. A wave-packet launched in one of the waveguides will oscillate back and forth between the two branches, as it passes through the region where the potential barrier is located. The electron can be switched between the two waveguides, which are therefore used to represent the two logic states of the qubit. An alternative approach, based on the use of plane waves, rather than wave packets, has also been proposed [23]. The electron-wave probability can be switched back and forth between the two waveguides, by controlling the length of the coupling window. Unfortunately, no practical implementations of the waveguide-based qubit have been demonstrated.

Quantum dot arrays is another class of mesoscopic devices aimed at implementation of "wireless" computing in solid-state electronics [33]. Quantum dot cell performs based on four phase adiabatic clocking scheme (computation, cell locked, cell relaxed, and inactive). It enables signal propagation within the arrays of quantum dots. However, the original idea of quantum dot arrays experiences major practical problems because of the unsolved questions of mesoscopic devices:

- Coherent hybridization of quantum dot levels, similar to that well studied in atomic and molecular physics results in tunneling from an island to bulk electrodes that brings a substantial dissipation into this system and suppresses the system coherence (see, e.g., [18, 21]). In general, energy quantization effects have not yet led to suggestions for any new practical applications, with the exception of single-electron spectroscopy, and some highly controversial logic device proposals.
- Coherence of the electron waves is disrupted by phase-randomizing scattering. Among the possible sources of this scattering are electron-phonon and electronelectron interaction. Another effect of increasing temperature is thermal smearing of the electron population near the Fermi level, and this effect combines with scattering-induced decoherence [21].

Reversible logic. An alternative class of machines intensively studied today is based on asymptotically reversible logic devices which entail no minimum entropy per operation [54]. They are organized in such a way as to permit arbitrary patterns of logically reversible information processing in 3-D space. Recently, a plenty of techniques for logic synthesis of reversible has been proposed [26, 42, 43], including multivalued reversible circuits [44].

So far, reversible computation has not gone past the theoretical stage. Nanoscale devices must, however, obey the fundamental constraints that the laws of physics impose on the manipulation of information, e.g.:

- Some physical constraints, such as the fact that the speed at which information can travel optically through free space is upperbounded by a constant (the speed of light, 299 792 458 m/s), are still present in nanodimensions. The electrical transmission of signals along wires is slower than light, so the current propagation delays along dissipative wires are significant.
- The time of transmission is proportional to the square of the distance, unless the signals are periodically regenerated by additional buffering logic.

Resonant tunneling has been first observed in 1974 [6]. Since then, the progress in heterostructure epitaxy has lead to quantum effect devices with adjustable peak current densities and peak voltages (e.g., MOBILE technology [3, 8]). Resonant-tunneling devices' application is primarily in the area of multivalued logic [7], and threshold logic gates [8]. A second way to implement a threshold gate on the device level is the neuron MOS transistor [55].

It must be emphasized that a circuit family based on threshold logic is able to compute every arbitrary switching function in a network with several layers. Linear threshold gates have been proposed in [3]. This gate consists of two resonant tunneling diodes in series whose peak current is modulated by a gate or by means of a parallel connected FET. Applying a bias voltage oscillating between the peak voltage and the double peak voltage, the output node is either monostable or bistable.

Rapid single-flux-quantum technology. This technology is rather submicron, however, these low-temperature devices precede single-electron devices [28]. Flux quantum devices have been first proposed and evaluated numerically in [37], including logic switches [70]. This SFQ D_2 flip-flop is switching the pulse (the messenger), the particle representing the state of the system, from the root into one of the branches, depending on its internal state.

An approach to designing rapid SFQ logic circuits based on binary decision diagrams has been proposed in [69]. It should be noted that BDD devices on SFQ are dual rail, and they are data driven and self-timed, so no clock signal distribution is required. The only requirement for the timing is that the messenger has to arrive at the gates after the transition caused by input data.

Molecular technology and other perspectives. The chemically directed self-assembly of molecular single-electron devices on prefabricated nanowires is considered by Fendler [14]. The single-molecule SETs operating at room temperature have been demonstrated experimentally [49, 71].

The main problem now is to synthesize molecules that would combine suitable device characteristics with the ability to self-assemble, with high yield, on a few *nm* gaps between prefabricated nanowires. The general idea of such CMOS/nanowire/MOLecular hybrid circuits is to combine the advantages of single-electron devices [61], on one side, and room temperature molecule scalability.

A three-terminal version of the inorganic molecule device earlier suggested by Collier et al., and Ellenbogen and Love [9, 12], has been proposed by Fölling [15]. The device is a simple combination of the SET and the trap, working together as a latching switch.

A three-terminal CMOL device is a simple combination of the SET and the trap, working together as a latching switch. The device consists of three small islands connected by four tunnel junctions. The first island, together with input and output wires serve as a source and drain, forming a single-electron transistor; other islands form a single-electron trap. If the effective source-to-drain voltage applied from two inputs of the device is low, the trap in equilibrium has no extra electrons, and the transistor remains in the Coulomb blockade state – input and output wires are disconnected. If the source-to-drain voltage increases beyond a certain threshold (which should be lower than the Coulomb blockade threshold voltage of the transistor), one electron is injected into the trap.

The latching switch is sensitive to random charge impurities, e.g., if the source-drain voltage is low for a long time, either thermal fluctuations or cotunneling forces the trapped electron out of the trap and the transistor closes, disconnecting the wires. Therefore, the device can be used in redundant circuits that may tolerate at least a small percentage of faulty devices.

A possible single-molecule implementation of such a device called *CrossNets* is discussed by Türel and Likharev [61]. Estimates show that neural cell density may be as high as 108 cm^2, implying that brain-scale systems could be implemented on a silicon area on the order of 10×10 cm^2. The estimated speed of scaling of this network is even more impressive, with the time of signal propagation between the neural cells on the order of 100 ns, at acceptable power dissipation, with the speed six orders of magnitude higher than that in biological neural networks. In addition, the system is capable of self-evolving at a high speed.

Nanowires and nanotrees. Nanowires can form building blocks to create logic gates [25]. Since the possibility of creating nanowires on silicon technology has been demonstrated (Harvard, Berkeley, and Lund University), they have been considered as reasonable candidates to move forward to the generation of well-defined nanowire network structures on almost any solid material, up to macroscopic sample sizes.

One recent method (Lund University) is to create nanowires with diameter <16 nm within cracks in a thin film; the cracks are straight, scalable down to nanometre size, and can be aligned via strain. Creation of patterns of branched "nanotrees" is controlled by seeding of multiple branching events: first, tiny nanoparticles of gold are created and placed on a semiconducting tray. Then reactive molecules are released that contain the atoms to form the nanowires. The reactive molecules seek out the catalytic gold particles and build crystals on the tiny contact surface directly under the gold. The wires are on the scale of a few micrometers (1,000ths of mm) in length and about 50 nanometers thick.

In the second step new gold particles are then sprayed onto the nanowires, and the procedure is repeated. Now new "branches" are grown at sites where the gold particles landed. The number of branches grown is determined by the crystal structure of the trunk. There can be three branches at 60-degree angles, four branches at 90-degree angles, and six branches at 30-degree angles. In this way veritable forests of nanotrees can be created.

Carbon nanotube devices. Derycke et al. and Martel et al. consider logic gates based on carbon nanotube (CNT) devices [11, 41]. CNTFETs with single- and multiwall carbon nanotubes as the channel were fabricated and tested by Bachtold et al. [4].

Scaling and fabrication. Another option is fabrication of discrete transistors with scanning probes, for example by nanooxidation of metallic films [45] or manipulation with carbon nanotubes [51]. For the former devices, single-electron addition energies as high as 1 *Volt* have been reached, but unfortunately, the current in these transistors is very low (below $10^{-11}A$). Fabrication of single-electron transistors is accomplished by the oxidation of a thin silicon channel until it breaks into one or several tunnel-coupled islands, see, for example, the work by Takahashi et al. [57] and recent results by Saitoh et al. [53].

CAD of ICs. Some models to describe 3-D submicron structure have been proposed in the past [24]. At the same time, a lot of 3-D graph-based topologies have been proposed for massively parallel computing (see [11] and also "Further Reading" in Chapter 5) on the system level, but low-level logic design. In the coming era of nanodimensions, spatial topology might be considered at the logic design level, since the entire design process is becoming horizontal rather than vertical in terms of consequent top-down steps [16]. Hypercube-like technologies have been reported for the first time by Shmerko and Yanushkevich [56]. 3-D aspects of nanoICs have not been given the appropriate attention as of yet, except for the results by Endoh et al., and Goldstein and Budiu [13, 17].

2.9 References

[1] Averin DV, and Likharev KK. Coulomb blockade of tunneling, and coherent oscillations in small junctions. *Journal Low Temperature Physics*, 62:345–372, Feb., 1986.

[2] Asahi N, Akazawa M, and Amemiya Y. Single-electron logic device based on the binary decision diagram. *IEEE Transactions on Electron Devices*, 44(7):1109–1116, 1997.

[3] Akeyoshi T, Meazawa K, and Mitzutani T. Weighted sum threshold logic operation of MOBILE (monostable-bistable transition logic element) using resonant-tunneling transistors. *IEEE Electron Device Letters.* 14(10):475-477, 1993.

[4] Bachtold A, Hadley P, Nakanishi T, and Dekker C. Logic circuits with carbon nanotube transistors. *Science*, 294:1317-1320, 2001.

[5] Bertoni A, Bordone P, Brunetti R, Jacobini C, and Reggiani S. Quantum logic gates based on coherent electron transport in quantum wires. *Physical Review Letters*, 84:5912–5915, 2000.

[6] Chang LL, Esaki R, Tsu R. Resonant tunneling in semiconductor double barriers. *Applied Physics Letters*, 24(12):1805–1806, 1974.

[7] Chen KJ, Waho T, Maezawa K, and Yamamoto M. An exclusive-OR logic circuit based on controlled quenching of series-connected negative differential resistance devices. *IEEE Electron Device Letters*, 17(6):309–311, 1996.

[8] Chen KJ, Maezawa K, and Yamamoto M. *InP*-based on high-performance monostable-bistable transition logic elements (MOBILE)'s using integrated multiple-input resonant-tunneling devices. *IEEE Electron Device Letters*, 17(3):127–129, 1996.

[9] Collier CP, Wong EW, Belohradsk M, Raymo FM, Stoddart JF, Kuekes PJ, Williams RS, and Heath JR. Electronically configurable molecular-based logic gates. *Science*, 285:391–394, July, 1999.

[10] Degawa K, Aoki T, Higuchi T, Inokawa H, and Takahashi Y. A single-electron transistor logic gate family and its application. In *Proceedings 34th IEEE International Symposium on Multiple-Valued Logic*, pp. 262–268, 2004.

[11] Derycke V, Martel R, Appenzeller J, and Avouris Ph. Carbon nanotube inter – and intramolecular logic gates. *Nano Letters*, 1(9):453–456, 2001.

[12] Ellenbogen JC, and Love JC. Architectures for molecular electronic computers: logic structures and an adder designed from molecular electronic diodes. In *Proceedings of the IEEE*, 88:386–426, Mar., 2000.

[13] Endoh T, Sakuraba H, Shinmei K, and Masuoka F. New three dimensional (3D) memory array architecture for future ultra high density DRAM. In *Proceedings 22nd International Conference on Microelectronics*, 442:447–450, 2000.

[14] Fendler JH. Chemical self-assembly for electronics applications. *Chemistry of Materials* 13:3196, 2001.

[15] Fölling S, Türel Ö, and Likharev KK. Single-electron latching switches as nanoscale synapses. In *Proceedings International Joint Conference on Neural Networks* (IEEE/Omnipress 2001), pp. 216–221, 2001.

[16] Frank MP, and Knight TF Jr. Ultimate theoretical models of nanocomputers. *Nanotechnology*, 9:162–176, 1998

[17] Goldstein SC, and Budiu M. NanoFabrics: spatial computing using molecular electronics. In *Proceedings 28th International Symposium on Computer Architecture*, pp. 178–189, IEEE, Goteborg, Sweden, 2001.

[18] Goser K, Pacha C, Kanstein A, and Rossmann ML. Aspects of systems and circuits for nanoelectronics. *Proceedings of the IEEE*, 85:558–573, April, 1997.

[19] Gerousis C, Goodnick SM, and Porod W. Toward nanoelectronic cellular neural networks. *International Journal of Circuits Theory and Applications* 28(6):523-535, 2000.

[20] Goddard W, Brenner D, Lyshevski S, and Iafrate G., Eds., *Handbook of Nanoscience, Engineering and Technology*, vol. 6, CRC Press, Boca Raton, FL, 2002.

[21] Goodnick SM, and Bird J. Quantum effect and single-electron devices. *IEEE Transactions on Nanotechnology*, 2(4):368–385, Dec., 2003.

[22] Goossens J, Ritskes J, Verhoeven C, and van Roermund A. Learning single electron tunneling neural nets, In *Proceedings ProRISC Workshop on Circuits, Systems, Signal Processing*, pp. 179–186, 1997.

[23] Harris J, Akis R, and Ferry DK. Magnetically switched quantum waveguide qubit. *Applied Physics Letters,* 79:2214–2215, 2001.

[24] Hitschefeld N, Conti P, and Fichtner W. Mixed element trees: a generalization of modified octrees for the generation of meshes for the simulation of complex 2-D semiconductor device structures. *IEEE Transactions on Computer-Aided-Design of Integrated Circuits and Systems*, 12(11):1714–1725, 1993.

[25] Huang Y, Duan X, Cui Y, Lauhon LJ, Kim KH, and Liber MC. Logic gates and computation from assembled nanowire building blocks. *Science*, 294:1313–1317, 2001.

[26] Hung W, Song X, Yang G, Yang J, and Perkowski M. Quantum logic synthesis by symbolic reachability analysis. In *Proceedings ACM IEEE Design Automation Conference*, pp. 838–841, San Diego, CA, USA, 2004.

[27] Inokawa H, and Takahashi Y. Experimental and simulation studies of single-electron transistor-based multiple-valued logic. In *Proceedings 33rd IEEE International Symposium on Multiple-Valued Logic*, pp. 259–266, Japan, 2003.

[28] Kadin AM. *Introduction to Superconducting Circuits.* John Wiley & Sons, New York, 1999.

[29] Kasai S, and Hasegawa H. A single-electron binary-decision-diagram quantum logic circuit based on Schottky wrap gate control of a GaAs nanowire hexagon. *IEEE Electron Device Letters*, 23(8), 2002.

[30] Korotkov A, Chen RH, and Likharev K. Possible performance of capacitively coupled single-electron transistors in digital circuits. *Journal of Applied Physics*, 78:2520-2530, 1995.

[31] Korotkov A, and Likharev K. Single-electron-parametron-based logic devices. *Journal of Applied Physics*, 84(11):6114–6126, 1998.

[32] Kulik IO, and Shekhter RI. Kinetic phenomena and charge discreteness effects in granular media. *Journal Eksp. Teor. Fiz. (Sov. Phys. JETP)*, 41:308–316 (62:623–640), Feb., 1975.

[33] Lent CS, Tougaw PD, Porod W, and Bernstein GH. Quantum cellular automata. *Nanotechnology*, 4(1):49-57, 1993.

[34] Likharev KK and Zorin AV. Theory of the bloch-wave oscillations in small Josephson junctions. *Journal of Low Temperature Physics*, 59(5/6):347–382, 1985.

[35] Likharev KK, and Averin DV. Coulomb blockade of tunneling, and coherent oscillations in tunnel junctions. *Journal of Low Temperature Physics*, 62(3/4):345–372, 1986.

[36] Likharev K, and Semenov V. Possible logic circuits based on the correlated single-electron tunnelling in ultrasmall junctions. In *International Superconductivity Electronics Conference*, pp. 128–131, Tokyo, 1987.

[37] Likharev K, and Semenov V. RSFQ logic/memory family: a new Josephson junction technology for sub-teraherz clock frequency digital systems. *IEEE Transactions on Applied Superconductivity*, 1:3–28, Mar., 1991.

[38] Likharev KK. Single-electron devices and their applications. In *Proceedings of the IEEE*, 87(4):606–632, 1999.

[39] Lyshevski SE. Devising and classification of nanodevices. In *Proceedings IEEE Conference on Nanotechnology*, pp. 471–476, Washington, DC, 2002.

[40] Lyshevski SE. *Nano- and Micro-Electromechanical Systems: Fundamental of Micro- and Nano-Engineering.* CRC Press, Boca Raton, FL, 2000.

[41] Martel R, et al. Carbon nanotube field effect transistors for logic applications. In *Proceedings Electron Devices Meeting, IEDM Technical Digest*, 7(5):1–4, 2001.

[42] Maslov D, Dueck GW, and Miller DM. Fredkin/Toffoli templates for reversible logic synthesis. In *Proceedings ICCAD*, pp. 256–261, San Jose, CA, 2003.

[43] Miller DM, Maslov D, and Dueck GW. A transformation based algorithm for reversible logic synthesis. In *Proceedings Design Automation Conference*, pp. 318–323, Anaheim, CA, 2003.

[44] Miller DM, Dueck GW, and Maslov D. A synthesis method for MVL reversible logic. In *Proceedings 34th International Symposium on Multiple-Valued Logic*, pp. 74–80, Canada, 2004.

[45] Matsumoto K, Ishii M, Segawa K, and Oka Y. Room temperature operation of a single electron transistor made by the scanning tunneling microscope nanooxidation process for the TiOx/Ti system. *Applied Physics Letters*, 68(1), Jan., 1996.

[46] Nakazato K. Blaikie RJ, and Ahmed H. Single electron memory. *Journal of Applied Physics*, 75:5123–5134, 1994.

[47] Ohring S, and Das SK. Incomplete hypercubes: embeddings of tree-related networks. *Journal of Parallel and Distributed Computing*, 126:36–47, 1995.

[48] Oya T, Asai T, Kukui T, and Amemiya Y. A majority logic device using an irreversible single-electron box. *IEEE Transactions on Nanoelectronics*, 2(1), Mar., 2003.

[49] Park J. Coulomb blockade and the Kondo effect in single-atom transistors. *Nature*, 417:722, 2002.

[50] Poole CP. Jr, and Owens FJ. *Introduction to Nanotechnology*. John Wiley & Sons, New York, 2003.

[51] Postma HWCh, Teepen TF, Yao Z, Grifoni M, and Dekker C. Carbon nanotubes single-electron transistors at room temperature. *Science*, 293:76–79, 2001.

[52] Reed MA, and Tour JM. Computing with molecules. *Scientific American*, 86, June, 2000.

[53] Saitoh M, Murakami T, and Hiramoto T. Large Coulomb blockade oscillations at room temperature in ultra-narrow wire channel MOSFETs formed by slight oxidation process. *IEEE Transactions on Nanotechnology* 2(4), Dec., 2003.

[54] Shende VV. Prasad AK, Markov IL, and Hayes JP. Synthesis of reversible logic circuits. *IEEE Transactions on Computer Aided Design of Integrated Systems*, 22:710–722, June, 2003.

[55] Shibata T, and Ohmi T. An intelligent MOS-transistor featuring gate-level weighted sum and threshold operations. In *Proceedings International Electron Device Meeting, Tech. Digest*, pp. 919–922, Washington DC, Dec. 8–11, 1991.

[56] Shmerko VP, and Yanushkevich SN. Three-dimensional feedforward neural networks and their realization by nano-devices. In: Yanushkevich SN., Ed., *Artificial Intelligence in Logic Design*. Kluwer, Dordrecht, pp. 313–334, 2004.

[57] Takahashi T, Yoshita M, and Sakaki H. Tunneling characterization of GaAs and InAs by scanning tunneling microscopy under laser irradiation. In *Proceedings 8th International Conference on Scanning Tunnel-*

ing Microscopy/Spectroscopy and Related Techniques, Colorado, July, 21–ThA10, 1995.

[58] Takahashi Y, Fujiwara A, Ono Y, and Murase K. Silicon single-electron devices and their application. In *Proceedings 30th IEEE International Symposium on Multiple-Valued Logic*, pp. 411–420, 2000.

[59] Tour JM, Zandt WLV, Husband CP, Husband SM, Wilson LS, Franzon PD, and Nackashi DP. Nanocell logic gates for molecular computing. *IEEE Transactions on Nanotechnology*, 1(2):100–109, 2002.

[60] Tour JM. Molecular electronics. Synthesis and testing of components. *Accounts of Chemical Research*, 33(11):791–804, Nov., 2000.

[61] Türel Ö, and Likharev K. CrossNets: possible neuromorphic networks based on nanoscale components. *International Journal of Circuit Theory and Applications* 31:37–53, 2003.

[62] Tucker JR. Complementary digital logic based on the Coulomb blockade. *Journal of Applied Physics*, 72:4399–4413, 1992.

[63] Wasshuber C. *Computational Single-Electronics*, Springer-Verlag, Heidelberg, 2001.

[64] Yamada T, Kinoshita Y, Kasai S, Hasegawa H, and Amemiya Y. Quantum-dot logic circuits based on shared binary decision diagram. *Journal of Applied Physics*, 40(7):4485–4488, 2001.

[65] Yamamura K, and Suda Y. Novel single-electron logic circuits using charge-induced signal transmission (CIST) structures. *IEEE Transactions on Nanotechnology*, 2(1):1–9, Mar., 2003.

[66] Yano K, Ishii T, Hashimoto T, Kobayashi T, Murai F, and Seki K. Room-temperature single-electron memory. *IEEE Transactions on Electronic Devices*, 41(9):1628–1638, 1994.

[67] Yano K, Ishii T, Sano T, Mine T, Murai F, Kure T, and Seki K. 128 MB early prototype for gigascale single-electron memories, In *Proceedings IEEE ISSCC*, pp. 344–345, 1996.

[68] Yanushkevich SN, Shmerko VP, Guy L, and Lu D. 3D Multiple-valued circuits design based on single-electron logic. In *Proceedings 34th IEEE International Symposium on Multiple-Valued Logic*, pp. 275–280, Canada, 2004.

[69] Yoshikawa N, Tago H, and Yoneyama K. New approach to RSFQ logic circuits based on the binary decision diagrams. *IEEE Transactions on Applied Superconductivity*, 9(2):3161–3164, June, 1999.

[70] Zinoviev D, and Likharev K. Feasibility study of RSFQ-based self-routing nonblocking digital switches. *IEEE Transactions on Applied Superconductivity*, 7(2):3155–3163, 1997.

[71] Zhitenev NB, Meng H, and Bao Z. Conductance of small molecular junctions. *Physical Review Letters* 88:226801, 2002.

3

Basics of Logic Design in Nanospace

In this chapter, selected methods of classical logic design are revised. The chapter consists of nine sections. In each section we deal with canonical forms for representation of switching functions. In general, the problem is to choose an appropriate form, which complies with particular properties of 3-D implementation. Such a formulation of the problem implies the following strategy for material representation:

- ► The general equation that generates a variety of forms of a switching function is introduced;
- ► The algebraic, matrix and hypercube-based methods of computing a switching function in a given form are described. Two aspects of the problem are the focus of discussion: construction of a given form (direct problem) and restoration of an initial function (inverse problem);
- ► Decision trees and decision diagram design are revisited, as a basis for 3-D embedding techniques.

In this chapter, we will focus on utilization of traditional hypercube structures. In this approach, each variable of a switching function is associated with one dimension in hyperspace. Manipulation of the function is based on special encoding of the vertices and edges in the hypercube. The hypercube is used as an effective algebraic model of a switching function.

The basics of multidimensional logic design include

- Topology design,
- Representations of switching functions (in algebraic and matrix form),
- Decision tree and decision diagram design in 2-D and 3-D space.

Six forms of representations of switching functions are revisited. These forms are divided into two groups: those for single-output functions, and word-level forms that represent multioutput functions. The attractive features of these forms are listed below:

- ► They are sums (logical, arithmetical, bitwise) of product terms that are uniform;

- ▶ They can be derived from the unified spectral as direct and inverse transforms in a given basis, and their transform matrices are uniform and factorizable, i.e., they can be used for computing in different forms;
- ▶ The algorithms to calculate switching functions are based on the above uniform and homogeneous (flowgraphs, cube-based, decision trees, decision diagrams) representations;
- ▶ This uniformity is a prerequisite for representation in 3-D space considered in Chapter 5.

The above data structures are represented in this chapter as follows. First, a brief introduction to the graphical and topological data structures is offered in Section 3.1. In Section 3.2, the general formal description for sum-of-products, Reed-Muller, arithmetic and word-level logics is introduced. Various forms of this general description are discussed: algebraic form, matrix form, cube form, and graphical form. Then, these forms are discussed in detail. Sections 3.3 and 3.4 focus on sum-of-products expression. In Sections 3.5 and 3.6, Reed-Muller representation and Davio decision trees and diagrams are introduced. Then, the technique for computing arithmetic expressions is given in Sections 3.7 and 3.8. Finally, after a brief summary, we provide problems and recommendations for "Further Reading."

3.1 Graphs

In this section the elements of graph theory are introduced. The focus is the design of decision trees and decision diagrams, as well as their structural and topological properties.

3.1.1 Definitions

A *graph* is defined as $G = (V, E)$, where V is the vertex (node) set and E is the edge set. We say $(i, j) \in E$, where $i, j \in V$. The terms *edge*, *link*, *connection* and *interconnection* are used here interchangeably, and the terms *graph* and *logic network* are considered synonyms. Functional elements (gates, circuits) correspond to nodes and communication links correspond to edges in the graph. The number of nodes in G is $n = |V|$. An edge in E between nodes v and u is written as an unordered pair (v, u), and v and u are said to be *adjacent* to each other or just *neighbors*. The *distance* between two nodes i and j in a graph G, is the number of edges in G on the shortest path connecting i and j. The *diameter* of a graph G is the maximum distance between two nodes in G. A graph G is *connected* if a path always exists between any pair of nodes i and j in G.

3.1.2 Directed graphs

A *directed* graph G consists of a finite vertex set V (nodes) and a set E of edges between the vertices. The *indegree* of a node (vertex) i in a graph G is the number of edges in G leading to i. The *outdegree* of a node (vertex) i in a graph G is the number of edges in G starting in i. A node is called a *sink* if it has an outdegree of 0. If the outdegree of v is bigger than 0, v is called an *internal* node. A node is called a *root* if it has an indegree of 0. The adjacency matrix $A = (a_{ij})$ of a graph G is defined as $|G| \times |G|$ size matrix such that $a_{ij} = 1$ if $(i, j) \in E$, and $a_{ij} = 0$ otherwise.

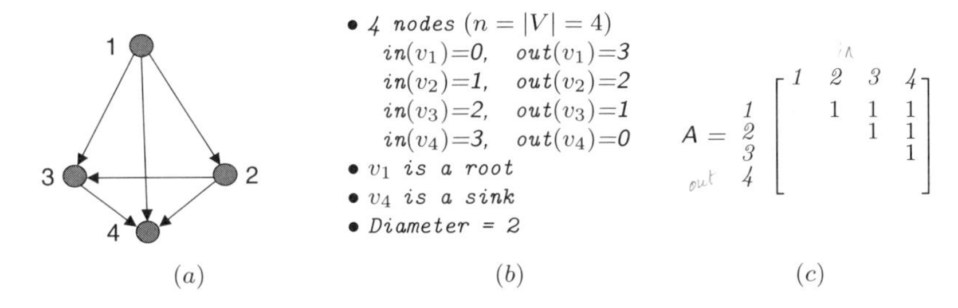

FIGURE 3.1
The directed graph (a), its properties (b), and an adjacency matrix (c) (Example 3.1).

Example 3.1 *The properties of the directed graph with four nodes and its adjacency matrix A are illustrated in Figure 3.1, where $in(v_i)$ and $out(v_i)$ are indegrees and outdegrees of a node v_i.*

3.1.3 Undirected graphs

In the case of undirected graphs the edges are considered unordered pairs and therefore have no distinguished direction. The *degree* of a node i in a graph G is the number of edges in G that are incident with i, i.e., where the outdegree and the indegree coincide.

Example 3.2 *Figure 3.2 illustrates the properties of the undirected graph. Its adjacency matrix A is equal to the transposed matrix A, $A = A^T$.*

3.1.4 Cartesian product graphs

Cartesian product graphs provide a framework in which it is convenient to analyze as well as to construct new graphs. Let $G_1 = (V_1, E_1)$ and $G_12 =$

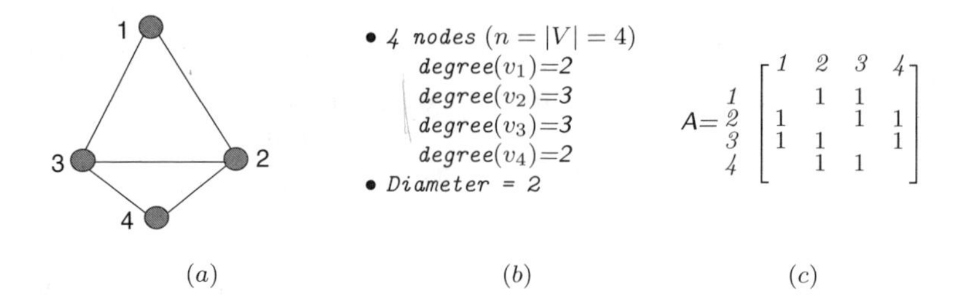

FIGURE 3.2
The undirected graph (a), its properties (b), and an adjacency matrix (c) (Example 3.2).

(V_2, E_2) are two graphs. The product of G_1 and G_2, denoted $G_1 \times G_2 = (V_1 \times V_2, E)$, is a graph where the set of nodes is the product set

$$V_1 \times V_2 = \{x_1x_2 | x_1 \in V_1 \text{ and } x_2 \in V_2\}$$
$$E = \{\langle x_1x_2, y_1, y_2\rangle | (x_1 = y_1 \text{ and } \langle x_2, y_2\rangle \in E_2) \text{ or } (x_2 = y_2 \text{ and } \langle x_1, y_1\rangle \in E_1)\}.$$

It can be shown that the hypercube can be defined as the product of n copies of the complete graph on two vertices, K_2. That is

$$H_n = H_{n-1} \times K_2.$$

3.1.5 Interconnection networks

The interconnection network organization depends on data structure.

Interconnection network organization. Let the set of processing elements be given. Suppose that this set of processing elements must be organized on the principle of massive parallel computing. To design the model, different interconnection organizations can be used.

Example 3.3 *In Figure 3.3a, this is a 7-node binary decision tree that models interconnection organization based on "1-input, 2-outputs" processing elements. The mesh interconnection network with 16 nodes shown in Figure 3.3b is based on the principle of nearest neighbor communication (each processing element has four neighbors). The ring network in Figure 3.3c includes 8 nodes: each processing element communicates with two neighbors. In the 8-node hypercube network, each processing element communicates with three neighbors (Figure 3.3d).*

Notice that meshes, rings, and binary trees can be mapped into a hypercube.

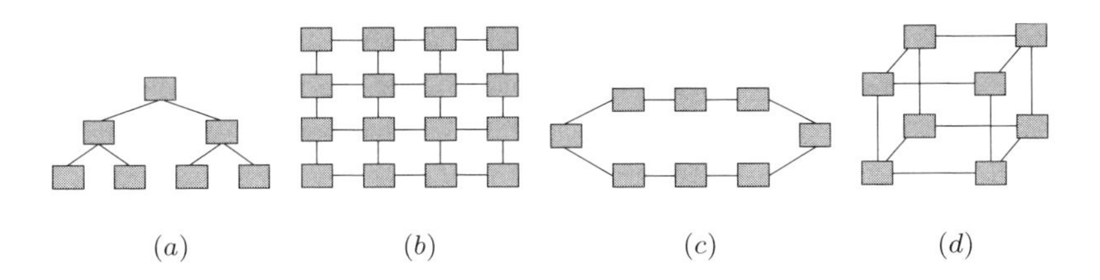

FIGURE 3.3
Interconnection network organization: a binary tree (a), mesh (b), ring (c), and hypercube (d) (Example 3.3).

Modeling a logic network by direct acyclic graphs. A logic network is modeled by a direct acyclic graph (DAG) $G(V, E)$, whose vertex set V is in one-to-one correspondence with primary inputs, local functions and primary outputs. In other words, a logic network is a finite collection of interconnected gates, network input terminals, and network output terminals with the following restrictions:

- ▶ No gate output terminal or network input terminal is connected to another gate output terminal or network input terminal,
- ▶ Every network output terminal or gate input terminal is wired (via one or more wires) to a constant value, a network input terminal, or a gate output terminal.

A combinational gate network is one in which the values of the signal present on its input terminals uniquely determine the signal values at its output terminals.

Example 3.4 *A logic network graph shown in Figure 3.4(b) has been derived from the circuit given in Figure 3.4(a).*

3.1.6 Decision tree

A graph is called *rooted* if there exists exactly one node with an indegree of 0, the root. A *tree* is a rooted acyclic graph in which every node but the root has an indegree of 1. This implies that in a tree, for every vertex v there exists a unique path from the root to v. The length of this path is called the *depth* or *level* of v. The *height* of a tree is equal to the largest depth of any node in the tree. A node with no children is a *terminal* (*external*) node or *leaf.* A nonleaf node is called an *internal* node. A *complete* $n-$level p-tree, is a tree with p^k nodes on level k for $k = 0, \dots, n-1$. A p^n-leaf complete tree has a level hierarchy (levels $0, 1, \dots, n$), the root is associated with level zero and its p children are on level one. This edge model describes the transmission of

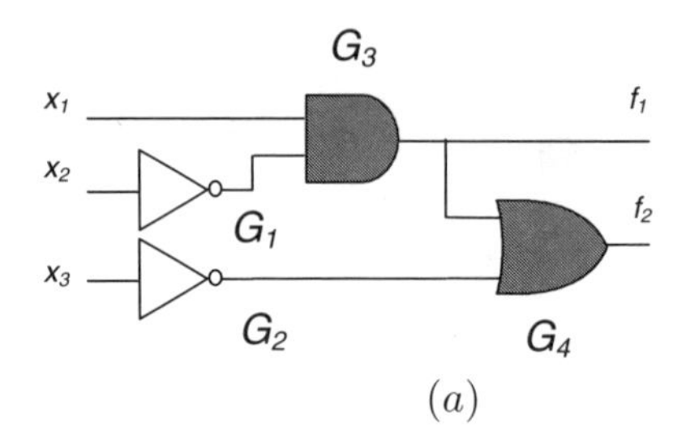

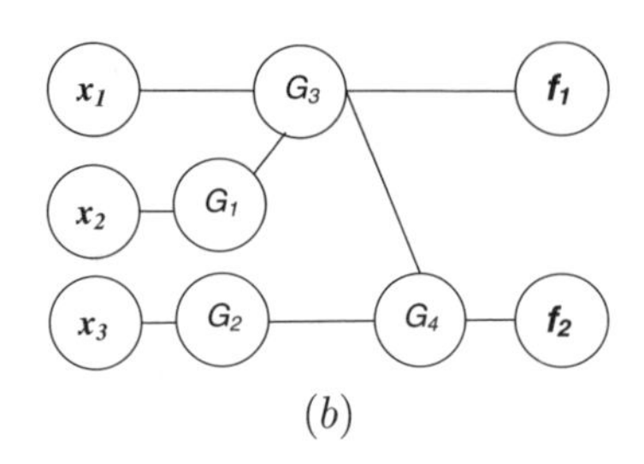

FIGURE 3.4
Logic circuit (a) and its logic network graph (b) (Example 3.4).

data from the child to the parent, so that data are sent in only one direction at a time, up or down.

A decision tree is characterized by a set of parameters:

- ▶ The *size*, the number of nodes;
- ▶ The *depth*, the number of levels;
- ▶ The *width*, the maximum number of nodes for a level; and
- ▶ The *area*, $Depth \times Width$.

Example 3.5 *In Figure 3.5, the complete ternary ($p = 3$) 3-level ($n = 3$) tree is given. The root corresponds to the level (depth) 0. Its three children are associated with level 1 ($3^1 = 3$). Level 2 includes $3^2 = 9$ nodes. Finally, there are $3^3 = 27$ terminal nodes.*

3.1.7 Embedding of a guest graph in a host graph

An *embedding*, $\langle \varphi, \alpha \rangle$, of a graph G in a graph H is a one-to-one mapping φ: $V(G) \rightarrow V(H)$, along with a mapping α that maps an edge $(u, v) \in E(G)$ to a path between $\varphi(u)$ and $\varphi(v)$ in H[11]. The embedding of a guest graph G in a host graph H is an injection (one-to-one mapping) of the nodes in G to the nodes in H. An embedding is characterized by a set of parameters:

- ▶ The *expansion* is the ratio $|V(H)|/|V(G)|$.
- ▶ The *dilation* cost of an embedding of G in H is the maximum distance in H between the images of any two neighboring nodes in G. This cost gives a measure of the proximity in H of the neighboring nodes in G under an embedding.
- ▶ The *congestion* of the embedding is the maximum of the congestions of all edges of H.

Example 3.6 *Details of embedding of graph G in a host graph H are given in Figure 3.6.*

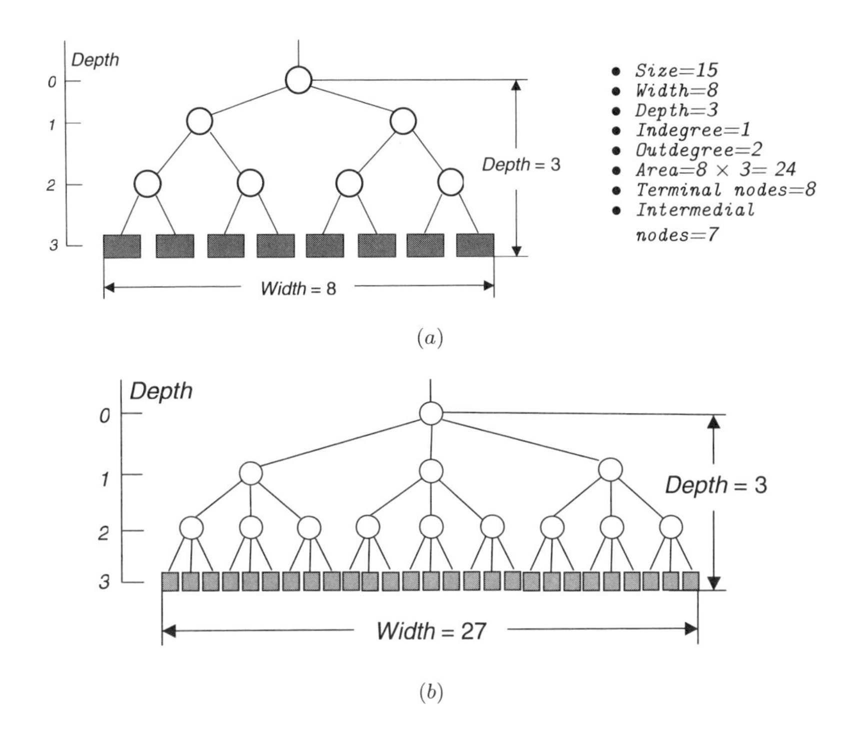

FIGURE 3.5
The complete binary (a) and ternary (b) tree (Example 3.5).

3.1.8 Binary decision diagrams

A decision tree is *reduced* if it does not contain any vertex v whose successors lead to the same node, and if it does not contain any distinct vertices v and v' such that the subgraphs rooted in v and v' are isomorphic. Reduced decision trees are usually referred to as decision diagrams. A binary decision diagram (BDD) is a directed acyclic graph with exactly one root, whose sinks are labeled by the constants 1 and 0, and whose internal nodes are labeled by a Boolean variable x_i and have exactly two outgoing edges, a 0-edge and 1-edge. BDD represents a switching function f in the following way:

- ▶ Each assignment to the input variables x_i defines a uniquely determined path from the root of the graph to one of the sinks.
- ▶ The label of the reached sink gives the function value of this input.

An *ordered* BDD (OBDD) is a rooted directed acyclic graph that represents a switching function. A linear variable order is placed on the input variables. The variables' occurrences on each path of this diagram have to be consistent with this order. A OBDD is called *reduced* if it does not contain any vertex

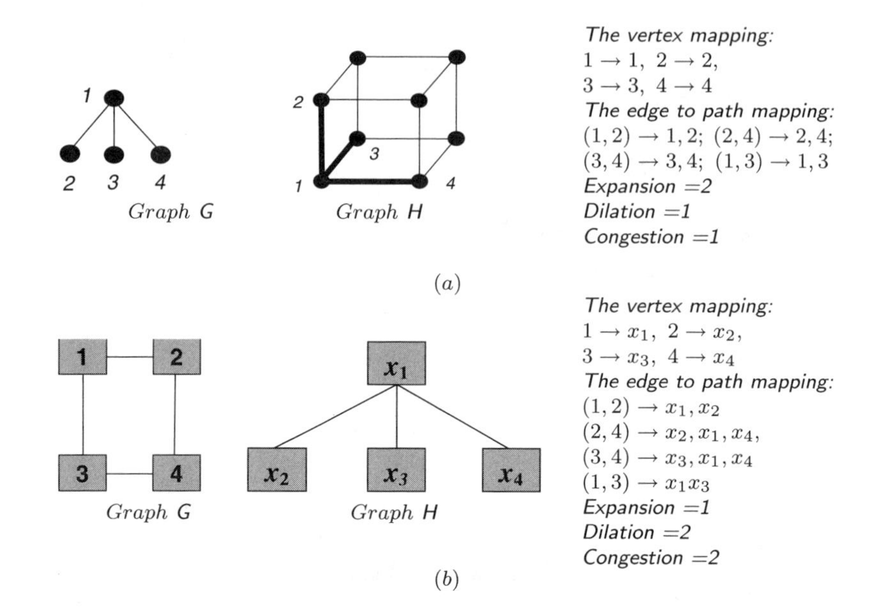

FIGURE 3.6
Embedding of graph G in a host graph H (Example 3.6).

v such that the 0-edge and 1-edge of v lead to the same node, and it does not contain any distinct vertices v and v' such that the subgraphs rooted in v and v' are isomorphic.

A decision diagram is characterized, similarly to a decision tree, by the *size*, *depth*, *width*, *area*, and the *efficiency of reduction* of two decision diagrams (trees) of size $Size_1$ and $Size_2$:

$$100 \times \frac{Size_1}{Size_2}\%$$

Example 3.7 *Figure 3.7 illustrates the efficiency of the reduction of decision tree for the switching function* $f_1 = x_1x_2 \vee x_3$.

A multioutput switching function is represented by a multirooted decision diagram, which is called a *shared* decision diagram.

Example 3.8 *A multioutput switching function is represented by a shared OBDD. Let there be two-output function*

$$f_1 = x_1x_2 \vee x_3,$$
$$f_2 = x_1 \vee x_2 \vee x_3.$$

The shared ROBDD is represented in Figure 3.7c.

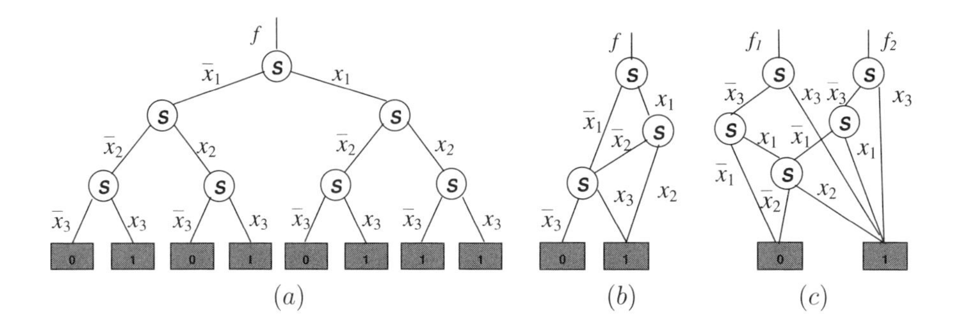

FIGURE 3.7
The complete (a) and reduced (b) decision tree for the switching function $f_1 = x_1x_2 \vee x_3$; the shared ROBDD (c) (Examples 3.7 and 3.8).

In a multiplexer tree (network), each internal tree node is represented as a 2 to 1 multiplexer controlled by the node variable and each leaf is implemented as a constant logical value (wired at 0 or wired at 1); the interconnection scheme is that of the decision tree (diagram). The evaluation of a function then proceeds from the leaves (the constant values) to the root multiplexer, the function variables, used as control variables, select a unique path from the root to one leaf, and the value assigned to that leaf propagates along the path to the output of the root multiplexer.

3.2 Data structures for switching functions

Circuit, switching net, netlist, fan-in, and fan-out. Switching functions have a corresponding implementation in terms of interconnected gates. This is called a *circuit*, or *schematic*. The composition of primitive components is accomplished by physically wiring the gates together. A collection of wires that always carry the same electrical signal is called a *switching net*. The tabulation of gate inputs and outputs and the nets to which they are connected is called the *netlist*. The *fan-in* of a gate is its number of inputs. The *fan-out* of a gate is the number of inputs to which the gate's output is connected.

Data structure is the term used to define an abstract data type. Data structure for switching functions is a collection of Boolean variables connected in various ways. In other words, it is a mathematical model of a switching function. *Data type* is a property of the mathematical model. Data structures for switching functions considered in this book are the following:

- Sum-of-products expressions,
- Reed-Muller expressions,
- Arithmetic expressions, and
- Word-levels (arithmetic, sum-of-products, and Reed-Muller).

These data structures for a switching function f of n variables $x_1, x_2, \ldots x_n$ are represented in different mathematical *forms* or *descriptions*:

- General (algebraic) expansion,
- Matrix form,
- Cube form representation,
- Flowgraph,
- Decision tree, and
- Decision diagram.

In addition, each of these forms can be modified with respect to so-called *polarity*. Choosing the appropriate

- data structure,
- Data type,
- Data description, and
- Polarity

for a switching function is the crucial point in circuit design in spatial dimensions.

General algebraic equation. This formal description carries information about the data structure through the algebraic relations between variables. Each variable in the product term is called a *space coordinate*, and the number of variables specifies the number of space dimensions. For example, a one-variable function corresponds to one-dimensional space, two-variables means two-dimensional space, etc. Coefficient Ω_i is 0 or 1 for sum-of-products and Reed-Muller expressions, and Ω_i is integer number for arithmetic and word-level representations. The product $(x_1^{i_1} \cdots x_n^{i_n})$ generates two kinds of expressions for $i = 0, 1, 2, \ldots, 2^n - 1$. If $x_j^{i_j} = \overline{x}_j$ for $i_j = 0$, and $x_j^{i_j} = x_j$ for $i_j = 1$, then

$$(x_1^{i_1} \cdots x_n^{i_n}) = \overline{x}_1 \ldots \overline{x}_n \odot \overline{x}_1 \overline{x}_2 \ldots \overline{x}_{n-1} x_n \odot \ldots \odot x_1 x_2 \ldots x_n.$$

If $x_j^{i_j} = 1$ for $i_j = 0$, and $x_j^{i_j} = x_j$ for $i_j = 1$, then

$$(x_1^{i_1} \cdots x_n^{i_n}) = 1 \odot x_n \odot x_{n-1} x_n \odot \ldots \odot x_1 x_2 \ldots x_{n-1} \odot x_1 x_2 \ldots x_n.$$

The general structure of the equation is illustrated in Figure 3.8 where

$$
\odot = \begin{cases} \vee, \text{ for sum-of-products expression;} \\ \oplus, \text{ for Reed-Muller expression;} \\ +, \text{ for arithmetic and word-level arithmetic expression;} \\ \overset{\frown}{\vee}, \text{ for word-level sum-of-products expression;} \\ \overset{\frown}{\oplus}, \text{ for word-level Reed-Muller expression.} \end{cases}
$$

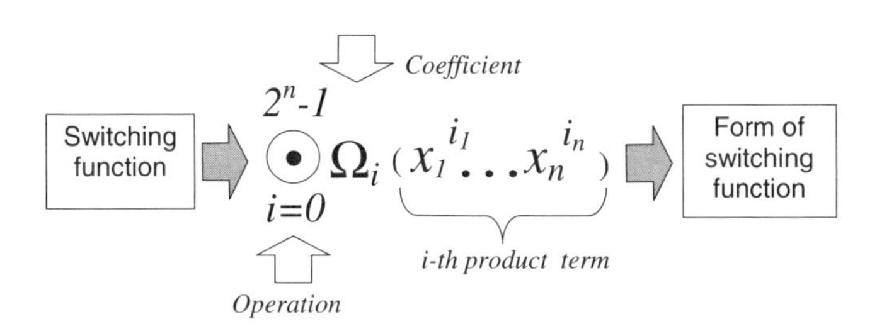

FIGURE 3.8
Structure of the equation to represent a switching function in various forms.

The matrix (spectral) representation. A data structure can be described in matrix form. This form often provides a better understanding of the data structure, for example, symmetric properties, or from a related field of view (spectral theory, parallel processing). Matrix form (Figure 3.9) is based on notation of truth-vector $\mathbf{F}$ of a given switching function f, vector of coefficients (spectrum) $\mathbf{\Omega}$, and transform matrix $\mathbf{\Omega}_{2^n}$. Given the truth-vector $\mathbf{F}$, the result of the direct transformation is the vector of coefficients $\mathbf{\Omega}$. Inverse transform is used to restore the truth-vector $\mathbf{F}$ given a vector of coefficients $\mathbf{\Omega}$. Matrix form is mapped into a flowgraph that carries useful algorithmic properties (parallel computing and complexity).

A flowgraph is the representation of the transform algorithm. The graph edges correspond to parallel streams of computing. The flowgraph is derived from *factorization* of the transform matrix $\mathbf{\Omega}_{2^n}$. The nodes of the flowgraph implement the operation Ω. For example, in the graph of Reed-Muller transform, the nodes implement modulo two operations. Correspondingly, arithmetic sum operation is implemented in the nodes of flowgraphs for arithmetic and word-level arithmetic transforms. For Reed-Muller and arithmetic transforms, the basic configuration of the flowgraph is "butterfly," well-known from the fast Fourier transform (FFT) used in digital signal processing.

The hypercube is a topological representation of a switching function by n-dimensional graph. This representation is aimed at:

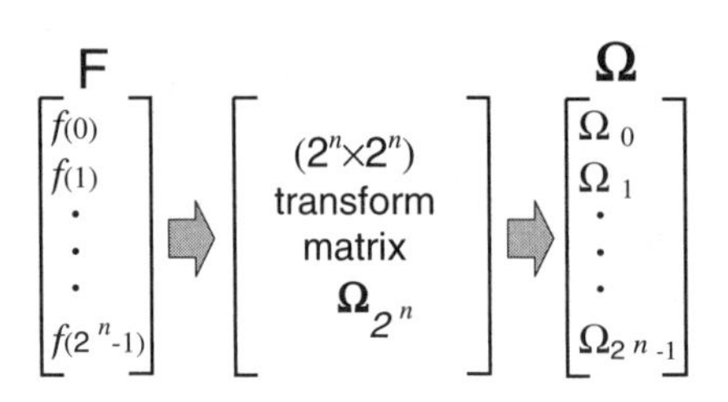

- *Truth-vector* **F** *of a switching function f*
- *Transform matrix* $\mathbf{\Omega}_{2^n}$
- *Vector of coefficients (spectrum)* $\mathbf{\Omega}$
- *The result of the direct transformation is the vector of coefficients* $\mathbf{\Omega}$
- *Inverse transform is used to restore the truth-vector* **F** *given a vector of coefficients* $\mathbf{\Omega}$

FIGURE 3.9
The matrix (spectral) representation of a switching function.

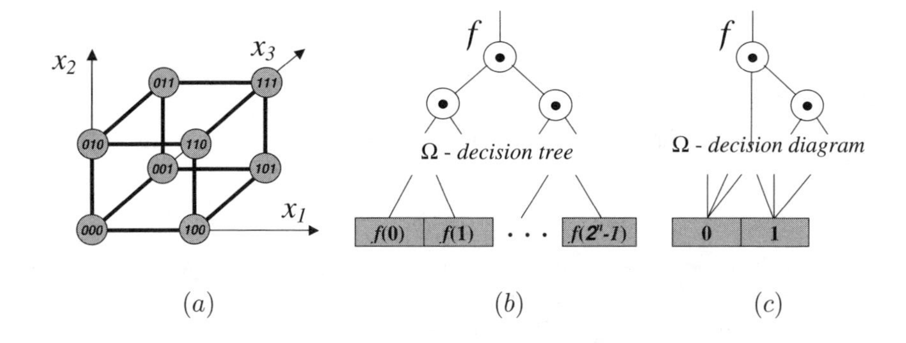

FIGURE 3.10
The data structures for representation of switching functions: hypercube (a), decision tree (b), and decision diagram (c).

- ▶ *Interpretation* (representation) of the function in a form useful for manipulation, and
- ▶ *Mapping* of the function to 3-D space.

In switching theory, a hypercube is defined as a collection of 2^m minterms, therefore, the vertices of the hypercube are assigned with the minterms, $m \leq n$. In Figure 3.11, the hypercubes for $n = 1, \ldots, 5$, carry information about the product terms in sum-of-products form. This information is encoded as shown in Figure 3.12. The design of the hypercube includes the encoding of switching function accordingly to the rules given in Figure 3.12, assigning the codes to 2^n vertices and $n2^{n-1}$ edges in the hypercube. The operations between two hypercubes produce a new hypercube (product) that is utilized in optimization problems.

The implementation problem is defined as embedding of a 2-D graphical structure (decision tree or decision diagram of a switching function) in a hypercube that carries *functional* and *topological* information about the switching function.

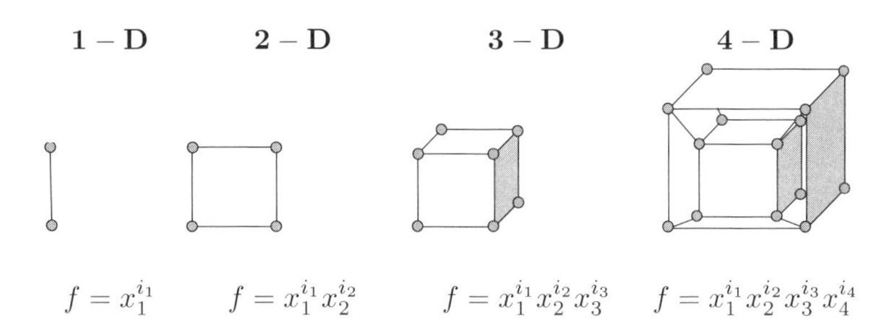

FIGURE 3.11
Product term of n variables and its spatial interpretation by n-dimensional hypercube, $n = 1, 2, 3, 4$.

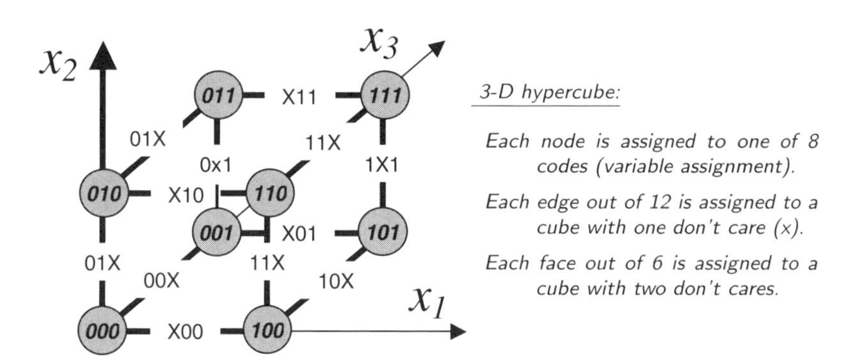

FIGURE 3.12
A hypercube data structure for representation and manipulation of switching functions.

Let us represent switching function f, given by a truth table, by hypercube (Figure 3.13a). In classical 2-D hypercube (Figure 3.13b):

► The four corners are called *vertices*. They correspond to the four rows of a truth table.

- Each vertex is identified by two coordinates. The horizontal coordinate is assumed to correspond to variable x_1, and vertical coordinate to x_2.
- The function f is equal to 1 for vertices 01, 10, and 11. The function f can be expressed as a set of vertices, $f = \{01, 10, 11\}$.
- The edge joins two vertices for which the labels differ in the value of only one variable. For example, $f = 1$ for vertices 10 and 11. They are joined by the edge that is labeled 1**x**. The letter **x** is used to denote the fact that the corresponding variable can be either 0 and 1. Hence 1**x** means that $x_1 = 1$, while x_2 can be either 0 or 1. Similar, vertices 01 and 11 are joined by the edge labeled **x**1. The edge 1**x** is the logical sum of vertices 10 and 11.
- The term x_1 is the sum of minterms $x_1\overline{x}_2$ and x_1x_2. It is follows that $x_1\overline{x}_2 \vee x_1x_2 = x_1$. The edges 1**x** and **x**1 define in a unique way the function f, hence we can write $f = \{1\mathbf{x}, \mathbf{x}1\}$. This corresponds to the function $f = x_1 \vee x_2$.

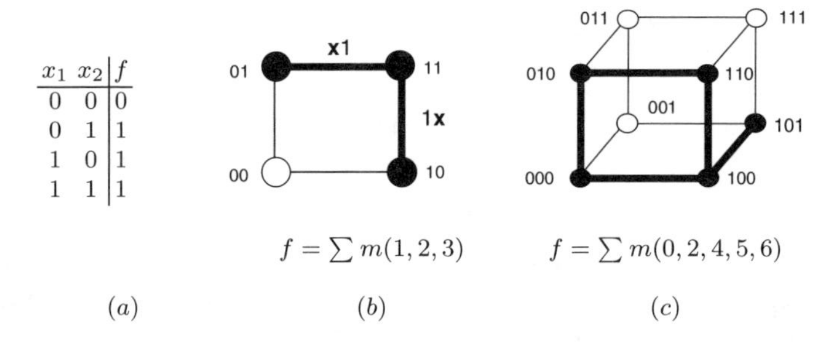

x_1	x_2	f
0	0	0
0	1	1
1	0	1
1	1	1

FIGURE 3.13
Representation of switching functions by the hypercube: truth table (a) and 2-D hypercube representation (b); 3-D hypercube for the function $f = \overline{x}_3 \vee x_1\overline{x}_2$ (c).

Let $f = \sum m(0, 2, 4, 5, 6)$. In 3-D hypercube (Figure 3.13c):

(*a*) There are five vertices that correspond $f = 1$: 000,010,100,101, and 110.

(*b*) These vertices are joined by five edges, $\mathbf{x}00, 0\mathbf{x}0, \mathbf{x}10, 1\mathbf{x}0$ and $10\mathbf{x}$.

(*c*) These vertices include all variations of x_1 and x_2, when x_3 is 0, and can be specified by the term **xx**0. This term means that $f = 1$ if $x_3 = 0$, regardless of the values of x_1 and x_2.

(d) The function f can be represented in several ways. Some of the possibilities are:

$$f = \{000, 010, 100, 101, 110\} = \{0\mathbf{x}0, 1\mathbf{x}0, 101\} = \{\mathbf{x}00, \mathbf{x}10, 101\}$$
$$= \{\mathbf{x}00, \mathbf{x}10, 10\mathbf{x}\} = \{\mathbf{xx}0, 10\mathbf{x}\}.$$

Obviously, the least-expensive circuit is obtained if $f = \{\mathbf{xx}0, 10\mathbf{x}\}$, which is equivalent to the logic expression $f = \overline{x}_3 \vee x_1\overline{x}_2$.

A 4-D hypercube consists of two 3-D hypercubes with their corners connected. The simplest way to visualize a 4-D hypercube is to have one hypercube placed inside the other hypercube.

The decision tree can be derived from the functional equation or its truth table. It provides a canonical representation of functions in graphical form, so that for a fixed variable order there is a bijection between switching functions and decision diagrams. Canonicity is important in two respects: it makes equivalence tests easy, and it increases efficiency. A node of the decision tree implements decomposition with respect to a variable. There are 2^n terminal nodes (exactly the same as the number of values in the truth vector) in the complete decision tree. It is a canonical data structure.

The decision diagram is constructed by reducing a decision tree. The decision diagram is a canonical form, however, it represents the optimal form of switching function.

A polarity of expression is one of the possible ways to represent the function. The polarity of a variable is an indication of complemented variable ($\overline{x}$ or polarity 1) or uncomplemented (x). All above forms can be interpreted and modified with respect to 2^n polarities. Moreover, in this modification the fixed and mixed polarity are distinguished.

In Table 3.1 three forms of switching function are given. In the next sections, we consider the techniques for computing the coefficients Ω_i.

3.3 Sum-of-products expressions

Sum-of-products expressions are derived as canonical polynomials of variables over OR, AND, NOT operations. Sum-of-products correspond to two-level AND-OR logic networks.

TABLE 3.1
Expansion of a switching function and corresponding decomposition rules for decision trees and diagrams.

Expansion	Formal description	$x_j^{i_j}$	Decomposition
Sum-of-products	$\bigvee_{i=0}^{2^n-1} s_i \cdot (x_1^{i_1} \cdots x_n^{i_n})$	$\begin{cases} \overline{x}_j, i_j = 0; \\ x_j, i_j = 1. \end{cases}$	Shannon
Reed-Muller	$\bigoplus_{i=0}^{2^n-1} r_i \cdot (x_1^{i_1} \cdots x_n^{i_n})$	$\begin{cases} 1, \ i_j = 0; \\ x_j, i_j = 1. \end{cases}$	Davio
Arithmetic	$\sum_{i=0}^{2^n-1} p_i \cdot (x_1^{i_1} \cdots x_n^{i_n})$	$\begin{cases} 1, \ i_j = 0; \\ x_j, i_j = 1. \end{cases}$	Arithmetic Davio

3.3.1 General form

Given a switching function f of n variables, the sum-of-products expression is specified by

$$f = \bigvee_{i=0}^{2^n-1} s_i \cdot \underbrace{(x_1^{i_1} \cdots x_n^{i_n})}_{i-th\ product}, \tag{3.1}$$

where $s_i \in \{0, 1\}$ is a coefficient, i_j is the j-th bit $(j = 1, 2, \ldots, n)$ in the binary representation of the index $i = i_1 i_2 \ldots i_n$, and $x_j^{i_j}$ is defined as

$$x_j^{i_j} = \begin{cases} \overline{x}_j, i_j = 0; \\ x_j, i_j = 1. \end{cases} \tag{3.2}$$

Example 3.9 *An arbitrary function of two variables ($n = 2$) is represented in sum-of-products form (3.1) and (3.2):*

$$f = s_0(\overline{x}_1\overline{x}_2) \vee s_1(\overline{x}_1 x_2) \vee s_2(\overline{x}_1 x_2) \vee s_3(x_1 x_2).$$

3.3.2 Computing the coefficients

Given a truth vector $\mathbf{F} = [f(0)\ f(1) \ldots f(2^n - 1)]^T$, the vector of coefficients $\mathbf{S} = [s_0\ s_1 \ldots s_{2^n-1}]^T$ is derived by the matrix equation specified on AND and OR operations

$$\mathbf{S} = \mathbf{S}_{2^n} \cdot \mathbf{F}, \tag{3.3}$$

where the $2^n \times 2^n$ matrix $\mathbf{S}_{2^n}$ is formed by the Kronecker (tensor) product $\mathbf{S}_{2^1}$,

$$\mathbf{S}_{2^n} = \bigotimes_{i=1}^{n} \mathbf{S}_{2^1}, \qquad \mathbf{S}_{2^1} = \begin{bmatrix} 1 & 0 \\ 0 & 1 \end{bmatrix}. \tag{3.4}$$

The Kronecker product is defined as follows. Let $\mathbf{A}_2$ and $\mathbf{B}_2$ be the 2×2 matrices

$$\mathbf{A}_2 = \begin{bmatrix} a_{11} & a_{12} \\ a_{21} & a_{22} \end{bmatrix}, \quad \mathbf{B}_2 = \begin{bmatrix} b_{11} & b_{12} \\ b_{21} & b_{22} \end{bmatrix}.$$

The Kronecker product of $\mathbf{A}_2$ and $\mathbf{B}_2$ is 4×4 matrix $\mathbf{C}_2$

$$\mathbf{C}_{2^2} = \mathbf{A}_2 \otimes \mathbf{B}_2 = \begin{bmatrix} a_{11}\mathbf{B}_2 & a_{12}\mathbf{B}_2 \\ a_{21}\mathbf{B}_2 & a_{22}\mathbf{B}_2, \end{bmatrix} = \begin{bmatrix} a_{11}b_{11} & a_{11}b_{12} & a_{12}b_{13} & a_{12}b_{14} \\ a_{11}b_{21} & a_{11}b_{22} & a_{12}b_{23} & a_{12}b_{24} \\ a_{21}b_{31} & a_{21}b_{32} & a_{22}b_{33} & a_{22}b_{34} \\ a_{21}b_{41} & a_{21}b_{42} & a_{22}b_{43} & a_{22}b_{44} \end{bmatrix}$$

Since, $\mathbf{S}_{2^n}$ is an identity matrix, $\mathbf{F} = \mathbf{S}$. This sum-of-products is the particular case of other forms. In Figure 3.14a, the general scheme of computing is represented.

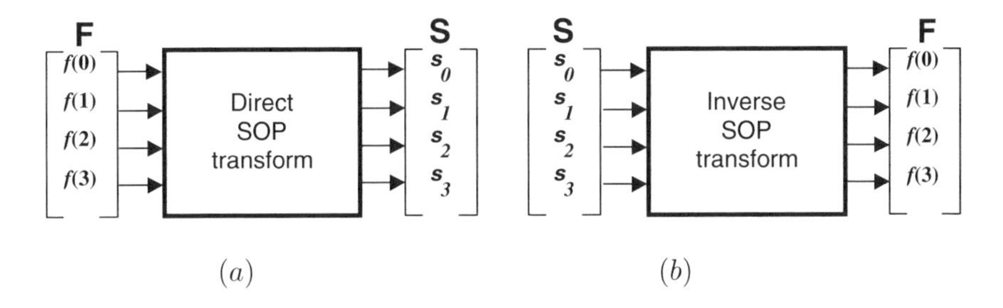

FIGURE 3.14
Direct (a) and inverse (b) sum-of-products (SOP) transforms for a switching function of two variables.

Example 3.10 *Computing the coefficients of the sum-of-products by Equation 3.3 and Equation 3.4 for the elementary function* $f = x_1 \oplus x_2$ *given by the truth vector* $\mathbf{F} = [0\ 1\ 1\ 0]^T$ *is illustrated in Figure 3.15.*

3.3.3 Restoration

The following matrix equation using AND and OR operations restores the truth-vector $\mathbf{F}$ from the vector of coefficients $\mathbf{S}$ (Figure 3.14b):

$$\mathbf{F} = \mathbf{S}_{2^n}^{-1} \cdot \mathbf{S}, \tag{3.5}$$

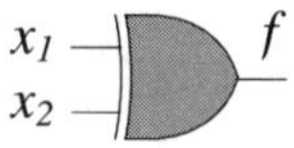

x_1, x_2, f

$f = x_1 \oplus x_2$

Computing the vector of coefficients

$$\mathbf{S} = \mathbf{S}_{2^3} \cdot \mathbf{F} = \left[\begin{array}{cc|cc} 1 & & & \\ & 1 & & \\ \hline & & 1 & \\ & & & 1 \end{array}\right] \begin{bmatrix} 0 \\ 1 \\ 1 \\ 0 \end{bmatrix} = \begin{bmatrix} 0 = s_0 \\ 1 = s_1 \\ 1 = s_2 \\ 0 = s_3 \end{bmatrix}$$

Sum-of-products expression

$$f = s_0(\overline{x}_1\overline{x}_2) \vee s_1(\overline{x}_1 x_2) \vee s_2(x_1\overline{x}_2) \vee s_3(x_1 x_2) = \overline{x}_1 x_2 \vee x_1\overline{x}_2$$

FIGURE 3.15
Computation of the sum-of-products expression for EXOR gate (Example 3.10).

where $\mathbf{S}_{2^1}^{-1} = \mathbf{S}_{2^1}$. Notice that the matrix $\mathbf{S}_{2^1}$ is self-inverse.

Example 3.11 *Restore the truth-vector* $\mathbf{F}$ *of sum-of-products of a function* f *given by the vector of coefficients* $\mathbf{S} = [0\ 1\ 1\ 0]^T$*:*

$$\mathbf{F} = \mathbf{S}_{2^3}^{-1} \cdot \mathbf{S} = \left[\begin{array}{cc|cc} 1 & & & \\ & 1 & & \\ \hline & & 1 & \\ & & & 1 \end{array}\right] \begin{bmatrix} 0 \\ 1 \\ 1 \\ 0 \end{bmatrix} = \begin{bmatrix} 0 \\ 1 \\ 1 \\ 0 \end{bmatrix}.$$

3.3.4 Useful rules

Because the vector of coefficients $\mathbf{S}$ and the truth-vector $\mathbf{F}$ are equal, the sum-of-products expression can be derived directly from the truth vector $\mathbf{F}$.

Example 3.12 *Given* $\mathbf{F} = [0\ 1\ 0\ 0]^T$*, we derive* $\mathbf{S} = [0\ 1\ 0\ 0]^T$*, i.e.,* $f = \overline{x}_1 x_2$.

3.3.5 Hypercubes

A topological representation of a switching function of n variables is the n-dimensional hypercube such that:

- The vertices of the hypercube denote the minterms, thus, the hypercube is a collection of minterms;
- The number of minterms is a power of two, 2^m, for some $m \leq n$;
- The number of edges in a hypercube is $3 \cdot s^{n-1}$.

In Figure 3.16, the hypercubes for some logic functions of three variables are represented.

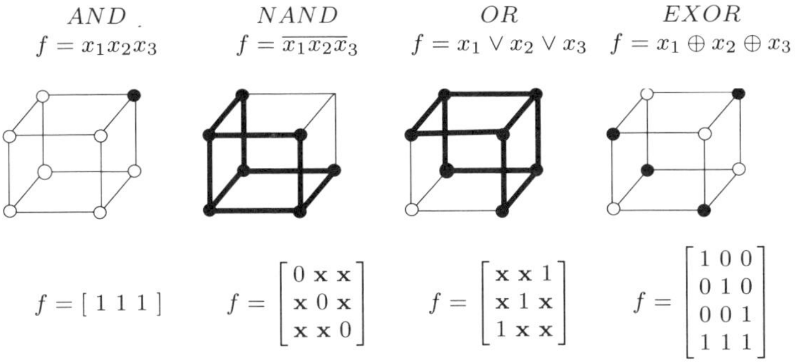

AND	*NAND*	*OR*	*EXOR*
$f = x_1x_2x_3$	$f = \overline{x_1x_2x_3}$	$f = x_1 \vee x_2 \vee x_3$	$f = x_1 \oplus x_2 \oplus x_3$

$$f = [\ 1\ 1\ 1\ ] \qquad f = \begin{bmatrix} 0\ \mathrm{x}\ \mathrm{x} \\ \mathrm{x}\ 0\ \mathrm{x} \\ \mathrm{x}\ \mathrm{x}\ 0 \end{bmatrix} \qquad f = \begin{bmatrix} \mathrm{x}\ \mathrm{x}\ 1 \\ \mathrm{x}\ 1\ \mathrm{x} \\ 1\ \mathrm{x}\ \mathrm{x} \end{bmatrix} \qquad f = \begin{bmatrix} 1\ 0\ 0 \\ 0\ 1\ 0 \\ 0\ 0\ 1 \\ 1\ 1\ 1 \end{bmatrix}$$

FIGURE 3.16
A three-D hypercube and cube-based representation of AND, NAND, OR and EXOR tree-input gates.

3.4 Shannon decision trees and diagrams

A binary decision tree that corresponds to a canonical sum-of-products representation of a switching function is called a Shannon decision tree. This tree is associated with the Shannon decision diagram (BDD, ROBDD).

3.4.1 Formal synthesis

A node in a Shannon decision tree of a switching function f corresponds to the Shannon decomposition of the function with respect to a variable x_i

$$f = \overline{x}_i f_0 \vee x_i f_1, \tag{3.6}$$

where $f_0 = f|_{x_i=0}$ and $f_1 = f|_{x_i=1}$. Here $f = f|_{x_i=a}$ denotes the cofactor of f after assigning the constant a to the variable x_i. Shannon decomposition is labeled by S (Figure 3.17).

In matrix notation, the transform in a node of the Shannon decision tree for a function of a single variable x_i given by the truth-vector $\mathbf{F} = [\ f(0)\ f(1)\ ]^T$ is given below

$$f = [\ \overline{x}_i\ x_i\ ] \begin{bmatrix} 1\ 0 \\ 0\ 1 \end{bmatrix} \begin{bmatrix} f_0 \\ f_1 \end{bmatrix} = [\ \overline{x}_i\ x_i\ ] \begin{bmatrix} f_0 \\ f_1 \end{bmatrix} = \overline{x}_i f_0 \vee x_i f_1,$$

where $f_0 = f|_{x_i=0}$, $f_1 = f|_{x_i=1}$. Recursive application of the Shannon expansion to f given by truth-vector $\mathbf{F} = [f(0)\ f(1) \ldots f(2^n - 1)]^T$ is described in the matrix notation as

$$f = \widehat{\mathbf{X}}\ \mathbf{S}_{2^n}\ \mathbf{F}, \tag{3.7}$$

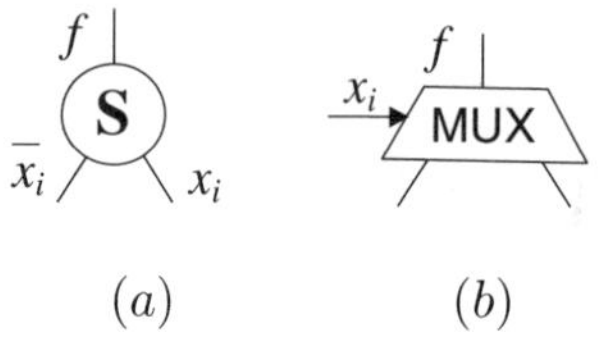

f, S, $\overline{x}_i$, x_i

(a)

f, x_i, MUX

(b)

$$f = \overline{x}_i f_0 \vee x_i f_1$$
$$f_0 = f|_{x_i=0}$$
$$f_1 = f|_{x_i=1}$$

(c)

$$f = [\ \overline{x}_i\ x_i\ ] \begin{bmatrix} 1 & 0 \\ 0 & 1 \end{bmatrix} \begin{bmatrix} f_0 \\ f_1 \end{bmatrix}$$

(d)

FIGURE 3.17
The node of the Shannon decision tree (a), its implementation by multiplexer (MUX) (b), algebraic (c), and matrix (d) descriptions.

where

$$\widehat{\mathbf{X}} = \bigotimes_{i=1}^{n} [\ \overline{x}_i\ x_i\ ], \quad \mathbf{S}_{2^n} = \bigotimes_{i=1}^{n} \mathbf{S}_2, \quad \mathbf{S}_2 = \begin{bmatrix} 1 & 0 \\ 0 & 1 \end{bmatrix},$$

and $\otimes$ denotes the Kronecker product.

Example 3.13 *Let us derive the Shannon decision tree for the switching function* $f = \overline{x}_1 \vee x_2$ *given truth table* $\mathbf{F} = [1\ 1\ 0\ 1]^T$. *The solution is shown in Figure 3.18. The minterms are generated by the Kronecker product* $\widehat{\mathbf{X}}$. *The* 4×4 *transform matrix* $\mathbf{S}$ *is formed by the Kronecker product of the basic matrix* $\mathbf{S}_{2^1}$. *The final result, sum-of-products, is directly mapped into the Shannon decision tree.*

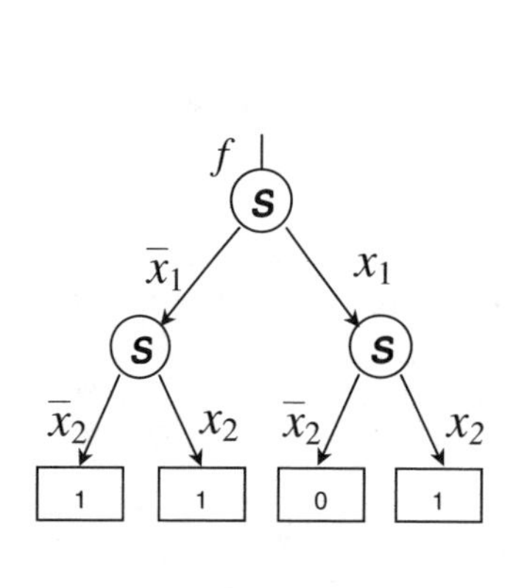

Transform matrix

$$\mathbf{S}_{2^2} = \mathbf{S}_2 \otimes \mathbf{S}_2 = \begin{bmatrix} 1 & 0 \\ 0 & 1 \end{bmatrix} \otimes \begin{bmatrix} 1 & 0 \\ 0 & 1 \end{bmatrix} = \left[\begin{array}{cc|cc} 1 & & & \\ & 1 & & \\ \hline & & 1 & \\ & & & 1 \end{array}\right]$$

Sum-of-products expression

$$f = \widehat{\mathbf{X}}\,\mathbf{S}_{2^2}\,\mathbf{F} = \widehat{\mathbf{X}} \left[\begin{array}{cc|cc} 1 & & & \\ & 1 & & \\ \hline & & 1 & \\ & & & 1 \end{array}\right] \begin{bmatrix} 1 \\ 1 \\ 0 \\ 1 \end{bmatrix} = \widehat{\mathbf{X}} \begin{bmatrix} 1 \\ 1 \\ 0 \\ 1 \end{bmatrix}$$

$$\widehat{\mathbf{X}} = [\ \overline{x}_1\ x_1\ ]^T \otimes [\ \overline{x}_2\ x_2\ ]^T = [\ \overline{x}_1\overline{x}_2,\ \overline{x}_1 x_2,\ x_1\overline{x}_2,\ x_1 x_2\ ]^T$$
$$f = \overline{x}_1\overline{x}_2 \vee \overline{x}_1 x_2 \vee x_1 x_2$$

FIGURE 3.18
Deriving the Shannon decision tree for the switching function $f = \overline{x}_1 \vee x_2$ (Example 3.13).

3.4.2 Structural properties

A Shannon decision tree is a canonical representation of f in graphical form with structural properties as follows:

- The nodes implement the Shannon expansion.
- Nonterminal nodes are distributed over levels, each level corresponding to a variable x_i in f, starting from the root node corresponding to x_i, or some other variable that was chosen first.
- Since the variables appear in a fixed order, such a tree is an ordered Shannon decision tree.
- Each path from the root node to the terminal nodes corresponds to a minterm in the sum-of-products representation of the function; the minterm is determined as the product of labels at the edges.
- The values of terminal nodes are the values of the represented functions. Thus, the terminal nodes are assigned with the truth values $\mathbf{F}=[f_{000}\ f_{001}...f_{111}]$ of switching function f, where "0" corresponds to the value $f_0 = f|_{x_i=0}$, and "1" corresponds to the value $f_1 = f|_{x_i=1}$.

Figure 3.19 illustrates graphical representations of sum-of-products for AND function. The flowgraph is degenerated because inputs and outputs are the same, i.e., $\mathbf{F} = \mathbf{S}$. The complete decision tree is reduced to a decision diagram. Finally, the decision tree is embedded into a 3-D structure.

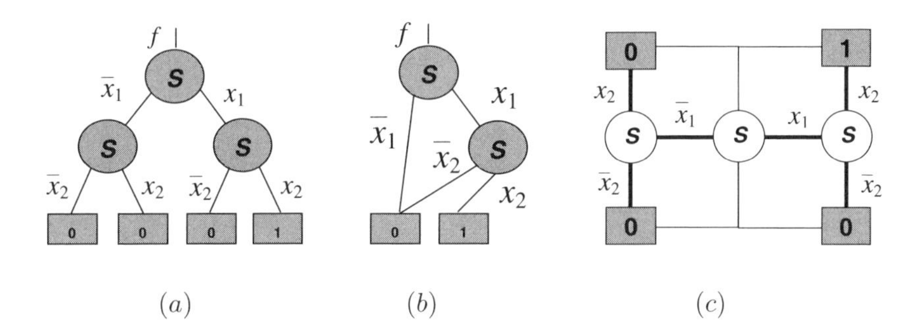

FIGURE 3.19
Graphical representation of an AND function of two variables in sum-of-products form: decision tree (a), decision diagram (b), and decision tree embedded in a hypercube (c).

Example 3.14 *An arbitrary switching function f of three variables can be represented by the Shannon decision tree shown in Figure 3.20: 3 levels ($n = 3$), 7 nodes, 8 terminal nodes, 2^k nodes at the k-th level, $k = 0, 1, 2$. To design* *this tree, the Shannon expansion is used as follows:*

- *with respect to variable* x_1: $f = \overline{x}_1 f_0 \vee x_1 f_1$;
- *with respect to variable* x_2: $f_0 = \overline{x}_2 f_{00} \vee x_2 f_{01}$ *and* $f_1 = \overline{x}_2 f_{10} \vee x_2 f_{11}$;
- *with respect to variable* x_3: $f_{00} = \overline{x}_3 f_{000} \vee x_3 f_{001}$, $f_{01} = \overline{x}_3 f_{010} \vee x_3 f_{011}$, $f_{10} = \overline{x}_3 f_{100} \vee x_3 f_{101}$, *and* $f_{11} = \overline{x}_3 f_{110} \vee x_3 f_{111}$.

Hence, the Shannon decision tree represents a switching function f *in the form of the sum-of-products* $f = f_{000}\overline{x}_1\overline{x}_2\overline{x}_3 \vee f_{001}\overline{x}_1\overline{x}_2 x_3 \vee f_{010}\overline{x}_1 x_2\overline{x}_3 \vee f_{011}\overline{x}_1 x_2 x_3 \vee f_{100} x_1\overline{x}_2\overline{x}_3 \vee f_{101} x_1\overline{x}_2 x_3 \vee f_{110} x_1 x_2\overline{x}_3 \vee f_{111} x_1 x_2 x_3$.

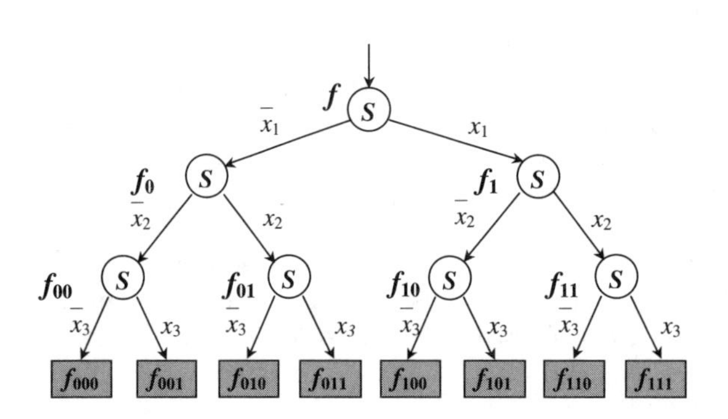

There are 8 paths from f *to the terminal nodes:*

Path 1: $m_1 = \overline{x}_1\overline{x}_2\overline{x}_3$
Path 2: $m_2 = \overline{x}_1\overline{x}_2 x_3$
Path 3: $m_3 = \overline{x}_1 x_2\overline{x}_3$
Path 4: $m_4 = \overline{x}_1 x_2 x_3$
Path 5: $m_5 = x_1\overline{x}_2\overline{x}_3$
Path 6: $m_6 = x_1\overline{x}_2 x_3$
Path 7: $m_7 = x_1 x_2\overline{x}_3$
Path 8: $m_8 = x_1 x_2 x_3$

Sum-of-products
$f = m_1 \vee m_2 \vee ... \vee m_8$

FIGURE 3.20
The Shannon decision tree for the sum-of-products of a switching function of three variables (Example 3.14).

3.4.3 Decision tree reduction

The Shannon decision diagram for a given function f is derived from the Shannon decision tree for f by deleting redundant nodes, and by sharing equivalent subgraphs. The reduction rules are as follows (Figure 3.21):

Elimination rule: If two descendent nodes of a node are identical, then delete the node and connect the incoming edges of the deleted node to the corresponding successor.

Merging rule: Share equivalent subgraphs.

In a tree, edges longer than 1, i.e., connecting nodes at nonsuccessive levels, can appear. For example, the length of an edge connecting a node at $(i-1)$-th level with a node at $(i+1)$-th level is two.

Elimination rule

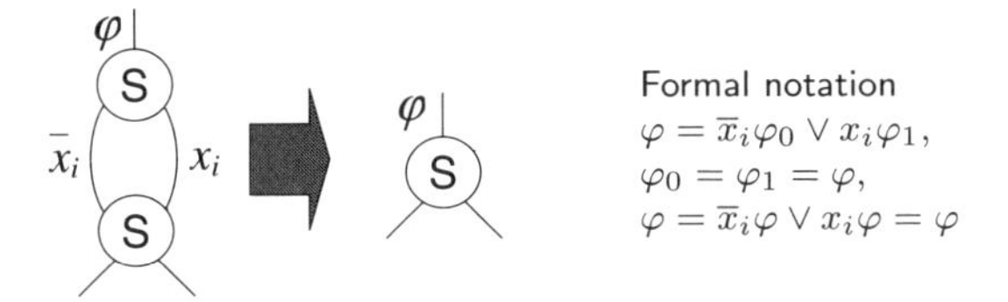

Merging rule

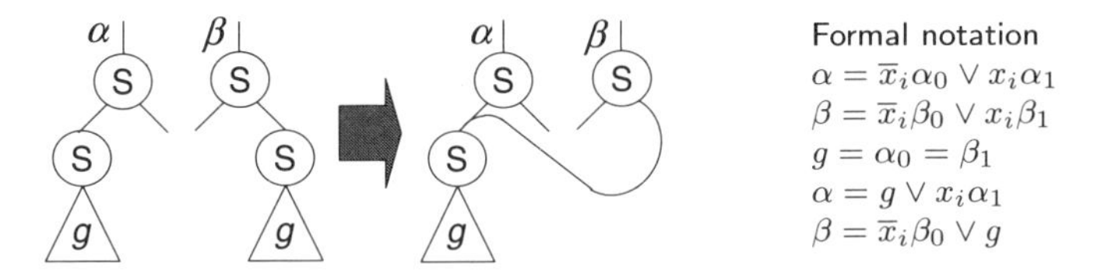

FIGURE 3.21
Reduction of the Shannon decision tree.

Example 3.15 *Application of the reduction rules to a decision tree of the 3-input NOR function is demonstrated in Figure 3.22.*

3.5 Reed-Muller expressions

Reed-Muller algebra is a universal basis that includes constant 1, EXOR, AND, and NOT operations over Boolean variables. Reed-Muller expressions are classified as fixed and mixed polarity expressions. In this section the fixed polarity Reed-Muller expressions are introduced.

3.5.1 General form

Given a switching function f of n variables, the Reed-Muller expression is specified by

$$f = \bigoplus_{i=0}^{2^n-1} r_i \cdot \underbrace{(x_1^{i_1} \cdots \ x_n^{i_n})}_{i-th\ product}, \tag{3.8}$$

where $r_i \in \{0, 1\}$ is a coefficient, i_j is the j-th bit $j = 1, 2, \ldots, n$, in the binary representation of the index $i = i_1 i_2 \ldots i_n$, and $x_j^{i_j}$ is defined as

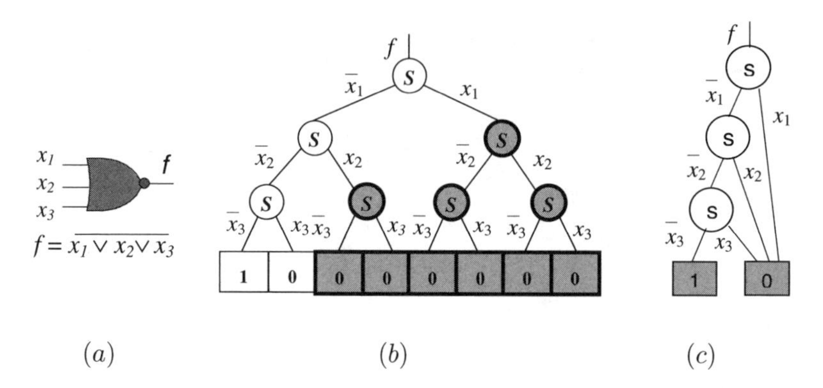

FIGURE 3.22
The three-input NOR function (a), Shannon decision tree (b), and decision diagram with the lexicographical order of variables (c) (Example 3.15).

$$x_j^{i_j} = \begin{cases} 1, & \text{if } i_j = 0; \\ x_j, & \text{if } i_j = 1. \end{cases} \tag{3.9}$$

Example 3.16 *An arbitrary switching function of two variables is represented by the Reed-Muller expression by Equation 3.8 and Equation 3.9*

$$\begin{aligned} f &= r_0(x_1^0 x_2^0) \oplus r_1(x_1^0 x_2^1) \oplus r_2(x_1^1 x_2^0) \oplus r_3(x_1^1 x_2^1) \\ &= r_0 \oplus r_1 x_2 \oplus r_2 x_1 \oplus r_3 x_1 x_2. \end{aligned}$$

Figure 3.23 illustrates how to assemble the expression given the coefficients r_j*, where the shadowed nodes implement the AND operations.*

3.5.2 Computing the coefficients

Given a truth vector $\mathbf{F} = [f(0)\ f(1) \ldots f(2^n - 1)]^T$, of a function f, the vector of Reed-Muller coefficients $\mathbf{R} = [r_0\ r_1 \ldots r_{2^n-1}]^T$ is derived by the matrix equation with respect to AND and EXOR operations

$$\mathbf{R} = \mathbf{R}_{2^n} \cdot \mathbf{F} \quad (mod\ 2), \tag{3.10}$$

the $2^n \times 2^n$ matrix $\mathbf{R}_{2^n}$ is formed by the Kronecker product

$$\mathbf{R}_{2^n} = \bigotimes_{j=1}^{n} \mathbf{R}_{2^j}, \qquad \mathbf{R}_{2^1} = \begin{bmatrix} 1 & 0 \\ 1 & 1 \end{bmatrix}. \tag{3.11}$$

In Figure 3.24a a general computing scheme is shown for $n = 2$.

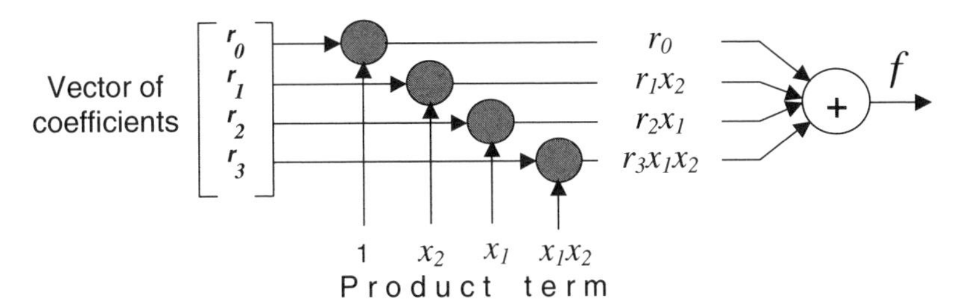

FIGURE 3.23
Deriving the Reed-Muller expression for a function of two variables (Example 3.16).

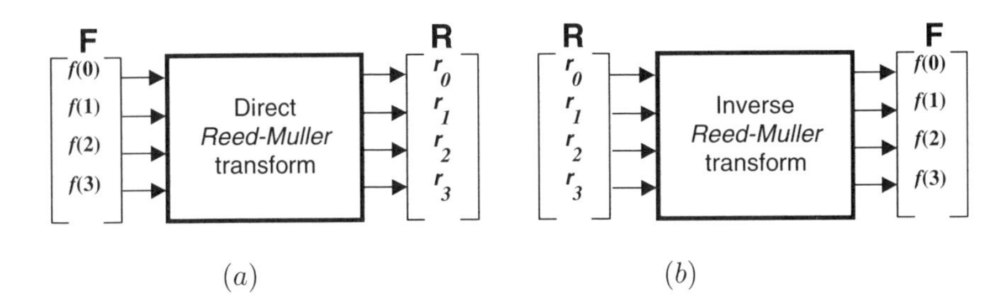

FIGURE 3.24
Direct (a) and inverse (b) Reed-Muller transforms for a switching function of two variables.

3.5.3 Flowgraphs

To design the flowgraph of the algorithm, the matrix $\mathbf{R}_{2^n}$ must be represented in the factorized form

$$\mathbf{R}_{2^n} = \mathbf{R}_{2^n}^{(1)}\mathbf{R}_{2^n}^{(2)} \cdots \mathbf{R}_{2^n}^{(n)}, \tag{3.12}$$

where $\mathbf{R}_{2^n}^{(i)}$, $i = 1, 2, \ldots, n$, is formed by the Kronecker product

$$\mathbf{R}_{2^n}^{(i)} = \mathbf{I}_{2^{n-i}} \otimes \mathbf{R}_{2^1} \otimes \mathbf{I}_{2^{i-1}}. \tag{3.13}$$

Hence, Reed-Muller coefficients are computed in n iterations.

Example 3.17 *Computing the Reed-Muller coefficients by Equation 3.10 and Equation 3.14 for the function $f = x_1 \vee \overline{x}_2$ is illustrated in Figure 3.25. The flowgraph includes two iterations accordingly to factorization relations (Equation 3.12 and Equation 3.13):*

$$
\begin{aligned}
\mathbf{R}_{2^2} &= \mathbf{R}_{2^2}^{(1)}\mathbf{R}_{2^2}^{(2)} = (\mathbf{I}_{2^{2-1}} \otimes \mathbf{R}_{2^1} \otimes \mathbf{I}_{2^{1-1}})(\mathbf{I}_{2^{2-2}} \otimes \mathbf{R}_{2^1} \otimes \mathbf{I}_{2^{2-1}}) \\
&= \underbrace{(\mathbf{I}_{2^1} \otimes \mathbf{R}_{2^1} \otimes 1)}_{1st\ iteration}\underbrace{(1 \otimes \mathbf{R}_{2^1} \otimes \mathbf{I}_{2^1})}_{2nd\ iteration} \\
&= \left[\begin{array}{c|c} \mathbf{R}_{2^1} & \\ \hline & \mathbf{R}_{2^1} \end{array}\right]\left[\begin{array}{c|c} \mathbf{I}_{2^1} & \\ \hline \mathbf{I}_{2^1} & \mathbf{I}_{2^1} \end{array}\right] = \left[\begin{array}{cc|cc} 1 & & & \\ 1 & 1 & & \\ \hline & & 1 & \\ & & 1 & 1 \end{array}\right]\left[\begin{array}{cc|cc} 1 & & & \\ & 1 & & \\ \hline 1 & & 1 & \\ & 1 & & 1 \end{array}\right].
\end{aligned}
$$

x_1, x_2, f

$f = x_1 \vee \overline{x}_2$

Vector of coefficients

$$
\mathbf{R} = \mathbf{R}_{2^2} \cdot \mathbf{F} = \left[\begin{array}{cc|cc} 1 & 0 & 0 & 0 \\ 1 & 1 & 0 & 0 \\ \hline 1 & 0 & 1 & 0 \\ 1 & 1 & 1 & 1 \end{array}\right]\begin{bmatrix} 1 \\ 0 \\ 1 \\ 1 \end{bmatrix} = \begin{bmatrix} 1 \\ 1 \\ 0 \\ 1 \end{bmatrix} \ (mod\ 2)
$$

Reed-Muller expression

$f = 1 \oplus x_2 \oplus x_1x_2$

Flow graph of the algorithm

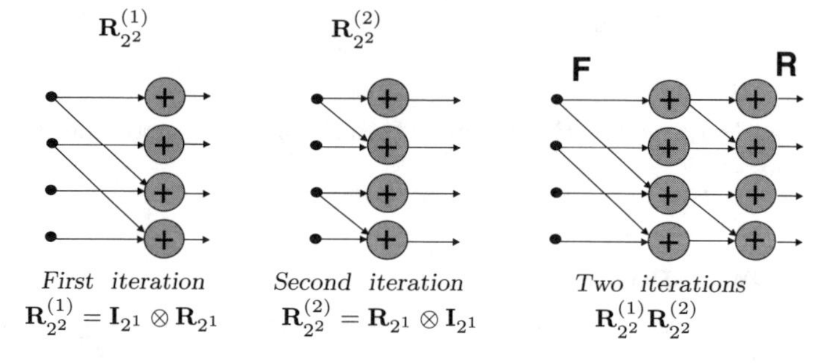

FIGURE 3.25
Computing the Reed-Muller expression for the two-input OR (Example 3.17).

3.5.4 Restoration

The following matrix equation with AND and EXOR operations restores the truth-vector $\mathbf{F}$ from the coefficients vector $\mathbf{R}$ (Figure 3.24b):

$$
\mathbf{F} = \mathbf{R}_{2^n}^{-1} \cdot \mathbf{R} \ \ (mod\ 2), \tag{3.14}
$$

where $\mathbf{R}_{2^1}^{-1} = \mathbf{R}_{2^1}$. Notice that the matrix $\mathbf{R}_{2^1}$ is a self-inverse matrix over

Galois field $GF(2)$, i.e., in terms of logic operations AND and EXOR.

Example 3.18 *Restore the truth-vector* $\mathbf{F}$ *of a switching function* f *given by the vector of Reed-Muller coefficients* $\mathbf{R} = [1\ 1\ 0\ 1]^T$*:*

$$\mathbf{F}{=}\mathbf{R}_{2^3}^{-1} \cdot \mathbf{R} = \left[\begin{array}{cc|cc} 1 & 0 & 0 & 0 \\ 1 & 1 & 0 & 0 \\ \hline 1 & 0 & 1 & 0 \\ 1 & 1 & 1 & 1 \end{array}\right] \begin{bmatrix} 1 \\ 1 \\ 0 \\ 1 \end{bmatrix} = \begin{bmatrix} 1 \\ 0 \\ 1 \\ 1 \end{bmatrix} \quad (mod\ 2).$$

3.5.5 Useful rules

Rule 1: To derive the Reed-Muller expression from the canonical sum-of-products, the OR operation must be replaced by the EXOR operation. For example, $x_1\overline{x}_2 \vee \overline{x}_1 x_2 {=} x_1\overline{x}_2 \oplus \overline{x}_1 x_2$.
Rule 2: To represent the expression in the NAND form, replace the complement variables $\overline{x}_i$ by $x_1 \oplus 1$, and simplify the obtained expression. For example,

$$\begin{aligned} x_1 \vee x_2 &= \overline{\overline{x_1 \vee x_2}} = \overline{\overline{x}_1\overline{x}_2} \\ &= (1 \oplus x_1)(1 \oplus x_2) \\ &= x_1 \oplus x_2 \oplus x_1 x_2. \end{aligned}$$

3.5.6 Hypercube representation

Let $x_j^{i_j}$ be a literal of a Boolean variable x_j such that $x_j^{i_j} = \overline{x}_j$ if $i_j = 0$, and $x_j^{i_j} = x_j$ if $i_j = 1$. A product of literals $x_1^{i_1} x_2^{i_2} \ldots x_n^{i_n}$ is called a *product term.* If the variable x_j is not present in a cube, $i_j = \mathbf{x}$ (don't care), i.e., $x_j^{\mathbf{x}} = 1$. In cube notation, a term is described by a cube that is a ternary n-tuple of components $i_j \in \{0,\ 1,\ \mathbf{x}\}$. A set of cubes corresponding to the true values of the switching function f represents the sum-of-products for this function.

A Reed-Muller expression consists of products combined by EXOR operation. For example, a sum-of-products form given by the cubes $[\mathbf{x}\ \mathbf{x}\ 0] \vee [1\ 0\ \mathbf{x}]$ can be written as exclusive sum-of-products (ESOP) $[\mathbf{x}\ \mathbf{x}\ 0] \oplus [0\ 1\ 0]$. The different cubes arise because of the different operation between the cubes in the expressions. Thus, the manipulation of the cubes involves OR, AND and EXOR operations, applied to the appropriate literals following the rules given in Table 3.2.

Example 3.19 *Given the cubes* $[1\ 1\ \mathbf{x}]$ *and* $[1\ 0\ \mathbf{x}]$*. The AND, OR, and EXOR operations with these cubes are shown in Figure 3.26.*

Suppose a sum-of-products expression for a function f is given by cubes. To represent this function in Reed-Muller form, we have to generate cubes based on the equation $x \vee y = x \oplus y \oplus xy$ that can be written in cube notation as

$$[C_1] \vee [C_2] = [C_1] \oplus [C_2] \oplus [C_1][C_2]. \tag{3.15}$$

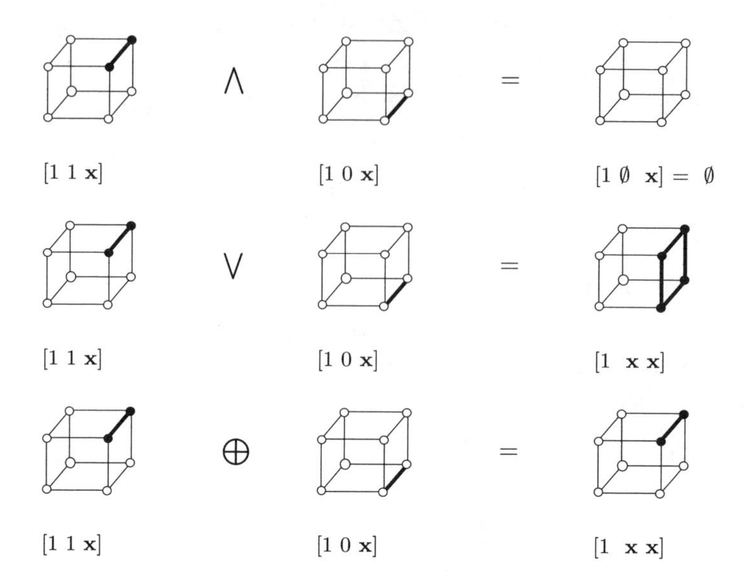

FIGURE 3.26
AND, OR, and EXOR operations over cubes (Example 3.19).

TABLE 3.2
AND, OR and EXOR operation with respect to literals of cubes.

AND C_i / C_j	0	1	x	*OR* C_i / C_j	0	1	x	*EXOR* C_i / C_j	0	1	x
0	0	∅	0	0	0	x	x	0	0	x	1
1	∅	1	1	1	x	1	1	1	x	1	0
x	0	1	x	x	x	1	x	x	1	0	x

Example 3.20 *A switching function is given in a sum-of-products form by four cubes,*

$$f = [\mathbf{x}\ 1\ 0\ 1] \vee [1\ 0\ 0\ \mathbf{x}] \vee [0\ \mathbf{x}\ \mathbf{x}\ 0] \vee [\mathbf{x}\ \mathbf{x}\ 1\ 0].$$

To find its ESOP expression we apply the cube generation operation, i.e., we replace $\vee$ *by* $\oplus$ *and compute AND for each cube distinguished by only one literal (the rules for AND are given in Table 3.2). We then obtain the cube representation*

$$f = [\mathbf{x}\ 1\ 0\ 1] \oplus [1\ 0\ 0\ \mathbf{x}] \oplus [0\ \mathbf{x}\ \mathbf{x}\ 0] \oplus [\mathbf{x}\ \mathbf{x}\ 1\ 0] \oplus [0\ \mathbf{x}\ 1\ 0]$$

that is

$$f = x_2\overline{x}_3x_4 \oplus x_1\overline{x}_2\overline{x}_3 \oplus \overline{x}_1\overline{x}_4 \oplus x_3\overline{x}_4 \oplus \overline{x}_1x_3\overline{x}_4.$$

Note that ESOP is a mixed polarity form where a variable entries both complemented and uncomplemented. The term *fixed polarity* is used to describe Reed-Muller expansions in which each variable appears uncomplemented or complemented, and never in both expressions.

3.5.7 Polarity

The polarity of a variable x_j can be:

$c_j = 1$, corresponding to the uncomplimented variable x_j, or
$c_j = 0$, corresponding to the complimented variable $\overline{x}_j$.

Let the polarity $c = c_1c_2\dots c_n$, $c \in \{0, 1, 2, \dots, 2^n - 1\}$, where c_j is the j-th bit in binary representation of c. For a switching function f of n variables, the Reed-Muller expression in a given polarity $c = c_1c_2\dots c_n$ of variables $x_1x_2\dots x_n$ is given by

$$f = \bigoplus_{i=0}^{2^n-1} r_i \cdot \underbrace{(x_1 \oplus c_1)^{i_1} \cdots \ (x_n \oplus c_n)^{i_n}}_{i-th\ product}, \tag{3.16}$$

where r_i is the i-th coefficient, and $(x_j \oplus c_j)^{i_j}$ is defined as

$$a_j^{i_j} = \begin{cases} 1, \text{ if } i_j = 0; \\ a, \text{ if } i_j = 1. \end{cases} \quad x_j \oplus c_j = \begin{cases} x_j, \text{ if } c_j = 0; \\ \overline{x}_j, \text{ if } c_j = 1. \end{cases} \tag{3.17}$$

Example 3.21 *In Example 3.16, a switching function of two variables has been represented by the zero-polarity Reed-Muller expression, thus for* $c = 0\ (c_1c_2 = 00)$. *We now represent this function in the polarity* $c = 2,\ c_1c_2 = 10$. *By Equation 3.16 and Equation 3.17:*

$$f = r_0(x_1 \oplus 1)^0(x_2 \oplus 0)^0 \oplus r_1(x_1 \oplus 1)^0(x_2 \oplus 0)^1 \oplus r_2(x_1 \oplus 1)^1(x_2 \oplus 0)^0$$
$$\oplus\, r_3(x_1 \oplus 1)^1(x_2 \oplus 0)^1 = r_0 \oplus r_1x_2 \oplus r_2\overline{x}_1 \oplus r_3\overline{x}_1x_2.$$

Let $f = x \vee y$, *then four fixed polarity Reed-Muller (FPRM) expressions can be derived as shown in Figure 3.27.*

Recall, in an FPRM expansion of a given switching function f, every variable appears either complemented or uncomplemented; never in both forms. If all variables are uncomplemented (complemented), the FPRM expansion is called a *positive (Negative) polarity* Reed-Muller form. FPRM expansions are *unique*. Thus, only one representation exists for the positive and negative FPRM or indeed any FPRM of f. FPRM expansions have been used to classify a given switching function into an equivalence class of switching functions, where two switching functions are equivalent if one is transformed into the other by permuting variables, complementing variables, and/or complementing the switching function, and is useful for determining library cells in computer aided design (CAD) tools. This is called *Boolean matching* and is important in the determination of library cells for use by computer-aided design tools.

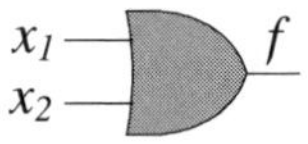

x_1, x_2, f

$f = x_1 \vee x_2$

$$
\begin{aligned}
0-polarity &: \ f = x_1 \oplus x_2 \oplus x_1 x_2 \\
1-polarity &: \ f = 1 \oplus \overline{x}_2 \oplus x_1 \overline{x}_2 \\
2-polarity &: \ f = 1 \oplus \overline{x}_1 \oplus \overline{x}_1 x_2 \\
3-polarity &: \ f = 1 \oplus \overline{x}_1 \overline{x}_2
\end{aligned}
$$

FIGURE 3.27
Representation of two-input OR gate by Reed-Muller forms of $2^2 = 4$ polarities (Example 3.21).

Given the truth table $\mathbf{F} = [f(0)\ f(1) \dots f(2^n-1)]^T$, the vector of Reed-Muller coefficients in the polarity c, $\mathbf{R^{(c)}} = [r_0^{(c)}\ r_1^{(c)} \dots r_{2^n-1}^{(c)}]^T$ is derived by the matrix eqaution

$$\mathbf{R^{(c)}} = \mathbf{R}_{2^n}^{(c)} \cdot \mathbf{F} \quad (mod\ 2), \tag{3.18}$$

where the $2^n \times 2^n$-matrix $\mathbf{R}_{2^n}^{(c)}$ is generated by the Kronecker product

$$\mathbf{R}_{2^n}^{(c)} = \bigotimes_{j=1}^{n} \mathbf{R}_{2^1}^{(c_j)}, \qquad \mathbf{R}_{2^1}^{(c)} = \begin{cases} \begin{bmatrix} 0 & 1 \\ 1 & 1 \end{bmatrix}, & c_j = 0; \\ \\ \begin{bmatrix} 0 & 1 \\ 1 & 1 \end{bmatrix}, & c_j = 1. \end{cases} \tag{3.19}$$

Example 3.22 *In the matrix form, the solution to Example 3.17 can be derived by the Equation 3.18 as follows:*

$$\mathbf{R^{(2)}} = \mathbf{R}_{2^2}^{(2)} \cdot \mathbf{F} = \left[\begin{array}{cc|cc} 0 & 0 & 0 & 1 \\ 0 & 0 & 1 & 1 \\ \hline 0 & 1 & 0 & 1 \\ 1 & 1 & 1 & 1 \end{array}\right] \begin{bmatrix} 1 \\ 0 \\ 1 \\ 1 \end{bmatrix} = \begin{bmatrix} 1 \\ 0 \\ 1 \\ 1 \end{bmatrix} \quad (mod\ 2)$$

where the matrix $\mathbf{R}_{2^2}^{(2)}$ *given* $c = 2$ *is generated by Equation 3.19 as*

$$\mathbf{R}_{2^2}^{(2)} = \mathbf{R}_{2^1}^{(1)} \otimes \mathbf{R}_{2^1}^{(0)} = \begin{bmatrix} 0 & 1 \\ 1 & 1 \end{bmatrix} \otimes \begin{bmatrix} 1 & 0 \\ 1 & 1 \end{bmatrix}.$$

The vector of coefficients $\mathbf{R^{(2)}} = [1\ 0\ 1\ 1]^T$ *corresponds to the expression* $f = 1 \oplus \overline{x}_1 \oplus \overline{x}_1 x_2$.

3.6 Decision trees and diagrams

A binary decision tree that corresponds to the Reed-Muller canonical representation of a switching function is called a *Davio decision tree*. A Davio (functional) decision diagram can be derived from a Davio decision tree.

3.6.1 Formal design

A node in a Davio decision tree of a switching function f corresponds to the Davio decomposition of the function with respect to variable x_i. There exists:

- *The positive Davio* expansion

$$f = f_0 \oplus x_i f_2, \tag{3.20}$$

  where $f_0 = f|_{x_i=0}$ and $f_2 = f|_{x_i=1} \oplus f|_{x_i=0}$, and

- *The negative Davio* expansion

$$f = \overline{x}_1 f_2 \oplus f_1, \tag{3.21}$$

  where $f_1 = f_1 = f|_{x_i=1}$.

Positive and negative Davio decomposition are labeled as pD and nD respectively (Figure 3.28).

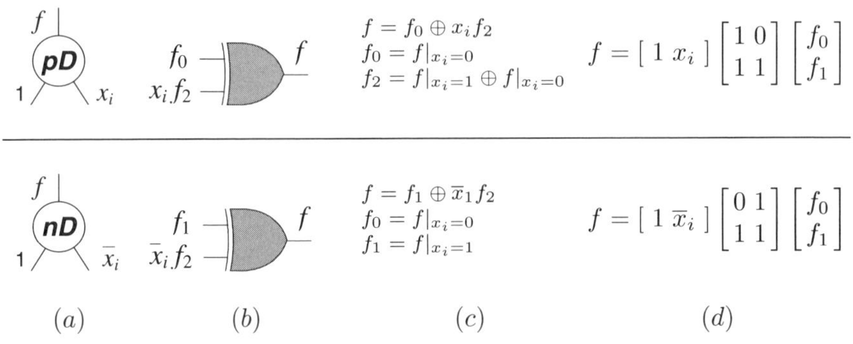

pD: f, 1, x_i; f_0, $x_i.f_2$, f

$f = f_0 \oplus x_i f_2$
$f_0 = f|_{x_i=0}$
$f_2 = f|_{x_i=1} \oplus f|_{x_i=0}$

$$f = [\,1\ x_i\,]\begin{bmatrix}1 & 0\\ 1 & 1\end{bmatrix}\begin{bmatrix}f_0\\ f_1\end{bmatrix}$$

nD: f, 1, $\overline{x}_i$; f_1, $\overline{x}_i.f_2$, f

$f = f_1 \oplus \overline{x}_1 f_2$
$f_0 = f|_{x_i=0}$
$f_1 = f|_{x_i=1}$

$$f = [\,1\ \overline{x}_i\,]\begin{bmatrix}0 & 1\\ 1 & 1\end{bmatrix}\begin{bmatrix}f_0\\ f_1\end{bmatrix}$$

(a) (b) (c) (d)

FIGURE 3.28
The node of a Davio decision tree (a), realization (b), algebraic (c) and matrix (d), descriptions.

In matrix notation, the function f of the node is a function of a single variable x_i given by truth-vector $\mathbf{F} = [\ f(0)\ f(1)\ ]^T$ and defined as

$$
\begin{aligned}
f &= [\ \overline{x}_i\ x_i\ ] \begin{bmatrix} 1 & 0 \\ 1 & 1 \end{bmatrix} \begin{bmatrix} f_0 \\ f_1 \end{bmatrix} = [\ \overline{x}_i\ x_i\ ] \begin{bmatrix} f_0 \\ f_1 \end{bmatrix} \\
&= \overline{x}_i f_0 \oplus x_i f_1 = (1 \oplus x_i) f_0 \oplus x_i f_1 = f_0 \oplus x_i f_2,
\end{aligned}
$$

where $f_0 = f|_{x_i=0}$, $f_2 = f_0 \oplus f_1$. Recursive application of the positive Davio expansion to the function f given by the truth-vector $\mathbf{F} = [f(0)\ f(1) \ldots f(2^n - 1)]^T$ can be expressed in matrix notation

$$
f = \widehat{\mathbf{X}}\ \mathbf{R}_{2^n}\ \mathbf{F}, \tag{3.22}
$$

where

$$
\widehat{\mathbf{X}} = \bigotimes_{i=1}^{n} [\ 1\ x_i\ ], \quad \mathbf{R}_{2^n} = \bigotimes_{i=1}^{n} \mathbf{R}_2, \quad \mathbf{R}_2 = \begin{bmatrix} 1 & 0 \\ 1 & 1 \end{bmatrix},
$$

and $\otimes$ denotes the Kronecker product.

Structure of Reed-Muller expression

$$
\begin{aligned}
\widehat{\mathbf{X}} &= [\ 1\ x_1\ ] \otimes [\ 1\ x_2\ ] \\
&= [\ 1,\ x_2,\ x_1,\ x_1 x_2\ ]
\end{aligned}
$$

Transform matrix

$$
\mathbf{R}_{2^2} = \mathbf{R}_2 \otimes \mathbf{R}_2 = \begin{bmatrix} 1 & 0 \\ 0 & 1 \end{bmatrix} \otimes \begin{bmatrix} 1 & 0 \\ 0 & 1 \end{bmatrix} = \left[\begin{array}{cc|cc} 1 & & & \\ 1 & 1 & & \\ \hline 1 & & 1 & \\ 1 & 1 & 1 & 1 \end{array}\right]
$$

Reed-Muller expression

$$
\begin{aligned}
f &= \widehat{\mathbf{X}}\ \mathbf{R}_{2^2}\ \mathbf{F} = \widehat{\mathbf{X}} \left[\begin{array}{cc|cc} 1 & & & \\ 1 & 1 & & \\ \hline 1 & & 1 & \\ 1 & 1 & 1 & 1 \end{array}\right] \begin{bmatrix} 1 \\ 1 \\ 0 \\ 1 \end{bmatrix} = \widehat{\mathbf{X}} \begin{bmatrix} 1 \\ 0 \\ 1 \\ 1 \end{bmatrix} \\
&= 1 \oplus x_1 \oplus x_1 x_2
\end{aligned}
$$

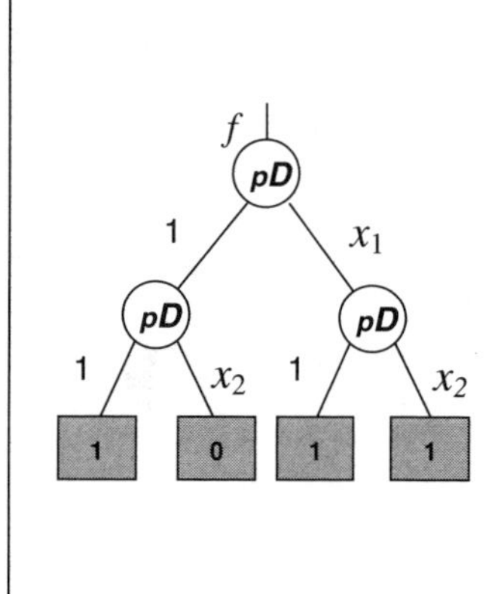

FIGURE 3.29
The Davio decision tree for the switching function $f = \overline{x}_1 \vee x_2$ and calculations in matrix form (Example 3.23).

Example 3.23 *Let us derive the Davio decision tree for the switching function $f = \overline{x}_1 \vee x_2$, given its truth-vector $\mathbf{F} = [1\ 1\ 0\ 1]^T$. We apply Equation 3.22, to get the solution shown in Figure 3.40. The product terms are generated by the Kronecker product $\widehat{\mathbf{X}}$. The 4×4 transform matrix $\mathbf{R}$ is generated by the Kronecker product of the basic matrix $\mathbf{R}_{2^1}$. The final result, the Reed-Muller expression, is directly mapped to the complete positive Davio decision tree.*

3.6.2 Structural properties

The most important structural properties of the positive Davio decision tree are described below:

- The Davio expansion is in the nodes of the decision tree.
- n-variable switching function is represented by an n-level Davio decision tree.
- The i-th level, $i = 1, \ldots, n$, includes 2^{i-1} nodes.
- Nodes at the n-th level are connected to 2^n terminal nodes, which take values 0 or 1.
- The nodes, corresponding to the i-th variable, form the i-th level in the Davio decision tree.
- In every path from the root node to a terminal node, the variables appear in a fixed order; it is said that this tree is ordered.
- The values of constant nodes are the values of the positive polarity Reed-Muller expression for the represented function. Thus, they are elements of the Reed-Muller coefficient vector $\mathbf{R} = [f_{000}\ f_{002}\ f_{020}\ f_{022}\ f_{200}\ f_{202}\ f_{220}\ f_{222}]$, where "0" corresponds to the value $f_0 = f|_{x_i=0}$, and "2" corresponds to the value $f_2 = f|_{x_i=1} \oplus f|_{x_i=0}$.

Figure 3.30 summarizes the useful sum-of-products graphical representations by the example of an AND function. The complete decision tree is reduced to a decision diagram. Finally, the decision tree or diagram is embedded in a hypercube.

Example 3.24 *An arbitrary switching function f of three variables can be represented by the Davio decision tree shown in Figure 3.31 (3 levels, 7 nodes, 8 terminal nodes). To design this tree, the positive Davio expansion (Equation 3.20) is used as follows:*

- *with respect to variable x_1: $f = f_0 \oplus x_1 f_2$;*
- *with respect to variable x_2: $f_0 = f_{00} \oplus x_2 f_{02}$ and $f_1 = f_{10} \oplus x_2 f_{22}$; and*
- *with respect to variable x_3: $f_{00} = f_{000} \oplus x_3 f_{002}$, $f_{02} = f_{020} \oplus x_3 f_{022}$, $f_{20} = f_{200} \oplus x_3 f_{202}$, $f_{22} = f_{220} \oplus x_3 f_{222}$.*

Hence, the Davio decision tree represents switching function f in the form of the Reed-Muller expression $f = f_{000} = \oplus f_{002} x_3 \oplus f_{020} x_2 \oplus f_{022} x_2 x_3 \oplus f_{200} x_1 \oplus f_{202} x_1 x_3 \oplus f_{220} x_1 x_2 \oplus f_{222} x_1 x_2 x_3$.

3.6.3 Decision tree reduction

The Davio decision diagram is derived from the Davio decision tree by deleting redundant nodes, and by sharing equivalent subgraphs. The rules below produce the reduced Davio decision diagram (Figure 3.32):

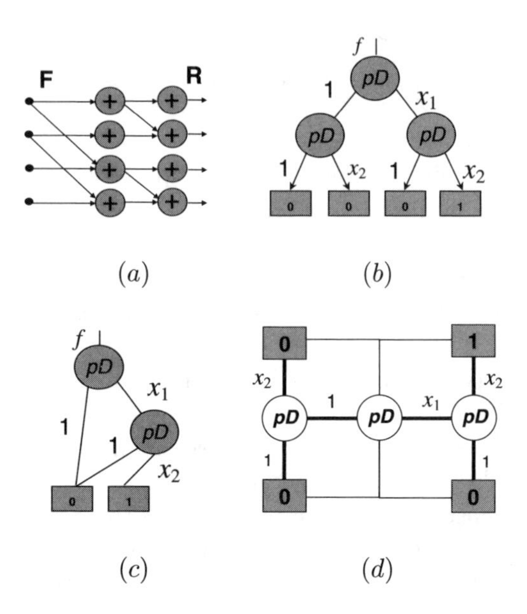

FIGURE 3.30
Graphical representation of AND function of two variables: flowgraph (a), decision tree (b), decision diagram (c), and decision tree embedded in the hypercube (d).

Elimination rule: If the outgoing edge of a node labeled with x_i and $\overline{x}_i$ points to the constant zero, then delete the node and connect the edge to the other subgraph directly.

Merging rule: Share equivalent subgraphs.

In a tree, edges longer than one, i.e., connecting nodes at nonsuccessive levels, can appear. For example, the length of an edge connecting a node at $(i-1)$-th level with a node at $(i+1)$-th level is two.

Example 3.25 *Application of the reduction rules to the three-variable NAND function is demonstrated in Figure 3.33.*

3.7 Arithmetic expressions

Arithmetic representation of switching functions is useful for the word-level description, and for linearization (Chapters 4 and 8). There are a number of similarities between arithmetic and Reed-Muller expressions. The main difference is that, in arithmetic expression, arithmetic addition is used in

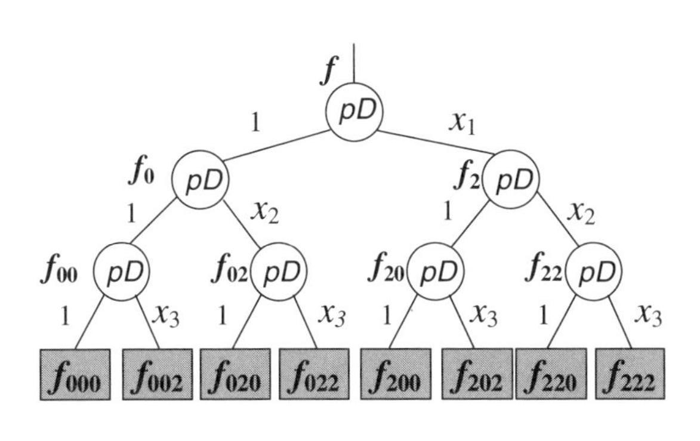

There are 8 paths from f textto the terminal nodes:

Path 1: $t_1 = 1$
Path 2: $t_2 = x_3$
Path 3: $t_3 = x_2$
Path 4: $t_4 = x_2x_3$
Path 5: $t_5 = x_1$
Path 6: $t_6 = x_1x_3$
Path 7: $t_7 = x_1x_2$
Path 8: $t_8 = x_1x_2x_3$

Reed-Muller expression
$f = f_{000}t_1 \oplus f_{002}t_2 \oplus \ldots \oplus f_{222}t_8$

FIGURE 3.31
Reed-Muller representation of a switching function of three variables by the Davio tree (positive polarity) (Example 3.24).

contrast to modulo-two sum in Reed-Muller polynomials.

3.7.1 General form

For two Boolean variables x_1 and x_2 the following is true:

$$\overline{x} = 1 - x, \qquad x_1 \vee x_2 = x_1 + x_2 - x_1x_2,$$
$$x_1 \wedge x_2 = x_1x_2, \qquad x_1 \oplus x_2 = x_1 + x_2 - 2x_1x_2.$$

The right part of the equation is called *arithmetic expressions.* A switching function of n variables is the mapping $\{0,1\}^n \to \{0,1\}$, while an integer-valued function in arithmetical logic denotes the mapping $\{0,1\}^n \to \{0, 1, \ldots, p-1\}$ where $p > 2$.

For a switching function f of n variables, the arithmetic expression is given by

$$f = \sum_{i=0}^{2^n-1} a_i \cdot \underbrace{(x_1^{i_1} \cdots x_n^{i_n})}_{i-th\ product} \tag{3.23}$$

where a_i is a coefficient (integer number), i_j is the j-th bit $1, 2, \ldots, n$, in the binary representation of the index $i = i_1 i_2 \ldots i_n$, and $x_j^{i_j}$ is defined as

$$x_j^{i_j} = \begin{cases} 1, & i_j = 0; \\ x_j, & i_j = 1. \end{cases} \tag{3.24}$$

Note that $\sum$ is the arithmetic addition.

Reduction rule 1

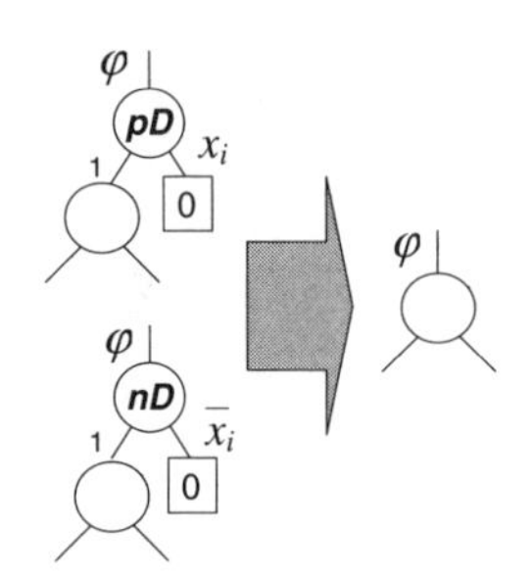

Formal notation

Positive Davio

pD: $\varphi = \varphi_0 \oplus x_i\varphi_2$

$\varphi_2 = 0$

$\varphi = \varphi_0$

Negative Davio

nD: $\varphi = \varphi_1 \oplus \overline{x}_i\varphi_2$

$\varphi_2 = 0$

$\varphi = \varphi_1$

Reduction rule 2

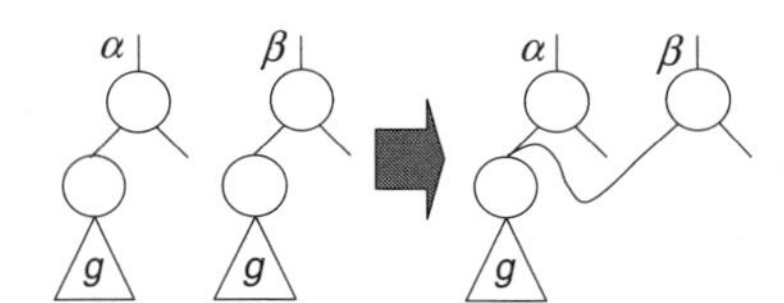

Formal notation

$\alpha = \alpha_0 \oplus x_i\alpha_2$

$\beta = \beta_0 \oplus x_i\beta_2$

$g = \alpha_0 = x_i\beta_0$

FIGURE 3.32
Reduction of a Davio decision tree.

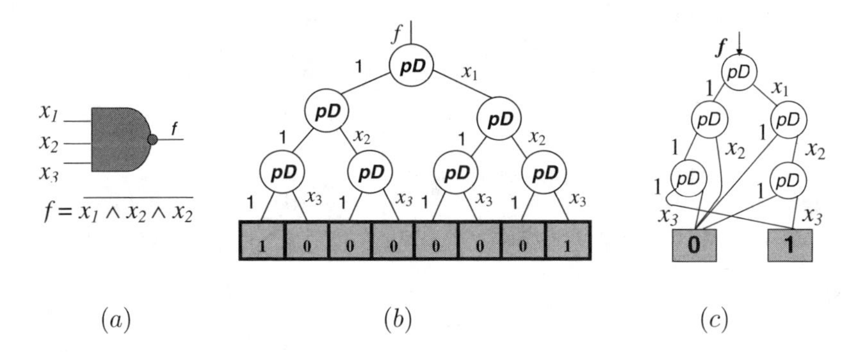

(*a*) (*b*) (*c*)

FIGURE 3.33
The three-variable NAND function, its Davio decision tree and reduced Davio decision diagram (Example 3.25).

Example 3.26 *An arbitrary switching function of three variables is represented in the arithmetic expression by Equation 3.23 and Equation 3.24:* $f = a_0 + a_1 \cdot x_3 + a_2 \cdot x_2 + a_3 \cdot (x_2x_3) + a_4 \cdot x_1 + a_5 \cdot (x_1x_3) + a_6 \cdot (x_1x_2) + a_7 \cdot (x_1x_2x_3)$.

Figure 3.34 illustrates the structure of arithmetic expression, where the shared nodes implement AND operations, and the sum is an arithmetic operation

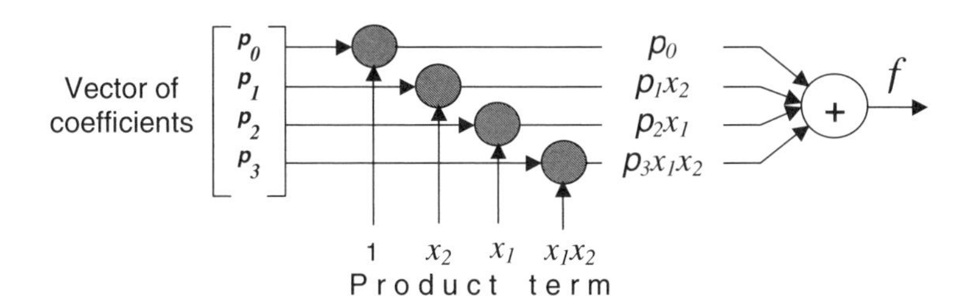

FIGURE 3.34
Deriving the arithmetic expression for a function of two variables (Example 3.26).

3.7.2 Computing the coefficients

Given the truth vector $\mathbf{F} = [f(0)\ f(1) \ldots f(2^n - 1)]^T$, the vector of arithmetic coefficients $\mathbf{P} = [p_0\ p_1 \ldots p_{2^n-1}]^T$ is derived by the matrix equation with respect to AND and arithmetic sum operations

$$\mathbf{P} = \mathbf{P}_{2^n} \cdot \mathbf{X} \tag{3.25}$$

where the $2^n \times 2^n$-matrix $\mathbf{P}_{2^n}$ is formed by the Kronecker product

$$\mathbf{P}_{2^n} = \bigotimes_{j=1}^{n} \mathbf{P}_{2^j}, \qquad \mathbf{P}_{2^j} = \begin{bmatrix} 1 & 0 \\ -1 & 1 \end{bmatrix}. \tag{3.26}$$

Notice, in the Reed-Muller matrix R_{2^n} the elements are logical values 0 and 1, and the calculation with R_{2^n} is performed in $GF(2)$. In arithmetic transform matrix P_{2^n}, the elements are the integers 0 and 1, and the calculation with P_{2^n} is performed on integers.

The general scheme of computing is shown in Figure 3.35a.

3.7.3 Flowgraphs

To design the flowgraph of the algorithm, the matrix $\mathbf{P}_{2^n}$ is represented in the factorized form

$$\mathbf{P}_{2^n} = \mathbf{P}_{2^n}^{(1)} \mathbf{P}_{2^n}^{(2)} \cdots \mathbf{P}_{2^n}^{(n)}, \tag{3.27}$$

where $\mathbf{P}_{2^n}^{(i)}$, $i = 1, 2, \ldots, n$, is formed by the Kronecker product

$$\mathbf{P}_{2^n}^{(i)} = \mathbf{I}_{2^{n-i}} \otimes \mathbf{P}_{2^1} \otimes \mathbf{I}_{2^{i-1}}. \tag{3.28}$$

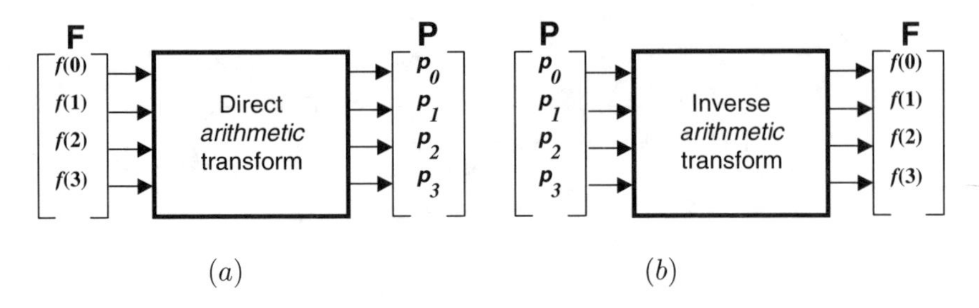

FIGURE 3.35
Direct (a) and inverse (b) arithmetic transform for a switching function of two variables.

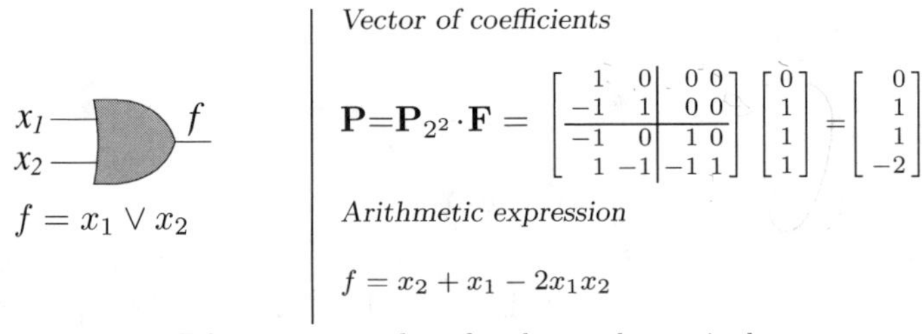

$$\mathbf{P}=\mathbf{P}_{2^2}\cdot\mathbf{F}=\begin{bmatrix}1&0&0&0\\-1&1&0&0\\-1&0&1&0\\1&-1&-1&1\end{bmatrix}\begin{bmatrix}0\\1\\1\\1\end{bmatrix}=\begin{bmatrix}0\\1\\1\\-2\end{bmatrix}$$

Flow graph of the algorithm

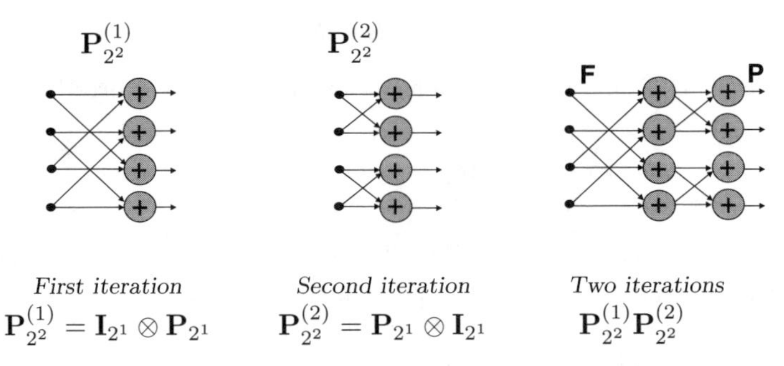

First iteration $\mathbf{P}_{2^2}^{(1)} = \mathbf{I}_{2^1} \otimes \mathbf{P}_{2^1}$ Second iteration $\mathbf{P}_{2^2}^{(2)} = \mathbf{P}_{2^1} \otimes \mathbf{I}_{2^1}$ Two iterations $\mathbf{P}_{2^2}^{(1)}\mathbf{P}_{2^2}^{(2)}$

FIGURE 3.36
Calculation of arithmetic expansion for OR gate (Example 3.27).

Hence, arithmetic coefficients are computed in n iterations.

Example 3.27 *Computing the arithmetic coefficients p_i by Equation 3.25 for the elementary function $f = x_1 \vee x_2$ given its truth-vector $\mathbf{F} = [0\ 1\ 1\ 1]^T$ is illustrated in Figure 3.36. The flowgraph includes two iterations, accordingly to factorization rules (Equation 3.27 and Equation 3.28). The matrix $\mathbf{P}_{2^2}$ is* *formed as follows:*

$$\mathbf{P}_{2^2} = \mathbf{P}_{2^2}^{(1)}\mathbf{P}_{2^2}^{(2)} = (\mathbf{I}_{2^{2-1}} \otimes \mathbf{P}_{2^1} \otimes \mathbf{I}_{2^{1-1}})(\mathbf{I}_{2^{2-2}} \otimes \mathbf{P}_{2^1} \otimes \mathbf{I}_{2^{2-1}})$$

$$= \underbrace{(\mathbf{I}_{2^1} \otimes \mathbf{P}_{2^1} \otimes 1)}_{1st\ iteration}\underbrace{(1 \otimes \mathbf{P}_{2^1} \otimes \mathbf{I}_{2^1})}_{2nd\ iteration}$$

$$= \left[\begin{array}{c|c} \mathbf{P}_{2^1} & \\ \hline & \mathbf{P}_{2^1} \end{array}\right]\left[\begin{array}{c|c} \mathbf{I}_{2^1} & \\ \hline -\mathbf{I}_{2^1} & \mathbf{I}_{2^1} \end{array}\right] = \left[\begin{array}{cc|cc} 1 & & & \\ -1 & 1 & & \\ \hline & & 1 & \\ & & -1 & 1 \end{array}\right]\left[\begin{array}{cc|cc} 1 & & & \\ & 1 & & \\ \hline -1 & & 1 & \\ & -1 & & 1 \end{array}\right].$$

3.7.4 Restoration

The following matrix equation over AND and arithmetic sum operations restore the truth-vector $\mathbf{F}$ from the vector of coefficients $\mathbf{P}$ (Figure 3.35)

$$\mathbf{F} = \mathbf{P}_{2^n}^{-1} \cdot \mathbf{P}, \tag{3.29}$$

where $2^n \times 2^n$ matrix $\mathbf{P}_{2^n}^{-1}$ is formed by the Kronecker product

$$\mathbf{P}_{2^n}^{-1} = \bigotimes_{i=0}^{n} \mathbf{P}_{2^1}^{-1}, \quad \mathbf{P}_{2^1}^{-1} = \begin{bmatrix} 1 & 0 \\ 1 & 1 \end{bmatrix} \tag{3.30}$$

Example 3.28 *To restore the truth-vector* $\mathbf{F}$ *given by its vector of arithmetical coefficients* $\mathbf{P} = [0\ 1\ 1\ -2]^T$, *Equation 3.29 can be applied:*

$$\mathbf{F} = \mathbf{P}_{2^3}^{-1} \cdot \mathbf{P} = \left[\begin{array}{cc|cc} 1 & 0 & 0 & 0 \\ 1 & 1 & 0 & 0 \\ \hline 1 & 0 & 1 & 0 \\ 1 & 1 & 1 & 1 \end{array}\right] \begin{bmatrix} 0 \\ 1 \\ 1 \\ -2 \end{bmatrix} = \begin{bmatrix} 0 \\ 1 \\ 1 \\ 1 \end{bmatrix}$$

3.7.5 Useful rules

The simple arithmetic expressions $\overline{x} = 1 - x$, $x_1 \vee x_2 = x_1 + x_2 - x_1x_2$, $x_1x_2 = x_1x_2$, $x_1 \oplus x_2 = x_1 + x_2 - 2x_1x_2$ can be generalized towards the switching functions f_1 and f_2:

- $\mathcal{P}\{\overline{f}\} = 1 - \mathcal{P}\{f\}$
- $\mathcal{P}\{f_1 \vee f_2\} = \mathcal{P}\{f_1\} + \mathcal{P}\{f_2\} - \mathcal{P}\{f_1\}\mathcal{P}\{f_2\}$
- $\mathcal{P}\{f_1f_2\} = \mathcal{P}\{f_1\}\mathcal{P}\{f_2\}$
- $\mathcal{P}\{f_1 \oplus f_2\} = \mathcal{P}\{f_1\} + \mathcal{P}\{f_2\} - 2\mathcal{P}\{f_1\}\mathcal{P}\{f_2\}$

where $\mathcal{P}\{\cdot\}$ denotes an arithmetic transform.

Example 3.29 *The rules above allow us to simplify the manipulations:*

- $\mathcal{P}\{x_1\overline{x}_2 \vee x_1\overline{x}_3\} = \mathcal{P}\{x_1\overline{x}_2\} + \mathcal{P}\{x_1\overline{x}_3\} - \mathcal{P}\{x_1\overline{x}_2\}\mathcal{P}\{x_1\overline{x}_3\}$
  $= x_1\overline{x}_2 + x_1\overline{x}_3 - x_1\overline{x}_2\overline{x}_3$
- $\mathcal{P}\{x_1 \oplus x_2\} = x_1 - 2x_1x_2 + x_2$
- $\mathcal{P}\{(x_1 \vee x_2) \oplus x_3\} = \mathcal{P}\{(x_1 \vee x_2) + \mathcal{P}\{x_3\} - 2\mathcal{P}\{(x_1 \vee x_2)\mathcal{P}\{x_3\}$
  $= x_1 + x_2 + x_3 - x_1x_2 - 2x_1x_3 - 2x_2x_3 + 2x_1x_2x_3$

3.7.6 Hypercube representation

Let a function be given by its cubes. To derive an arithmetic expression of the function, we can employ an algorithm similar to the one used to derive its ESOP form. However, it must be taken into account that operations over cubes in an arithmetic form are specific. The generation of a new cube is based on the equation $x \vee y = x + y - xy$. Given the cubes $[C_1]$ and $[C_2]$ of the sum-of-products expression, the cubes to be included in the arithmetical expression are derived by the equation

$$[C_1] \vee [C_2] = [C_1] + [C_2] - [C_1][C_2]. \tag{3.31}$$

Example 3.30 *Let* $f = \overline{x}_2 x_3 + x_1 x_3$, *i.e.,* $[C_1] = [\mathbf{x}\ 0\ 1]$ *and* $[C_2] = [1\ \mathbf{x}\ 1]$. *To derive arithmetic form, Equation 3.31 is applied and three cubes are produced:* $[C_1]$, $[C_2]$, *and new cube* $-[C_1][C_2] = -[1\ 0\ 1]$. *Thus,* $f = \overline{x}_2 x_3 + x_1 x_3 - x_1 \overline{x}_2 x_3$. *Figure 3.37 illustrates this calculation.*

A cube that corresponds to a product in the arithmetic expression of a switching function is composed of the components: $\{0,\ 1,\ \mathbf{x},\ a, b\}$, where $a = -\overline{x}_i + x_i = (-1)^{\overline{x}_i}$ and $b = -x_i + \overline{x}_i = (-1)^{x_i}$.

Example 3.31 *Given the arithmetic expression* $f = -\overline{x}_1 x_2 \overline{x}_3 + x_1 x_2 \overline{x}_3$, *its cube form is derived as follows:*

$$f = -\overline{x}_1 x_2 \overline{x}_3 + x_1 x_2 \overline{x}_3 = x_2 \overline{x}_3(-\overline{x}_1 + x_1) = (-1)^{\overline{x}_1} x_2 \overline{x}_3,$$

which corresponds to $f = [a\ 1\ 0]$.

3.7.7 Polarity

The polarity of a variable x_j can take the values:

(*i*) $c_j = 1$, corresponding to the uncomplemented variable x_j, or
(*ii*) $c_j = 0$, corresponding to the complemented variable $\overline{x}_j$.

Let the polarity $c = c_1, c_2, \ldots, c_n$, $c \in \{0, 1, 2, \ldots, 2^n - 1\}$, where c_j is the j-th bit of binary representation of c. For a switching function f of n variables, the arithmetic expression, given the polarity $c = c_1, c_2, \ldots, c_n$ of variables $x_1, x_2, \ldots, x_n$ is as follows

$$f = \sum_{i=0}^{2^n-1} p_i \cdot \underbrace{(x_1 \oplus c_1)^{i_1} \cdots (x_n \oplus c_n)^{i_n}}_{i-th\ product}, \tag{3.32}$$

where p_i is the coefficient, and $(x_j \oplus c_j)^{i_j}$ is defined as

$$a_j^{i_j} = \begin{cases} 1, \text{ if } i_j = 0; \\ a, \text{ if } i_j = 1. \end{cases} \quad x_j \oplus c_j = \begin{cases} x_j, \text{ if } c_j = 0; \\ \overline{x}_j, \text{ if } c_j = 1. \end{cases} \tag{3.33}$$

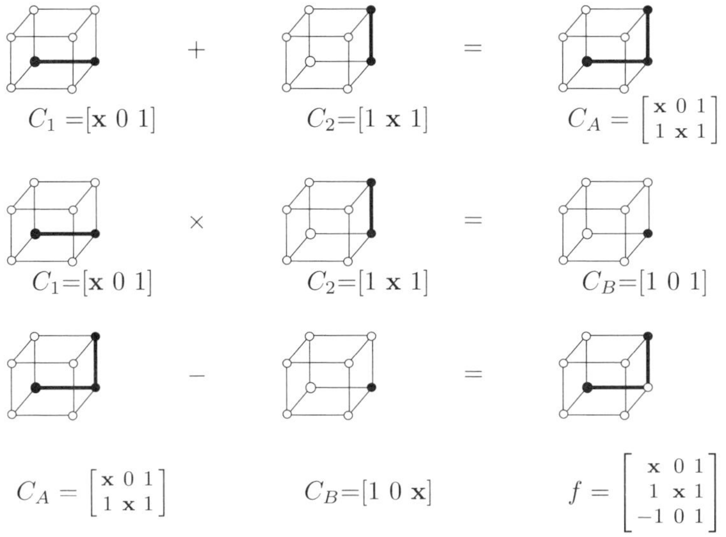

$C_1 = [\mathbf{x}\ 0\ 1]$ + $C_2 = [1\ \mathbf{x}\ 1]$ = $C_A = \begin{bmatrix} \mathbf{x} & 0 & 1 \\ 1 & \mathbf{x} & 1 \end{bmatrix}$

$C_1 = [\mathbf{x}\ 0\ 1]$ × $C_2 = [1\ \mathbf{x}\ 1]$ = $C_B = [1\ 0\ 1]$

$C_A = \begin{bmatrix} \mathbf{x} & 0 & 1 \\ 1 & \mathbf{x} & 1 \end{bmatrix}$ − $C_B = [1\ 0\ \mathbf{x}]$ = $f = \begin{bmatrix} \mathbf{x} & 0 & 1 \\ 1 & \mathbf{x} & 1 \\ -1 & 0 & 1 \end{bmatrix}$

FIGURE 3.37
Computing the cubes of a switching function of three variables by the rule $f = [C_1] + [C_2] - [C_1][C_2]$ (Example 3.30).

Example 3.32 *Let* $c = 2$, $c_1, c_2 = 1, 0$. *The representation of a switching function of two variables by arithmetic expression of the polarity* $c = 2$ *can be derived by Equation 3.32 and Equation 3.33:*

$$f = p_0(x_1 \oplus 1)^0(x_2 \oplus 0)^0 + p_1(x_1 \oplus 1)^0(x_2 \oplus 0)^1 + p_2(x_1 \oplus 1)^1(x_2 \oplus 0)^0$$
$$+ p_3(x_1 \oplus 1)^1(x_2 \oplus 0)^1 = p_0 + p_1x_2 + p_2\overline{x}_1 + p_3\overline{x}_1x_2.$$

The coefficients p_i can be derived from the function's truth values by the technique presented below.

Given the truth vector $\mathbf{F} = [f(0)\ f(1) \ldots f(2^n - 1)]^T$, the vector of arithmetic word-level coefficients of polarity c, $\mathbf{P^{(c)}} = [p_0^{(c)}\ p_1^{(c)} \ldots p_{2^n-1}^{(c)}]^T$ is derived by the matrix equation

$$\mathbf{P^{(c)}} = \mathbf{P}_{2^n}^{(c)} \cdot \mathbf{F}, \tag{3.34}$$

where the $2^n \times 2^n$-matrix $\mathbf{P}_{2^n}^{(c)}$ is generated by the Kronecker product

$$\mathbf{P}_{2^n}^{(c)} = \bigotimes_{j=1}^{n} \mathbf{P}_{2^1}^{(c_j)}, \qquad \mathbf{P}_{2^1}^{(c)} = \begin{cases} \begin{bmatrix} 1 & 0 \\ -1 & 1 \end{bmatrix}, & c_j = 0; \\ \\ \begin{bmatrix} 0 & 1 \\ 1 & -1 \end{bmatrix}, & c_j = 1. \end{cases} \tag{3.35}$$

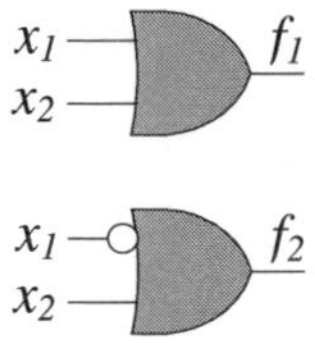

x_1, x_2 → f_1 (OR gate); $\overline{x}_1$, x_2 → f_2 (OR gate with inverted input x_1)

$$f_1 = x_1 \vee x_2$$
$$f_2 = \overline{x}_1 \vee x_2$$

$$\mathbf{F} = \left[\mathbf{F_1} | \ \mathbf{F_2}\right] = \left[\begin{array}{c|c} 1 & 0 \\ 1 & 1 \\ 0 & 1 \\ 1 & 1 \end{array}\right] = \begin{bmatrix} 2 \\ 3 \\ 1 \\ 3 \end{bmatrix}$$

$$\mathbf{P^{(2)}} = \mathbf{P}_{2^2}^{(2)} \cdot \mathbf{F}$$
$$= \left[\begin{array}{cc|cc} 0 & 0 & 1 & 0 \\ 0 & 0 & -1 & 1 \\ \hline 1 & 0 & -1 & 0 \\ -1 & 1 & 1 & -1 \end{array}\right] \begin{bmatrix} 2 \\ 3 \\ 1 \\ 3 \end{bmatrix} = \begin{bmatrix} 1 \\ 2 \\ 1 \\ -1 \end{bmatrix},$$

$$f = 1 + 2x_2 + \overline{x}_1 - \overline{x}_1 x_2$$

Alternatively,

$$\mathbf{P_1^{(2)}} = \begin{bmatrix} 1 \\ 0 \\ -1 \\ 1 \end{bmatrix}, \quad \mathbf{P_2^{(2)}} = \begin{bmatrix} 0 \\ 1 \\ 1 \\ -1 \end{bmatrix}$$

$$\mathbf{P^{(2)}} = 2^1 \mathbf{P_2^{(2)}} + 2^0 \mathbf{P_1^{(2)}} = \begin{bmatrix} 2 \\ 3 \\ 1 \\ 3 \end{bmatrix}$$

FIGURE 3.38
Computing the arithmetic expression of polarity $c = 2$ for the two gates (Example 3.33).

Example 3.33 *Figure 3.38 demonstrates derivation of coefficients p_i to Example 3.32 by matrix Equation 3.34. Here, the matrix $\mathbf{P}_{2^2}^{(2)}$ for the polarity $c = 2$ is generated by Equation 3.35 as follows:*

$$\mathbf{P}_{2^2}^{(2)} = \mathbf{P}_{2^1}^{(1)} \otimes \mathbf{P}_{2^1}^{(0)} = \begin{bmatrix} 0 & 1 \\ 1 & -1 \end{bmatrix} \otimes \begin{bmatrix} 1 & 0 \\ -1 & 1 \end{bmatrix}.$$

Note that in this example the two-output function is presented by the truth vector **F** *that is a word-level interpretation of two functions, f_1 and f_2. The matrix manipulation of word-level vectors is the same as of binary ones. Word-level forms will be considered in Chapter 4.*

3.8 Decision trees and diagrams

A binary decision tree that corresponds to the arithmetic canonical representation of a switching function is called an *arithmetic* decision tree. Arithmetic decision trees are associated with the arithmetic decision diagrams, also called *spectral* or *functional*, diagram.

3.8.1 Formal design

A node in an arithmetic decision tree of a switching function f corresponds to the arithmetic analog of Davio decomposition with respect to a variable x_i.

There exist:

- *The arithmetic analog of the positive Davio* expansion

$$f = f_0 + x_i f_2, \tag{3.36}$$

where $f_0 = f|_{x_i=0}$ and $f_2 = f|_{x_i=1} + f|_{x_i=0}$, and
- *The arithmetic analog of the negative Davio* expansion

$$f = \overline{x}_1 f_2 + f_1, \tag{3.37}$$

where $f_1 = f|_{x_i=1}$.

Arithmetic analogs of positive and negative Davio expansion are labeled as pD_A and nD_A correspondingly (Figure 3.39).

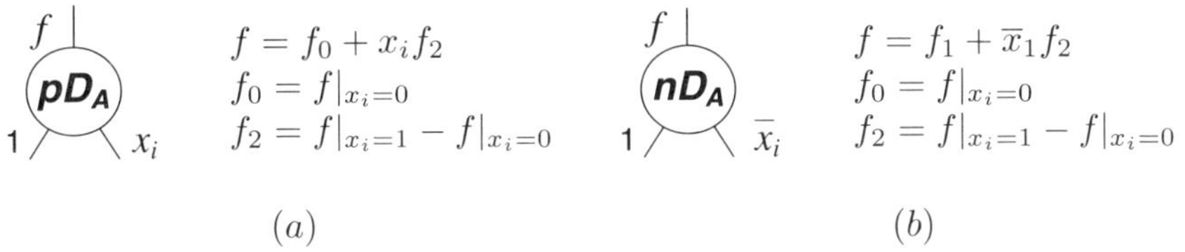

f, pD_A, 1, x_i

$$f = f_0 + x_i f_2$$
$$f_0 = f|_{x_i=0}$$
$$f_2 = f|_{x_i=1} - f|_{x_i=0}$$

(a)

f, nD_A, 1, $\overline{x}_i$

$$f = f_1 + \overline{x}_1 f_2$$
$$f_0 = f|_{x_i=0}$$
$$f_2 = f|_{x_i=1} - f|_{x_i=0}$$

(b)

FIGURE 3.39
The positive (a) and negative (b) Davio node of an arithmetic decision tree and its formal description.

In matrix notation, the expansion of switching function f given by truth-vector $\mathbf{F} = [\ f(0)\ f(1)\ ]^T$, implemented in the node is defined as

$$f = [\ \overline{x}_i\ x_i\ ] \begin{bmatrix} 1 & 0 \\ 1 & 1 \end{bmatrix} \begin{bmatrix} f_0 \\ f_1 \end{bmatrix} = [\ \overline{x}_i\ x_i\ ] \begin{bmatrix} f_0 \\ f_1 \end{bmatrix}$$
$$= \overline{x}_i f_0 \oplus x_i f_1 = (1 \oplus x_i) f_0 \oplus x_i f_1 = f_0 \oplus x_i f_2,$$

where $f_0 = f|_{x_i=0}$, $f_2 = f_0 \oplus f_1$. Recursive application of the arithmetic analog of positive Davio expansion to a function f given by its truth-vector $\mathbf{F} = [f(0)\ f(1) \ldots f(2^n - 1)]^T$ is expressed in matrix notation as

$$f = \widehat{\mathbf{X}}\ \mathbf{P}_{2^n}\ \mathbf{F}, \tag{3.38}$$

where

$$\widehat{\mathbf{X}} = \bigotimes_{i=1}^{n} [\ 1\ x_i\ ], \quad \mathbf{P}_{2^n} = \bigotimes_{i=1}^{n} \mathbf{P}_2, \quad \mathbf{P}_2 = \begin{bmatrix} 1 & 0 \\ 1 & 1 \end{bmatrix},$$

and $\otimes$ denotes the Kronecker product.

Example 3.34 *Let us derive the arithmetic decision tree of the switching function* $f = \overline{x}_1 \vee x_2$ *given by the truth-vector* $\mathbf{F} = [1\ 1\ 0\ 1]^T$. *The solution to Equation 3.38 is shown in Figure 3.40. The product terms are generated by the Kronecker product* $\widehat{\mathbf{X}}$. *The* 4×4 *transform matrix* $\mathbf{P}$ *is generated by the Kronecker product over the basic matrix* $\mathbf{P}_{2^1}$. *The final result, the arithmetic coefficients, is directly mapped into the complete arithmetic decision tree.*

$$\widehat{\mathbf{X}} = [\ 1\ x_1\ ] \otimes [\ 1\ x_2\ ] = [\ 1,\ x_2,\ x_1,\ x_1x_2\ ]$$

$$\mathbf{P}_{2^2} = \mathbf{P}_2 \otimes \mathbf{P}_2 = \begin{bmatrix} 1 & 0 \\ -1 & 1 \end{bmatrix} \otimes \begin{bmatrix} 1 & 0 \\ -1 & 1 \end{bmatrix} = \left[\begin{array}{cc|cc} 1 & & & \\ -1 & 1 & & \\ \hline -1 & & 1 & \\ 1 & -1 & -1 & 1 \end{array}\right]$$

$$f = \widehat{\mathbf{X}}\ \mathbf{P}_{2^2}\ \mathbf{F} = \widehat{\mathbf{X}} \left[\begin{array}{cc|cc} 1 & & & \\ -1 & 1 & & \\ \hline -1 & & 1 & \\ 1 & -1 & -1 & 1 \end{array}\right] \begin{bmatrix} 1 \\ 1 \\ 0 \\ 1 \end{bmatrix} = \widehat{\mathbf{X}} \begin{bmatrix} 1 \\ 0 \\ -1 \\ 1 \end{bmatrix} = 1 - x_1 + x_1x_2$$

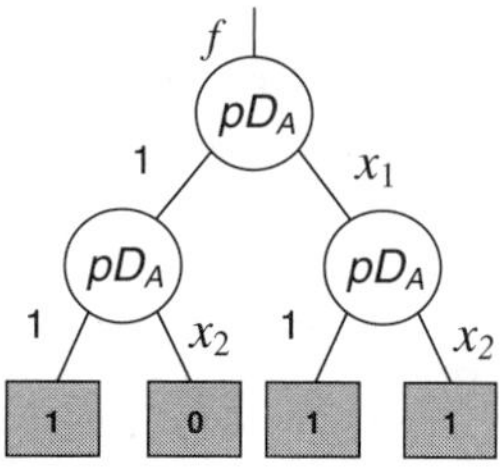

FIGURE 3.40
Derivation of the arithmetic decision tree for the switching function $f = \overline{x}_1 \vee x_2$ (Example 3.34).

3.8.2 Structural properties

Structural properties of the arithmetic decision tree are similar to the Reed-Muller decision tree except:

- The values of terminal nodes are integer numbers and correspond to the coefficients of the arithmetic expression.
- Each path from the root to a terminal node corresponds to a product in the arithmetic expression.
- The values of constant nodes are the values of the arithmetic spectrum in the positive polarity for the represented functions. Thus, they are elements of the vector of arithmetic coefficients $\mathbf{P}_f = [f_{000}\ f_{002}\ f_{020}\ f_{022}\ f_{200}\ f_{202}\ f_{220}\ f_{222}]$, where "0" corresponds to the value of $f_0 = f|_{x_i=0}$, and "2" corresponds to the value of $f_2 = f|_{x_i=1} \oplus f|_{x_i=0}$.

Example 3.35 *An arbitrary switching function* f *of three variables can be represented by the arithmetic decision tree shown in Figure 3.41 (3 levels, 7*

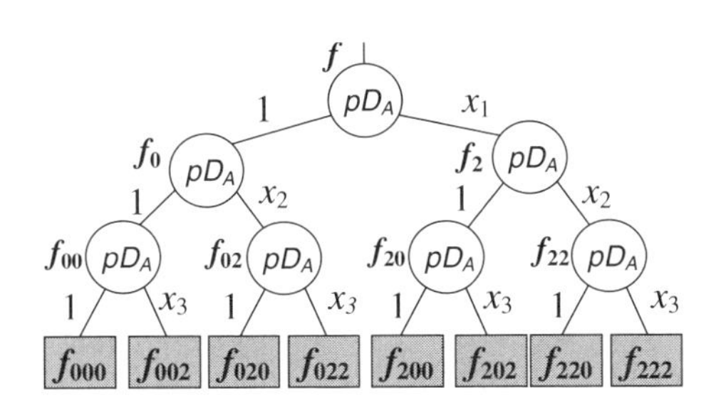

There are 8 paths from f to the terminal nodes:

Path 1: $t_1 = 1$
Path 2: $t_2 = x_3$
Path 3: $t_3 = x_2$
Path 4: $t_4 = x_2x_3$
Path 5: $t_5 = x_1$
Path 6: $t_6 = x_1x_3$
Path 7: $t_7 = x_1x_2$
Path 8: $t_8 = x_1x_2x_3$

Arithmetic expression
$f = f_{000}t_1 + f_{002}t_2 + \ldots$
$+ f_{222}t_8$

FIGURE 3.41
Arithmetic representation of a switching function of three variables by the arithmetic tree ($c = 000$) (Example 3.35).

nodes, 8 terminal nodes). To design this tree, the positive arithmetic expansion (Equation 3.20) is used as follows:

(a) *with respect to the variable* x_1*:* $f = f_0 + x_1 f_2$

(b) *with respect to the variable* x_2*:* $f_0 = f_{00} + x_2 f_{02}$ *and* $f_1 = f_{10} + x_2 f_{22}$

(c) *with respect to the variable* x_3*:*

$$f_{00} = f_{000} + x_3 f_{002}, \qquad f_{02} = f_{020} + x_3 f_{022},$$
$$f_{20} = f_{200} + x_3 f_{202}, \qquad f_{22} = f_{220} + x_3 f_{222}.$$

Hence, the arithmetic decision tree represents switching function f in the form of the logic expression $f = f_{000} = \oplus f_{002}x_3 + f_{020}x_2 + f_{022}x_2x_3 + f_{200}x_1 + f_{202}x_1x_3 + f_{220}x_1x_2 + f_{222}x_1x_2x_3$.

3.8.3 Decision tree reduction

The arithmetic decision diagram is derived from the arithmetic decision tree by deleting redundant nodes, and by sharing equivalent subgraphs. Reduction rules produce the reduced arithmetic decision diagram (see "Further Reading" Section in this chapter).

3.9 Summary

1. To represent a circuit in spatial dimensions with respect to the requirements of the technology, an appropriate data structure must be chosen. In this chapter, three types of representations of switching functions, sum-of-products, Reed-Muller, and arithmetic, are revisited. State-of-the-art techniques are used for the interpretation of each of the above representations: algebraic, matrix, and graph-based (flowgraphs, hypercubes, decision trees, and decision diagrams).

2. Each form of data structure corresponds to a certain level of abstraction and is useful at certain phases of logic design:

- ► *Algebraic equations* are useful for analyzing the general properties of logic functions.
- ► *Matrix* (spectral) representation is used to study the properties of the functions through computing so-called *spectra* of the functions. For each form, matrix interpretation is based on two transforms: direct and inverse. The unique property of the transform matrices is their representability in factorized form. This provides the possibility for studying signal flows.
- ► *The flowgraph*, formally, is derived from the factorized matrix transformation of switching functions. The flowgraph reveals the possibilities for the parallel processing and the relationship between matrix notation and hardware implementation.
- ► *The hypercube* is useful for manipulation of switching functions. However, this topological model cannot be considered as a prototype ready for direct implementation.
- ► *The decision tree* is an interpretation of the computing of a switching function. There are various kinds of decision tree, different in the interpretation of the node function. The complete decision tree is a canonical representation of the switching function. The key reason to focus on the decision tree representation is that it can be embedded into hypercube structure. Later, in Chapters 5 and 6, this property will be utilized to represent a circuit in spatial dimensions.
- ► *The decision diagram*, the reduced decision tree, is useful for manipulation of switching functions. There are a variety of decision diagrams, most of them, in particular OBDD, provide canonical representations of switching functions.

3.10 Problems

Problem 3.1 Derive the Shannon decision tree and BDD for the switching functions given below. Use Examples 3.13, 3.14, and 3.15 for reference.

(*a*) $f = \overline{x}_1 \vee x_2 \vee x_3$
(*b*) $f = x_1 \vee (x_2 \oplus \overline{x}_3)$
(*c*) $f = x_1x_2 \vee \overline{x}_2x_3 \vee x_1\overline{x}_3$
(*d*) $f = x_1x_2x_3$

Problem 3.2 Represent the switching function given below by the Reed-Muller expression of polarities $c = 1$ and $c = 2$ (follow Example 3.21).

(*a*) $f = \overline{x}_1x_2 \vee x_1x_2\overline{x}_3 \vee x_3$
(*b*) $f = x_1 \vee x_2 \vee \overline{x}_1x_3$
(*c*) $f = x_1 \vee x_1x_2x_3 \vee x_1x_3$
(*d*) $f = x_1 \vee (x_2 \oplus x_3)$

Problem 3.3 Consider the Reed-Muller expressions given below.

(*a*) $f = \overline{x}_1x_2 \oplus x_1x_2\overline{x}_3 \oplus x_3$
(*b*) $f = x_1 \oplus x_1x_2x_3 \oplus x_1x_3$
(*c*) $f = x_1 \oplus x_2 \oplus x_3$
(*d*) $f = 1 \oplus x_1 \oplus x_2 \oplus \overline{x}_1x_3$

- ▶ Represent the expressions in polarity $c = 0$ (follow Example 3.21)
- ▶ Derive the flowgraph for a switching function of three variables (follow Example 3.17)
- ▶ Represent the expressions by the Davio decision tree and Davio decision diagram. Use Examples 3.23, 3.24 and 3.25 for reference

Problem 3.4 Represent the switching functions given below by an arithmetic expression (follow the rules given in subsection 3.7.5)

(*a*) $f = \overline{x}_1x_2 \vee x_2\overline{x}_3$
(*b*) $f = x_1 \oplus x_1x_2x_3$
(*d*) $f = x_1 \vee x_2 \vee x_3$
(*e*) $f = (x_1 \vee x_2) \oplus x_3$

Problem 3.5 Below some arithmetic expressions are given.

- ▶ Represent the function by an arithmetic form of polarity $c = 3$ (follow Example 3.32)
- ▶ Derive the flowgraph of arithmetic transform for the 3-variable switching function (follow Example 3.27)
- ▶ Represent the arithmetic expression by a decision tree and decision diagram

Use Examples 3.34 and 3.35 for reference

(a) $x_3 + x_1 - x_1x_3 - x_1x_2$

(b) $x_1 - x_1x_3 + x_1x_2x_3$

(c) $x_2 + x_3 - x_2x_3 - x_1x_2x_3$

(d) $x_1 + x_2x_3 - x_1x_3 - x_1x_2 + 2x_1x_2x_3$

Problem 3.6 Using the appropriate inverse transform, find the truth-vector for the function given by the vector of coefficients

(a) $\mathbf{S} = [10101011]^T$ (follow Example 3.11)

(b) $\mathbf{R} = [10101011]^T$ (follow Example 3.17)

(c) $\mathbf{P} = [10101011]^T$ (follow Example 3.27)

Problem 3.7 Prove that the following expressions represent the same switching function, $f_1 = x_1 \cdot x_3 \vee x_2 \cdot \overline{x}_3 \vee x_2 \cdot \overline{x}_4$:

(a) $f_1 = 1 \oplus \overline{x}_2 \cdot \overline{x}_3 \cdot x_4 \cdot x_5 \oplus \overline{x}_2 \cdot \overline{x}_4 \cdot \overline{x}_5 \oplus \overline{x}_2 \cdot \overline{x}_4 \oplus x_2 \cdot x_3 \cdot x_4$

(b) $f_2 = x_2 \oplus x_5 \oplus x_2 \cdot x_5 \oplus x_2 \cdot x_3 \cdot x_4 \oplus x_3 \cdot x_4 \cdot x_5 \oplus x_2 \cdot x_3 \cdot x_4 \cdot x_5$

(c) $f_3 = 1 \oplus \overline{x}_2 \cdot \overline{x}_5 \oplus x_3 \cdot x_4 \oplus \overline{x}_2 \cdot x_3 \cdot x_4 \cdot \overline{x}_5$

(d) $f_4 = x_2 \oplus \overline{x}_2 \cdot x_5 \oplus x_2 \cdot x_3 \cdot x_4 \oplus x_2 \cdot x_3 \cdot x_4 \cdot \overline{x}_5$

(e) $f_5 = \overline{x}_2 \cdot \overline{x}_5 \oplus \overline{x}_2 \cdot x_3 \cdot x_4 \cdot x_5 \oplus x_2 \cdot x_3 \cdot x_4$

(f) $f_6 = x_1 \cdot x_3 \oplus x_2 \cdot \overline{x}_3 \oplus x_3 \cdot \overline{x}_4 \oplus \overline{x}_1 \cdot \overline{x}_2 \cdot x_3 \cdot \overline{x}_4$

Problem 3.8 Given an n-variable switching function, 2^n different Reed-Muller expansions of polarities $0, 1, \ldots, 2^n - 1$ can be generated. Among them, an optimal expansion can be defined as an expansion with the minimum total number of literals. There exist a significant number of algorithms to for finding an optimal Reed-Muller expansion, including exhaustive search, evolutionary minimization algorithm, and decision diagram based approach. Find a method of manipulation of an $\mathcal{N}$-hypercube (with Shannon or pD expansion in the nodes) given a function of 3 variables (you can also propose how to extend it to other number of variables).

3.11 Further reading

The basics of logic synthesis. There are many excellent textbooks on the theory of switching functions and very large scale integration (VLSI) design of logic circuits, in particular, [7, 11, 13, 18, 22, 23, 24].

Decision trees and diagrams. The fundamentals of decision diagrams go back to Lee (1959), Akers (1977) and Bryant (1986). Lee called decision trees *binary decision programs* [16]. This result attracted interest of specialists in complexity and algorithmic theory. Akers has developed the BDD-based methodology for generation tests [2]. Bryant's paper [4] has stimulated much research in developing the technique of decision diagrams for logic design problems.

Decision tree techniques are introduced in [3, 2, 20] and in the textbooks [18, 23]. In a number of books, various aspects of the decision trees and decision diagrams have been discussed, including manipulation and reduction [6, 4, 26, 28]. Technique of application of decision diagrams is introduced in many monographs and textbooks, in particular, [8, 18].

Reed-Muller representation of switching functions. There is a long history of research on sum-of-products expressions in which *sum* corresponds to Exclusive-OR and *product* corresponds to the AND of variables or complements of variables. It is known that such EXOR expressions require, on the average, fewer product terms than OR sum-of-products. Such structures are known to be easily testable. AND-EXOR circuits have been used in arithmetic, error correcting, and telecommunications applications. Fundamental aspects of the technique of manipulation with Reed-Muller expressions are discussed by Davio et al. [6] and in the textbook by Sasao [23]. In a number of papers and books, the usefulness of manipulation with different polarities has been shown [11, 25, 30, 31]. Matrix (spectral) computing of Reed-Muller expressions is discussed in [6, 26, 28]. Computing the Reed-Muller expressions is presented in detail by Yanushkevich [32]. The reader can find a very detailed study on fixed and mixed polarity Reed-Muller expressions in the papers of *International Workshops on Applications of the Reed-Muller Expansion in Circuit Design*. The most important results in this field are discussed in the paper by Stanković et al. [27].

Symmetry detection. There have been many studies of symmetric functions dating back to the early history of switching theory. A number of fundamental results on this problem are presented by Davio et al. [6]. In particular, the properties of Lagrange, Newton and Nyquist expansions of symmetric functions are investigated.

Arithmetic representation of switching functions. Apparently, the first attempts to present logic operations by arithmetical ones were taken by the founder of Boolean algebra Boolé (1854). He did not used the Boolean operators well known today. Rather he used arithmetic expressions. It is interesting to note that: Aiken first found that arithmetic expressions can be useful to design circuits and used it in the Harvard MARK 3 and MARK 4 computers [1].

Arithmetic expressions are closely related to Reed-Muller expressions. However, with variables and function values interpreted as integers 0 and 1 instead of logic values. In this way, arithmetic expressions can be considered as integer counterparts of Reed-Muller expressions.

Arithmetic logic has many applications in contemporary logic design, for example in the computation of signal probabilities for test generations, and switching activities for power and noise analysis. In a number of papers the usefulness of manipulation with arithmetic expressions is shown [10, 4, 21, 32]. Matrix (spectral) computing of word-level expressions is discussed in [26, 28, 32]. Jain introduces probabilistic computing of arithmetic transforms [14].

Walsh representation and other spectral forms. Discrete Walsh functions are a discrete version of the functions introduced by Walsh in 1923 for solving some problems in approximation of square-integrable functions on the interval 0, 1. The basic Walsh matrix is defined as $W_2 = \begin{bmatrix} 1 & 1 \\ 1 & -1 \end{bmatrix}$. The Walsh transform matrix is constructed using Kronecker product as $W_{2^i} = \begin{bmatrix} W_{2^{i-1}} & W_{2^{i-1}} \\ W_{2^{i-1}} & -W_{2^{i-1}} \end{bmatrix}$. This is a so-called *Hadamard-ordered* Walsh functions. Walsh functions possess the symmetry properties, and due to this the Walsh matrix is orthogonal, symmetric and self-inverse (with normalization constant 2^{-n}). The Walsh functions take two values, +1 and −1, and in that respect are compatible with switching functions, which are also two-valued. Discrete Walsh transform is defined as $\mathbf{W} = W_{2^n}\mathbf{F}$.

The reader can find more details in [9, 15, 26, 32, 33], including classification and relations to other orthogonal transforms.

3.12 References

[1] Aiken HH. Synthesis of electronic computing and control circuits. *Ann. Computation Laborarатory of Harvard University*, XXVII, Harvard University, Cambridge, MA, 1951.

[2] Akers S. Binary decision diagrams. *IEEE Transactions on Computers*, 27(6):509–516, 1978.

[3] Atallah MJ. *Algorithms and Theory of Computation Handbook.* CRC Press, Boca Raton, FL, 1999.

[4] Bryant RE. Graph-based algorithms for Boolean function manipulation. *IEEE Transactions on Computers*, 35(6):677–691, 1986.

[5] Cormen TH, Leiserson CE, Rivest RL, and Stein C. *Introduction to Algorithms.* MIT Press, Cambridge, MA, 2001.

[6] Davio MJ, Deschamps P, and Thayse A. *Discrete and Switching Functions.* McGraw-Hill, New York, 1978.

[7] DeMicheli G. *Synthesis and Optimization of Digital Circuits.* McGraw-Hill, New York, 1994.

[8] Drechsler R, and Becker B. *Binary Decision Diagrams: Theory and Implementation.* Kluwer, Dordrecht, 1999.

[9] Falkowski BJ. A note on the polynomial form of Boolean functions and related topics. *IEEE Transactions on Computers*, 48(8):860–863, 1999.

[10] Falkowski BJ, and Stanković RS. Spectral interpretation and applications of decision diagrams. *VLSI Design International Journal of Custom Chip Design, Simulation and Testing*, 11(2):85–105, 2000.

[11] Green D. *Modern Logic Design.* Addison-Wesley, Reading, MA, 1986.

[12] Green DH. Families of Reed-Muller canonical forms. *International Journal of Electronics*, 2:259–280, 1991.

[13] Hachtel G, and Somenzi F. *Logic Synthesis and Verification Algorithms.* 3rd ed., Kluwer, New York, 2000.

[14] Jain J. Arithmetic transform of Boolean functions. In Sasao T. and Fujita M, Eds., *Representations of Discrete Functions*, Kluwer, Dordrecht, 1996, pp. 55–92.

[15] Janković D, Stanković RS, and Drechsler R. Decision diagram method for calculation of pruned Walsh transform. *IEEE Transactions on Computers*, 50(2):147–157, 2001.

[16] Lee CY. Representation of switching circuits by binary-decision programs. *Bell System Technical Journal*, 38:985–999, 1959.

[17] Malyugin VD. Representation of Boolean functions by arithmetical polynomials. *Automation and Remote Control*, Kluwer/Plenum Publishers, 43(4):496–504, 1982.

[18] Meinel C, and Theobald T. *Algorithms and Data Structures in VLSI Design.* Springer-Verlag/Heidelberg, 1998.

[19] Minato S. *Binary Decision Diagrams and Applications for VLSI Design.* Kluwer, Dordrecht, 1996.

[20] Moret BME. Decision trees and diagrams. *Computing Surveys*, 14(4):593–623, 1982.

[21] Papaioannou SG. Optimal test generation in combinational networks by pseudo-Boolean programming. *IEEE Transactions on Computers*, 26:553–560, 1977.

[22] Roth JP. *Mathematical Design. Building Reliable Complex Computer Systems.* IEEE Press, New York, 1999.

[23] Sasao T. *Switching Theory for Logic Synthesis.* Kluwer, Dordrecht, 1999.

[24] Sasao T, and Fujita M., Eds., *Representations of Discrete Functions.* Kluwer, Dordrecht, 1996.

[25] Sasao T. Representation of logic functions using EXOR operators. In Sasao T, and Fujita M., Eds., *Representations of Discrete Functions,* Kluwer, Dordrecht, 1996, pp. 29–54.

[26] Stanković RS, and Astola JT. *Spectral Interpretation of Decision Diagrams.* Springer-Verlag/Heidelberg, 2003

[27] Stanković RS, Moraga C, and Astola JT. Reed-Muller expressions in the previous decade. In *Proceedings 5th International Workshop on Applications of the Reed-Muller Expansion in Circuit Design*, pp. 7–26, Mississippi State University, MS, 2001.

[28] Thornton MA, Dreschler R, and Miller DM. *Spectral Techniques in VLSI CAD.* Kluwer, Dordrecht, 2002.

[29] Thornton M, and Nair V. Efficient calculation of spectral coefficients and their applications. *IEEE Transactions on Computer-Aided Design of Integrated Circuits and Systems*, 14(1):1328–13411, 1995.

[30] Tsai CC, and Marek-Sadowska M. Boolean functions classification via fixed polarity Reed-Muller forms. *IEEE Transactions on Computers,* 46(2):173–186, 1997.

[31] Tsai CC, and Marek-Sadowska M. Generalized Reed-Muller forms as a tool to detect symmetries. *IEEE Transactions on Computers*, 45:33–40, 1996.

[32] Yanushkevich SN. Arithmetical canonical expansions of Boolean and MVL functions as generalized Reed-Muller series. In *Proceedings of the IFIP WG 10.5 Workshop on Applications of the Reed-Muller Expansions in Circuit Design*, pp. 300–307, Japan, 1995.

[33] Yanushkevich SN. Multiplicative properties of spectral Walsh coefficients of the Boolean function. *Automation and Remote Control,* Kluwer/Plenum Publishers, 64(12):1938–1947, 2003.

4

Word-Level Data Structures

In this chapter, selected methods of word-level logic design are introduced. Word-level and manipulation of switching function is the key point of parallel and homogeneous computing. Therefore, word-level processing is a suitable candidate for calculations on nanostructures that are parallel and distributed systems. This is the main motivation to discuss these methods of computing in spatial dimensions. All methods for word-level representation and manipulation discussed here are divided into three groups:

- Computing with *words of assignments*,
- Computing with *words of functions*, and
- Computing with words of assignments and words of functions.

These methods, as well as the basics of the word-level approach, are revised in Section 4.1. The rest of this chapter focuses on: (i) word-level arithmetic expressions (Section 4.2), (ii) word-level sum-of-products expressions (Section 4.3), and (iii) word-level Reed-Muller expressions (Section 4.4). The technique of computation is introduced in each section by general equations for word-level representations, and computing methods including algebraic, matrix, and decision diagrams.

4.1 Word-level data structures

The state-of-the-art approaches to word-level representation and manipulation of switching functions operate on the following groups of data structures:

- Word-level set of assignments,
- Word-levels, and
- Word-levels and set of assignments.

In this section, the general strategy for synthesis and analysis of these data structures is discussed.

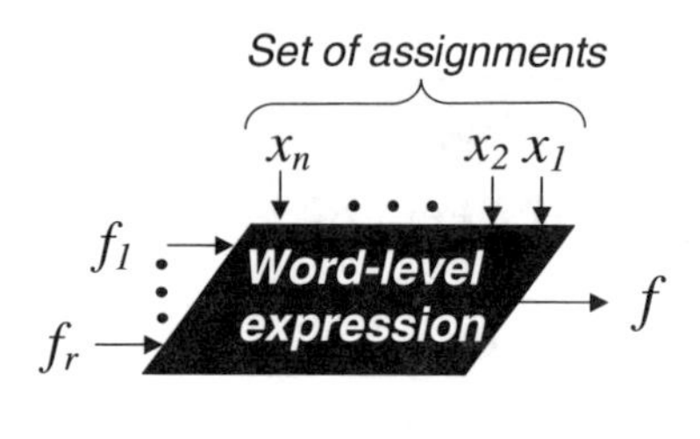

Let f be a switching function of two variables x_1, x_2 $(n = 2)$

The set of assignments

$u = \{00, 10, 11\}$

Vectors of assignments:

$$\mathbf{x}_1 = \begin{bmatrix} 0 \\ 1 \\ 1 \end{bmatrix}, \quad \mathbf{x}_2 = \begin{bmatrix} 0 \\ 0 \\ 1 \end{bmatrix}$$

FIGURE 4.1
Word-level set of assignments for computing the switching function f given a set of word-level assignments (Example 4.1).

4.1.1 Computing by word-level set of assignments

The problem is stated as follows: given a set of n, $u \in 0, 1, \ldots, 2^n$ assignments of n variables $x_1, x_2, \ldots, x_n$ for a switching function f, calculate the function f for these u assignments. Denote the columns $x_1, x_2, \ldots, x_n$ of the truth table of f given u assignments, or $u \times 1$ vectors $\mathbf{x}_1, \mathbf{x}_2, \ldots, \mathbf{x}_n$ called the *vectors of assignments.*

Example 4.1 *For function f of two variables x_1, x_2 $(n = 2)$, the set of assignments is $u = \{00, 10, 11\}$. Vectors of assignments are given in Figure 4.1.*

Denote the values of a switching function f given the vectors of assignments $\mathbf{x}_1, \mathbf{x}_2, \ldots, \mathbf{x}_n$ by $\mathbf{f}$. In its turn, $\mathbf{f}$ is expressed by the equation $\mathbf{f} = \mathbf{f}(\mathbf{x}_1, \mathbf{x}_2, \ldots, \mathbf{x}_n)$. The algorithm of computing the switching function f given a set of word-level assignments is shown in Figure 4.2. This approach is illustrated below with the example for a single output function.

Example 4.2 *Given the switching function $f = x_1 \vee x_2 x_3$, calculate its values for assignments* 000, 010, 110, 111 *(Figure 4.3).*

Step 1. Determine the vectors of assignments $\mathbf{x}_1, \mathbf{x}_2$ and $\mathbf{x}_3$.
Step 2. Calculate $\mathbf{f} = \mathbf{x}_1 \vee \mathbf{x}_2 \mathbf{x}_3$. Figure 4.3 illustrates the operations over vectors $\mathbf{x}_1, \mathbf{x}_2$ and $\mathbf{x}_3$ to derive $\mathbf{f}$.

These calculations can be represented by the bitwise over integer numbers (words): $\mathbf{x}_1 = 12,\ \mathbf{x}_2 = 14, \mathbf{x}_3 = 8$,

$$\mathbf{f} = \mathbf{x}_1 \mathbin{\widehat{\vee}} \mathbf{x}_2 \mathbin{\widehat{\wedge}} \mathbf{x}_3 = 12 \mathbin{\widehat{\vee}} 14 \mathbin{\widehat{\wedge}} 8 = 12$$

4.1.2 Computing by word-level expressions

The forms of switching functions considered in this chapter are the following:

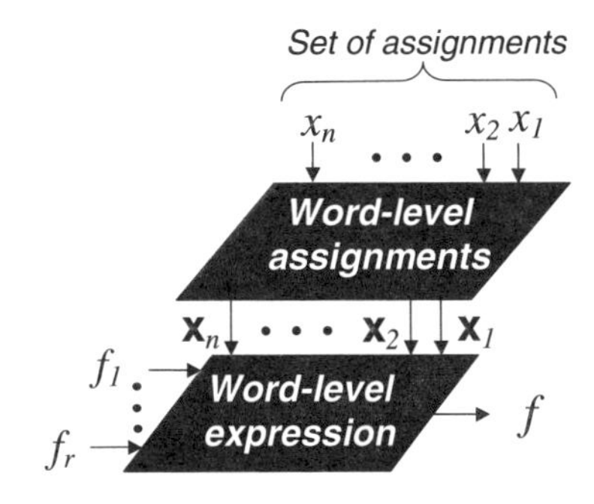

Given :
A set of u assignments, and a word-level expression of switching function f.

(*i*) *Represent the set of u assignments by $u \times 1$ vectors of assignments* $\mathbf{x}_1, \mathbf{x}_2, \ldots, \mathbf{x}_n$.

(*ii*) *Represent the word-level expression of f* $\mathbf{f} = \mathbf{f}(\mathbf{x}_1, \mathbf{x}_2, \ldots, \mathbf{x}_n)$.

(*iii*) *Calculate the outputs of* $\mathbf{f} = (\mathbf{x}_1, \mathbf{x}_2, \ldots, \mathbf{x}_n)$.

FIGURE 4.2
An algorithm for computing the switching function f given a set of word-level assignments.

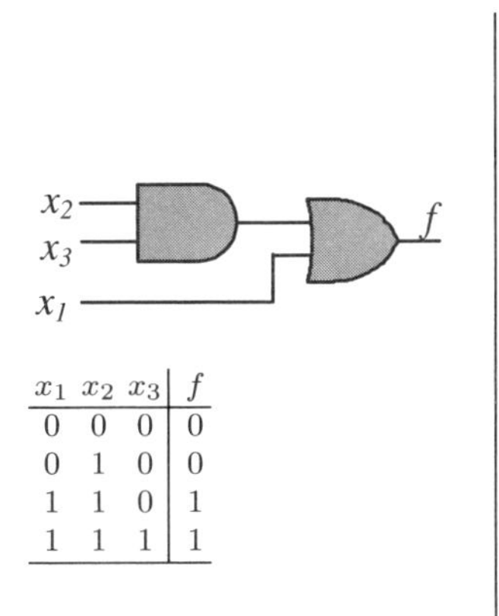

x_1	x_2	x_3	f
0	0	0	0
0	1	0	0
1	1	0	1
1	1	1	1

The switching function

$$f = x_1 \vee x_2 x_3$$

Vectors of assignments

$$\mathbf{x}_1 = \begin{bmatrix} 0 \\ 0 \\ 1 \\ 1 \end{bmatrix}, \ \mathbf{x}_2 = \begin{bmatrix} 0 \\ 1 \\ 1 \\ 1 \end{bmatrix}, \ \mathbf{x}_3 = \begin{bmatrix} 0 \\ 0 \\ 0 \\ 1 \end{bmatrix}$$

Equation for assignments

$$\mathbf{f} = \mathbf{x}_1 \vee \mathbf{x}_2 \mathbf{x}_3$$

Output :

$$\mathbf{f} = \begin{bmatrix} 0 \\ 0 \\ 1 \\ 1 \end{bmatrix} \vee \begin{bmatrix} 0 \\ 1 \\ 1 \\ 1 \end{bmatrix} \begin{bmatrix} 0 \\ 0 \\ 0 \\ 1 \end{bmatrix} = \begin{bmatrix} 0 \\ 0 \\ 1 \\ 1 \end{bmatrix}$$

FIGURE 4.3
Derivation of the switching function $f = x_1 \vee x_2 x_3$ given by word-level assignments (Example 4.2).

- ▶ Word-level arithmetic expressions,
- ▶ word-level sum-of-products expressions, and
- ▶ word-level Reed-Muller expressions.

These forms of a switching function f of n variables $x_1, x_2, \ldots x_n$ have the common data structure in formal expressions as shown in Table 4.1, where

$$\odot = \begin{cases} +, \text{ for word-level arithmetic expression;} \\ \frown{\vee}, \text{ for word-level sum-of-products expression;} \\ \frown{\oplus}, \text{ for word-level Reed-Muller expression.} \end{cases}$$

TABLE 4.1
Word-level expressions for a switching function and decomposition rules for relevant decision trees and diagrams.

Expression	Formal description	$x_j^{i_j}$	Decomposition
Word-level arithmetic	$\sum_{i=0}^{2^n-1} d_i \cdot (x_1^{i_1} \cdots x_n^{i_n})$	$\begin{cases} 1, & i_j = 0; \\ x_j, & i_j = 1. \end{cases}$	Davio arithmetic
Word-level sum-of-products	$\frown{\bigvee}_{i=0}^{2^n-1} v_i \cdot (x_1^{i_1} \cdots x_n^{i_n})$	$\begin{cases} \overline{x}_j, & i_j = 0; \\ x_j, & i_j = 1. \end{cases}$	Shannon
Word-level Reed-Muller	$\frown{\bigoplus}_{i=0}^{2^n-1} w_i \cdot (x_1^{i_1} \cdots x_n^{i_n})$	$\begin{cases} 1, & i_j = 0; \\ x_j, & i_j = 1. \end{cases}$	Davio

The expressions are consistent with the relevant data structures – matrix representation, flowgraph, decision tree and decision diagram – that will be considered later in this chapter. We observe that these expressions are the same algebraic structure.

4.2 Word-level arithmetic expressions

Word-level arithmetic expansion serves to represent a logic function whose outputs are grouped to build a bitstring (word). In other words, the advantage of arithmetic logic is a possibility to represent an r-output function f with outputs $f_1, f_2, ..., f_r$, so that the outputs can be restored in a unique way.

4.2.1 General form

For an r-output switching function f of n variables, the arithmetic word-level expression is the weighted sum of arithmetic expressions of f_j, $j = 1, \ldots, r$,

$$f = 2^{r-1} f_r + \cdots + 2^1 f_2 + 2^0 f_1. \tag{4.1}$$

This expression (Equation 4.1) can be rewritten in the form

$$f = \sum_{i=0}^{2^n-1} d_i \cdot \underbrace{(x_1^{i_1} \cdots \, x_n^{i_n})}_{i-th \ product}, \tag{4.2}$$

where coefficient d_i is an integer number, i_j, $j = 1, 2, \ldots, n$, is the j-th bit in the binary representation of the index $i = i_1 i_2 \ldots i_n$, and $x_j^{i_j}$ is defined as

$$x_j^{i_j} = \begin{cases} 1, & i_j = 0; \\ x_j, & i_j = 1. \end{cases} \tag{4.3}$$

Example 4.3 *An arbitrary r-output switching function of two variables is represented by the arithmetic word-level expression according to Equation 4.2 and Equation 4.3 by*

$$\begin{aligned} f &= d_0(x_1^0 x_2^0) + d_1(x_1^0 x_2^1) + d_2(x_1^1 x_2^0) + d_3(x_1^1 x_2^1) \\ &= d_0 + d_1 x_2 + d_2 x_1 + d_3 x_1 x_2. \end{aligned}$$

Figure 4.4 illustrates the structure of this expression, where the nodes implement the AND operations.

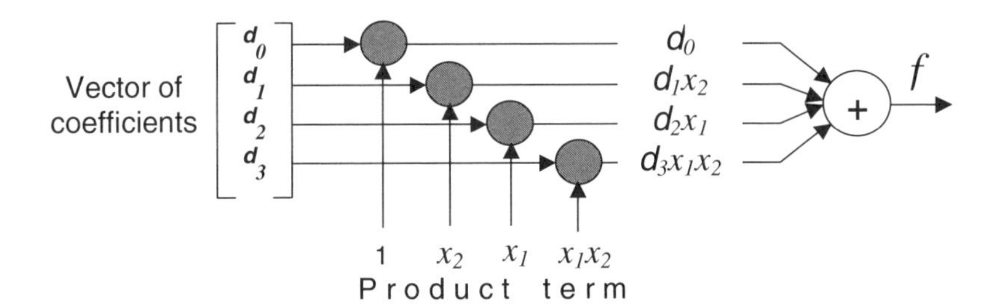

FIGURE 4.4
Deriving the arithmetic word-level expression for an r-output switching function of two variables (Example 4.3).

4.2.2 Masking operator

The *masking operator* $\Xi^{\tau}\{f\}$ is used to recover the single function f_{τ} from the word-level representation of a r-output switching function f:

$$f_{\tau} = \Xi^{\tau}\{f\}.$$

Example 4.4 *Let the switching function be given in 4-bit word format. Recover:*

(a) the output f_2 and f_4,
(b) the output f_1 and f_4.

In Figure 4.5, the solutions (a) and (b) are given.

Example 4.5 *In Example 4.3, $f = x_1 + x_2$, which corresponds to the coefficient vector* $\begin{bmatrix} 0 \\ 1 \\ 1 \\ 0 \end{bmatrix}$ *and to the truth vector* $\begin{bmatrix} 0 \\ 1 \\ 1 \\ 2 \end{bmatrix} = \begin{bmatrix} 0\ 0 \\ 0\ 1 \\ 0\ 1 \\ 1\ 0 \end{bmatrix}$, *which indeed describes two-output function. The masking operator recovers two switching functions* $\Xi^1\{x_1 + x_2\} = x_1 \oplus x_2$ *and* $\Xi^2\{x_1 + x_2\} = x_1 x_2$.

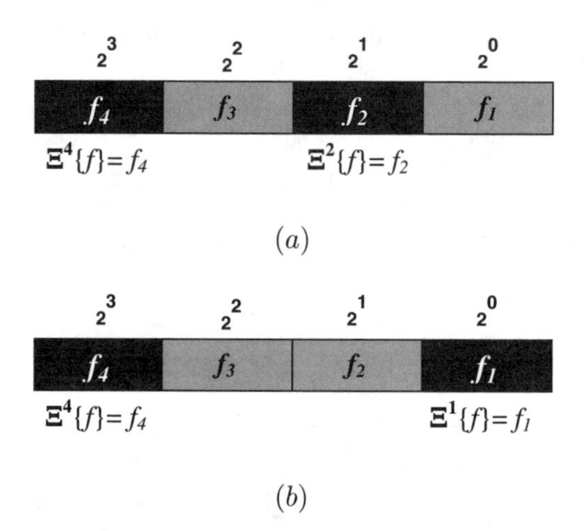

FIGURE 4.5
The masking operator $\Xi^{\tau}\{f\}$, $\tau = 1, 2, 3, 4$, recovers the function f_{τ} from a word-level representation of the switching function f (Example 4.4).

4.2.3 Computing the coefficients

Given a truth vector $\mathbf{F} = [f(0)\ f(1) \ldots f(2^n - 1)]^T$, the vector of coefficients $\mathbf{D} = [d_0\ d_1 \ldots d_{2^n-1}]^T$ is derived by the matrix eqaution with AND and arithmetic sum operations

$$\mathbf{D} = \mathbf{P}_{2^n} \cdot \mathbf{F}, \tag{4.4}$$

where the $2^n \times 2^n$ matrix $\mathbf{P}_{2^n}$ is formed by the Kronecker product

$$\mathbf{P}_{2^n} = \bigotimes_{j=1}^{n} \mathbf{P}_{2^j}, \quad \mathbf{P}_{2^1} = \begin{bmatrix} 1 & 0 \\ -1 & 1 \end{bmatrix} \tag{4.5}$$

The general scheme of computing is shown in Figure 4.6a.

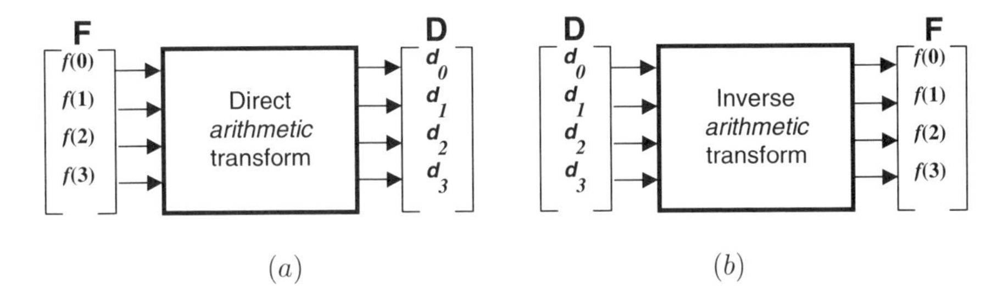

FIGURE 4.6
Direct (a) and inverse (b) arithmetic word-level transform for a r-output switching function of two variables.

Example 4.6 *Computing the coefficients by Equation 4.4 and Equation 4.6 for the two-output switching function $f_1 = x_1 \vee x_2$, $f_2 = \overline{x}_1 \vee x_2$ with truth-vector $\mathbf{F_1}$ and $\mathbf{F_2}$ is illustrated by Figure 4.7. The truth vector $\mathbf{F}$ is defined as*

$$\mathbf{F} = [\mathbf{F_2}|\mathbf{F_1}] = \begin{bmatrix} 1 & 0 \\ 1 & 1 \\ 0 & 1 \\ 1 & 1 \end{bmatrix} = \begin{bmatrix} 2 \\ 3 \\ 1 \\ 3 \end{bmatrix}.$$

The same result can be obtained by algebraic Equation 4.1 given

$$f_1 = x_1 \vee x_2 = x_2 + x_1 - x_1x_2,$$
$$f_2 = \overline{x}_1 \vee x_2 = 1 - x_1 + x_1x_2,$$

i.e.,

$$f = 2^1 f_2 + 2^0 f_1 = 2^1 \underbrace{(1 - x_1 + x_1x_2)}_{Output\ f_2} + 2^0 \underbrace{(x_2 + x_1 - x_1x_2)}_{Output\ f_1}$$
$$= 2 + x_2 - x_1 + x_1x_2.$$

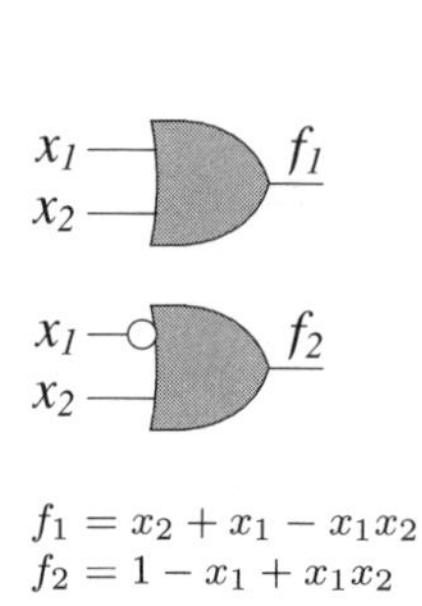

Direct arithmetic transform

$$\mathbf{D} = \mathbf{P}_{2^2} \cdot \mathbf{F} = \left[\begin{array}{rr|rr} 1 & 0 & 0 & 0 \\ -1 & 1 & 0 & 0 \\ \hline -1 & 0 & 1 & 0 \\ 1 & -1 & -1 & 1 \end{array}\right] \begin{bmatrix} 2 \\ 3 \\ 1 \\ 3 \end{bmatrix} = \begin{bmatrix} 2 \\ 1 \\ -1 \\ 1 \end{bmatrix}$$

Arithmetic expression

$f = 2 + x_2 - x_1 + x_1 x_2$

The equation is equivalent to

$$\mathbf{F} = 2^1 \mathbf{F_2} + 2^0 \mathbf{F_1} = 2^1 \begin{bmatrix} 1 \\ 1 \\ 0 \\ 1 \end{bmatrix} + 2^0 \begin{bmatrix} 0 \\ 1 \\ 1 \\ 1 \end{bmatrix} = \begin{bmatrix} 2 \\ 3 \\ 1 \\ 3 \end{bmatrix}$$

Applying transformation (Equation 4.4) implies:

$\mathbf{D} = \mathbf{P_{2^2}} \mathbf{F} = [\,2\;1\;-1\;1\,]^T$

i.e., $f = 2 + x_2 - x_1 + x_1 x_2$

FIGURE 4.7
Computing the arithmetic word-level expression for a two-output circuit (Example 4.6).

4.2.4 Restoration

Given the matrix eqaution with AND and arithmetic sum operations, restore the truth-vector **F** of switching function f from the vector of coefficients **D** (Figure 4.6b):

$$\mathbf{F} = \mathbf{P}_{2^n}^{-1} \cdot \mathbf{D} \tag{4.6}$$

where the $2^n \times 2^n$ matrix $\mathbf{P}_{2^n}^{-1}$ is formed by the Kronecker product

$$\mathbf{P}_{2^n}^{-1} = \bigotimes_{j=0}^{n} \mathbf{P}_{2^j}^{-1}, \qquad \mathbf{P}_{2^1}^{-1} = \begin{bmatrix} 1 & 0 \\ 1 & 1 \end{bmatrix}. \tag{4.7}$$

Example 4.7 *Restore the truth-vector* **F** *of a switching function* f *given by the vector of coefficients* $\mathbf{D} = [2\;1\;-1\;1]^T$. *Using Equation 4.6 implies*

$$\mathbf{F} = \mathbf{P}_{2^3}^{-1} \cdot \mathbf{D} = \left[\begin{array}{cc|cc} 1 & 0 & 0 & 0 \\ 1 & 1 & 0 & 0 \\ \hline 1 & 0 & 1 & 0 \\ 1 & 1 & 1 & 1 \end{array}\right] \begin{bmatrix} 2 \\ 1 \\ -1 \\ 1 \end{bmatrix} = \begin{bmatrix} 2 \\ 3 \\ 1 \\ 3 \end{bmatrix}.$$

This means $\begin{bmatrix} 2 \\ 3 \\ 1 \\ 3 \end{bmatrix} = \begin{bmatrix} 1 & 0 \\ 1 & 1 \\ 0 & 1 \\ 1 & 1 \end{bmatrix}$, *i.e.,* $f_1 = x_1 \vee x_2,\; f_2 = \overline{x}_1 \vee x_2$.

4.2.5 Useful properties

The linearity property of a word-level sum of truth-vectors $\mathbf{F}_i$, $i = 1, 2, \ldots, r$ (Figure 4.8a)

$$\mathbf{F} = \underbrace{2^{r-1}\mathbf{F}_r + \cdots + 2^1\mathbf{F}_2 + 2^0\mathbf{F}_1}_{r\ truth\ vectors},$$

and linearity property of arithmetic transform over this sum yields

$$\begin{aligned}\mathbf{D} &= \mathbf{P}_{2^n}\mathbf{F}\\ &= \underbrace{2^{r-1}\mathbf{P}_{2^n}\mathbf{F}_r + \cdots + 2^1\mathbf{P}_{2^n}\mathbf{F}_2 + 2^0\mathbf{P}_{2^n}\mathbf{F}_1}_{r\ arithmetic\ transforms}.\end{aligned}$$

Therefore, in a word-level representation, the arithmetic transform of the truth-vector $\mathbf{F}$ can be replaced with r arithmetic transforms of the truth-vectors $\mathbf{F}_i$.

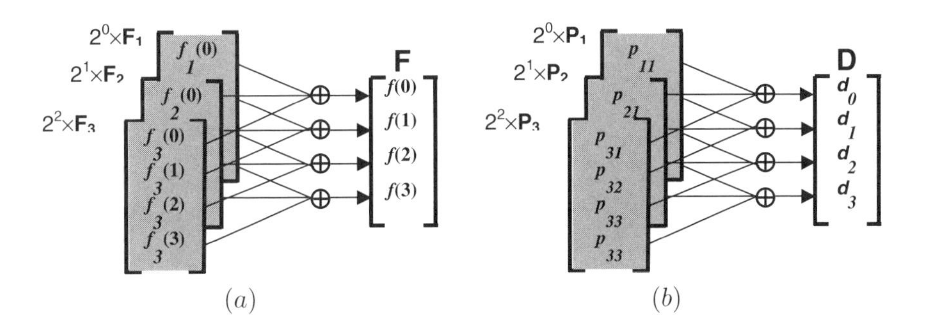

FIGURE 4.8
Linear properties of arithmetic transform: word-level sum of truth-vectors (a) and word-level sum of vectors of coefficients (b).

By analogy, a word-level sum of vectors of coefficients (Figure 4.8b)

$$\mathbf{D} = \underbrace{2^{r-1}\mathbf{D}_r + \cdots + 2^1\mathbf{D}_2 + 2^0\mathbf{D}_1}_{r\ vectors},$$

then the inverse arithmetic transform of a vector of coefficients $\mathbf{D}$ can be replaced by r arithmetic transforms of the vectors $\mathbf{D}_i$

$$\begin{aligned}\mathbf{F} &= \mathbf{P}_{2^n}^{-1} \cdot \mathbf{D}\\ &= \underbrace{2^{r-1}\mathbf{P}_{2^n}\mathbf{D}_r + \cdots + 2^1\mathbf{P}_{2^n}\mathbf{D}_2 + 2^0\mathbf{P}_{2^n}\mathbf{D}_1}_{r\ inverse\ arithmetic\ transforms}.\end{aligned}$$

Figure 4.9 illustrates the final stage of the transformation.

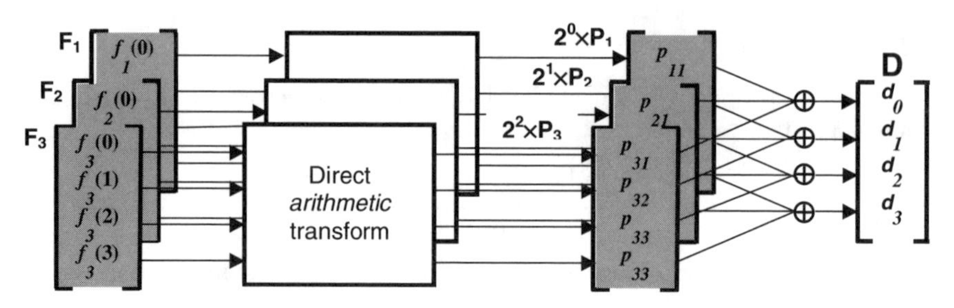

FIGURE 4.9
The r-iteration direct arithmetic transform.

4.2.6 Polarity

The polarity of a variable x_j can be:

$c_j = 0$ that corresponds to the uncomplemented variable x_j, or
$c_j = 1$ that corresponds to the complemented variable $\overline{x}_j$.

Let the polarity $c = c_1, c_2, \ldots, c_n$, $c \in \{0, 1, 2, \ldots, 2^n - 1\}$, where c_j is the j-th bit of binary representation of c. For a r-output switching function f of n variables, the arithmetic word-level form in a given polarity $c = c_1, c_2, \ldots, c_n$ of variables $x_1, x_2, \ldots, x_n$ is defined as follows

$$f = \sum_{i=0}^{2^n-1} d_i \cdot \underbrace{(x_1 \oplus c_1)^{i_1} \cdots \ (x_n \oplus c_n)^{i_n}}_{i-th\ product}, \tag{4.8}$$

where d_i is the coefficient, and $(x_j \oplus c_j)^{i_j}$ is defined as

$$a_j^{i_j} = \begin{cases} 1, \text{ if } i_j = 0; \\ a, \text{ if } i_j = 1. \end{cases} \quad x_j \oplus c_j = \begin{cases} x_j, \text{ if } c_j = 0; \\ \overline{x}_j, \text{ if } c_j = 1. \end{cases} \tag{4.9}$$

Example 4.8 *Represent the r-output switching function of two variables by arithmetic word-level expression of the polarity $c = 2$, $c_1, c_2 = 1, 0$. By Equation 4.8 and Equation 4.9,*

$$f = d_0 \underbrace{(x_1 \oplus 1)^0 (x_2 \oplus 0)^0}_{0-st\ product} + d_1 \underbrace{(x_1 \oplus 1)^0 (x_2 \oplus 0)^1}_{1-st\ product} + d_2 \underbrace{(x_1 \oplus 1)^1 (x_2 \oplus 0)^0}_{2-nd\ product}$$
$$+ d_3 \underbrace{(x_1 \oplus 1)^1 (x_2 \oplus 0)^1}_{3-rd\ product} = d_0 + d_1 x_2 + d_2 \overline{x}_1 + d_3 \overline{x}_1 x_2.$$

Given the truth vector $\mathbf{F} = [f(0)\ f(1) \ldots f(2^n - 1)]^T$, the vector of arithmetic word-level coefficients of polarity c, $\mathbf{D^{(c)}} = [d_0^{(c)}\ d_1^{(c)} \ldots d_{2^n-1}^{(c)}]^T$, is derived by the matrix eqaution

$$\mathbf{D^{(c)}} = \mathbf{P}_{2^n}^{(c)} \cdot \mathbf{F}, \tag{4.10}$$

where the $2^n \times 2^n$ matrix $\mathbf{P}_{2^n}^{(c)}$ is generated by the Kronecker product

$$\mathbf{P}_{2^n}^{(c)} = \bigotimes_{j=1}^{n} \mathbf{P}_{2^1}^{(c_j)}, \qquad \mathbf{P}_{2^1}^{(c)} = \begin{cases} \begin{bmatrix} 1 & 0 \\ -1 & 1 \end{bmatrix}, & c_j = 0; \\ \begin{bmatrix} 0 & 1 \\ 1 & -1 \end{bmatrix}, & c_j = 1. \end{cases} \tag{4.11}$$

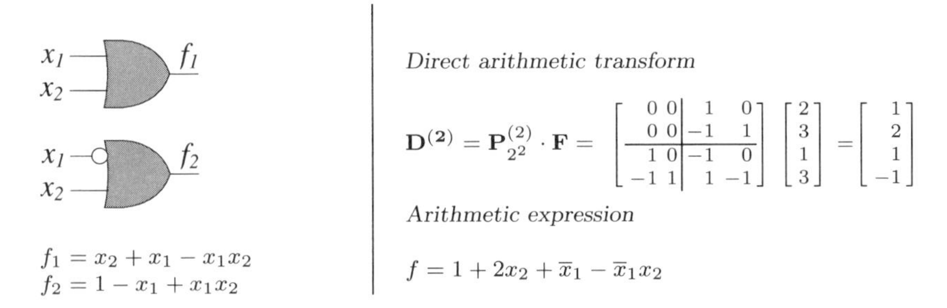

FIGURE 4.10
Computing the arithmetic word-level expression of polarity $c = 2$ for the two-output circuit (Example 4.9).

Example 4.9 *In the matrix form, the solution to Example 4.6 using Equation 4.8 is given in Figure 4.10, where the matrix* $\mathbf{P}_{2^2}^{(2)}$ *for the polarity* $c = 2$ *is generated by Equation 4.9 as*

$$\mathbf{P}_{2^2}^{(2)} = \mathbf{P}_{2^1}^{(1)} \otimes \mathbf{P}_{2^1}^{(0)} = \begin{bmatrix} 0 & 1 \\ 1 & -1 \end{bmatrix} \otimes \begin{bmatrix} 1 & 0 \\ -1 & 1 \end{bmatrix}$$

4.2.7 Computing for a word-level set of assignments

In Section 4.1, the word-level techniques have been distinguished by using words of assignments and words of functions. The implementation aspects are discussed in this section.

Algebraic form.

Example 4.10 *Calculate the outputs of the circuit depicted in Figure 4.11 given the word-level arithmetic expression of the switching function:*

$$f = 6x_4 + 2x_3\overline{x}_4 - 4x_1\overline{x}_2x_4 + 5x_1\overline{x}_2,$$

for the set of assignments $u \in \{1000, 0010, 0001\}$. The solution is shown in Figure 4.11. Vectors of assignments $\mathbf{x}_1, \mathbf{x}_2, \mathbf{x}_3$, and $\mathbf{x}_4$ are derived from the truth table. Calculations over these vectors are performed by simulation the arithmetic expression of f.

In algebraic form, this calculation is performed over integers $\mathbf{x}_1 = 4$, $\mathbf{x}_2 = 2$, $\mathbf{x}_3 = 6$, $\mathbf{x}_4 = 5$.

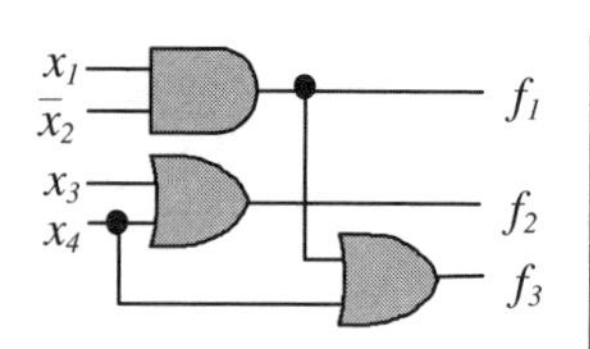

x_1	x_2	x_3	x_4	$f_3\ f_2\ f_1$
0	0	0	1	*Find*
0	1	1	0	*Find*
1	0	1	1	*Find*

Three-output switching function:

$$f = 6x_4 + 2x_3\overline{x}_4 - 4x_1\overline{x}_2x_4 + 5x_1\overline{x}_2$$

Vectors of assignments:

$$\mathbf{x}_1 = \begin{bmatrix} 0 \\ 0 \\ 1 \end{bmatrix}, \quad \mathbf{x}_2 = \begin{bmatrix} 0 \\ 1 \\ 0 \end{bmatrix}, \quad \mathbf{x}_3 = \begin{bmatrix} 0 \\ 1 \\ 1 \end{bmatrix}, \quad \mathbf{x}_4 = \begin{bmatrix} 1 \\ 0 \\ 1 \end{bmatrix}$$

Equation for assignments in algebraic form:

$$\mathbf{f} = 6\mathbf{x}_4 + 2\mathbf{x}_3\overline{\mathbf{x}}_4 - 4\mathbf{x}_1\overline{\mathbf{x}}_2\mathbf{x}_4 + 5\mathbf{x}_1\overline{\mathbf{x}}_2$$

Equation for assignments in matrix form:

$$\mathbf{f} = 6\begin{bmatrix} 1 \\ 0 \\ 1 \end{bmatrix} + 2\begin{bmatrix} 1 \\ 0 \\ 1 \end{bmatrix}\begin{bmatrix} 0 \\ 1 \\ 0 \end{bmatrix} - 4\begin{bmatrix} 0 \\ 0 \\ 1 \end{bmatrix}\begin{bmatrix} 0 \\ 0 \\ 1 \end{bmatrix}\begin{bmatrix} 0 \\ 0 \\ 1 \end{bmatrix} + 5\begin{bmatrix} 0 \\ 0 \\ 1 \end{bmatrix}\begin{bmatrix} 0 \\ 0 \\ 1 \end{bmatrix}$$

$$= \begin{bmatrix} 6 \\ 2 \\ 7 \end{bmatrix} = \begin{bmatrix} 1 & 1 & 0 \\ 0 & 1 & 0 \\ 1 & 1 & 1 \end{bmatrix}$$

Outputs: $\mathbf{f_1} = \begin{bmatrix} 0 \\ 0 \\ 1 \end{bmatrix}, \quad \mathbf{f_2} = \begin{bmatrix} 1 \\ 1 \\ 1 \end{bmatrix}, \quad \mathbf{f_3} = \begin{bmatrix} 1 \\ 0 \\ 1 \end{bmatrix}$

FIGURE 4.11
Computing a three-output function via word-level assignments (Example 4.10).

Decision trees and diagrams. Word-level decision trees and diagrams are analogs of Davio trees and diagrams whereas the nodes correspond to the arithmetic analog of positive Davio expansion

$$f = 1 \cdot f_0 + x_i(-f_0 + f_1), \tag{4.12}$$

where $f_0 = f(x_i = 0)$, $f_1 = f(x_i = 1)$. The difference is that f_0 and f_1 take not integer but binary values, and thus, the edges of the word-level Davio tree are multibit, or wordwise.

4.3 Word-level sum-of-products expressions

Sum-of-products expressions describe single-output switching functions. In this section, we introduce so-called *word-level sum-of-products expressions* that describe multi/output switching functions.

4.3.1 General form

For an r-output switching function f of n variables, the word-level sum-of-products expression is the bitwise of sum-of-products expressions of f_j, $j = 1, \ldots, r$,

$$f = 2^{r-1} f_r \mathbin{\widehat{\vee}} \cdots \mathbin{\widehat{\vee}} 2^1 f_2 \mathbin{\widehat{\vee}} 2^0 f_1, \tag{4.13}$$

where $\widehat{\vee}$ denotes a bitwise operation. Equation 4.13 can be rewritten in the form

$$f = \widehat{\bigvee}_{i=0}^{2^n-1} v_i \cdot \underbrace{(x_1^{i_1} \cdots x_n^{i_n})}_{i-th\ product}, \tag{4.14}$$

where coefficient v_i is a positive integer number, i_j is j-th bit in the binary representation of the index $i = i_1 i_2 \ldots i_n$, and $x_j^{i_j}$ is defined as

$$x_j^{i_j} = \begin{cases} \overline{x}_j, \; i_j = 0; \\ x_j, \; i_j = 1. \end{cases} \tag{4.15}$$

It can be observed from Equation 4.14 and Equation 4.15 that:

- Word-level sum-of-products include the product terms of n literals. For example, for $n = 2$ variables, the expression includes no product terms of two literals.
- A coefficient v_i carries information about the distribution of product terms over the switching functions: 1s in its binary representation indicate the index of function that comprises the product $(x_1^{i_1} \cdots x_n^{i_n})$. For example, $v_2 = 6 = 110$ carries information about the product $x_1\overline{x}_2$: this product is included in the functions f_3, $x_1\overline{x}_2 \in f_3$, and f_2, and not in the function f_1, $x_1\overline{x}_2 \notin f_3$.

Example 4.11 *The calculations below are based on Equation 4.14 and Equation 4.15.*

(*i*) *Let* $n = 1$, *then*

$$f = v_0\overline{x}_1 \mathbin{\widehat{\vee}} v_1 x_1 = v_0\overline{x}_1 \vee v_1 x_1.$$

If $v_0 = v_1 = 1$, *this expression represents constant 1:* $f = \overline{x}_1 \vee x_1 = 1$.

(*ii*) *Let* $n = 1$, $v_0 = 1$, *and* $v_1 = 2$. *The expression*

$$\overline{x}_1 \mathbin{\widehat{\vee}} 2x_1$$

implies that there are two functions: $f_1 = \overline{x}_1$ *and* $f_2 = x_1$.

(*iii*) *An arbitrary* $r-$*output switching function of two variables* ($n = 2$) *is represented by the word-level sum-of-products*

$$f = v_0(\overline{x}_1\overline{x}_2) \mathbin{\widehat{\vee}} v_1(\overline{x}_1x_2) \mathbin{\widehat{\vee}} v_2(x_1\overline{x}_2) \mathbin{\widehat{\vee}} v_3(x_1x_2).$$

Figure 4.12 illustrates the structure of the above expression, where the nodes implement AND function.

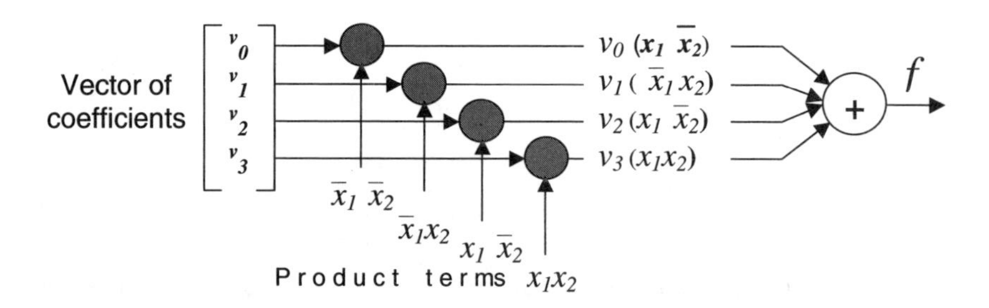

FIGURE 4.12
Deriving the word-level sum-of-profucts expression for a r-output switching function of two variables (Example 4.11).

4.3.2 Masking operator

The *masking operator* is used to recover a switching function from the word-level representation. However, for the word-level sum-of-products representation, the masking operator degenerates to lexicographical order of functions. In Figure 4.13 the masking operator recovers subfunctions from the four-output switching function f.

4.3.3 Computing the coefficients

Given a truth vector $\mathbf{F} = [f(0)\ f(1) \ldots f(2^n - 1)]^T$, the vector of coefficients $\mathbf{V} = [v_0\ v_1 \ldots v_{2^n-1}]^T$ is derived by the matrix equation with AND and OR operations

$$\mathbf{V} = \mathbf{S}_{2^n} \cdot \mathbf{F}, \tag{4.16}$$

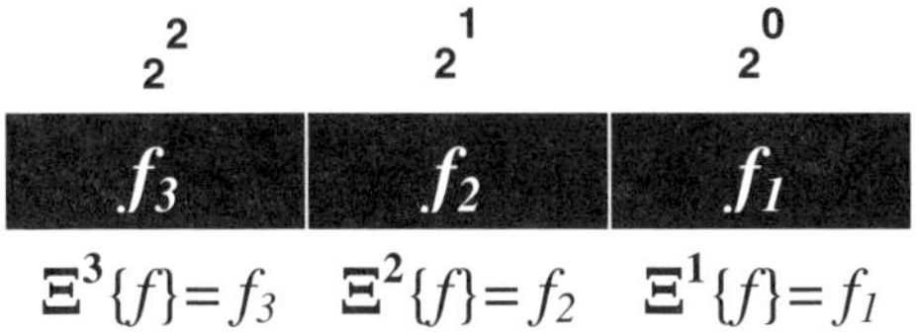

2^2 2^1 2^0

f_3 f_2 f_1

$\Xi^3\{f\}=f_3$ $\Xi^2\{f\}=f_2$ $\Xi^1\{f\}=f_1$

FIGURE 4.13
Masking operator $\Xi^\tau\{f\}$, $\tau = 1, 2, 3$, recovers the switching function f_τ from a word-level representation.

where the $2^n \times 2^n$ matrix $\mathbf{S}_{2^n}$ is formed by the Kronecker product

$$\mathbf{S}_{2^n} = \bigotimes_{j=1}^{n} \mathbf{S}_{2^j}, \quad \mathbf{S}_{2^1} = \begin{bmatrix} 1 & 0 \\ 0 & 1 \end{bmatrix}. \tag{4.17}$$

The general computing scheme is shown in Figure 4.14.

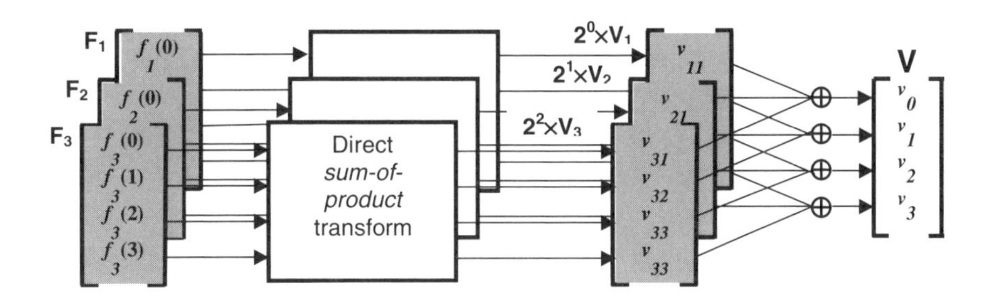

FIGURE 4.14
Direct arithmetic word-level transform over a two-output switching function of two variables.

Example 4.12 *Compute the coefficients from Equation 4.16 and Equation 4.17 for the two-output switching function*

$$f_1 = x_1 \oplus x_2$$
$$f_2 = x_1 x_2$$

given the truth-vectors $\mathbf{F_1}$ *and* $\mathbf{F_2}$. *Figure 4.15 is an illustration. The truth vector* **F** *is defined as*

$$\mathbf{F} = [\mathbf{F_2}|\mathbf{F_1}] = \begin{bmatrix} 0\,0 \\ 0\,1 \\ 0\,1 \\ 1\,0 \end{bmatrix} = \begin{bmatrix} 0 \\ 1 \\ 1 \\ 2 \end{bmatrix}.$$

Notice, the same result can be obtained by algebraic Equation 4.13 given $f_1 = x_1 \oplus x_2$ *and* $f_2 = \overline{x}_1 x_2$*:*

$$\begin{aligned} f &= 2^1 f_2 \overset{\frown}{\vee} 2^0 f_1 = 2^1(x_1 x_2) \overset{\frown}{\vee} 2^0(x_1\overline{x}_2 \vee \overline{x}_1 x_2) \\ &= 2x_1x_2 \overset{\frown}{\vee} x_1\overline{x}_2 \overset{\frown}{\vee} \overline{x}_1 x_2. \end{aligned}$$

Two-output function

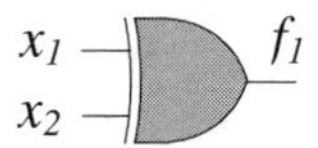

$f_1 = x_1\overline{x}_2 \vee \overline{x}_1 x_2$
$f_2 = x_1 x_2$

The direct transform (Equation 4.16) to compute vector **V**

$$\mathbf{V} = \mathbf{S}_{2^2} \cdot \mathbf{F} = \begin{bmatrix} 1 & & & \\ & 1 & & \\ & & 1 & \\ & & & 1 \end{bmatrix} \begin{bmatrix} 0 \\ 1 \\ 1 \\ 2 \end{bmatrix} = \begin{bmatrix} 0 \\ 1 \\ 1 \\ 2 \end{bmatrix}$$

Word-level expression of the two-output switching function

$f = \overline{x}_1 x_2 \overset{\frown}{\vee} x_1\overline{x}_2 \overset{\frown}{\vee} 2x_1x_2$

FIGURE 4.15
Computing the arithmetic word-level expression for the two-output circuit (Example 4.12).

4.3.4 Restoration

Restoration from the algebraic and matrix eqaution possesses some specific features.

Restoration from the algebraic equation. Let the coefficients of the word-level expression be $v_0 = 1,\ v_1 = 2,\ v_2 = 3,\ v_3 = 4$ (see Example 4.11), that is

$$f = \overline{x}_1\overline{x}_2 \overset{\frown}{\vee} 2\overline{x}_1 x_2 \overset{\frown}{\vee} 3x_1\overline{x}_2 \overset{\frown}{\vee} 4x_1x_2.$$

This equation yields the matrix of coefficients

$$f = \begin{bmatrix} v_0 \\ v_1 \\ v_2 \\ v_3 \end{bmatrix} = \begin{bmatrix} 1 \\ 2 \\ 3 \\ 4 \end{bmatrix}, \quad \text{C=} \begin{matrix} \\ v_0,\ \{x_1x_2\} \\ v_1,\ \{x_1x_2\} \\ v_2,\ \{x_1x_2\} \\ v_3,\ \{x_1x_2\} \end{matrix} \begin{matrix} f_3 & f_2 & f_1 \\ \left[\begin{matrix} 0 \\ 0 \\ 0 \\ 1 \end{matrix}\right. & \begin{matrix} 0 \\ 1 \\ 1 \\ 0 \end{matrix} & \left.\begin{matrix} 1 \\ 0 \\ 1 \\ 0 \end{matrix}\right] \end{matrix}$$

Therefore,

- ► f generates three switching functions f_1, f_2, f_3, since the maximal coefficient $v_3 = 4$ is represented by three bits.
- ► f_1 includes the products $\overline{x}_1\overline{x}_2$ and $x_1\overline{x}_2$ because the least significant bit of the coefficients $v_0 = 1 = 001$ and $v_2 = 3 = 011$ is equal to 1. Hence, $f_1 = \overline{x}_1\overline{x}_2 \vee x_1\overline{x}_2$.
- ► f_2 includes the products $\overline{x}_1x_2$ and $x_1\overline{x}_2$, because the second bit of the coefficients $v_1 = 2$ and $v_2 = 3$ is equal to 1. Hence, $f_2 = \overline{x}_1x_2 \vee x_1\overline{x}_2$.
- ► f_3 includes the product x_1x_2 because the most significant bit of the coefficient $v_3 = 4$ is equal to 1. Hence, $f_3 = x_1x_2$.

Restoration from the matrix equation. The following matrix equation with AND and arithmetic sum operations restores the truth-vector $\mathbf{F}$ from the vector of coefficients $\mathbf{V}$ (Figure 4.15b):

$$\mathbf{F} = \mathbf{S}_{2^n}^{-1} \cdot \mathbf{V} \tag{4.18}$$

where the $2^n \times 2^n$ matrix $\mathbf{S}_{2^n}^{-1} = \mathbf{S}_{2^n}$.

Example 4.13 *(Continuation Example 4.12) Restore the truth-vector* $\mathbf{F}$ *of a switching function* f *given by the vector of coefficients* $\mathbf{V}$ *(Figure 4.16).*

4.3.5 Computing for a word-level set of assignments

Word-level forms offer a convenient way to compute the values of switching functions given variable assignments.

Example 4.14 *Calculate the switching function*

$$f = 5x_1\overline{x}_2 \stackrel{\frown}{\vee} 2x_3 \stackrel{\frown}{\vee} 6x_4,$$

given the set of assignments $u \in \{0001, 0110, 1011\}$. *The logic circuit described by this expression is shown in Figure 4.17. First, vectors of assignments* $\mathbf{x}_1, \mathbf{x}_2, \mathbf{x}_3$, *and* $\mathbf{x}_4$ *are derived from* u. *Second, the bitwise calculations are performed on these vectors according to the word-level sum-of-products.*

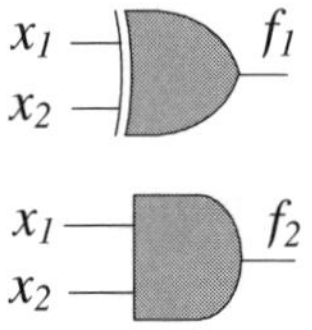

x_1, x_2 → f_1

x_1, x_2 → f_2

Given:
vector of coefficients

$$\mathbf{V} = \begin{bmatrix} 0 \\ 1 \\ 1 \\ 2 \end{bmatrix}$$

Applying Equation 4.18 implies

$$\mathbf{F} = \mathbf{S}_{2^2}^{-1} \cdot \mathbf{V} = \left[\begin{array}{cc|cc} 1 & & & \\ & 1 & & \\ \hline & & 1 & \\ & & & 1 \end{array}\right] \begin{bmatrix} 0 \\ 1 \\ 1 \\ 2 \end{bmatrix} = \left[\begin{array}{c|c} 0 & 0 \\ 0 & 1 \\ 0 & 1 \\ 1 & 0 \end{array}\right] = [\mathbf{F}_2 | \mathbf{F}_1]$$

$$\begin{aligned} f &= \widehat{\mathbf{X}} \cdot \mathbf{V} \\ &= 0(\overline{x}_1\overline{x}_2) \,\widehat{\vee}\, 1(\overline{x}_1 x_2) \,\widehat{\vee}\, 1(x_1\overline{x}_2) \,\widehat{\vee}\, 2(x_1 x_2) \\ &= \overline{x}_1 x_2 \,\widehat{\vee}\, x_1\overline{x}_2 \,\widehat{\vee}\, 2x_1x_2, \end{aligned}$$

where

$$\begin{aligned} \widehat{\mathbf{X}} &= [\,\overline{x}_1 \; x_1\,] \otimes [\,\overline{x}_2 \; x_2\,] \\ &= [\,\overline{x}_1\overline{x}_2,\; \overline{x}_1 x_2,\; x_1\overline{x}_2,\; x_1x_2\,] \end{aligned}$$

Functions f_1 and f_2 are recovered from the least and the most significant bits of the word representation v_i, $i = 0, 1, 2, 3$,

$$\begin{aligned} f_1 &= 0(\overline{x}_1\overline{x}_2) \vee 0(\overline{x}_1 x_2) \vee 1(\overline{x}_1 x_2) \vee 1(x_1\overline{x}_2) \vee 0(x_1x_2) \\ &= \overline{x}_1 x_2 \vee x_1\overline{x}_2, \\ f_2 &= 0(\overline{x}_1\overline{x}_2) \vee 0(\overline{x}_1 x_2) \vee 0(\overline{x}_1 x_2) \vee 0(x_1\overline{x}_2) \vee 1(x_1x_2) \\ &= x_1x_2 \end{aligned}$$

FIGURE 4.16
Restoration of truth-vector of a two-output switching function f given by the vector of coefficients (Example 4.13).

In the above example, the coefficients $5 = 101$, $2 = 010$ and $6 = 110$ indicate the scheme of computing. For example, $\mathbf{f_1}$ includes the product $\mathbf{x_1\overline{x}_2}$ only (the least significant bit equal to 1 in coefficient 5 only).

In algebraic form this calculation is performed by analogy with $\mathbf{x}_1 = 4$, $\mathbf{x}_2 = 2$, $\mathbf{x}_3 = 6$, $\mathbf{x}_4 = 5$. For example, given $u = 0001$, $f = (5 \cdot 0 \cdot \overline{0}) \,\widehat{\vee}\, (2 \cdot 0) \,\widehat{\vee}\, (6 \cdot 1) = 6$, and $f_3, f_2, f_1 = 1, 1, 0$.

4.3.6 Word-level Shannon decision trees and diagrams

Sum-of-products are useful for word-level representation for three reasons. First of all, the manipulation of the word-level form is very simple. Second, restoration procedure is simple, compared to restoration by calculation of arithmetic word-level form. Finally, the linearity property can be utilized in order to decrease the computational cost.

A node in a word-level Shannon decision tree of a r-output switching function f corresponds to the word-level Shannon expansion with respect to a

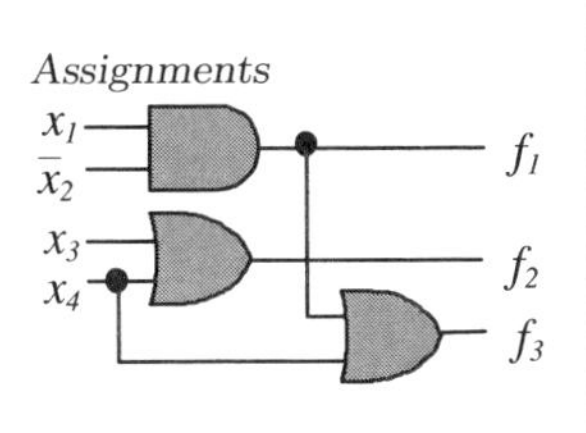

x_1	x_2	x_3	x_4	$f_3\ f_2\ f_1$
0	0	0	1	*Find*
0	1	1	0	*Find*
1	0	1	1	*Find*

Word-level expression:

$$f = 5x_1\overline{x}_2 \,\widehat{\vee}\, 2x_3 \,\widehat{\vee}\, 6x_4 = \begin{bmatrix} 5 \\ 2 \\ 6 \end{bmatrix} = \begin{bmatrix} 101 \\ 010 \\ 110 \end{bmatrix} = [f_3|f_2|f_1]$$

$$f_1 = x_1\overline{x}_2, \quad f_2 = x_3 \vee x_4, \quad f_3 = x_1\overline{x}_2 \vee x_4$$

Vectors of assignments:

$$\mathbf{x_1} = \begin{bmatrix} 0 \\ 0 \\ 1 \end{bmatrix} \quad \mathbf{x_2} = \begin{bmatrix} 0 \\ 1 \\ 0 \end{bmatrix}$$

$$\mathbf{x_3} = \begin{bmatrix} 0 \\ 1 \\ 1 \end{bmatrix} \quad \mathbf{x_4} = \begin{bmatrix} 1 \\ 0 \\ 1 \end{bmatrix}$$

Equations for assignments:

$$\mathbf{f} = 5\mathbf{x_1}\overline{\mathbf{x}}_\mathbf{2} \,\widehat{\vee}\, 2\mathbf{x_3} \,\widehat{\vee}\, 6\mathbf{x_4}$$
$$\mathbf{f_1} = \mathbf{x_1}\overline{\mathbf{x}}_\mathbf{2}, \quad \mathbf{f_2} = \mathbf{x_3} \vee \mathbf{x_4}, \quad \mathbf{f_3} = \mathbf{x_1}\overline{\mathbf{x}}_\mathbf{2} \vee \mathbf{x_4}$$

Outputs (vectors):

$$\mathbf{f_1} = \begin{bmatrix} 0 \\ 0 \\ 1 \end{bmatrix}\begin{bmatrix} 1 \\ 0 \\ 1 \end{bmatrix} = \begin{bmatrix} 0 \\ 0 \\ 1 \end{bmatrix}, \; \mathbf{f_2} = \begin{bmatrix} 0 \\ 1 \\ 1 \end{bmatrix} \vee \begin{bmatrix} 1 \\ 0 \\ 1 \end{bmatrix} = \begin{bmatrix} 1 \\ 1 \\ 1 \end{bmatrix}, \; \mathbf{f_3} = \begin{bmatrix} 0 \\ 0 \\ 1 \end{bmatrix}\begin{bmatrix} 1 \\ 0 \\ 1 \end{bmatrix} \vee \begin{bmatrix} 1 \\ 0 \\ 1 \end{bmatrix} = \begin{bmatrix} 1 \\ 0 \\ 1 \end{bmatrix}$$

FIGURE 4.17

Computing the three-output switching function $f = 5x_1\overline{x}_2 \,\widehat{\vee}\, 2x_3 \,\widehat{\vee}\, 6x_4$ by word-level assignments (Example 4.14).

variable x_i

$$f_j = f_j(x_i = 0) \vee x_i f_j(x_i = 1), \tag{4.19}$$

where $j = 1, 2, \ldots, r$, and $f_j = f|_{x_i=a}$ denotes the subfunctions of f_j that are derived from f after assigning the constant $a \in \{0, 1\}$ to the argument denoted by the variable x_i. The Shannon expansion of f is labeled as $\widehat{S}$ (Figure 7.18).

Word-level Shannon decision tree is a canonical representation of switching function f in graph-based form with structural properties similar to the Shannon decision tree for a single-output switching function.

Example 4.15 *A three-output switching function*

$$f = 7\overline{x}_1\overline{x}_2\overline{x}_3 \vee 3x_1x_2\overline{x}_3 \vee 6x_1x_2x_3$$

of three variables x_1, x_2 and x_3, where

$$f_1 = \overline{x}_1\overline{x}_2\overline{x}_3 \vee x_1x_2\overline{x}_3$$
$$f_2 = \overline{x}_1\overline{x}_2\overline{x}_3 \vee x_1x_2\overline{x}_3 \vee x_1x_2x_3$$
$$f_3 = \overline{x}_1\overline{x}_2\overline{x}_3 \vee x_1x_2x_3$$

can be represented by the word-level Shannon decision diagram shown in Figure 4.18a. The terms $7\overline{x}_1\overline{x}_2\overline{x}_3$, $3x_1x_2\overline{x}_3$ *and* $6x_1x_2x_3$ *are shown in Figure 4.18b. The bit-level representation would require three binary decision diagrams as shown in Figure 4.19.*

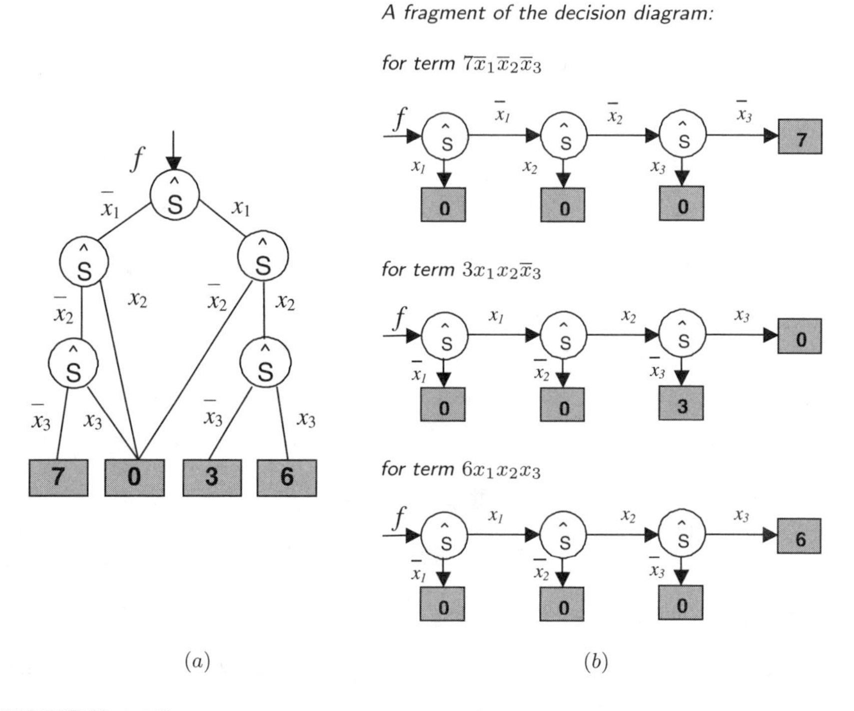

FIGURE 4.18
Word-level sum-of-products representation of the three-output switching function $7\overline{x}_1\overline{x}_2\overline{x}_3 \vee 3x_1x_2\overline{x}_3 \vee 6x_1x_2x_3$ (Example 4.15): Shannon tree (a) and the fragments of the tree corresponding to the terms of f (b)

4.4 Word-level Reed-Muller expressions

Given a multioutput switching function, it is possible to calculate Reed-Muller expression for the participating functions using bitwise on the whole word at once.

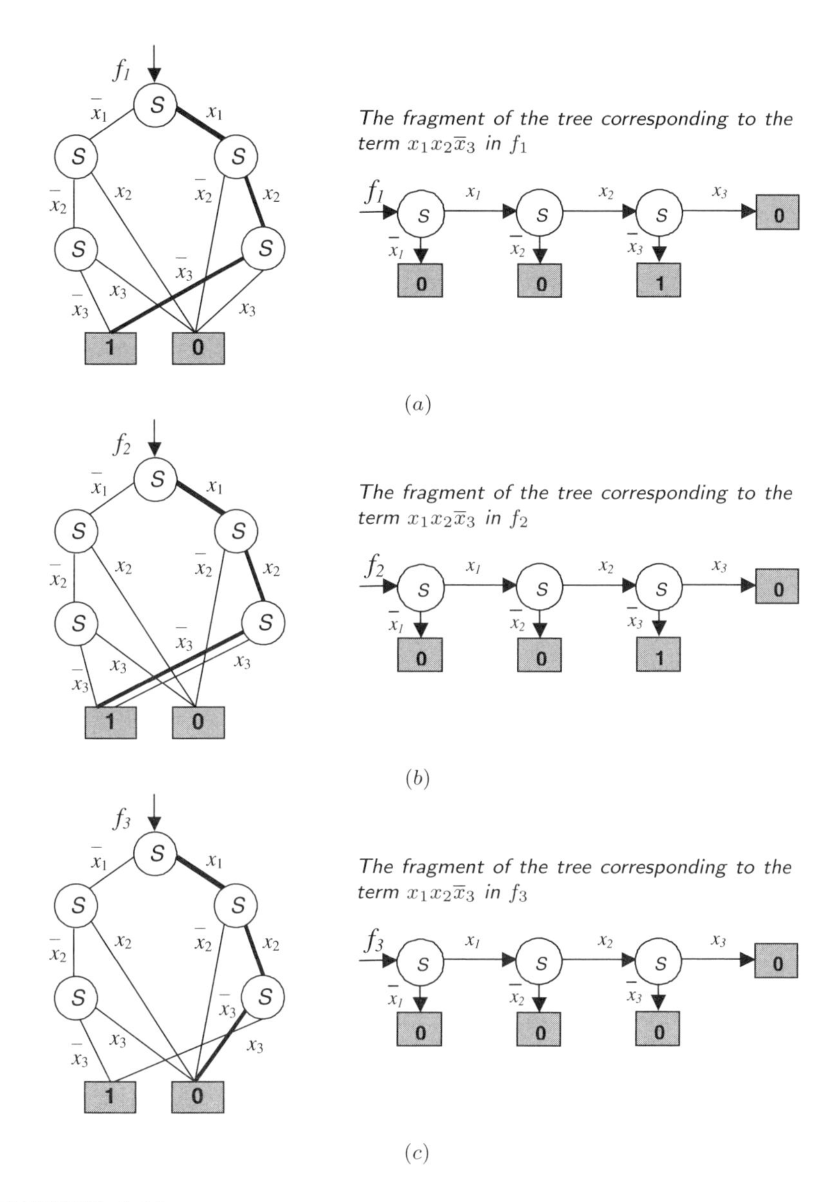

FIGURE 4.19
Shannon diagrams of the switching subfunctions f_1 (a), f_2 (b) and f_3 (c) of the three-output switching function $7\overline{x}_1\overline{x}_2\overline{x}_3 \vee 3x_1x_2\overline{x}_3 \vee 6x_1x_2x_3$, and the paths corresponding to the term $x_1x_2\overline{x}_3$ in each of the diagrams (Example 4.15).

4.4.1 General form

Given a r-output switching function f of n variables, the word-level Reed-Muller expression is the bitwise sum of Reed-Muller expressions of f_j, $j = 1, \ldots, r$,

$$f = 2^{r-1} f_r \mathbin{\widehat{\oplus}} \cdots \mathbin{\widehat{\oplus}} 2^1 f_2 \mathbin{\widehat{\oplus}} 2^0 f_1. \tag{4.20}$$

where $\widehat{\oplus}$ denotes a bitwise operation. Equation 4.20 can be rewritten in the form

$$f = \widehat{\bigoplus_{i=0}^{2^n-1}} w_i \cdot \underbrace{(x_1^{i_1} \cdots x_n^{i_n})}_{i-th\ product}, \tag{4.21}$$

where coefficient w_i is the positive integer number, i_j is j-th bit in the binary representation of the index $i = i_1 i_2 \ldots i_n$, and $x_j^{i_j}$ is defined as

$$x_j^{i_j} = \begin{cases} 1, & i_j = 0; \\ x_j, & i_j = 1. \end{cases} \tag{4.22}$$

Example 4.16 *An arbitrary r-output switching function of two variables ($n = 2$) is represented by the word-level EXOR expression, accordingly Equation 4.21 and Equation 4.22*

$$f = w_0 \mathbin{\widehat{\oplus}} w_1 x_2 \mathbin{\widehat{\oplus}} w_2 x_1 \mathbin{\widehat{\oplus}} w_3 x_1 x_2.$$

Figure 4.20 illustrates forming of this expression; the nodes in the graph implement AND function.

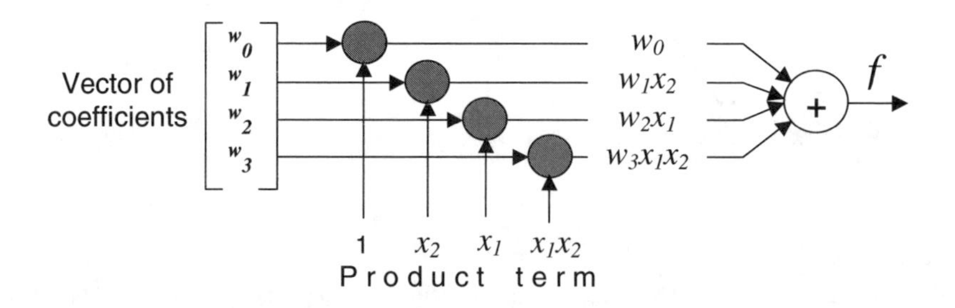

FIGURE 4.20
Deriving the Reed-Muller word-level expression for a r-output switching function of two variables (Example 4.16).

4.4.2 Masking operator

The *masking operator* is used to recover a switching function from the word-level representation. However, similar to word-level sum-of-products, in the word-level Reed-Muller representation, the masking operator degenerates to lexicographical order.

4.4.3 Computing the coefficients

Given a word-level truth vector $\mathbf{F} = [f(0)\ f(1) \dots f(2^n-1)]^T$, the vector of coefficients $\mathbf{W} = [w_0\ w_1 \dots w_{2^n-1}]^T$ is derived by the matrix equation over AND and EXOR operations

$$\mathbf{W} = \mathbf{R}_{2^n} \cdot \mathbf{F}, \tag{4.23}$$

where the $2^n \times 2^n$ - matrix $\mathbf{R}_{2^n}$ is formed by the Kronecker product

$$\mathbf{R}_{2^n} = \bigotimes_{j=1}^{n} \mathbf{R}_{2^j}, \qquad \mathbf{R}_{2^1} = \begin{bmatrix} 1 & 0 \\ 1 & 1 \end{bmatrix}. \tag{4.24}$$

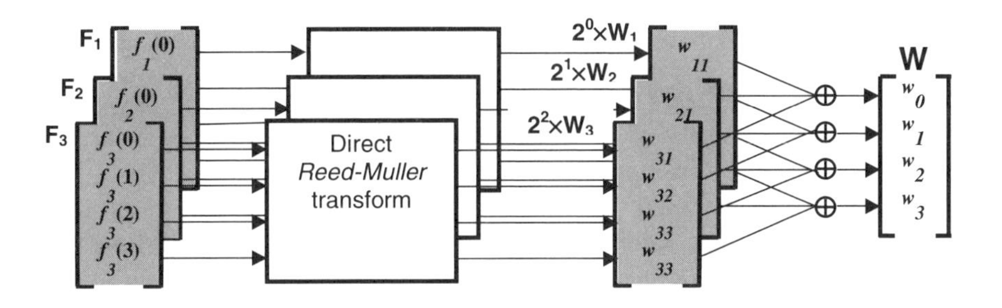

FIGURE 4.21
Direct word-level Reed-Muller transform for a two-output switching function of two variables.

The general scheme of computing is shown in Figure 4.21a.

Example 4.17 *Computing the word-level Reed-Muller coefficients by Equation 4.23 and Equation 4.24 for the two-output switching function* $f_1 = x_1 \oplus x_2$, $f_2 = \overline{x}_1 \oplus x_2$ *given by the truth-vectors* $\mathbf{F_1}$ *and* $\mathbf{F_2}$ *correspondingly is illustrated by Figure 4.22. The truth vector* $\mathbf{F}$ *is defined as*

$$\mathbf{F} = [\mathbf{F_2}|\mathbf{F_1}] = \begin{bmatrix} 1 & 0 \\ 0 & 1 \\ 0 & 1 \\ 1 & 0 \end{bmatrix} = \begin{bmatrix} 2 \\ 1 \\ 1 \\ 2 \end{bmatrix}.$$

The calculation using Equation 4.23 results to

$$\mathbf{W} = [\mathbf{W_2}|\mathbf{W_1}] = \begin{bmatrix} 1 & 0 \\ 1 & 1 \\ 1 & 1 \\ 0 & 0 \end{bmatrix} = \begin{bmatrix} 2 \\ 3 \\ 3 \\ 0 \end{bmatrix}.$$

Notice, that the same result can be obtained by the algebraic Equation 4.20 given $f_1 = x_1 \oplus x_2$ *and* $f_2 = \overline{x}_1 \oplus x_2 = 1 \oplus x_1 \oplus x_2$*:*

$$f = 2^1 f_2 \overset{\frown}{\oplus} 2^0 f_1 = 2^1(1 \oplus x_1 \oplus x_2) \overset{\frown}{\oplus} 2^0(x_1 \oplus x_2) = 2 \overset{\frown}{\oplus} 3x_2 \overset{\frown}{\oplus} 3x_1.$$

Two-output function

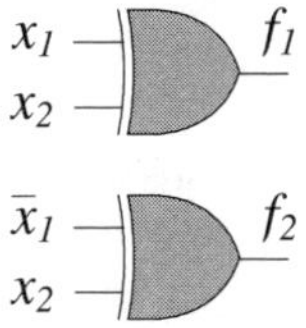

x_1, x_2 → f_1

$\overline{x}_1$, x_2 → f_2

$f_1 = x_1 \oplus x_2$
$f_2 = 1 \oplus x_1 \oplus x_2$

Calculation of vectors of coefficients

$$\mathbf{W_1} = \mathbf{R}_{2^2} \cdot \mathbf{F_1} = \left[\begin{array}{cc|cc} 1 & & & \\ 1 & 1 & & \\ \hline 1 & & 1 & \\ 1 & 1 & 1 & 1 \end{array}\right] \begin{bmatrix} 0 \\ 1 \\ 1 \\ 0 \end{bmatrix} = \begin{bmatrix} 0 \\ 1 \\ 1 \\ 0 \end{bmatrix} \quad (mod\ 2)$$

$$\mathbf{W_2} = \mathbf{R}_{2^2} \cdot \mathbf{F_2} = \left[\begin{array}{cc|cc} 1 & & & \\ 1 & 1 & & \\ \hline 1 & & 1 & \\ 1 & 1 & 1 & 1 \end{array}\right] \begin{bmatrix} 1 \\ 0 \\ 0 \\ 1 \end{bmatrix} = \begin{bmatrix} 1 \\ 1 \\ 1 \\ 0 \end{bmatrix} \quad (mod\ 2)$$

Vector of Reed-Muller coefficients of two-output function

$\mathbf{W} = [\mathbf{W_2}|\mathbf{W_1}] = [2\ 3\ 3\ 0]^T$

Word-level Reed-Muller expression

$f = 2 \overset{\frown}{\oplus} 3x_2 \overset{\frown}{\oplus} 3x_1$

FIGURE 4.22
Computing the word-level Reed-Muller expression of a two-output switching function (circuit) (Example 4.17).

4.4.4 Restoration

Restoration from the algebraic and matrix equations have a number of specific features.

Restoration from the algebraic equation. Let us consider the function from Example 4.16, $w_0 = 1,\ w_1 = 2,\ w_2 = 3,\ w_3 = 4$, that is

$$f = 1 \overset{\frown}{\oplus} 2x_2 \overset{\frown}{\oplus} 3x_1 \overset{\frown}{\oplus} 4x_1x_2.$$

It follows from this equation that function f generates three switching functions f_1, f_2, f_3 because the coefficient $w_3 = 4$ is represented by three bits. Therefore,

$$C = \begin{bmatrix} w_0 \\ w_1 \\ w_2 \\ w_3 \end{bmatrix} = \begin{bmatrix} 1 \\ 2 \\ 3 \\ 4 \end{bmatrix} = \begin{array}{l} w_0 \ \{1\} \\ w_1 \ \{x_2\} \\ w_2 \ \{x_1\} \\ w_3 \ \{x_1 x_2\} \end{array} \begin{array}{c} \begin{array}{ccc} f_3 & f_2 & f_1 \end{array} \\ \begin{bmatrix} 0 & 0 & 1 \\ 0 & 1 & 0 \\ 0 & 1 & 1 \\ 1 & 0 & 0 \end{bmatrix} \end{array}$$

$f_1 = 1 \oplus x_1, \ f_2 = x_2 \oplus x_1, \ f_3 = x_1 x_2$, i.e.,

- ▶ f_1 includes the products 1 and x_1 because the least significant bit of the coefficients $w_0 = 1$ and $w_2 = 3$ is equal to 1. Hence, $f_1 = 1 \oplus x_1$.
- ▶ f_2 includes the products x_2 and x_1 because the second bit of the coefficients $w_1 = 2$ and $w_2 = 3$ is 1. Hence, $f_2 = x_2 \oplus x_1$.
- ▶ f_3 includes the product $x_1 x_2$ because the most significant bit of the coefficient $w_3 = 4$ is 1. Hence, $f_3 = x_1 x_2$.

Restoration from the matrix equation. The following matrix equation over AND and Reed-Muller operations restores the truth-vector $\mathbf{F}_t$ from the vector of coefficients $\mathbf{W}_t, \ t = 1, 2, \ldots, r$:

$$\mathbf{F}_t = \mathbf{R}_{2^n}^{-1} \cdot \mathbf{W}_t \ \ (mod \ 2), \tag{4.25}$$

where $\mathbf{R}_{2^n}^{-1} = \mathbf{R}_{2^n}$.

Example 4.18 *Restore the truth-vector* $\mathbf{F}_t, \ t = 1, 2,$ *of a function* f *given by the vector of coefficients* $\mathbf{W} = [2\ 3\ 3\ 0]^T$ *(Example 4.17)*

$$\mathbf{F}_1 = \mathbf{R}_{2^2}^{-1} \cdot \mathbf{W}_1 = \left[\begin{array}{cc|cc} 1 & & & \\ 1 & 1 & & \\ \hline 1 & & 1 & \\ 1 & 1 & 1 & 1 \end{array}\right] \begin{bmatrix} 0 \\ 1 \\ 1 \\ 0 \end{bmatrix} = \begin{bmatrix} 0 \\ 1 \\ 1 \\ 0 \end{bmatrix}.$$

By analogy, $\mathbf{F}_2 = [1\ 0\ 0\ 1]^T$. *Hence,*

$$f_1 = \widehat{\mathbf{X}} \cdot \mathbf{F}_1 = \widehat{\mathbf{X}} \cdot [0\ 1\ 1\ 0]^T = x_1 \oplus x_2$$
$$f_2 = \widehat{\mathbf{X}} \cdot \mathbf{F}_2 = \widehat{\mathbf{X}} \cdot [1\ 0\ 0\ 1]^T = 1 \oplus x_1 \oplus x_2.$$

4.4.5 Computing for a word-level set of assignments

This section is devoted to implementation of word assignments in word-level Reed-Muller expression.

Example 4.19 *Calculate the outputs of the circuit (Figure 4.23) given the word-level arithmetic expression*

$$f = 5x_1\overline{x}_2 \ \widehat{\oplus} \ 2x_3\overline{x}_4 \ \widehat{\oplus} \ 6x_4,$$

and the set of assignments $u \in \{1000, 0010, 0001\}$. *First, the vectors of assignments* $\mathbf{x}_1, \mathbf{x}_2, \mathbf{x}_3$, *and* $\mathbf{x}_4$ *are derived from the truth-table. Second, the bitwise calculations are performed over these vectors according to the word-level sum-of-products.*

In this example, the coefficients $5 = 101$, $2 = 010$ and $6 = 110$ indicate that:

- $\mathbf{x}_1\overline{\mathbf{x}}_2 \in \mathbf{f}_1$ (the least significant bit is equal to 1 in the coefficient 5),
- $\mathbf{x}_3\overline{\mathbf{x}}_4 \in \mathbf{f}_2$ and $\mathbf{x}_4 \in \mathbf{f}_2$ (the second bit is equal to 1 in the coefficient 2 and coefficient 6),
- $\mathbf{x}_1\overline{\mathbf{x}}_2 \in \mathbf{f}_3$ and $\mathbf{x}_4 \in \mathbf{f}_3$ (the most significant bit is equal to 1 in the coefficients 4 and 6).

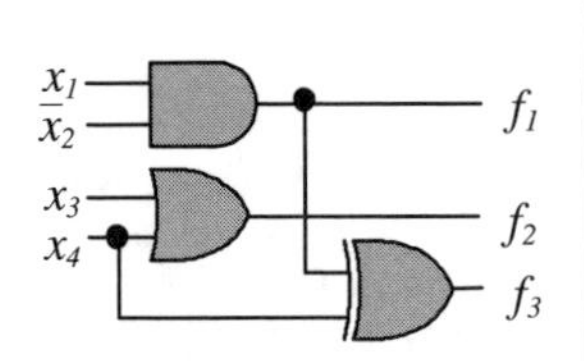

Assignments:

x_1	x_2	x_3	x_4	$f_3\ f_2\ f_1$
0	0	0	1	*Find*
0	1	1	0	*Find*
1	0	1	1	*Find*

Word-level Reed-Muller expression:

$$f = 5x_1\overline{x}_2 \,\widehat{\oplus}\, 2x_3\overline{x}_4 \,\widehat{\oplus}\, 6x_4 = \begin{bmatrix} 5 \\ 2 \\ 4 \end{bmatrix} = \begin{bmatrix} 101 \\ 010 \\ 100 \end{bmatrix} = [f_3|f_2|f_1]$$

Outputs:

$f_1 = x_1\overline{x}_2,\quad f_2 = x_3 \vee x_4,\quad f_3 = x_1\overline{x}_2 \oplus x_4$

Vectors of assignments:

$$\mathbf{x}_1 = \begin{bmatrix} 0 \\ 0 \\ 1 \end{bmatrix}, \quad \mathbf{x}_2 = \begin{bmatrix} 0 \\ 1 \\ 0 \end{bmatrix}, \quad \mathbf{x}_3 = \begin{bmatrix} 0 \\ 1 \\ 1 \end{bmatrix}, \quad \mathbf{x}_4 = \begin{bmatrix} 1 \\ 0 \\ 1 \end{bmatrix}$$

Equation for assignments:

$$\mathbf{f} = 5\mathbf{x}_1\overline{\mathbf{x}}_2 \,\widehat{\oplus}\, 2\mathbf{x}_3\overline{\mathbf{x}}_4 \,\widehat{\oplus}\, 6\mathbf{x}_4$$

Output vectors:

$$\mathbf{f}_1 = \mathbf{x}_1\overline{\mathbf{x}}_2 = \begin{bmatrix} 0 \\ 0 \\ 1 \end{bmatrix}\begin{bmatrix} 1 \\ 0 \\ 1 \end{bmatrix} = \begin{bmatrix} 0 \\ 0 \\ 1 \end{bmatrix}, \quad \mathbf{f}_2 = \mathbf{x}_3\overline{\mathbf{x}}_4 \oplus \mathbf{x}_4 = \begin{bmatrix} 0 \\ 1 \\ 1 \end{bmatrix}\begin{bmatrix} 0 \\ 1 \\ 0 \end{bmatrix} \oplus \begin{bmatrix} 1 \\ 0 \\ 1 \end{bmatrix} = \begin{bmatrix} 1 \\ 1 \\ 1 \end{bmatrix},$$

$$\mathbf{f}_3 = \mathbf{x}_1\overline{\mathbf{x}}_2 \oplus \mathbf{x}_4 = \begin{bmatrix} 0 \\ 0 \\ 1 \end{bmatrix}\begin{bmatrix} 1 \\ 0 \\ 1 \end{bmatrix} \oplus \begin{bmatrix} 1 \\ 0 \\ 1 \end{bmatrix} = \begin{bmatrix} 1 \\ 0 \\ 0 \end{bmatrix}$$

FIGURE 4.23

Computing the three-output switching function $f = 5x_1\overline{x}_2 \,\widehat{\oplus}\, 2x_3\overline{x}_4 \,\widehat{\oplus}\, 6x_4$ by word-level assignments (Example 4.19).

4.4.6 Word-level Davio decision trees and diagrams

At the node of a word-level Davio decision tree, the *positive* Davio expansion of each function f_j, $j = 1, 2, \ldots, r$, is implemented in parallel:

$$f = f_j(x_i = 0) \oplus x_i f_j(x_i = 1). \tag{4.26}$$

Word-level Davio decomposition is labeled as $\widehat{pD}$. The *negative* Davio expansion of f_j is specified as in Chapter 3.

Note that decision trees and diagrams can include nodes corresponding to pD, nD and S expansions if the Reed-Muller expression contain variables entering uncomplemented, complemented or in both forms.

Example 4.20 *Represent the word-level Davio decision tree of a three-output switching function given in Figure 4.23. The final result, Reed-Muller expression, is directly mapped into the decision tree in which $\widehat{pD}$ is applied to the uncomplemented variables x_1 and x_3, $\widehat{nD}$is applied to complemented variable x_2 and $\widehat{S}$ is applied to x_4 which enters in the expression in both complemented and uncomplemented forms (Figure 4.24). The decision trees and the fragments for one term are given for subfunctions f_1, f_2 and f_3 in Figure 4.25a, b, and c accordingly.*

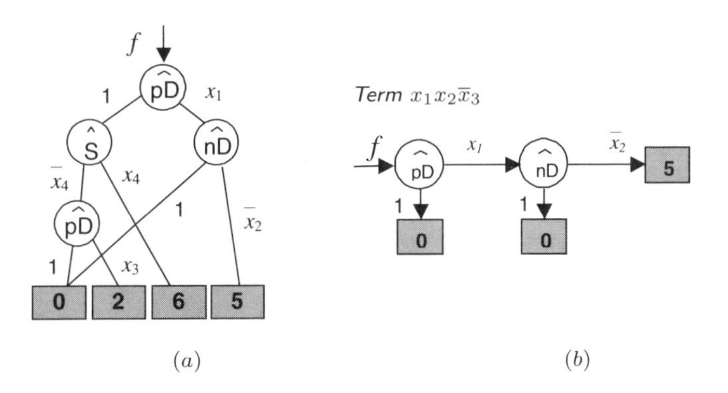

FIGURE 4.24

Word-level Davio decision tree of the function $f = 5x_1\overline{x}_2 \widehat{\oplus} 2x_3\overline{x}_4 \widehat{\oplus} 6x_4$ (a) and a fragment of the tree for the term $5x_1\overline{x}_2$ (b) (Example 4.20).

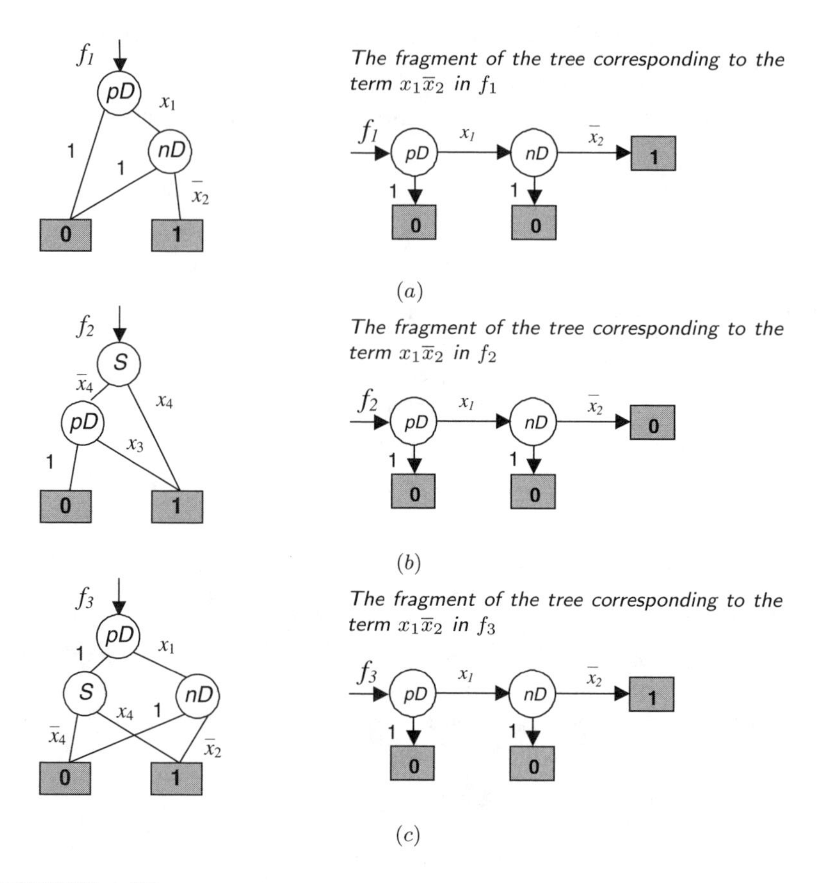

FIGURE 4.25
Decision diagrams for subfunctions f_1 (a), f_2 (b), and f_3 (c) of a tree-output function $f = 5x_1\overline{x}_2 \overset{\frown}{\oplus} 2x_3\overline{x}_4 \overset{\frown}{\oplus} 6x_4$ (Example 4.20).

4.5 Summary

In this chapter the advanced topics in word-level representation of switching functions are introduced. Word-level data structures are the integer counterparts of bit-level data structures. In contrast to bit-level representations with terminal nodes values 0 and 1, in the word-level expressions, decision trees and diagrams, integers or in general complex numbers are assigned to the terminal nodes. Word-level data structures provide more compact representation, and manipulation needs less time.

1. Word-level representation is useful for representation of multioutput switching functions. To compose the switching functions in a word, different forms of representation can be used. In this chapter, the word-level technique is introduced by arithmetic, sum-of-products, and Reed-Muller expressions. The result is a word-level representation of a multioutput switching function. The word-level expression can be in algebraic form, matrix form, or in graphical form (decision trees and diagrams).

2. In contrast to Boolean expression that takes values 1 and 0 through computing, the word-level expression takes integer values if assign the values of binary variables. To recover the values of switching functions a simple scheme of decoding is used (masking operator).

3. Word-level expressions have a number of common properties. However, they differ in implementation, in particular:

- ▶ In the word-level arithmetic representation, the coefficients are integer numbers (both positive and negative). Word-level arithmetic representation is useful when the linear properties of truth-vectors and vectors of coefficients are important. The drawbacks are the large coefficients and arithmetic operations in the nodes of decision tree and diagram. However, in Chapter 8, it will be shown that it is possible to alleviate this problem by the linearization technique and special encoding.
- ▶ Word-level sum-of-products representation is an attractive form of description because of optimal coefficients. This is the main advantage of this word-level form. The coefficients are positive integer numbers, and can be calculated for a large functions. Additional benefits follow from the structure of the coefficients: it is easy to calculate the distribution of the products between the output functions.
- ▶ Word-level Reed-Muller representation is useful for description of a multiple-output function given in the Reed-Muller form. Grouping and restoration procedures, and the properties, are the same as in the word-level sum-of-products expression. For example, from the binary codes of the coefficients (positive integer numbers), it is easy to represent the output functions. These functions are represented in Reed-Muller form.

4. Representation of word-level expressions by decision trees and decision diagrams provides the possibility to manipulate large multioutput functions. In the word-level decision trees and decision diagrams, the constant (terminal) nodes are integers. Different types of a word-level decision diagrams are developed for the optimization of switching functions (see "Further Reading" Section).

4.6 Problems

Problem 4.1 Represent the two-output switching function given below by a word-level arithmetic expression (follow Example 4.6)

$(a)\ f_1 = \overline{x}_1 x_2 \vee x_1 x_2 \overline{x}_3 \vee x_3;\ f_2 = x_1 \vee x_2 \vee x_3$
$(b)\ f_1 = x_1 \oplus x_1 x_2 x_3 \oplus x_1 x_3;\ f_2 = x_1 \oplus x_2 \oplus x_3$
$(c)\ f_1 = x_1 \vee x_2 \vee x_3;\ f_2 = x_1 \oplus x_2 \oplus x_3$
$(d)\ f_1 = x_1 \vee \overline{x}_2 f_2 = x_1 \oplus x_2$

Problem 4.2 Consider the word-level arithmetic expressions given below.

(A) Restore the switching function (follow Example 4.7)
(B) Represent the arithmetic expressions in polarity $c = 0$ (follow Example 4.8)
(C) Calculate the arithmetic expressions for assignments $000, 011, 110$ (follow Example 4.10)
(D) Represent the arithmetic expressions by the decision tree and decision diagram

$(a)\ f = 5 + 2x_1 x_2 - x_2 x_3 - x_3$
$(b)\ f = 3 + x_1 + x_2 x_3 - x_1 x_3$
$(c)\ f = 4 - x_1 - \overline{x}_2 - x_3$

Problem 4.3 Represent the switching function given below by a word-level sum-of-products (follow Example 4.12)

$(a)\ f_1 = \overline{x}_1 \vee x_2 \vee \overline{x}_3;\ f_2 = x_1 \vee x_2 \vee x_3$
$(b)\ f_1 = x_1 \overline{x}_2;\ f_2 = x_1 x_2 x_3$
$(c)\ f_1 = x_1 \vee x_2 \vee x_3;\ f_2 = x_1 \vee \overline{x}_2$

Problem 4.4 Below the word-level sum-of-products are given.

(A) Restore the switching functions (follow Example 4.13)
(B) Calculate the word-level sum-of-products expression for assignments $000, 011, 1$ (follow Example 4.14)
(C) Represent the word-level sum-of-products expressions by the decision tree and decision diagram

$(a)\ x_1 \widehat{\vee} 3x_2 \widehat{\vee} 2\overline{x}_1 \widehat{\vee} 3\overline{x}_3$
$(b)\ 2x_1 \widehat{\vee} 2x_2 \widehat{\vee} \overline{x}_1 \widehat{\vee} \overline{x}_2 \widehat{\vee} 3\overline{x}_3$
$(c)\ 7x_1 \widehat{\vee} 4x_2 \widehat{\vee} x_3 \widehat{\vee} 2\overline{x}_3$

Problem 4.5 Represent the switching function given below by a word-level Reed-Muller expression (follow Example 4.17)

(a) $f_1 = \overline{x}_1 \oplus x_2 \oplus x_3$; $f_2 = x_1 \oplus \overline{x}_2 \oplus x_3$
(b) $f_1 = x_1 \oplus x_2 \oplus x_3$; $f_2 = x_1 \oplus x_3$; $f_3 = \overline{x}_1 \oplus \overline{x}_2$
(c) $f_1 = \overline{x}_1 \vee \overline{x}_2 \vee x_3$; $f_2 = x_1 \oplus x_2$; $f_3 = x_3$

Problem 4.6 Below the word-level Reed-Muller expressions are given.

(A) Restore the switching functions (follow Example 4.18)
(B) Calculate the word-level Reed-Muller expressions for assignments $001, 01, 111$ (follow Example 4.19)
(C) Represent the word-level Reed-Muller expression by the decision tree and decision diagram (follow Example 4.20)

(a) $5x_1 \widehat{\oplus} 3x_2 \widehat{\oplus} 7x_3 \widehat{\oplus} \overline{x}_1$
(b) $3x_1 \widehat{\oplus} 2x_2 \widehat{\oplus} \overline{x}_3 \widehat{\oplus} 3\overline{x}_3$
(c) $x_1 \widehat{\oplus} 6x_2 \widehat{\oplus} x_3 \widehat{\oplus} 2\overline{x}_1 \widehat{\oplus} \overline{x}_2 \widehat{\oplus} \overline{x}_3$

Problem 4.7 Consider a two-level combinational circuit shown in Figure 6.10. Find:

(a) Linear arithmetic expression for each level of the circuit
(b) Sum-of-products
(c) Derive decision diagram
(d) Derive a set of word-level linear decision diagrams

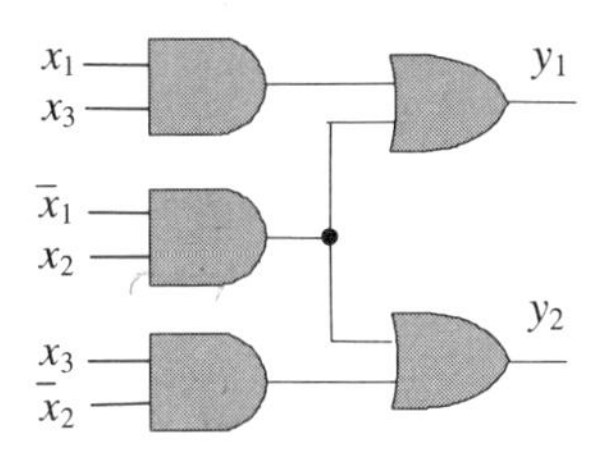

FIGURE 4.26
Two-output logic circuit for Problem 4.7.

Problem 4.8 Compare shared binary decision diagram (in terms of total nodes) that represent functions f_1, f_2 and word-level linear decision diagram

4.7 Further reading

Arithmetic word-level expressions and decision diagrams. Formal aspects of word-level arithmetic expressions are given in [3, 4, 5] and [6]. The technique of application of Word-level decision diagrams has been introduced in the references of Chapter 3. Spectral theory provides quite simple formal methods of design of word-level structures, for example, word-level Walsh and Haar representation.

Edge-valued binary decision diagram were introduced by Lai et al. [2] to improve efficiency of representation of multioutput switching functions. This is a class of decision diagrams with attributed edges. There is one constant node in EVBDDs, and its value is set to zero irrespective of the represented switching function. In contrast to BDDs, the algorithms for assigning switching function to the decision trees, and for determining this function from decision tree are related to the edges and weighting coefficients at the edges (not to the nodes). EVBDDs represent multioutput switching functions in the form of algebraic polynomials. The node of an EVBDD implements the particular case of arithmetic Davio expansion. The EVBDDs with a fixed order of variables provides a canonical representation of a switching function. Extensions to EVBDD for multiple-valued functions have been developed too. Spectral theory provides quite simple formal methods of design of EVBDDs.

Binary moment decision diagrams (BMDs) [1]. These are different from EVBDDs in the order of the decompositions. In BMDs, weights at the edges are factors in the arithmetic transform coefficients.

4.8 References

[1] Bryant R, and Chen Y. Verification of arithmetic functions using binary moment diagrams. In *Proceedings of Design Automation Conference*, pp. 535–541, 1995.

[2] Lai Y, Pedram M, and Vrudhula S. EVBDD-based algorithms for integer linear programming, spectral transformation, and function decomposition. *IEEE Transactions on Computer Aided Design of Integrated Circuits and Systems*, 13(8):959–975, 1994.

[3] Malyugin VD. Representation of Boolean functions by arithmetical polynomials. *Automation and Remote Control*, Kluwer/Plenum Publishers, 43(4):496–504, 1982.

[4] Minato S. *Binary Decision Diagrams and Applications for VLSI Design.* Kluwer, Dordrecht, 1996.

[5] Stanković RS, Sasao T, and Moraga C. Spectral transform decision

diagrams. In Sasao T. and Fujita M., Eds., *Representations of Discrete Functions*, Kluwer, Dordrecht, 1996, pp. 55–92.

[6] Stanković RS, Stanković M, Astola J, and Egiazarian K. Bit-level and word-level polynomial expressions for functions in Fibonacci interconnection topologies. In *Proceedings IEEE 31st International Symposium on Multiple-Valued Logic*, pp. 305–310, 2001.

[7] Malyugin VD. Realization of Boolean function's corteges by means of linear arithmetical polynomial. *Automation and Remote Control*, Kluwer/Plenum Publishers, 45(2):239–245, 1984.

[8] Shmerko VP. Synthesis of arithmetic forms of Boolean functions using the Fouier transform. *Automation and Remote Control*, Kluwer/Plenum Publishers, 50(5):684–691, 1989.

5

Nanospace and Hypercube-Like Data Structures

Logic design of nanoICs in spatial dimensions is based on selected methods of advanced logic design, and appropriate spatial topologies. This chapter focuses on properties of hypercube data structure, and hypercube-like topology, $\mathcal{N}$-hypercube, obtained by extension of the hypercube.

To introduce the main topic of this chapter, that is, the $\mathcal{N}$-hypercube data structure, classical hypercube structure is given. Classical hypercube structure is the base for $\mathcal{N}$-hypercube design and inherits most of the properties of classical hypercubes. Traditionally, hypercube topology is used in

- ► Logic design for switching function manipulation (minimization, representation),
- ► Communication problems for traffic representation and optimization, and
- ► Multiprocessor systems design for optimization of parallel computing.

In the first approach, each variable of a switching function is associated with one dimension in hyperspace. Manipulation of the function is based on a special encoding of the vertices and edges in the hypercube. The hypercube is used as an effective algebraic model of a switching function. In the second approach that is used in communication problems and multiprocessor systems design, the hypercube is the computational model. To design this model, a decision tree or a decision diagram must be constructed and embedded into a hypercube. In this approach the hypercube is utilized as a topological structure for computing in 3-D space.

The problem of assembling a hypercube from a number of topological components is introduced. This approach is especially useful when hypercube and hypercube-like structures carry information about switching functions. An alternative approach based on embedding graphs in hypercubes is also discussed.

Based on the above, the new, hypercube-like topology called the $\mathcal{N}$-hypercube is introduced. There are many reasons to develop $\mathcal{N}$-hypercube-based topologies:

- ► $\mathcal{N}$-hypercubes ideally reflect all properties of decision trees and decision diagrams, popular in advanced logic design data structure, enhancing them to more than two dimensions.

- $\mathcal{N}$-hypercubes inherit the classical hypercube's properties.
- $\mathcal{N}$-hypercubes satisfy a number of nanotechnology requirements.

Several of features distinguish the $\mathcal{N}$-hypercube from a hypercube: in particular, additional nodes, including a unique node called the root. Thanks to this, a distribution of information flows that is suitable from the point of view of technology can be achieved. Moreover, for the $\mathcal{N}$-hypercube distinguished configurations can be obtained by the rotation.

The structure of this chapter is as follows. A brief overview of various spatial configurations is given in Section 5.1. A classical hypercube data structure is introduced in Section 5.2. Section 5.3 is the brief introduction to assembling hypercubes. In Section 5.4, the basic properties of the $\mathcal{N}$-hypercube are discussed. Section 5.5 focuses on particular properties of $\mathcal{N}$-hypercube: degree of freedom and rotation and their relations with polarity and order of variables. The coordinate description of $\mathcal{N}$-hypercubes is given in Section 5.6. Sections 5.7, 5.8, and 5.9 represent the basics of technique for design of $\mathcal{N}$-hypercube structure. The measures in hypercube-like structures are discussed in Section 5.10. After the Summary (Section 5.11), we provide Problems (Section 5.12) and recommendation for "Further Reading" (Section 5.13).

5.1 Spatial structures

Several network topologies have been developed to fit different styles of computation. However, most of the known topologies for massive parallel computing have not been considered relevant to spatial logic design.

5.1.1 Requirement for representation in spatial dimensions

Ideally, a spatial network topology in space intended for switching function representation should possess the following characteristics:

- Minimal degree,
- Ability to extend the size of structure with minimal changes to the existing configuration,
- Ability to increase reliability and fault tolerance with minimal changes to the existing configuration,
- Good embedding capabilities,
- Flexibility of design methods, and
- Flexibility of technology.

5.1.2 Topologies

Based on the above criteria, a number of topologies can be considered relevant to the problems of spatial logic design, namely:

- Hypercube topology,
- Cube-connected cycles known as *CCC*-topology (hypercube),
- Pyramid topology,
- X-hypercube topology,
- Hybrid topologies, and
- Specific topologies (hyper-Peterson, hyper-star, Fibonacci cube, etc.)

Hypercube topology (Figure 5.1a) has received considerable attention in classical logic design due mainly to its ability to interpret logic formulas and logic computation (small diameter, regularity, high connectivity, symmetry). Hypercube-based structures are at the forefront of massive parallel computation because of the unique characteristics of hypercubes (fault tolerance, ability to efficiently permit the embedding of various topologies, such as lattices and trees).

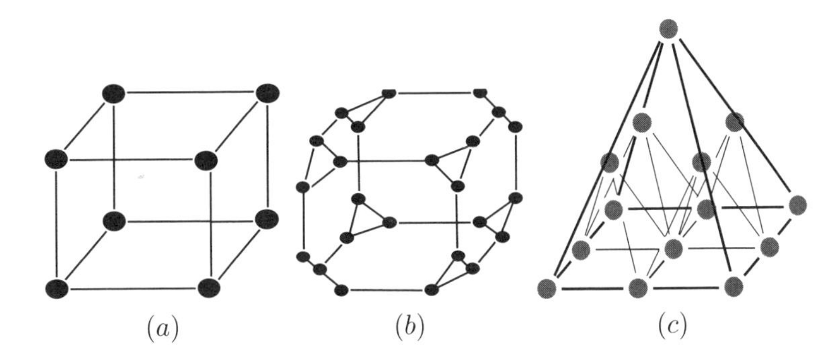

FIGURE 5.1
Spatial configurations: hypercube (a), *CCC*-hypercube (b), and pyramid (c).

The binary $n-$hypercube is a special case of the family of k-ary $n-$hypercubes, which are hypercubes with n dimensions and k nodes in each dimension. The total number of nodes in such a hypercube is $N = k^n$. Parallel computers with direct-connect topologies are often based on k-ary $n-$cubes or isomorphic structures such as rings, meshes, tori, and direct binary n-cubes. A node in a k-ary $n-$hypercube can be represented by an n-digit address $d_{n-1} \ldots d_0$ in radix k. The i-th digit, d_i, of the address represents the node's position in the i-th dimension, where $0 \leq d_i \leq k - 1$. Two nodes with addresses

$(d_{n-1}, d_{n-2}, \ldots, d_0)$ and $(d'_{n-1}, d'_{n-2}, \ldots, d'_0)$ are neighbors in the ith dimension if, and only if, either

$$d_i = (d'_i + 1) \; mod \; k, \qquad \text{or}$$
$$d_j = d'_j, \quad \forall j \neq i.$$

The CCC-hypercube is created from a hypercube by replacing each node with a cycle of s nodes (Figure 5.1b). It hence increases the total number of nodes from 2^n to $s \cdot 2^n$ and preserves all features of the hypercube. The CCC-hypercube is closely relevant to the butterfly network. As has been shown in the previous sections, "butterfly" flowgraphs are the nature of most transforms of switching functions in matrix form.

Pyramid topology (Figure 5.1c) is suitable for many computations based on the principle of hierarchical control, for example, decision trees and decision diagrams. An arbitrary large pyramid can be embedded into the hypercube with a minimal load factor. The dilation is two and the congestion is $\Theta(\sqrt{l})$ where l is the load factor. Pyramid P_n has nodes on levels $0, 1, \ldots, n$. The number of nodes on level i is 4^i. So, the number of nodes of P_n is $(4^{n+1}/3)$. The unique node on level 0 is called the *root* of pyramid. The subgraph of P_n induced by the nodes on level i is isomorphic to mesh $2^i \times 2^i$ The subgraph of P_n induced by the edges connecting different levels is isomorphic to a 4^n-leaf quad-tree. This structure is very flexible for extension. Pyramid topology is relevant also to fractal-based computation that effective for symmetric functions and is used in digital signal processing, image processing and pattern recognition.

Hybrid topology combines different structures. This topology is effective when local computation is reasonable to perform in different topological models. For example, a pyramid can be embedded into a hypercube with the minimal load factor (the maximum number of nodes of G mapped onto a node of H). In Figure 5.2, the example of extension of hypercube topology by hypercubes and extension of CCC-topology by hypercubes is given.

Some topologies are effective in particular cases, for example, for symmetric functions, partially specified functions, threshold functions, etc. In this chapter, hypercube topology and its special extension is the focus of consideration.

5.2 Hypercube data structure

In this section, the hypercube is described in a formal way, through definitions, characteristics, topological parameters and components.

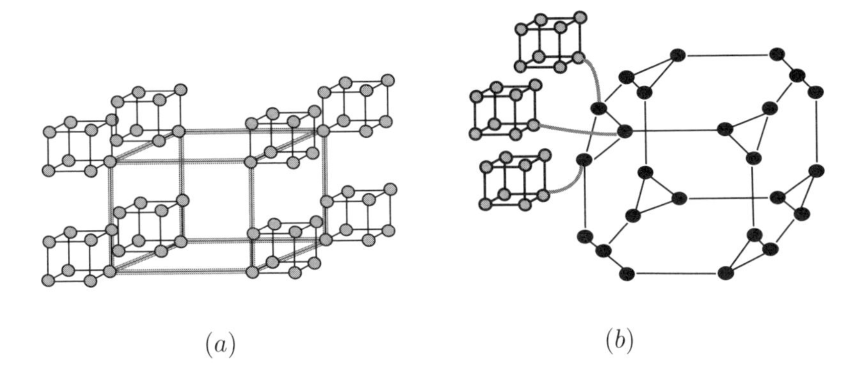

FIGURE 5.2
Hybrid topology design: extension of a hypercube by hypercubes (a) and extension of CCC topology by hypercubes (b).

5.2.1 Hypercube definition and characteristics

The hypercube is characterized by the features below (Figure 5.3):

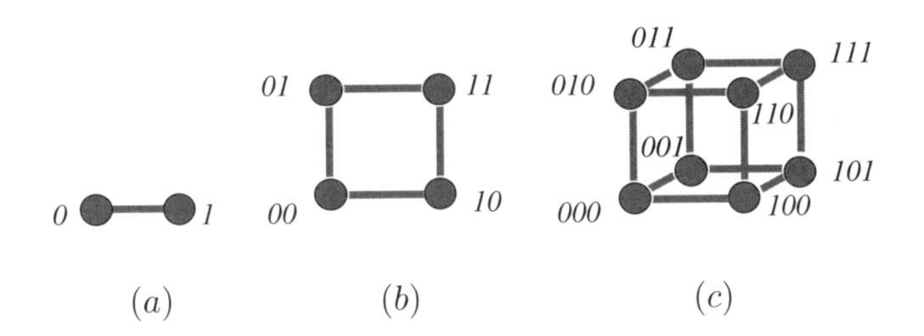

FIGURE 5.3
1-D hypercube (a), 2-D hypercube (b), and 3-D hypercube (c).

- ▶ A hypercube is an extension of a graph. The dimensions are specified by the set $\{0, 1, \ldots, n-1\}$. An n-dimensional binary hypercube is a network with $N = 2^n$ nodes and diameter n. There are $d \times 2^{d-1}$ edges in a hypercube of d dimensions.
- ▶ Each node of an n-dimensional hypercube can be specified by the binary address $(g_{n-1}, g_{n-2}, \ldots, g_0)$ of length n, where the bit g_i corresponds to the i-th dimension in a Boolean space. Two nodes with addresses $(g_{n-1}, g_{n-2}, \ldots, g_0)$ and $(g'_{n-1}, g'_{n-2}, \ldots, g'_0)$ are connected by an edge (or link) if and only if their addresses differ by exactly one bit.

- There are $\binom{n}{x}$ nodes at Hamming distance of x from a given node, and n node-disjoint paths between any pair of nodes of the hypercube.
- Hypercube Q_n can be defined recursively as the graph product.
- The *fan-out* (i.e. degree) of every node is n, and the total number of communication links is $\frac{1}{2}N \log N$.
- A k-dimensional subcube (k-subcube) of hypercube Q_n, $k \leq n$, is a subgraph of Q_n that is a k-dimensional hypercube. A k-subcube of Q_n is represented by a ternary vector $A = a_1a_2\ldots, a_n$, where $a_i \in \{0, 1, *\}$, and $*$denotes an element that can be either 0 or 1.
- Given two subcubes $A = a_1a_2\ldots, a_n$ and $B = b_1b_2\ldots, b_n$, the *intercube distance* $D_i(A, B)$ between A and B along the i-th dimension is 1 if $\{a_i, b_i = \{0, 1\}\}$; otherwise, it is 0. The distance between two subcubes A, B is given by $D(A, B) = \sum_{i=1}^{n} D_i(A, B)$.
- A *path* P of length l is an ordered sequence of nodes $x_{i_0}, x_{i_1}, x_{i_2}, \ldots, x_{i_l}$, where the nodes are labeled with x_{i_j}, $0 \leq j \leq l$, and $x_{i_k} \neq x_{i_{k+1}}$, for $0 \leq k \leq l-1$.

Example 5.1 *The string* $A = a_1a_2a_3a_4 = \{01**\}$ *represents the 2-subcube of* Q_4 *with the node set* $\{0100, 0101, 0110, 0111\}$.

5.2.2 Gray code

Gray code is used for encoding the indexes of the nodes. There are several reasons to encode the indexes. The most important of them is to simplify analysis, synthesis and embedding of topological structures. Gray code is referred to as *unite-distance* codes. Let $b_n...b_1b_0$ be a binary representation of an integer positive number B and $g_n...g_1g_0$ be its Gray code.

$$\texttt{Binary representation } b_n...b_1b_0 \iff \texttt{Gray code } g_n...g_1g_0$$

Binary code to Gray code. Suppose that $B = b_n...b_1b_0$ is given, then the corresponding binary Gray code representation

$$g_i = b_i \oplus b_{i+1} \tag{5.1}$$

where $b_{n+1} = 0$.

Gray code to binary code. Given Gray code $G = g_n...g_1g_0$, the corresponding binary representation is derived by

$$\begin{aligned} b_i &= g_0 \oplus g_1 \oplus \ldots g_{n-i} \\ &= \bigoplus_{i=0}^{n-i} g_i. \end{aligned} \tag{5.2}$$

Table 5.1 illustrates the above transformation for $n = 3$.

TABLE 5.1
Relationships for binary and Gray code for $n = 3$.

Binary code	Gray code		Binary code	Gray code
000	000		000	000
001	001		001	001
010	011		011	010
011	010		010	011
100	110		110	100
101	111		111	101
110	101		101	110
111	100		100	111

Example 5.2 *Binary to Gray and Gray to binary transformation is illustrated by Figure 5.4 for $n = 3$.*

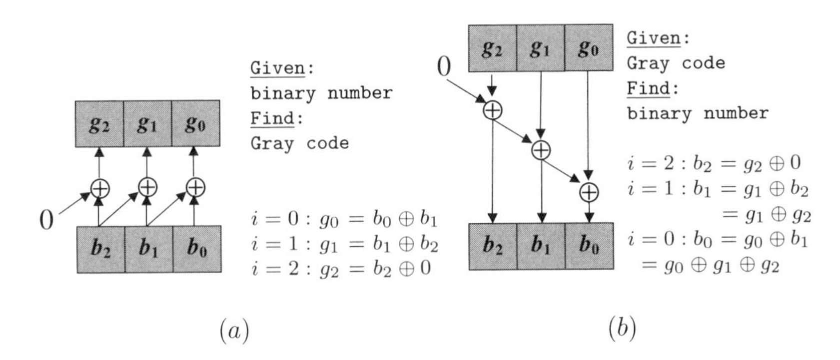

FIGURE 5.4
Flowgraph and formal equation for binary to the Gray code (a) and inverse transformation (b) (Example 5.2),

Useful rule. To build a Gray code for d dimensions, one takes the Gray code for $d - 1$ dimensions, reflects it top to bottom across a horizontal line just below the last element, and adds a leading one to each new element below the line of reflection.

5.2.3 Hamming distance

The *Hamming distance* is a useful measure in hypercube topology. The Hamming sum is defined as the bitwise operation

$$(g_{d-1} \cdots g_0) \oplus (g'_{d-1} \cdots g'_0) = (g_{d-1} \oplus g'_{d-1}), \ldots, (g_1 \oplus g'_1), (g_0 \oplus g'_0) \quad (5.3)$$

where $\oplus$ is an exclusive or operation.

In the hypercube, two nodes are connected by a link (edge) if and only if they have labels that differ by exactly one bit. The number of bits by which labels g_i and g_j differ is denoted by $h(g_i, g_j)$; this is the Hamming distance between the nodes.

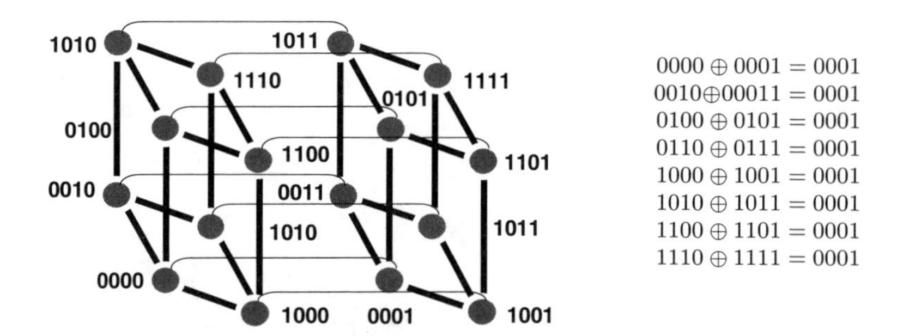

FIGURE 5.5
Hamming sum operation (Example 5.3).

Example 5.3 *Hamming sum operation on two hypercubes for 3-variable switching functions is illustrated in Figure 5.5.*

5.2.4 Embedding in a hypercube

Rings and chains. Let G be a ring (chain) with 2^d vertices. The ring (chain) vertices are numbered 0 through $2^d - 1$ in Gray code (Figure 5.6a). Map vertices of G to vertices of H, and map i, j-th edge of G to the unique edge in H_d that connects the corresponding vertices. The expansion, dilation, and congestion of this embedding are all equal to 1. Note that rings and chains are related to linear arrays.

Meshes. Denote the size of mesh G by $X \times Y$, where X and Y are both powers of two. Let $X = 2^x$ and $Y = 2^y$. Let $d = x + y$ in hypercube H_d (Figure 5.6b, $x = 1$, $y = 2$). The embedding of mesh G into H_d is characterized by expansion, dilation, and congestion of 1.

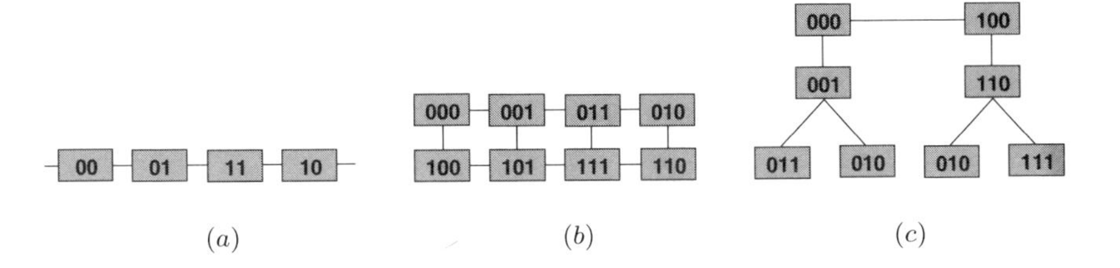

FIGURE 5.6
Ring (chain) (a), mesh (b), and binary tree (c) representation for embedding in hypercube.

Complete binary trees. Let T_i be the complete binary tree of height i. This tree consists of $2^i - 1$ vertices. The following statements are useful for embedding:

- Let $i > 0$, then there is dilation 1 embedding of complete binary tree T_i into the hypercube H_{i+1}.
- Let $i > 0$, then there is dilation 2 embedding of complete binary tree T_i into the hypercube H_i.
- Let $i > 2$, then there is no dilation 1 embedding of complete binary tree T_i into the hypercube H_i. One can justify this statement by construction: T_1 is trivially embedded into H_1, T_2 is embedded into H_2 with a dilation and congestion of one, and for $i > 2$ there is no dilation. Note that the condition of embedding with dilation 1 includes two requirements:

  (a) vertices of T_i that are on an odd level of tree hierarchy ($V_{odd\ level} = 2^0 + 2^2 + 2^4 + \cdots + 2^{i-1} = \frac{2^{i+1}-1}{3}$) are mapped to hypercube H_i vertices that have an even number of ones, and

  (b) vertices of T_i that are on an even level of tree ($V_{even\ level} = 2^0 + 2^3 + 2^5 + \cdots + 2^{i-1} = \frac{2(2^i-1)}{3}$) are mapped to hypercube H_i vertices that have an odd number of ones.

  Because $V_{odd\ level} > 2^{i-1}$ for $i > 1$ and $V_{even\ level} > 2^{i-1}$ for $i > 2$, the hypercube h_i does not have enough vertices with odd and even numbers of ones to host the vertices of tree T_i on odd and even levels of tree hierarchy.

More references to the above problem are given in "Further Reading" Section.

5.3 Assembling of hypercubes

Assembling is the basic topological operation that we apply to synthesize hypercube and hypercube-like data structure. Assembling is the first phase of the development of self-assembling, that is, the process of construction of a unity from components acting under forces/motives internal or local to the components themselves.

To apply assembly procedure, the following items must be defined:

- ▶ The structural topological components,
- ▶ Formal interpretation of the structural topological components in terms of the problem, and
- ▶ The rules of assembly.

The assembling is a key philosophy of building complex systems. For example, assembling a circuit after configuration. In this section, the assembling of classical hypercubes is considered.

5.3.1 Topological representation of products

Assembling a hypercube of switching functions is accomplished by:

- ▶ Generating the products as enumerated points (nodes) in the plane,
- ▶ Encoding the nodes by Gray code,
- ▶ Generating links using Hamming distance,
- ▶ Assembling the nodes and links, and
- ▶ Joining a topology of hypercube in n dimensions.

Let $x_j^{i_j}$ be a literal of a Boolean variable x_j such that $x_j^0 = \overline{x}_j$, and $x_j^1 = x_j$, and $x_1^{i_1}x_2^{i_2}\ldots x_n^{i_n}$ is a product of literals. Topologically, it is a set of points on the plane numerated by $i = 0, 1, \ldots, n$. To map this set into the hypercube, the numbers must be encoded by Gray code and represented by the corresponding graphs based on Hamming distance. The example below demonstrates the assembly procedure.

Example 5.4 *Figure 5.7a demonstrates the assembly of hypercubes for products of one, two and three variables:*

Product $x_1^{i_1} \Longleftrightarrow$ *2 points* $\Longleftrightarrow$ *1-D hypercube* $(n = 1)$
Product $x_1^{i_1}x_2^{i_2} \Longleftrightarrow$ *4 points* $\Longleftrightarrow$ *2-D hypercube* $(n = 2)$
Product $x_1^{i_1}x_2^{i_2}x_3^{i_3} \Longleftrightarrow$ *8 points* $\Longleftrightarrow$ *3-D hypercube* $(n = 3)$

Products with more than three variables are represented by assembling 3-D hypercubes (Figure 5.7b):

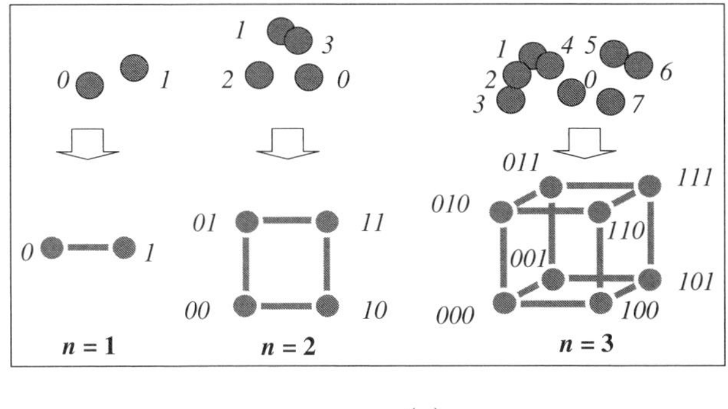

(a)

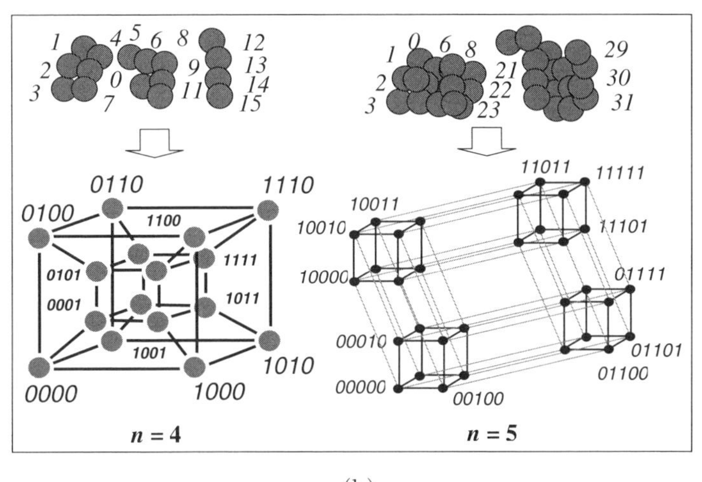

(b)

FIGURE 5.7
Assembling a hypercube for representation of product terms of a single variable $n = 1$, two variables $n = 2$, three variables $n = 3$ (a); assembling 3-D hypercubes to represent the product term of four $n = 4$ and five $n = 5$ variables (b) (Example 5.4).

$$\text{Product } x_1^{i_1} x_2^{i_2} x_3^{i_3} x_4^{i_4} \iff \text{16 points} \iff \text{4-D hypercube } (n = 4)$$
$$\text{Product } x_1^{i_1} x_2^{i_2} x_3^{i_3} x_4^{i_4} x_5^{i_5} \iff \text{32 points} \iff \text{5-D hypercube } (n = 5)$$

Notice that the 0-dimensional hypercube ($n = 0$) represents constant 0. The line segment connects vertices 0 and 1, and these vertices are called the *face* of 1-D hypercube and denoted by **x**. A 2-D hypercube has four faces, 0**x**, 1**x**, **x**0, and **x**1. The total 2-D hypercube can be denoted by **xx**.

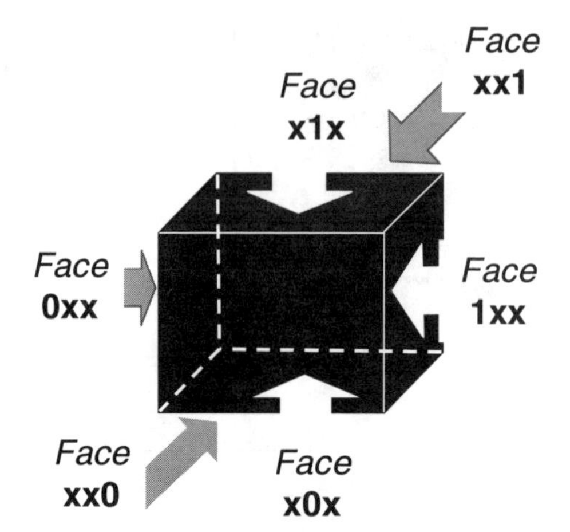
Face xx1
Face x1x
Face 0xx
Face 1xx
Face xx0
Face x0x
Faces of the hypercube are carriers of switching functions of three variables:
Face xx0: $\overline{x}_3(\overline{x}_1\overline{x}_2 \vee \overline{x}_1 x_2 \vee x_1\overline{x}_2 \vee x_1 x_2)$
Face xx1: $x_3(\overline{x}_1\overline{x}_2 \vee \overline{x}_1 x_2 \vee x_1\overline{x}_2 \vee x_1 x_2)$
Face 0xx: $\overline{x}_1(\overline{x}_2\overline{x}_3 \vee \overline{x}_2 x_3 \vee x_2\overline{x}_3 \vee x_2 x_3)$
Face 1xx: $x_3(\overline{x}_2\overline{x}_3 \vee \overline{x}_2 x_3 \vee x_2\overline{x}_3 \vee x_2 x_3)$
Face x0x: $\overline{x}_2(\overline{x}_1\overline{x}_3 \vee \overline{x}_1 x_3 \vee x_1\overline{x}_3 \vee x_1 x_3)$
Face x1x: $x_2(\overline{x}_1\overline{x}_3 \vee \overline{x}_1 x_3 \vee x_1\overline{x}_3 \vee x_1 x_3)$

FIGURE 5.8
Faces of the hypercube interpretation in the sum-of-products of a switching function of three variables (Example 5.5).

Example 5.5 *Six faces of the hypercube,* **xx**0, **xx**1, 0**xx**, 1**xx**, **x**1**x**, *and* **x**0**x** *(Figure 5.8) represent 1-term products for a switching function of three variables.*

5.3.2 Assembling hypercubes for switching functions

The assembly of the hypercube is illustrated for switching functions given by sum-of-products. The most useful property of a sum-of-products for the hypercube assembly is that it can be derived directly from the switching function. If the variable x_j is not present in the hypercube, then $c_j = \mathbf{x}$ (don't care), i.e., $x_j^{\mathbf{x}} = 1$. In hypercube notation, a term is described by a hypercube that is a ternary vector $[i_1 i_2 \dots i_n]$ of components $i_j \in \{0,\ 1,\ \mathbf{x}\}$. A set of cubes corresponding to the true values of a switching function f represents the sum-of-products for this function. The hypercube $[i_1 i_2 \dots i_n]$ is called n-hypercube or n-cube to specify the size of the hypercube. Cube algebra includes a set of elements as $C = \{1, 0, *\}$, and basic operations to find new cubes.

Directly adjacent elements of the on-set are called *adjacency plane.* Each adjacency plane corresponds to a product term.

Example 5.6 *Let a switching function of three variables $f = x_1\overline{x}_2 \vee \overline{x}_3$ be given by its truth table (Figure 5.9). Figure 5.10 illustrates assembling a hypercube.*

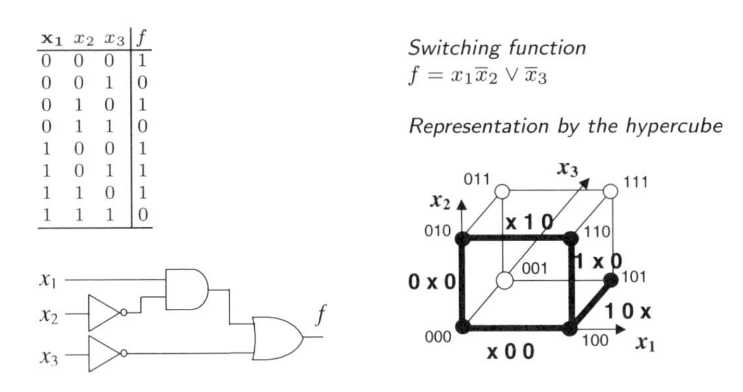

x_1	x_2	x_3	f
0	0	0	1
0	0	1	0
0	1	0	1
0	1	1	0
1	0	0	1
1	0	1	1
1	1	0	1
1	1	1	0

FIGURE 5.9
Truth table of the switching function f, the corresponding logic circuit, and the hypercube representation (Example 5.6).

5.3.3 Assembling hypercubes for state assignments of finite state machines

A finite state machine represents the behavior of a sequential network, that is a logic network with memory. Sequential networks are represented by state diagrams or state tables with input variables, state variables, and output variables. Two states s_i and s_j are distinguishable if they produce different output sequences for the same input sequence. Otherwise, s_i and s_j are equivalent. A design of a sequential network includes the minimization of the number of states, assigning a binary code to each state, and implementation of the network. Each state is represented by a binary vector of state variables and s state variables represent at most 2^s internal states. A state assignment assigns each state to a binary vector. The complexity of a sequential network depends on the method of state assignment. The problem of state assignment or *encoding problem* is to find a state assignment that simplifies the network.

Let a finite state machine be given by s states $Q_0, \cdots Q_{s-1}$, each of them is represented by a binary vector. The hypercube assembly procedure includes the following steps.

Step 1. Generate s states as points in the plane (0-D hypercubes). If s can not be represented by two to an integer power, it should be increased to 2^n by virtual states, where $n =]log_2 s[$ is the smallest integer that is equal to or greater than $log_2 s$.

Step s. The set of $(s-1)-$dimensional hypercubes is considered. They join into pairs, and an s-dimensional hypercube is assembled from each pair by adding edges properly. As far as it possible, those vertices i and j are chosen for being connected with an edge, which have the greatest value of corresponding w_{ij}.

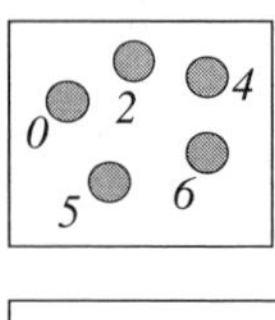

Step 1:

Generation of the set of points
$V = \{0, 2, 4, 5, 6\} = \{000, 010, 100, 101, 110\}$

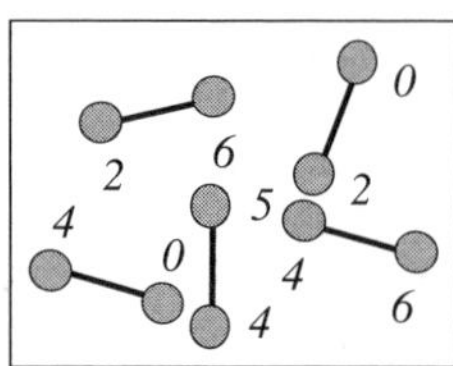

Step 2:

Generation of graphs of degree $d = 1$
for which the Hamming distance is equal to one:
$000 \oplus 010 = 010,\ 000 \oplus 100 = 100$
$010 \oplus 110 = 100,\ 100 \oplus 110 = 010$
$100 \oplus 101 = 001$

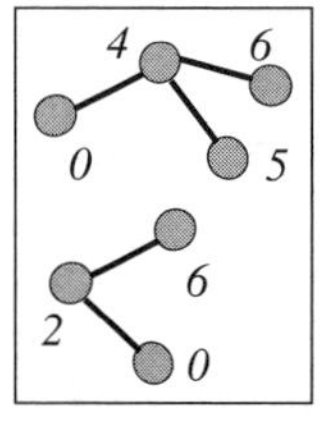

Step 3:

Generation of connected graphs with degree of $d > 1$
more than one:
combine vertices 100
combine vertices 010

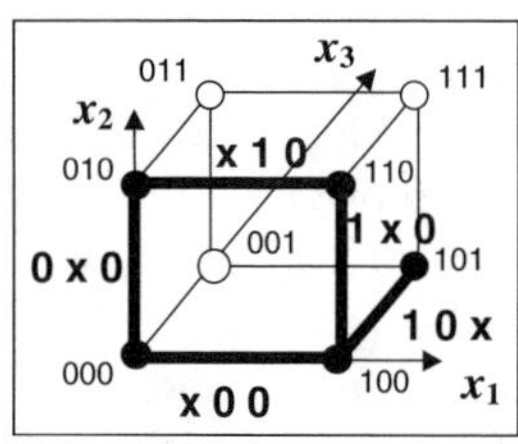

Step 4:

Embed the graphs in a hypercube.
The final representation of the switching function is

$$\begin{bmatrix} \mathbf{x} & \mathbf{x} & 0 \\ 1 & 0 & \mathbf{x} \end{bmatrix}$$

FIGURE 5.10
Assembling a hypercube (Example 5.6) for a switching function.

Step n. A n-dimensional hypercube is assembled. The n-component Boolean vectors are assigned to the vertices of the hypercube where the neighborhood relation between the vectors should be represented by the edges of the hypercube.

For two hypercubes of dimension $(s-1)$ represented by sequences $S^{'}$ and $S^{''}$, the sum $\sum \omega_{ij}$ is calculated where adding is performed over all pairs i, j of indices of vertices that take the same places in the sequences. This sum varies with permutations of vertices of one of the sequences, say $S.^{''}$ Only those permutations may be taken into consideration which preserve the adjacency relation among vertices.

The pair of hypercubes is chosen for which $\sum \omega_{ij}$ is maximal and they are joined into an s-dimensional hypercube by edges between vertices that are in related places of $S^{'}$ and $S^{''}$ (after the proper permutation). The sequence that represents the composed hypercube is formed by concatenation of the sequences S' and S'' one of which changes its order for the reverse one.

Example 5.7 *The state machine is given by 16 states. In Table 5.2, the machine is represented by* $w_{i,j}$. *The solution is given in Figure 5.11.*

TABLE 5.2
The values of $w_{i,j}$ of the state machine (Example 5.7).

2	3	4	5	6	7	8	9	10	11	12	13	14	
0	0	0	0	16	16	88	32	80	32	0	0	0	1
	0	0	0	0	0	0	0	0	0	0	32	0	2
		32	0	0	0	0	0	0	0	0	0	0	3
			0	0	0	0	0	0	0	0	16	0	4
				0	16	0	0	0	0	0	16	0	5
					48	16	48	16	32	0	0	16	6
						16	32	16	48	0	0	0	7
							32	80	32	0	0	0	8
								32	64	0	0	0	9
									32	32	0	0	10
										0	0	0	11
											0	0	12
												0	13

Some comments on Example 5.7 will be useful.

- Since the number of states must be equal to 2^4, states 15 and 16 are classified as virtual states.
- The states are interpreted as isolated vertices (hypercubes of zero dimension) in the plane (Figure 5.11, step 1).
- In the i-th step of assembling procedure, two vertices are connected with an edge if and only if the places taken by them in the sequence S correspond to the places of neighbor codes in the Gray code sequence of the same length as the length of s.

In addition,

$$w_{ij} = w_{ij}^{'} + w_{ij}^{''} + w_{ij}^{'''}$$

where $w_{ij}^{'}$ is the number of pairs of transitions in the given machine from the same state to i-th and j-th states at the neighbouring input states, $w_{ij}^{''}$ is the number of pairs of transitions of the machine from i-th and j-th states to the same state at the same input states, $w_{ij}^{'''}$ is the sum of output variables equal to 1. The sum is taken over all pairs of transitions of machine from i-th and j-th states at the same input state.

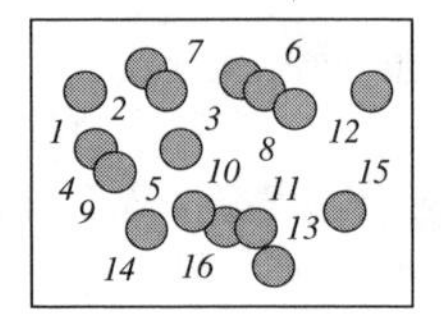

Step 1:

Generation of the set of points

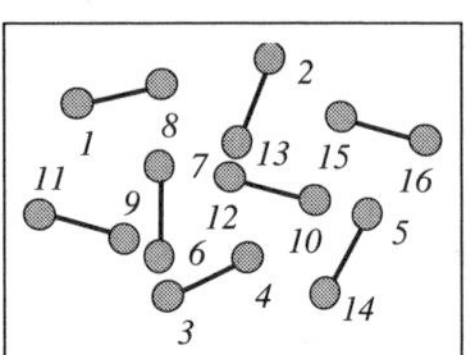

Step 2:

Generation of graphs of degree $d = 1$
accordingly to the criteria of Hamming distance

Step 3:

Generation of graphs of degree $d = 2$

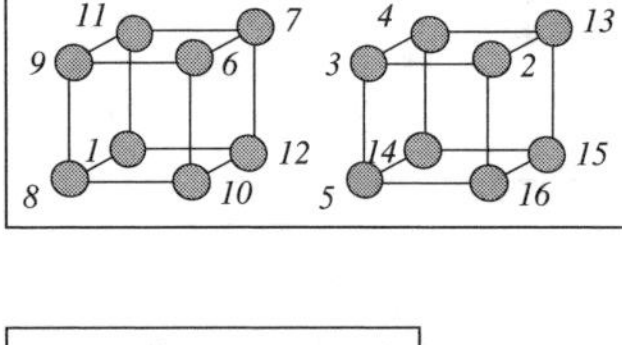

Step 4:

Generation of graphs of degree $d = 3$

Step 5:

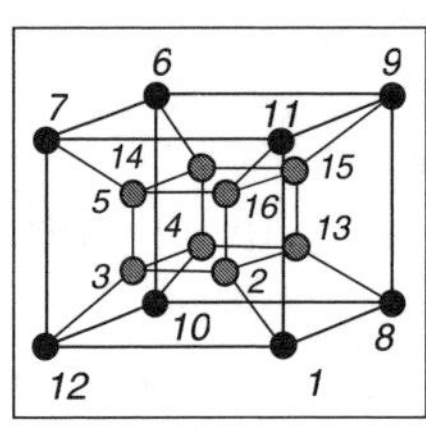

Embed hypercube in the hypercube.
Assign Boolean vectors to the states
of a finite state machine according to
their neighborhood relation:
1 - 0000, 2 - 0001, 3 - 0101, 4 - 1101,
5 - 0111, 6 - 0001, 7 - 0110, 8 - 1000,
9 - 1101, 10 - 1100, 11 - 0010, 12 - 0100,
13 - 1010, 14 - 1111

FIGURE 5.11
Assembling a hypercube to state assignment of finite state machine (Example 5.7).

5.4 $\mathcal{N}$-hypercube definition

In this section the extension of the traditional hypercube is considered. This extension is called the $\mathcal{N}$-hypercube.

5.4.1 Extension of a hypercube

Extension of a hypercube is made by:

- Embedding additional nodes,
- Distinguishing the types of nodes,
- Special space coordinate distribution of the additional nodes,
- New link assignments.

Additional embedded nodes and links assignments correspond to embedding decision trees in a hypercube and thus, convert a hypercube from the passive representation of a function to a connection-based structure, i.e., a structure in which calculations can be accomplished. In other words, information connectivity is introduced into the hypercube. Distinguishing the types of the nodes satisfies the requirements of graphical data structures of switching functions.

5.4.2 Structural components

An $\mathcal{N}$-hypercube includes:

- *Intermediate* nodes embedded into the edges and faces of a singular hypercube,
- *Terminal* nodes, which are the nodes of a singular hypercube,
- A *Root* node embedded into singular hypercube, and
- *Links* between nodes.

Intermediate nodes are the nodes of a decision diagram, i.e., the processing elements that operate on logical signals (Figure 5.12). The node of decision diagram is a demultiplexor element, i.e., the node performs Shannon expansion. In addition, each intermediate node is associated with a so-called *degree of freedom.* Terminal nodes carry information about the results of computing. The root node resembles the root node of a decision tree. The nodes (their functions and coordinates), and links carry information about the function implemented by the $\mathcal{N}$-hypercube.

5.5 Degree of freedom and rotation

The degree of freedom of each intermediate node can be used for variable order manipulation, as the order of variables is a parameter to adjust in decision trees and diagrams. Additional intermediate nodes, the root node and corresponding links in the $\mathcal{N}$-hypercube are associated with

- Polarity of variables in switching function representation,
- Structure of decision tree and decision diagram, and variable order, and
- Degree of freedom and rotation of the $\mathcal{N}$-hypercube based topology.

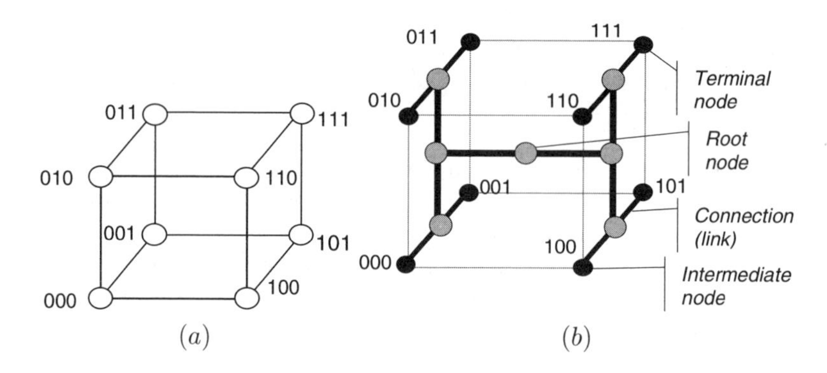

FIGURE 5.12
Classical hypercube (a) and $\mathcal{N}$-hypercube (b).

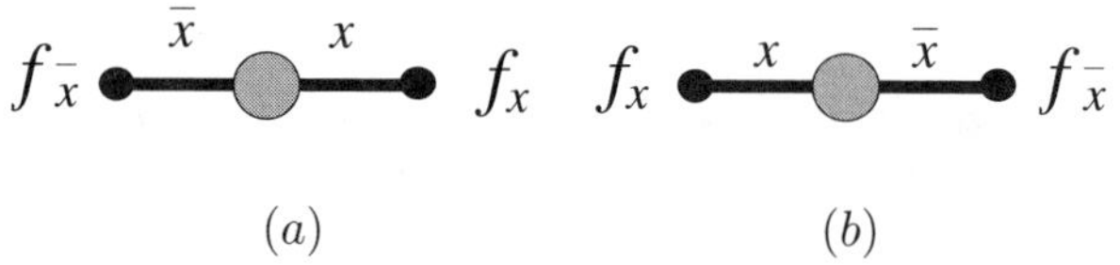

$f_{\bar{x}}$ — $\bar{x}$ — ● — x — f_x (a) f_x — x — ● — $\bar{x}$ — $f_{\bar{x}}$ (b)

FIGURE 5.13
The switching function of an uncomplement and complement variable x, and corresponding 1-D $\mathcal{N}$-hypercubes (a) and (b).

Consider the switching function of a single variable x. Corresponding 1-D $\mathcal{N}$-hypercubes are shown in Figure 5.13.

The term "degree of freedom" is related to the order of variables in decomposition, and hence, to the order of variables in decision trees and diagrams. The polarity of variables influences the above characteristics too.

Only intermediate and root nodes in $\mathcal{N}$-hypercube can be characterized by a degree of freedom. An intermediate node in the 1-D $\mathcal{N}$-hypercube has two degrees of freedom (Figure 5.14). The 2-D $\mathcal{N}$-hypercube is assembled from two 1-D $\mathcal{N}$-hypercubes, and includes three intermediate nodes. The $\mathcal{N}$-hypercube in 2-D has $2 \times 2 \times 2 = 8$ degrees of freedom. There are four decision trees with different orders of variables.

Consider an $\mathcal{N}$-hypercube in 3-D. This $\mathcal{N}$-hypercube is assembled of two two-dimensional $\mathcal{N}$-hypercubes and includes seven intermediate nodes and has $8 \times 8 \times 2 = 128$ degrees of freedom. The degree of freedom of an intermediate node at i-th dimension, $i = 2, 3, \ldots, n$, is equal to $DF_i = 2^{n-i} + 1$. In general, the degree of freedom of the n-dimensional $\mathcal{N}$-hypercube is defined as

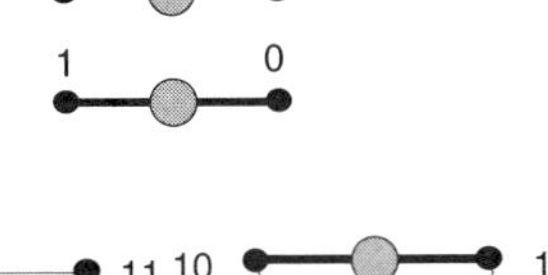

An intermediate node in the 1-D $\mathcal{N}$-hypercube has two degrees of freedom.

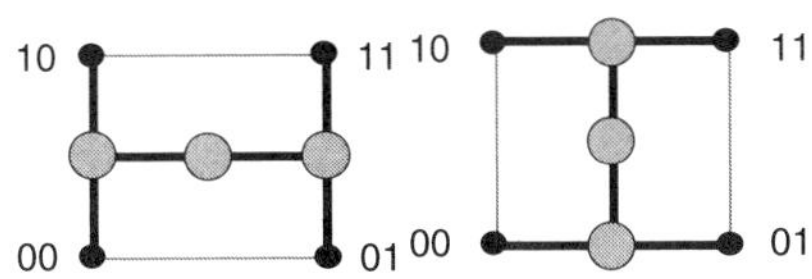

The 2-D $\mathcal{N}$-hypercube is assembled from two 1-D $\mathcal{N}$-hypercubes, and includes three intermediate nodes The $\mathcal{N}$-hypercube in 2-D has

$$2 \times 2 \times 2 = 8 \text{ degrees of freedom.}$$

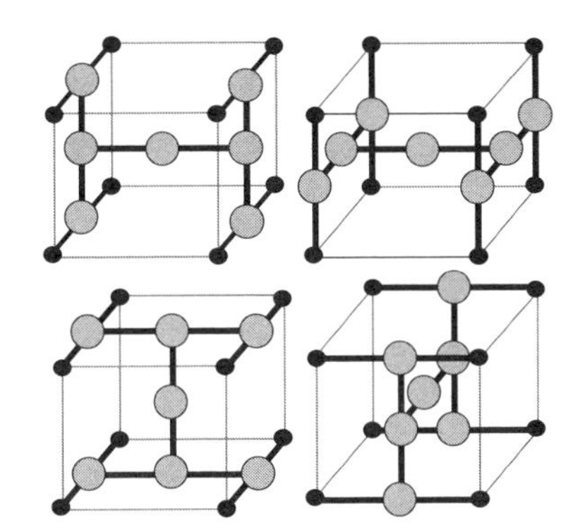

This $\mathcal{N}$-hypercube is assembled of two two-dimensional $\mathcal{N}$-hypercubes and includes 7 intermediate nodes and has

$$8 \times 8 \times 2 = 128 \text{ degrees of freedom.}$$

The degree of freedom of an intermediate node at i-th dimension, $i = 2, 3, \ldots n$, is equal to $DF_i = 2^{n-i} + 1$. In general, the degree of freedom DF of the n-dimensional $\mathcal{N}$-hypercube is defined as

$$DF = \sum_i DF_i = \sum_i (2^{n-i} + 1).$$

FIGURE 5.14
Degree of freedom and rotation of the $\mathcal{N}$-hypercube in 1-D and 2-D.

$$Degree\ of\ freedom = \sum_i DF_i = \sum_i (2^{n-i} + 1). \tag{5.4}$$

5.6 Coordinate description

There are two possible configurations of the intermediate nodes. The first configuration (planes) is defined as (Figure 5.15a)

x00 $\Longleftrightarrow$ x01 $\Longleftrightarrow$ $\boxed{\text{xx1}}$ $\Longleftrightarrow$ x11 $\Longleftrightarrow$ x10 $\Longleftrightarrow$ $\boxed{\text{xx0}}$ $\Longleftrightarrow$ x00.

The second possible configuration (planes) is related to the symmetric faces (Figure 5.15b)

00x $\Longleftrightarrow$ $\boxed{\text{0xx}}$ $\Longleftrightarrow$ 01x $\Longleftrightarrow$ 11x $\Longleftrightarrow$ $\boxed{\text{1xx}}$ $\Longleftrightarrow$ 10x $\Longleftrightarrow$ 00x.

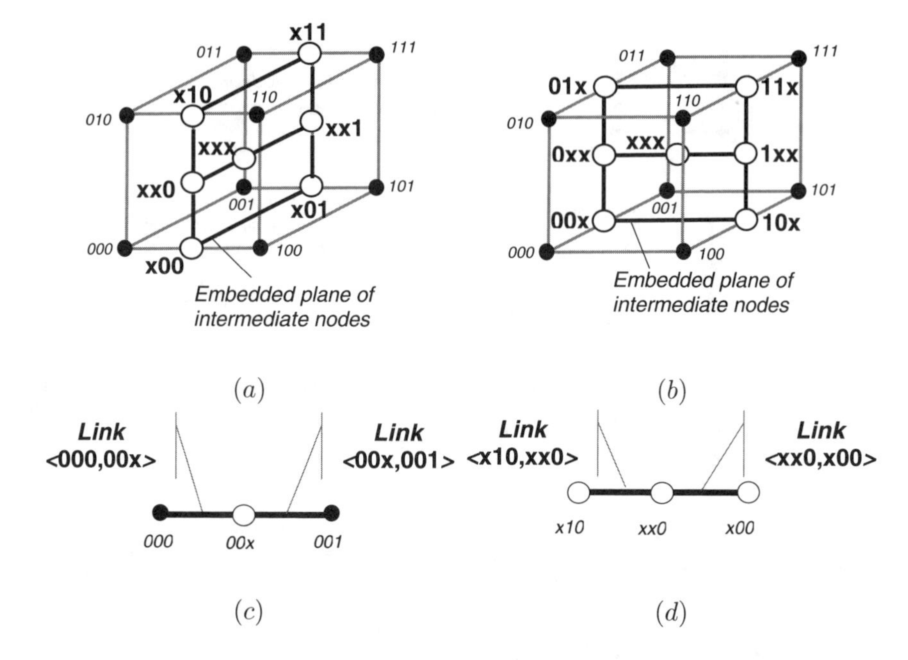

FIGURE 5.15
Coordinate description of the $\mathcal{N}$-hypercube: (a) the first plane, (b) the second plane, (c) the links of an intermediate node, and (d) links of the root.

An $\mathcal{N}$-hypercube includes two types of links with respect to the root node:

$$\text{Link 1: } \boxed{\text{xx0}} \Longleftrightarrow \text{xxx} \Longleftrightarrow \boxed{\text{xx1}},$$
$$\text{Link 2: } \boxed{\text{0xx}} \Longleftrightarrow \text{xxx} \Longleftrightarrow \boxed{\text{1xx}}.$$

The root node coordinate is xxx. There are two types of link in an $\mathcal{N}$-hypercube: links between terminal nodes and intermediate nodes, and links between intermediate nodes, including the root node.

Example 5.8 *In Figure 5.15, link <000,00x> indicates the connection of the terminal node 000 and intermediate node 00x. By analogy, if two intermediate nodes x10 and xx0 are connected, we indicate this fact by <x10,xx0> (Figure 5.15d).*

The number of terminal nodes in the $\mathcal{N}$-hypercube is always equal to the number of nodes in the hypercube. Therefore, the classical hypercube can be considered as the basic data structure for representation of a switching function, in which the $\mathcal{N}$-hypercube can be embedded. The obtained $\mathcal{N}$-hypercube brings the element of connectivity utilized for calculation of switching functions.

There are plain relationships between the hypercube and $\mathcal{N}$-hypercube: coordinate of a link (face) in the hypercube corresponds to the coordinate of an intermediate node located in the middle of this link (face in the $\mathcal{N}$-hypercube).

TABLE 5.3
Relationship between hypercube and $\mathcal{N}$-hypercube.

Hypercube	**$\mathcal{N}$-hypercube**
Link	Intermediate node
Face	Intermediate node

Example 5.9 *Relationship between coordinate description of the hypercube and the $\mathcal{N}$-hypercube is given in Table 5.4.*

TABLE 5.4
Relationship between the hypercube and $\mathcal{N}$-hypercube (Example 5.9).

Hypercube	**$\mathcal{N}$-hypercube**
Links x00, 0x0, x10, 10x	Intermediate nodes x00, 0x0, x10,10x
Faces xx0, xx1, 0xx, 1xx, x1x	Intermediate nodes 0xx, 1xx, x1x,x0x

5.7 $\mathcal{N}$-hypercube design for $n > 3$ dimensions

Consider two 3-D $\mathcal{N}$-hypercubes. For example, let us use the configuration shown in Figure 5.15b. To design a 4-D $\mathcal{N}$-hypercube, two $\mathcal{N}$-hypercubes must be joined by links. There are seven possibilities for connecting these $\mathcal{N}$-hypercubes because the links are allowed between intermediate nodes, intermediate node and the root node, and between the root nodes. The new root node is embedded in the link <xxx0, xxx1>.

Therefore, the number of bits in the coordinate description of both $\mathcal{N}$-hypercubes must be increased by one bit. Suppose that $\mathcal{N}$-hypercubes are connected via link <xxx0, xxx1> between the root nodes xxx0 and xxx1. The resulting topological structure is called a 4-D $\mathcal{N}$-hypercube.

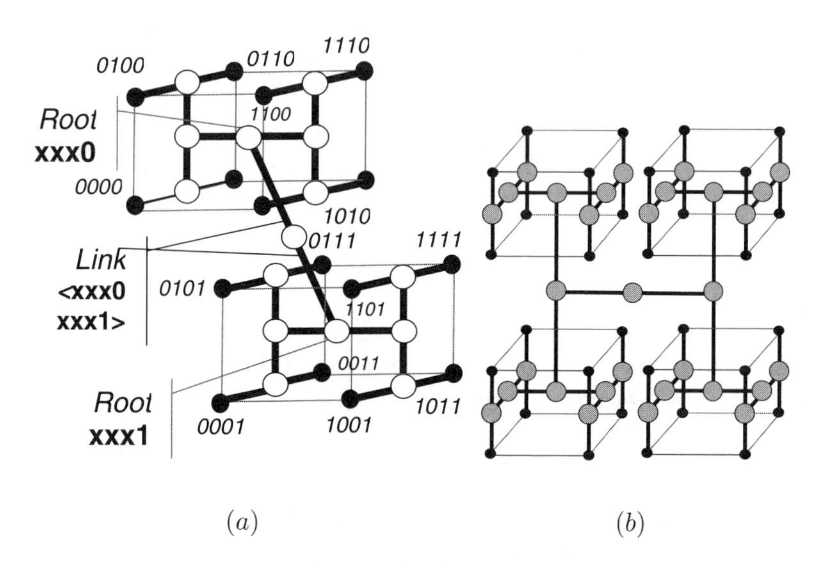

FIGURE 5.16
Connections between $\mathcal{N}$-hypercubes in n-dimensional space (Example 5.10).

Example 5.10 *Figure 5.16 shows the possibilities for connecting a given $\mathcal{N}$-hypercube to another $\mathcal{N}$-hypercube. This connection property follows from the properties of intermediate nodes.*

Summarizing the above characteristics, the $\mathcal{N}$-hypercube can be specified as an undirected hypercube with the following properties:

- ► $\mathcal{N}$-hypercube corresponds to an n-level 2^n-leaves complete binary tree.
- ► $k = 2^n$ terminal nodes labelled from 0 to $2^n - 1$ so that there is an edge between any two vertices if and only if the binary representations of their labels differ by one and only one bit.
- ► Each edge is assigned with an intermediate node which corresponds to binary representation of both edge-ending constant nodes with the don't care value for the only different bit.
- ► There is an edge between two intermediate nodes if and only if the binary representations of their labels differ by one and only one bit.

5.8 Embedding a binary decision tree in $\mathcal{N}$-hypercube

Rings, meshes, pyramids, shuffle-exchange networks, and complete binary trees can be embedded into hypercubes. For example, the leaf vertex, or node of the complete binary decision tree with q levels can be embedded into a hypercube with 2^q vertices and $q \times 2^{q-1}$ edges. This is because the complete binary decision tree with q levels has 2^q leaves. This is exactly the number of nodes in the hypercube structure, where each node is connected to $q-1$ neighbors and assigned the q-bit binary code that satisfies the Hamming encoding rule, and, thus, has $q \times 2^{q-1}$ edges.

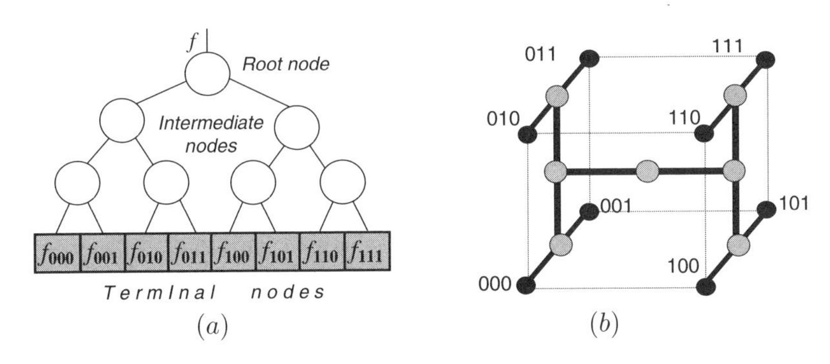

FIGURE 5.17
Correspondence of the attributes of binary tree and $\mathcal{N}$-hypercube.

Lemma 5.1 *The number of vertices of the binary tree embedded into the middle of each edge of the $\mathcal{N}$-hypercube is equal to 2^{q-1} whereas the possible number of such embeddings (the number of all wires) is $q \times 2^{q-1}$.*

PROOF The proof follows from the fact that the number of vertices in the second last level of the binary tree is 2^{q-1}. The number of possible variable orders is equal to the number of variables, q, so the possible number of embeddings is $q \times 2^{q-1}$. ∎

Lemma 5.2 *The number of inner nodes of the binary tree embedded into the middle of each edge of the $\mathcal{N}$-hypercube is equal to 2^{q-2} while the possible number of such embeddings (the number of all edges) is $q \times 2^{q-2}$.*

The proof is adequate to the proof for Lemma 5.1.

Recurrence is a general strategy to generate spatial and homogeneous structures. This strategy can be used for embedding a binary decision trees into a $\mathcal{N}$-hypercube:

Step 1. Embed 2^q leaves of the binary tree (nodes of the level $q-1$) into 2^q-node $\mathcal{N}$-hypercube; assign a code to the node so that each node is connected to q Hamming-compatible nodes.

Step 2. Embed 2^{q-1} inner nodes of the binary tree (nodes of the level $q-2$) into edges connecting the existing nodes of the $\mathcal{N}$-hypercube, taking into account the polarity of the variable.

Step 3. Embed 2^{q-2} inner nodes of the binary tree (nodes of the level $q-3$) into edges connecting the existing nodes of the $\mathcal{N}$-hypercube, taking into account the polarity of variable.

Step 4. Repeat recursively till we embed the root of the tree into the center of the $\mathcal{N}$-hypercube.

Example 5.11 *Let $q=1$, a binary decision tree, represent a switching function of one variable (Figure 5.18). The function takes value 1 while $x=0$ and value 0 while $x=1$. These values assign two leaves of the binary decision diagram.*

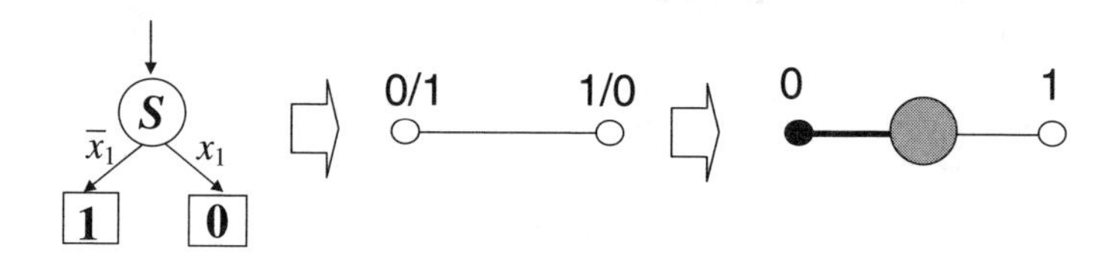

FIGURE 5.18
Embedding a binary decision tree of one variable into the $\mathcal{N}$-hypercube (Example 5.11).

Example 5.12 *Let $q=2$, the four leaf nodes of the complete binary decision tree of a two-variable function can be embedded into a $\mathcal{N}$-hypercube with four nodes (Figure 5.19).*

In more detail, the embedding is as follows (Figure 5.19:

Step 1. Embed 4 leaves of the binary tree into a 4-node $\mathcal{N}$-hypercube; assign the codes 00,01,10,11 so that each node is Hamming-compatible with neighbor nodes. The Hamming distance between the neighbor nodes is equal to one.

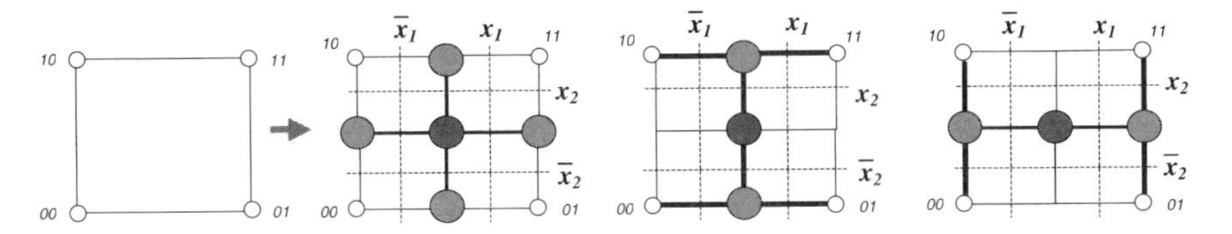

FIGURE 5.19
Embedding a binary decision tree into a 2-D $\mathcal{N}$-hypercube (Example 5.12).

Step 2. Embed 2 inner nodes of the binary tree into edges connecting the existing nodes of the $\mathcal{N}$-hypercube; note that two of the edge-embedded nodes must be considered at a time (Figure 5.19, the left figure corresponds to the order x_2, x_1, and the right figure describes the order x_1, x_2; the axes are associated with the polarity of variables (complemented, uncomplemented) and explain the meaning of the bold edges).

Step 3. Embed the root of the tree into the center of the facet of the $\mathcal{N}$-hypercube and connect it to the edge-embedded nodes.

Example 5.13 *Let $q = 3$, the 8 leaf nodes of the complete binary decision tree of a two-variable function are embedded into the 3-D $\mathcal{N}$-hypercube (Figure 5.20).*

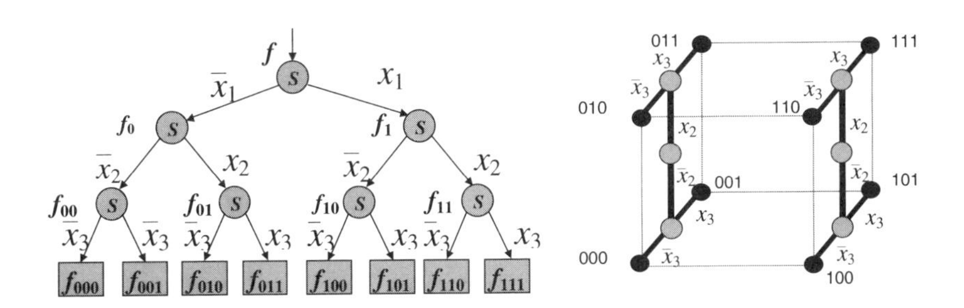

FIGURE 5.20
Embedding the complete binary tree into the 3-D $\mathcal{N}$-hypercube (Example 5.13).

The total number of nodes and the total number of edges (connections) between nodes in the $\mathcal{N}$-hypercube is specified as below

$$N_d = \sum_{i=0}^{n} 2^{n-i} C_i^n \tag{5.5}$$

$$N_c = \sum_{i=0}^{n} 2i \cdot 2^{n-i} C_i^n \tag{5.6}$$

It is obvious that the total number of internal nodes is equal to the number of all nodes except leaves in the complete binary decision tree that represents a switching function of n variables, and the total number of edges is equal to the number of edges in the complete binary decision tree.

5.9 Assembling

Assembling is one of the possible approaches to $\mathcal{N}$-hypercube design. The second approach, based on embedding decision trees into the $\mathcal{N}$-hypercube is discussed later in this chapter.

There are two assembling procedures for the $\mathcal{N}$-hypercube design:

- Assembling a $\mathcal{N}$-hypercube onto $\mathcal{N}$-hypercubes of smaller dimensions; this is a recursive procedure based on several restrictions (rules), and
- Assembling a shared the $\mathcal{N}$-hypercube based structures; in this approach, some extensions of the above mentioned rules are used.

The following rules are the basis of assembly procedure.

Rule 1 (Connections). A terminal node is connected to one intermediate node only. In Figure 5.21, there are 32 terminal nodes connected to 16 intermediate nodes.

Rule 2 (Connections). The root node is connected to two intermediate nodes located symmetrically in opposite faces. Figure 5.21 explains this for 5-dimensional structure.

Rule 3 (Symmetry). Configurations of the terminal and intermediate nodes on the opposite faces are symmetric. The two faces are connected via the root node. In Figure 5.21, two pairs of 3-D $\mathcal{N}$-hypercubes are connected via their root nodes forming two new root nodes, and then two pairs are symmetrically connected via a new root nodes of the 5-D $\mathcal{N}$-hypercube. Two symmetric planes include terminal nodes only.

If values of some terminal nodes in two $\mathcal{N}$-hypercubes assigned with the same codes are equal, then these $\mathcal{N}$-hypercubes can share some nodes. The $\mathcal{N}$-hypercubes are called *shared* $\mathcal{N}$-hypercubes

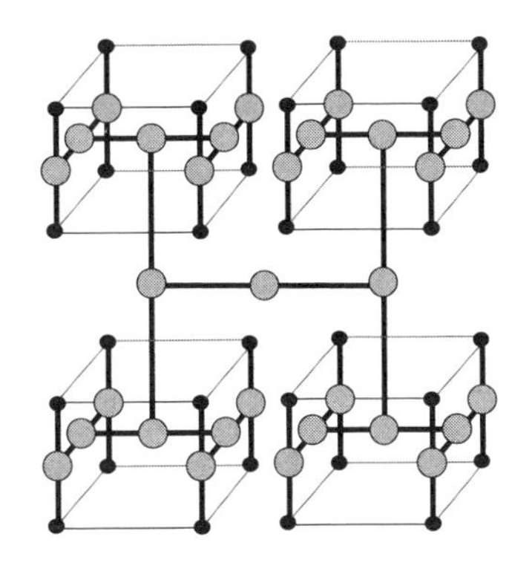

Rule 1 *A terminal node is connected to one intermediate node only.*

Rule 2 *The root node is connected to two intermediate nodes located symmetrically in opposite face.*

Rule 3 *Configurations of the terminal and intermediate nodes on the opposite face are symmetric. Two symmetric planes include terminal nodes only.*

FIGURE 5.21
Assembly rules for $\mathcal{N}$-hypercube design.

In Figure 5.22, assembling the shared $\mathcal{N}$-hypercube based structures is illustrated in 2-D and 3-D space. Two 2-D $\mathcal{N}$-hypercubes are combined accordingly to the merging rule by their two terminal nodes (Figure 5.22a). This type of connection is called *strong merging* in contrast to *weak merging* shown in Figure 5.22b. By analogy, in Figures 5.22c, d the merging procedure for 3-D $\mathcal{N}$-hypercubes is illustrated.

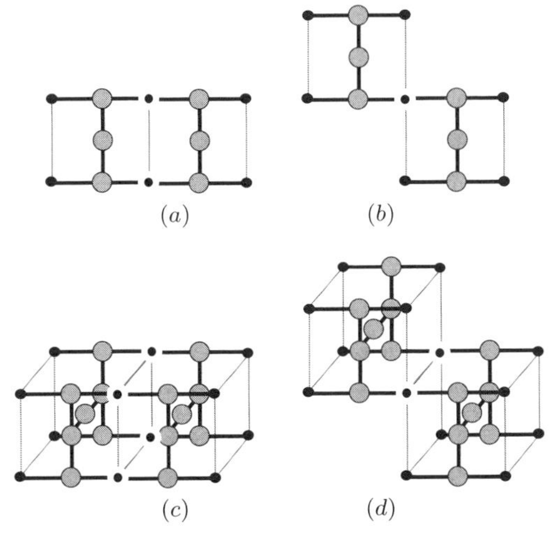

Two 2-D $\mathcal{N}$-hypercubes are combined by their two terminal nodes (strong merging). If one terminal node of each 3-D $\mathcal{N}$-hypercube is merged, it is weak merging.

The merging procedure for 3-D $\mathcal{N}$-hypercubes: strong merging means merging by four nodes, and and weak merging means merging by two nodes.

FIGURE 5.22
Assembling $\mathcal{N}$-hypercube based structures: strong (a, c) and weak (b, d) merging in 2-D and 3-D space.

Example 5.14 *Figure 5.23 illustrates the 3-D $\mathcal{N}$-hypercube assembling.*

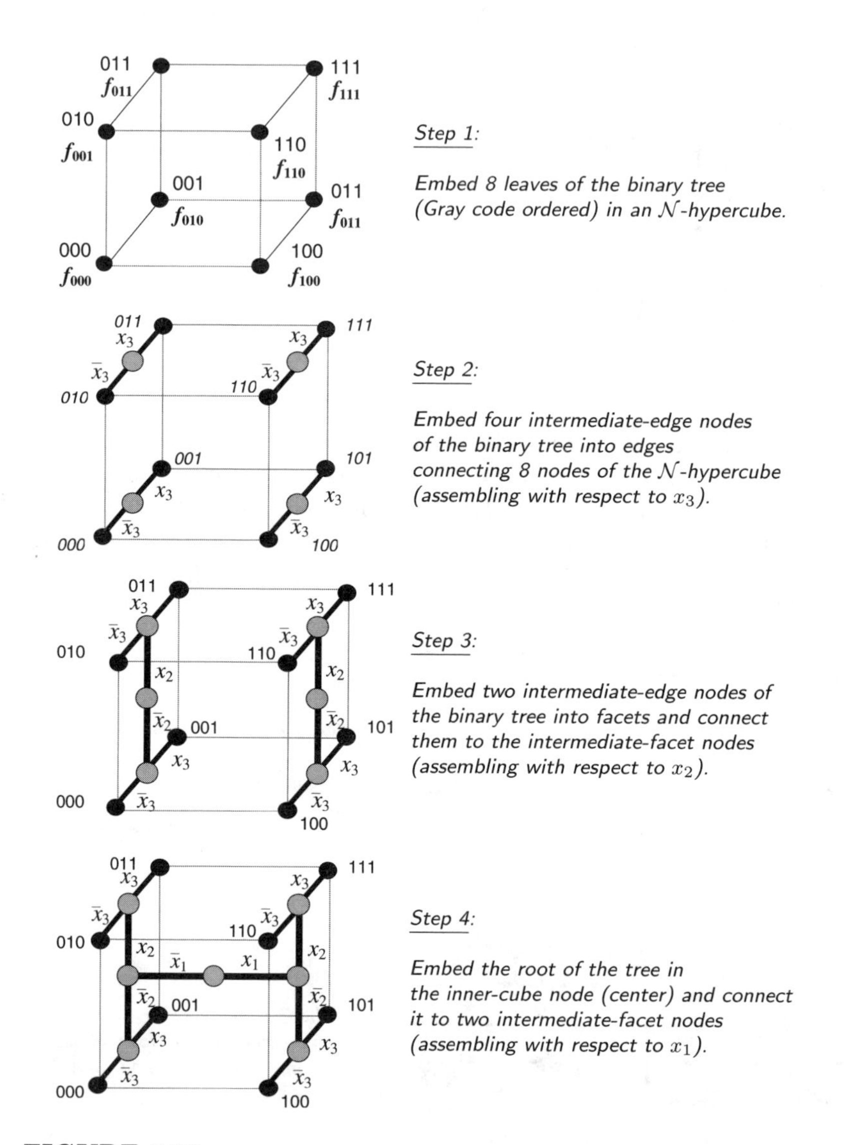

FIGURE 5.23
$\mathcal{N}$-hypercube design by embedding a complete binary tree in the $\mathcal{N}$-hypercube (Example 5.14).

5.10 Spatial topological measurements

There exists a number of basic measures for estimation of the $\mathcal{N}$-hypercube structure of a circuit (Table 5.5).

Diameter and link complexity. The diameter of a network is defined as the maximum distance between any two nodes in the network. Thus, the diameter determines the maximum number of hops that the carrier may have to do on the path from the root to a terminal node.

Link complexity or node degree is defined as the number of physical links per node. For a regular network, where all nodes have the same number of links, the node degree of the network is that of a node. In an $\mathcal{N}$-hypercube, the node degree is 3, except the terminal nodes whose degree is one.

Distance between sub$\mathcal{N}$-hypercubes. Given two sub$\mathcal{N}$-hypercubes $A = a_1a_2\ldots,a_n$ and $B = b_1b_2\ldots,b_n$, the *intercube distance* $D_i(A,B)$ between A and B along the i-th dimension is 1 if $\{a_i, b_i = \{0,1\}\}$; otherwise, it is 0. The distance between two sub$\mathcal{N}$-hypercubes A, B is given by $D(A,B) = \sum_{i=1}^{n} D_i(A,B)$.

Dilation. The *dilation* of an edge $e \in E(G)$ is the length of the path $\alpha(e)$ in H. The dilation of an embedding is the maximum dilation over all edges in G.

Bisection width. The bisection width of a network is defined as the minimum number of links that have to be removed to partition the network into two equal halves. The bisection width indicates the volume of communication allowed between any two halves of the network with an equal number of nodes. The bisection width of a n-dimensional hypercube is $2^{n-1} = N/2$ since many links are connected between two $(n-1)$-dimensional $\mathcal{N}$-hypercubes to form an n-dimensional $\mathcal{N}$-hypercube.

Granularity of size scaling. The granularity of size scaling is the ability of the system to increase in size with minor or no change to the existing configuration, and with an expected increase in performance proportional to the extent of the increase in size. The size of a hypercube can only be increased by doubling the number of nodes; that is, the granularity of size scaling in an n-dimensional hypercube is 2^n.

Average message distance. The average distance in a network is defined as the average number of links that a carrier (electron, dot) should travel

TABLE 5.5
Metrics on $\mathcal{N}$-hypercube based structures.

Metric	Characteristic
Diameter	The maximum distance between any two nodes in the network
Link complexity	The number of physical links per node
Dilation of an edge	The length of the path $\alpha(e)$ in H. The dilation of an embedding is the maximum dilation over all edges in G
Average message distance	The average number of links that a carrier should travel between any two nodes
Total number of primitives	The number of $\mathcal{N}$-hypercube primitives in the network
Effectiveness	The average number of variables that represent a $\mathcal{N}$-hypercube
Active nodes	The nodes connected to nonzero terminals through a path
Connectivity	The number of paths from the root
Average path length	The number of links connecting the root node to a nonzero terminal
Bisection width	The minimum number of links to be removed in order to partition the network into two equal halves
The granularity of size scaling	The ability of the system to increase in size with minor or no change to the existing configuration

between any two nodes. Let N_i represent the number of nodes at a distance i, then the average distance $\overline{L}$ is defined as

$$\overline{L} = \frac{1}{N-1}\sum_{i=1}^{n} iN_i$$

where N is the total number of nodes, and n is the degree.

Fault tolerance. In the probability fault model, the reliability of each node at time t is a random variable. The probability that sub-$\mathcal{N}$-hypercube is operational is represented by the reliability of the data processing in the sub-$\mathcal{N}$-hypercubes. The $\mathcal{N}$-hypercube reliability can be formulated as the union of probabilistic events that all the possible $\mathcal{N}$-hypercubes are operational.

5.11 Summary

1. The $\mathcal{N}$-hypercube is a data structure obtained by extension of the classical hypercube. Additional (intermediate) nodes and links contribute to the information connectivity of a hypercube. Intermediate nodes are associated with

- ▶ Degree of freedom, a new characteristic of a hypercube structure,
- ▶ Polarity of variables, closely related to the degree of freedom,
- ▶ Intermediate nodes of the embedded decision tree.

2. There are two approaches to design of $\mathcal{N}$-hypercubes

- ▶ Embedding of a complete binary decision tree in an $\mathcal{N}$-hypercube,
- ▶ Assembling the $\mathcal{N}$-hypercube of the $\mathcal{N}$-hypercube of smaller dimensions.

The basic procedure for both approaches is the representation of a switching function by a complete binary decision tree.

3. Hypercube-like topology is a reasonable model for various aspects of logic design:

- ▶ Representation and manipulation of switching functions. In this classical application the carriers of information are coordinates of a hypercube (Figure 5.24a).
- ▶ Representation of state assignments in sequential circuits. The carriers of information in this task are marks of nodes (Figure 5.24b).
- ▶ A hypercube-like structure can be a topological description of devices, for example, FPGA. In Figure 5.24c, the terminal nodes represent logic blocks and intermediate nodes correspond to switch blocks.
- ▶ A hypercube-like structure, $\mathcal{N}$-hypercube, can be used for switching functions implementation in spatial dimensions. The information is processed in the intermediate nodes and transmitted by links. In Figure 5.24d, the terminal nodes carry information about the values of function.

5. Spatial topological measurements include diameter and link complexity, distance between sub-$\mathcal{N}$-hypercubes, dilation, bisection width, granularity of size scaling, average message distance, fault tolerance, degree of freedom, and Voronoi diagrams.

6. Due to the additional nodes and their topological relationships, new properties are achieved, namely:

- ▶ Creating a hierarchy of information flows in space dimensions (this hierarchy satisfies the requirements of different data structures).

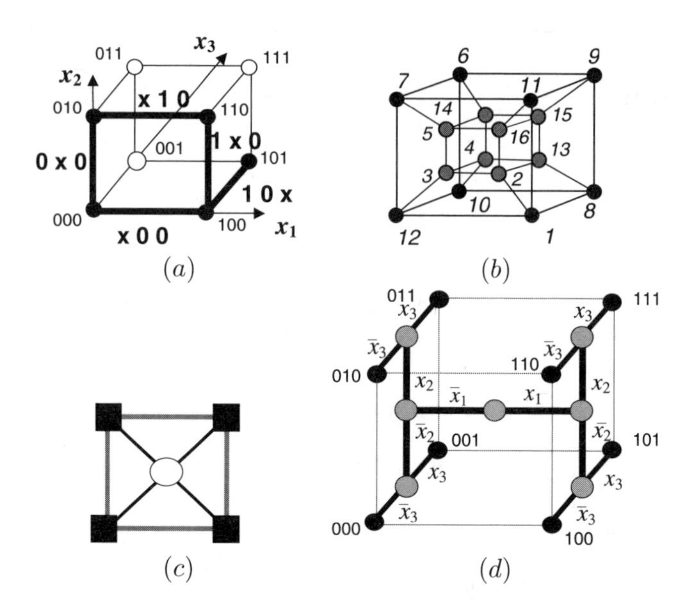

FIGURE 5.24
Carriers of information in topological structures.

- ▶ Embedding graphical data structures such as decision trees and decision diagrams in three dimensions.
- ▶ Assembling topological structures; this means new possibilities for flexible spatial logic design.

5.12 Problems

Problem 5.1 Use the assembling method to represent the switching functions given below by the hypercubes

(a) $f = \overline{x}_1 x_2 \vee x_1 x_2 \overline{x}_3 \vee x_3$
(b) $f = x_1 \oplus x_1 x_2 x_3 \oplus x_1 x_3$
(c) $f = (x_1 \vee x_2)(\overline{x}_2 \vee x_3)$
(d) $f = x_1 + x_3 - 2x_1 x_3$
(e) $f = 2x_1 \widehat{\vee} x_2 \widehat{\vee} 2x_3 \widehat{\vee} 3\overline{x}_3$

Problem 5.2 Given the $\mathcal{N}$-hypercube:

(a) Derive the switching function in the form of sum-of-products given the $\mathcal{N}$-hypercube (Figure 5.25a).

(*b*) Derive the switching function in Reed-Muller form given the $\mathcal{N}$-hypercube (Figure 5.25b).
(*c*) Derive the switching function in arithmetic form given the $\mathcal{N}$-hypercube (Figure 5.25c).

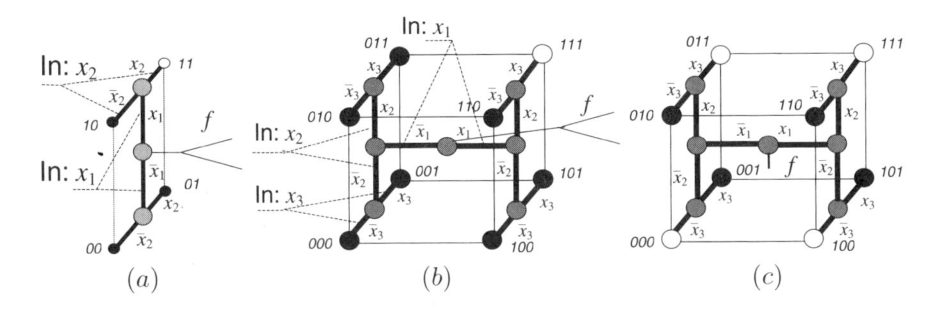

FIGURE 5.25
$\mathcal{N}$-hypercubes for Problem 5.2.

Problem 5.3 Represent the two-output switching function

$$f_1 = x_1x \vee \overline{x}_2x_3 \vee x_2\overline{x}_3$$
$$f_2 = x_1 \vee \overline{x}_2 \vee x_3$$

(*a*) By two complete decision trees
(*b*) By an embedded $\mathcal{N}$-hypercube
(*c*) By two decision diagrams
(*d*) By a shared decision diagram

Note, that a is used to represent several functions that can share equivalent subgraphs in the diagram.

Problem 5.4 Construct a binary decision tree of the switching functions given below and embed it in the $\mathcal{N}$-hypercube

(*a*) 2-input majority circuit
(*b*) 3-input majority circuit
(*c*) Full adder
(*d*) 2-input adder

Problem 5.5 Propose a way to implement a shared $\mathcal{N}$-hypercube for the two-output function represented in Example 5.3.

Problem 5.6 Chapter 2 discusses various nanotechnologies. Among them, the SET technology is the appropriate technology to implement decision trees and decision diagrams in $\mathcal{N}$-hypercube. However, other possibilities can be found. Propose an appropriate nanoelectronic implementation of the $\mathcal{N}$-hypercube structure.

5.13 Further reading

Hypercube architecture. An approach to design a supercomputer with a 3-D hypercube architecture has been proposed by Hayes et al. [3]. References to 3-D processing algorithms can be found in [1, 6]. Problems of 3-D representation and simulation of data have been considered in [9].

Hypercube-like structures. Hypercube data structures have been studied for a long time in the area of parallel and distributed computing. Lai and White studied the effectiveness of the pyramid topology [5]. Sources [13, 15, 16, 17] are useful for studying the characteristics of hypercube-like topologies and embedding graphs in these topologies.

Embedding complete binary trees in hypercubes has been studied by Leiss and Reddy [7]. In [18], the conditions of embedding of a full binary tree in the hypercube are given. The reader can find rigorous mathematical results in this field in [13]. Properties of incomplete hypercubes have been studied by Öhring and Das [11]. In [1, 8], a hypercube-like structure has been used for modeling semiconductor device structures. Measurements in hypercube-like structures are considered in [8, 10, 14].

Geometry. Geometric models such as meshes and octrees have been applied to simulation of 3-D semiconductor device structure [1, 4].

Assembling of hypercubes have been studied by Pottosin [12].

5.14 References

[1] Conti P, Hitschfeld N, and Fichtner W. Ω – an octree-based mixed element grid allocator for the simulation of complex 3-D device structures. *IEEE Transactions on Computer-Aided Design of Integrated Circuits and Systems*, 10(10):1231–1241, 1991.

[2] Cormen TH, Leiserson CE, Riverst RL, and Stein C. *Introduction to Algorithms.* MIT Press, Cambridge, MA, 2001.

[3] Hayes JP, Mudge TN, Stout QF, Colley S, and Palmer J. A microprocessor-based hypercube supercomputer. *IEEE Micro,* 6:6–17, 1986.

[4] Hitschfeld N, Conti P, and Fichtner W. Mixed element trees: a generalization of modified octrees for the generation of meshes for the simulation of complex 3-D semiconductor device structures. *IEEE Transactions on Computer-Aided Design of Integrated Circuits and Systems,* 12(11):1714–1725, 1993.

[5] Lai TH, and White W. Mapping pyramid algorithm into hypercubes. *Journal of Parallel and Distributed Computing*, 9(1):42–54, 1990.

[6] Leighton FT. *Introduction to Parallel Algorithms and Architectures: Arrays, Trees, and Hypercubes.* Morgan Kaufmann, San Mateo, CA, 1991.

[7] Leiss EL, and Reddy HN. Embedding complete binary trees into hypercubes. *International Parallel Letters*, 38:197–199, 1991.

[8] Leitner E, and Selberherr S. Mixed-element decomposition method for three-dimensional grid adaptation. *IEEE Transactions on Computer-Aided Design of Integrated Circuits and Systems*, 17(7):561–572, 1998.

[9] Lorenz J., Ed., *Three-Dimensional Process Simulation.* Springer-Verlag, Heidelberg, 1995.

[10] Okabe A, Boots B, and Sugihara K. *Spatial Tessellations. Concept and Applications of Voronoi Diagrams.* John Wiley & Sons, New York, 1992.

[11] Öhring S, and Das SK. Incomplete hypercubes: embeddings of tree-related networks. *Journal of Parallel and Distributed Computing*, 26:36–47, 1995.

[12] Pottosin YV. "Assembling" a Boolean hypercube: an approach to assignment of finite state machines. In *Proceedings 2nd International Conference on Computer-Aided Design of Discrete Devices*, Minsk, Republic of Belarus, vol. 1, pp. 54–59, 1997.

[13] Saad Y, and Schultz MH. Topological properties of hypercubes. *IEEE Transactions on Computers*, 37(7):867–872, 1988.

[14] Schroeder WJ, and Shephard MS. A combine octree/Delaunay method for fully automatic 3-D mesh generation. *International Journal Numerical Methods in Engineering*, 29:37–55, 1990.

[15] Shen X, Hu Q, and Liang W. Embedding k-ary complete trees into hypercubes. *Journal of Parallel and Distributed Computing*, 24:100–106, 1995.

[16] Varadarajan R. Embedding shuffle networks into hypercubes. *Journal of Parallel and Distributed Computing*, 11:252–256, 1990.

[17] Wagner AS. Embedding the complete tree in hypercube. *Journal of Parallel and Distributed Computing*, 26:241–247, 1994.

[18] Wu AY. Embedding a tree networks into hypercubes. *Journal of Parallel and Distributed Computing*, 2:238–249, 1985.

6

Nanodimensional Multilevel Circuits

Developed models of two- and multilevel circuits are derived from and closely dependent on data structure. Data structures such as $\mathcal{N}$-hypercube and hypercube-like have been chosen to represent the switching function after careful analysis of the principle of data processing and technological characteristics of existing and predictable nanodevices (see Chapter 2). However, combination of "nano"-compatible models with very well developed traditional (mostly gate-level) models of switching circuits is considered as well, as such an approach is acceptable for some nanotechnologies.

This chapter presents the methodology for spatial logic design of combinational circuits. The spatial topological structure is defined as an $\mathcal{N}$-hypercube. Two approaches to building $\mathcal{N}$-hypercube structures are introduced:

- ▶ A symbolic model of a switching circuit, i.e., sum-of-products expression, and corresponding decision trees or diagrams followed by embedding these structures in an $\mathcal{N}$-hypercube,
- ▶ A combined approach where a circuit as a network of gates is represented by a direct acyclic graph (DAG) in which each node is modeled by $\mathcal{N}$-hypercube; the DAG is embedded in a hypercube-like structure in 3-D.

The first approach is of exponential complexity, and the second allows reduction of the complexity. Both approaches are relevant to design of $\mathcal{N}$-hypercubes of elementary functions, or elementary gates. These primitive $\mathcal{N}$-hypercubes are constructed by embedding decision trees of elementary functions.

$\mathcal{N}$-hypercube structure is characterized by topological metrics (space coordinates and space size, diameter, etc.). The recombination of the $\mathcal{N}$-hypercube structure in space makes it possible to change the order and polarity of variables. Transformations of $\mathcal{N}$-hypercube structure (dimension reduction, $\mathcal{N}$-hypercube deleting or merging, etc.) are possible, and resemble the rules of manipulation of Boolean expressions and decision trees.

The material is introduced as follows. After having formulated the problem in Section 6.1, we describe a library of standard of combinational gates in Section 6.2. The principles of multilevel circuit design are presented in 6.3. The rules for transformation of networks are introduced in Section 6.4. Numerical evaluation of 3-D structures is given in Section 6.5.

6.1 Graph-based models in logic design of multilevel networks

A multilevel circuit is defined as interconnections of single-output combinational logic gates under the assumption that the interconnection provides a unidirectional (no feedback) flow of signals from primary inputs to primary outputs. Multilevel network implementations can be restricted to a gate type (NAND, NOR) with a fixed fan-in.

A multiinput multioutput switching function can be represented by

- A gate-level model, or network of gates, described by a direct acyclic graph (DAG), and
- A functional level model, that is a two-level sum-of-products expression, decision tree or decision diagram.

6.1.1 DAG-based representation of multilevel circuits

The well known techniques that are used in today's multilevel circuit design utilize DAG construction and optimization. In design at the physical level, Boolean mapping is required, which means covering of the DAG by DAGs of elementary gates from the library of gates. A DAG is the way contemporary multilevel circuits are represented and it is a gate-level model that is compatible with a library of traditional gates.

6.1.2 Decision diagram based representation of circuits

Decision diagrams correspond to multiplexer (MUX)-based implementation that is not widely used in today's logic design. The situation is quite opposite in the design of new devices: as we have shown in Chapter 2, the appropriate logic for nanowire wrap-gate single-electron devices is based on MUX-gate or T-gate implementation. This technique suffers from exponential complexity, which is reduced by transforming decision trees to decision diagrams. In nanodimensions, however, the regularity of the tree structures may hold more benefits than the compactness of decision diagrams.

6.1.3 $\mathcal{N}$-hypercube model of multilevel circuits

In Chapter 5, we considered an $\mathcal{N}$-hypercube to represent arbitrary n-input functions. It has been shown that an $\mathcal{N}$-hypercube can be generated by embedding a decision tree that represents a switching function. It has been mentioned that both DAG and $\mathcal{N}$-hypercube models can be combined so that a hybrid structure can be derived. This is a solution to the exponential complexity problem in decision trees mentioned above.

6.2 Library of $\mathcal{N}$-hypercubes for elementary logic functions

We will focus on

- Generation of elementary $\mathcal{N}$-hypercubes, and
- Evaluation of elementary $\mathcal{N}$-hypercube structures.

In this section, we introduce the library of $\mathcal{N}$-hypercubes that implement elementary switching and multivalued functions. This representation is based on:

- The two-level form of a switching function of a gate, corresponding to a two-level tree, and
- Characterization, analysis and study of recombination (while the order of variables is changed).

6.2.1 Structure of the library

In Boolean mapping, a library is understood as a set of logical elements. Design of a logic network over a given library of gates is accomplished by covering a DAG by DAGs of elementary gates from the library of gates. The library contains the set of logic gates that are available in the desired design style. Each element is characterized by its function, inputs, outputs, and some parameters such as area, delay and capacity load. The library used below is characterized by the following properties:

- A variety of models of gates that represent elementary switching functions (AND, OR, NOR, NAND, EXOR, etc.).
- Flexibility of modification of gates to meet requirements of technology (for example, extension of inputs and outputs designs of macrocells).
- Flexibility in choosing implementation models: decision trees (Shannon and Davio), linear word-level representation.
- Flexibility for multivalued logic.

6.2.2 Metrics of $\mathcal{N}$-hypercube

Let us define a *primitive* as a 1-dimensional $\mathcal{N}$-hypercube of (Figure 6.1). This will be used below as a unit to evaluate $\mathcal{N}$-hypercube models of gates.

The simplest topological characteristics of the $\mathcal{N}$-hypercube gate include:

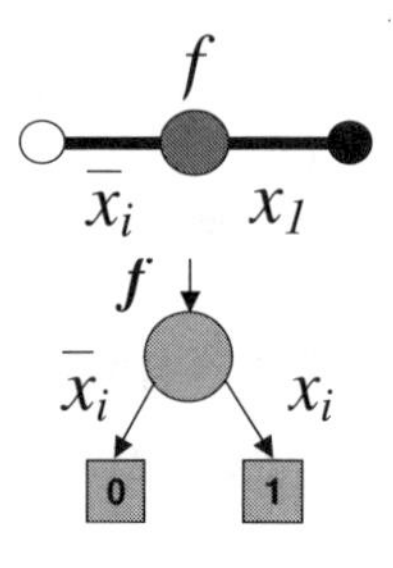
f
x̄ᵢ
x₁
f
x̄ᵢ
xᵢ
0
1
• A primitive 𝒩-hypercube is a 1-dimensional 𝒩-hypercube (a root node and two terminal nodes).
• The primitive 𝒩-hypercube corresponds to the decision tree of a single variable (node).
• The primitive is a unit metric unit in the 𝒩-hypercube structure.

FIGURE 6.1
The primitive 1-dimension $\mathcal{N}$-hypercube, and the corresponding node of a decision tree.

Number of primitives. Because an arbitrary $\mathcal{N}$-hypercube can be assembled on an $\mathcal{N}$-hypercube of lesser dimensions, the primitive is the most reasonable unit of topological complexity.

Number of nodes (terminal and intermediate). The active terminal nodes are used to evaluate power dissipation characteristics. The number of *active* nodes is equal to the number of 1's in the truth table of the implemented switching function, which is the current state of the $\mathcal{N}$-hypercube structure. The intermediate nodes implement the switching functions and, therefore, can be used for evaluation of circuit complexity.

Connectivity is the characteristic that describes the complexity in a number of links between root, terminal and intermediate nodes. Given the coordinates of a link, it is possible to measure the sizes of links, compare them, etc. Based on the notation of connectivity, the topological distributions of links are calculated, as are the distances from an arbitrary point to a set of points in space.

Space size. For a given sets of links, the space size $V = X \times Y \times Z$ can be calculated, for example, in the *primitive* units. The space size is used for local and global power characteristic evaluation.

Example 6.1 *The space size for the $\mathcal{N}$-hypercube given in Figure 6.4 can be calculated in units of the $\mathcal{N}$-hypercube primitives and links.*
(a) $V = X \times Y = 1 \times 1 = 1$ *primitive*^{2}.
(b) $V = X \times Y = 2 \times 2 = 2^2$ *links*^{2}.

Diameter is the global characteristic of a circuit in space. Diameter is the maximum number of links between any two nodes in the $\mathcal{N}$-hypercube.

Example 6.2 *In Table 6.1, the topological characteristics of the $\mathcal{N}$-hypercube model of two- and three-input $\mathcal{N}$-hypercube gates are given.*

We observe from Table 6.1, that

- Different gates can be characterized by the same topological characteristics. Hence, this metric is universal in this sense.
- 2- and 3-input gates are represented by 2-D and 3-D $\mathcal{N}$-hypercubes respectively.

TABLE 6.1
Metrics of 2-input and 3-input $\mathcal{N}$-hypercube gates.

Metrics	2-input	3-input
Total number of primitives, $\# \mathcal{N}$	3	7
Number of terminal nodes, N_T	4	8
Number of intermediate nodes, $N_I + 1(root)$	2+1	6+1
Total number of nodes, N_{Total}	3	14
Connectivity, C *links*:	6	14
Space size, V *links*^{2}:	2^2	2^3
Diameter, D *links*	4	6

6.2.3 Signal flowgraphs on an $\mathcal{N}$-hypercube

Diagrams that describe computation in terms of the flow of signals between operators are called *signal flowgraphs*. Signal flowgraphs of an $\mathcal{N}$-hypercube represent information streams between the nodes. Suppose that the direction of the information stream is from the root to the terminal nodes. Then the root node is associated with the source of information, and the terminal nodes correspond to the receivers of information.

Example 6.3 *Figure 6.2 shows one of possible ways to derive signal flowgraphs.*

(a) *In the 1-D $\mathcal{N}$-hypercube (Figure 6.2a), the directed stream of information from the source f reaches the terminal 0 that corresponds to the switching function $f = \overline{x}$.*

(b) *In the 2-D $\mathcal{N}$-hypercube (Figure 6.2b), the stream of information from the source f reaches the terminals 01 and 10 that correspond to the switching function given by the truth vector* $\mathbf{F} = \begin{bmatrix} f(00) \\ f(01) \\ f(10) \\ f(11) \end{bmatrix} = [0110]^T$, *that is* $f = x_1 \oplus x_2$.

(c) *By analogy, in the 3-D $\mathcal{N}$-hypercube (Figure 6.2c), the truth vector is* $\mathbf{F} = [11011000]^T$ *that is* $f = \overline{x}_1\overline{x}_2\overline{x}_3 \vee \overline{x}_1\overline{x}_2 x_3 \vee \overline{x}_1 x_2 x_3 \vee x_1\overline{x}_2\overline{x}_3 = \overline{x}_1\overline{x}_2 \vee \overline{x}_2\overline{x}_3 \vee \overline{x}_1 x_3$.

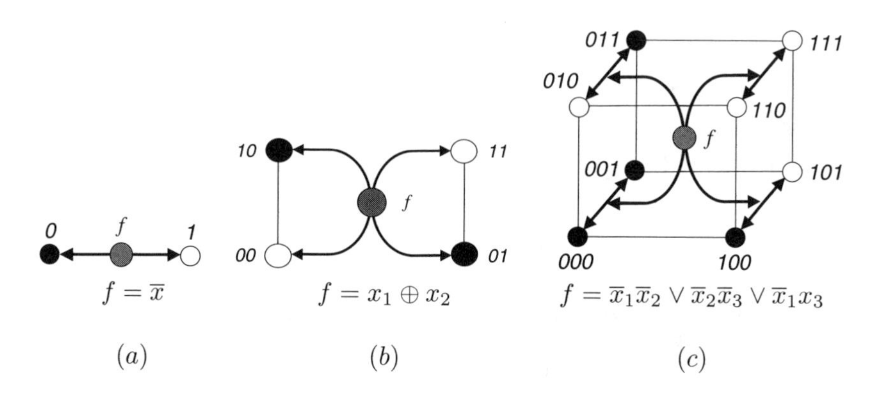

FIGURE 6.2
Signal flowgraph in the $\mathcal{N}$-hypercube: 1-D (a), 2-D (b), and 3-D (c) (Example 6.3).

The signal flow in the $\mathcal{N}$-hypercube is different from the signal graph in the gate-level model, or DAG. The input/output relationship of a circuit is described by equations (models). The signal propagation can be interpreted in various ways. For example, in decision trees and diagrams, the direction of a signal stream is usually chosen from the root node (f) to terminal nodes (the values of variables x_i). In the logic network, the signals are propagated from the inputs (the values of variables x_i) to the outputs (f_i). In physical implementation, the direction of a signal flow is defined by the direction of carriers of information. Hence, in mathematical models, we can choose a direction that better introduces the properties of the model.

Some nonformal rules that help to choose the direction of a signal flowgraph are as follows:

- Traditionally, in the models where input/output notation is used, the direction of a flow graph is from input to output.

- In technology-dependent models, the direction of information flow is chosen based on physical carriers of information. For example, in T-gate based nanowires with wrap-gate devices, the electron is injected to the root node.

6.2.4 Manipulation of $\mathcal{N}$-hypercube

Two attributes of a $\mathcal{N}$-hypercube can be changed by reconfiguration:

- Polarity of variables, and
- The order of variables in a decision tree and decision diagram.

The reconfiguration of the $\mathcal{N}$-hypercube is defined as *rotation.*

Example 6.4 *Two orders of variables are possible in the decision tree derived for the two-variable function NAND (Figure 6.3). The level exchange is required in the case we change the order. In the embedded $\mathcal{N}$-hypercube, a change of the order does not influence the levels, it results in a change of orientation of axes, i.e., rotation. Mapping is implemented in such a way that we do not need to change the configuration or assign variables to other edges, – we have to choose the proper edges (that exist already or a priori) depending on the given variable order.*

However, this structure must be configured with respect to the order of variables in the decision tree. There are six possible orders, in particular:

- The left figure corresponds to the order x_1, x_2,
- The right figure describes the order x_2, x_1, and

the axes are associated with the polarity of variables. It also follows from Figure 6.3, that rotation changes the configuration of flowgraphs. The above results can be developed for a network of two- and three-input gates, i.e., multilevel circuits.

6.2.5 Library-based design paradigm

An approach to creating a library of $\mathcal{N}$-hypercube gates is based on the following technique:

`Switching function (gate)` $\Longleftrightarrow$
`Decision tree` $\Longleftrightarrow$
$\mathcal{N}$`-hypercube`

This means that any standard gate of the conventional combinational logic, i.e., an n-input AND, OR, NOR, NAND, NOT and EXOR gate, can be represented by the $\mathcal{N}$-hypercube model in two steps:

- A decision tree of the gate is derived;
- The tree is embedded in an $\mathcal{N}$-hypercube of n-dimensions.

An $\mathcal{N}$-hypercube of a two-input gate includes the root node, two intermediate nodes, four terminal nodes, and the edges (connections).

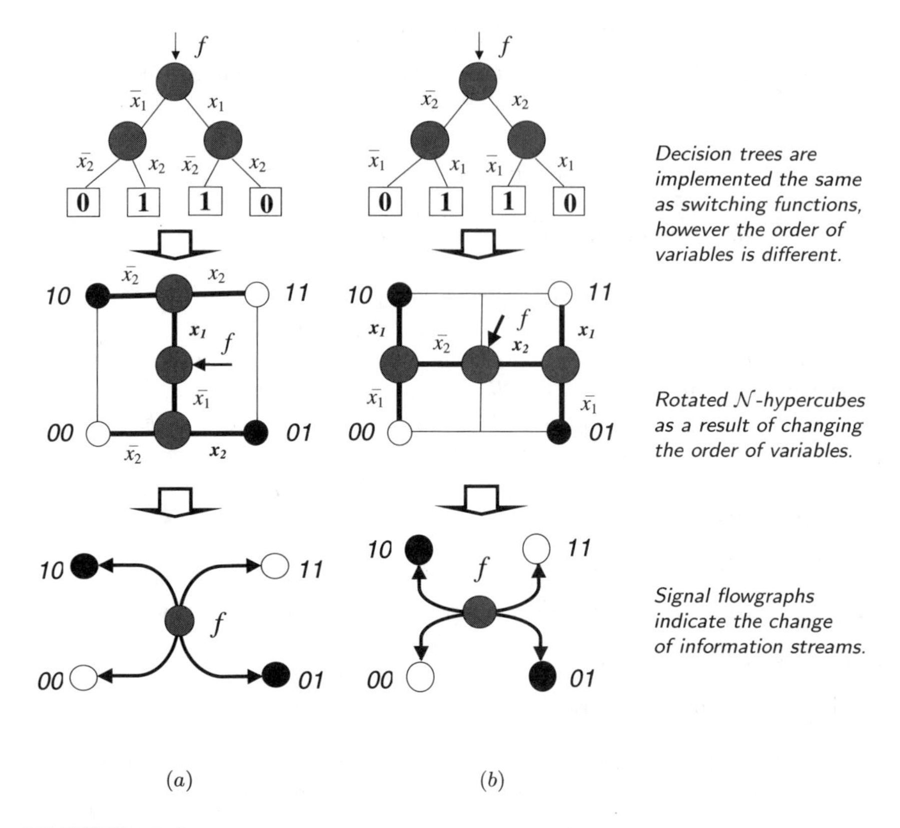

FIGURE 6.3
Embedding of a 2-variable binary tree into the $\mathcal{N}$-hypercube for variable order $\{x_1, x_2\}$ (a) and $\{x_2, x_1\}$ (b) (Example 6.4).

Example 6.5 *Figure 6.4 illustrates designing an $\mathcal{N}$-hypercube model for the two-input NAND gate. Given the two-input NAND function $f = \overline{x_1 x_2}$ (a), the Shannon decision tree is derived (b). Next, this tree is embedded in an $\mathcal{N}$-hypercube (c).*

6.2.6 Useful denotation

To simplify the graphical representation and visualization of a circuit in spatial dimensions, we use a uniform denotation as follows. A switching function of two-, three-, four-, or five variables is represented by a two-, three-, four-, or five-dimensional $\mathcal{N}$-hypercube respectively.

The $\mathcal{N}$-hypercube in 2-D, 3-D, 4-D, and 5-D is denoted by a cube with two, three, four, and five inputs respectively (Figure 6.5). Notice that from this denotation, it is easy to recognize:

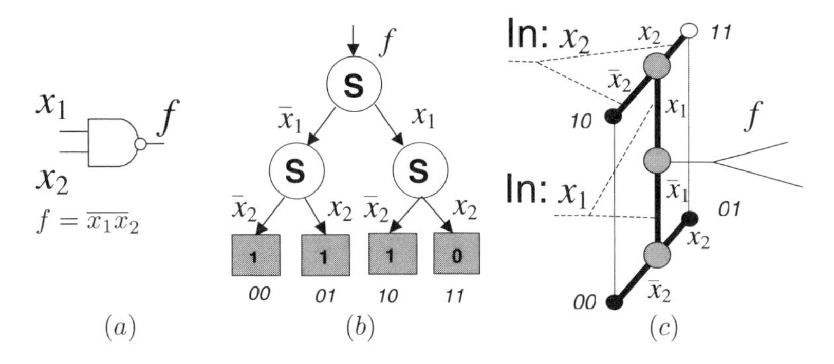

FIGURE 6.4
Fragment of a library of $\mathcal{N}$-hypercube gates: a two-input NAND gate (a), its decision tree (b), and the corresponding $\mathcal{N}$-hypercube (c) (Example 6.5).

- The function's dimensions.
- General topology configuration, and
- Structural properties.

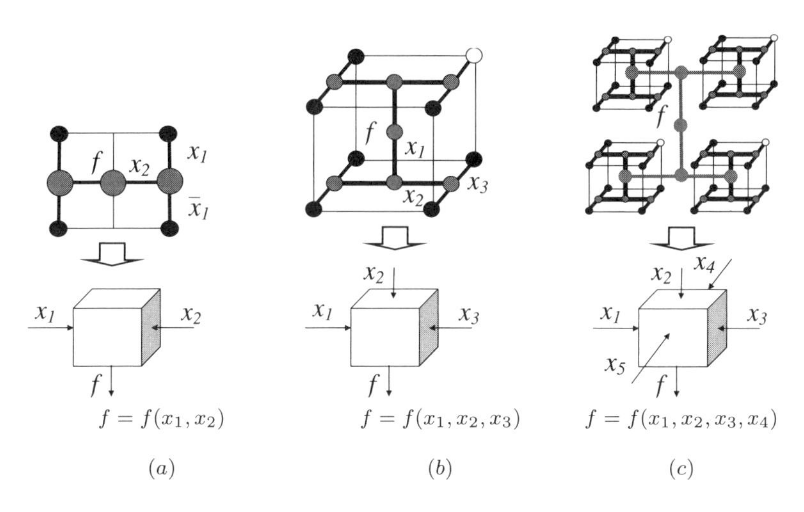

FIGURE 6.5
Denotation of a multidimensional $\mathcal{N}$-hypercube: (a) 2-D, (b) 3-D, and (c) 5-D.

6.3 Hybrid design paradigm: $\mathcal{N}$-hypercube and DAG

An arbitrary switching function can be represented by an $\mathcal{N}$-hypercube derived from the decision tree of a function. However, for a traditional approach based on covering the DAG of a circuit by the library of gates, netlist must be also considered. In terms of implementation on nanodevices (nanowire networks), the problem of interconnections arises. It is not an unsolved problem for all nanoelectronic implementations (see Chapter 2).

We focus on two approaches for deriving 3-D data structures for switching functions:

- ▶ A logic network (direct acyclic graph) presented, i.e., netlist is treated as a tree; and
- ▶ Embedding the tree in an $\mathcal{N}$-hypercube.

6.3.1 Embedding a DAG in $\mathcal{N}$-hypercube

Deriving a 3-D structure given a circuit netlist is implemented in two steps:

Step 1. The circuit netlist is represented by a DAG and levelized to obtain an incomplete decision tree;

Step 2. The tree is embedded in an $\mathcal{N}$-hypercube.

Thus, there are two levels of embedding:

- ▶ Embedding decision trees in elementary $\mathcal{N}$-hypercubes of gates, and
- ▶ Embedding a tree, i.e., DAG, in a "macro" $\mathcal{N}$-hypercube.

6.3.2 Levelization and cascading

The DAG of a switching circuit must be levelized in order to be treated as a binary decision tree. Given a multioutput function, each output is scanned from output to inputs, and the result is a DAG. Next, it is levelized.

The algorithm for levelization is based on the representation of i-th, $i = 1, 2, \ldots, k$, level of a k-level circuit by $\mathcal{N}$-hypercube structures. The result is a set of k $\mathcal{N}$-hypercubes with corresponding links. This algorithm can be used for different data structures; for example, word-level linear word-level representation. Levelized DAG is treated as a tree and embedding of the tree into an $\mathcal{N}$-hypercube goes ahead.

The algorithm for cascading r-output logic circuit is represented by a set of r $\mathcal{N}$-hypercubes. This partition (cascading) is made by traversing from each output to the inputs of the circuit.

6.4 Manipulation of $\mathcal{N}$-hypercubes

The technique of transformation of circuit models in spatial dimensions is based on

- ► Algebraic simplifications of switching functions,
- ► Topological simplifications, and
- ► Logic-topological transformations.

Algebraic simplifications can modify the topology, and topological simplifications can change the switching function described by this topology. Therefore, algebraic and topological transformations must be carefully combined. In this section, the rules for manipulation of an $\mathcal{N}$-hypercube are introduced. These rules allow reduction of dimensions and simplify circuit representation in spatial dimensions, and are also useful in verification of $\mathcal{N}$-hypercubes.

Dimension reduction. If the input i of an $\mathcal{N}$-hypercube to implement an n-input OR function is equal to 0, reduce this input and replace this hypercube with the $(n-1)$-dimensional $\mathcal{N}$-hypercube. Figure 6.6a illustrates this property for $i = 3$. By analogy, the dimensions of the $\mathcal{N}$-hypercube are decreased by 1 for an n-input AND function if the input is equal to 1 (Figure 6.6b).

$\mathcal{N}$-hypercube deleting. If the input of an $\mathcal{N}$-hypercube implementing an AND function is equal to 0, replace the hypercube with the constant 0 (Figure 6.6c). By analogy, replace an OR $\mathcal{N}$-hypercube with constant 1, if one of the outputs is equal to 1 (Figure 6.6d).

$\mathcal{N}$-hypercube merging. Two AND (OR) $\mathcal{N}$-hypercubes of dimensions i and j correspondingly connected in a series can be merged into one AND (OR) $\mathcal{N}$-hypercube of $i+j$ dimensions (Figure 6.7).

Deleting of duplicated $\mathcal{N}$-hypercubes. If there are two $\mathcal{N}$-hypercubes whose inputs and outputs are the same, remove one and create a fan-out (Figure 6.7c)

Reduction of redundant connections. Consider the manipulation of two $\mathcal{N}$-hypercubes connected in a series, A and B, with respect to the input x_j:

$$f = \underbrace{\overline{\underbrace{(\overline{x_j x_2 \overline{x}_3})}_{Hypercube\ A} \; x_j x_4 x_5}}_{Hypercube\ B} = \underbrace{\overline{\underbrace{(\overline{x_2 \overline{x}_3})}_{Hypercube\ A} \quad x_j x_4 x_5}}_{Hypercube\ B}$$

It follows from the above that the input x_j can be deleted from $\mathcal{N}$-hypercube A (Figure 6.7d).

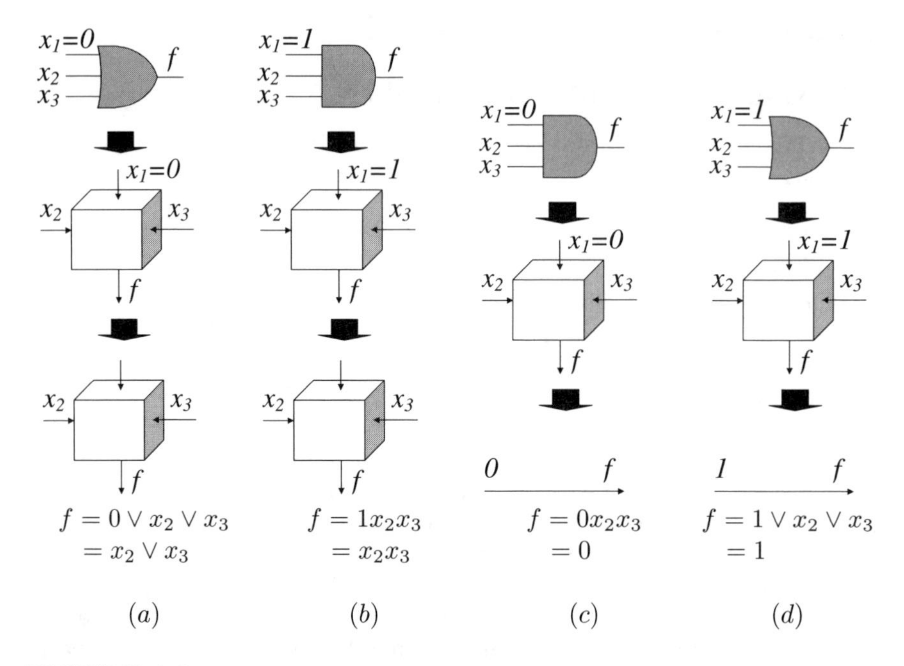

FIGURE 6.6
Reduction of dimensions to implement 3-input OR function when $x_i = 0$ (a), 3-input AND function when $x_i = 1$ (b), and $\mathcal{N}$-hypercube degeneration: 3-input OR function while $x_i = 0$ (c), and 3-input AND function while $x_i = 1$(d).

Based on the properties of switching functions, the remaining rules for simplification and manipulation of $\mathcal{N}$-hypercube topology structures can be derived.

Example 6.6 *In Figure 6.8, the extension of EXOR function with respect to inputs is given. The resulting function is a 5-input EXOR over a library of 2-input EXOR gates. To implement this function using $\mathcal{N}$-hypercube structure, four 2-D $\mathcal{N}$-hypercubes are used.*

Of course, it is possible to use the embedding or cascading method here.

Example 6.7 *In Figure 6.9a, the initial circuit is cascaded into two subcircuits with outputs f_1 and f_2. Since the gates can be considered as nodes, the subcircuits are equivalent to the two logic networks. The embedding of these networks results in two incomplete $\mathcal{N}$-hypercubes over the library of 3-D gates (Figure 6.9b).*

Hence, instead of representation of this circuit by a complete decision tree, that is, a tree with $\mathbf{2^6}$ terminal nodes, **6** levels, **62** intermediate nodes and a

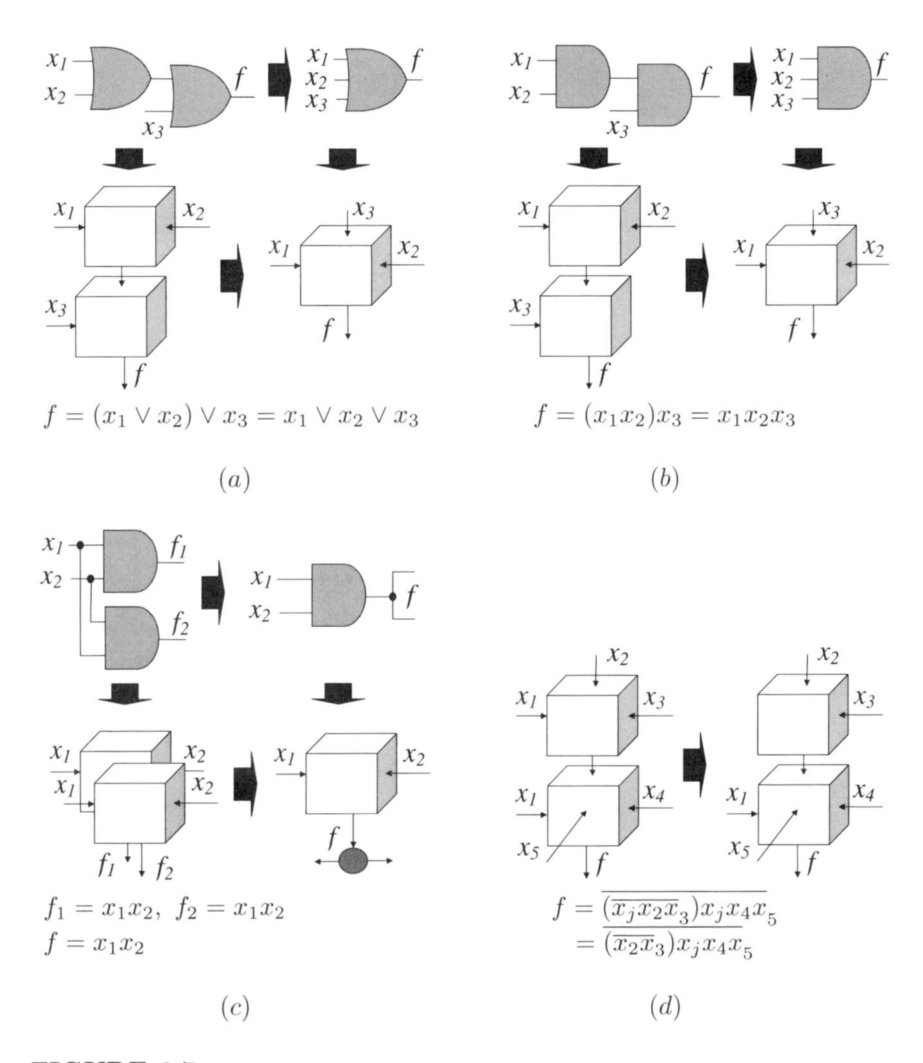

FIGURE 6.7
Merging of AND (a) and OR (b) $\mathcal{N}$-hypercubes, reduction of duplicated $\mathcal{N}$-hypercubes (c) and redundant connections (d).

root, or by 6-dimensional $\mathcal{N}$-hypercube, the circuit is described by two logic networks with the following characteristics:

- ▶ The first network includes 5 terminal nodes and 4 intermediate nodes, and
- ▶ The second network includes 4 terminal nodes and 3 intermediate nodes

that correspond to two incomplete 3-D $\mathcal{N}$-hypercubes.

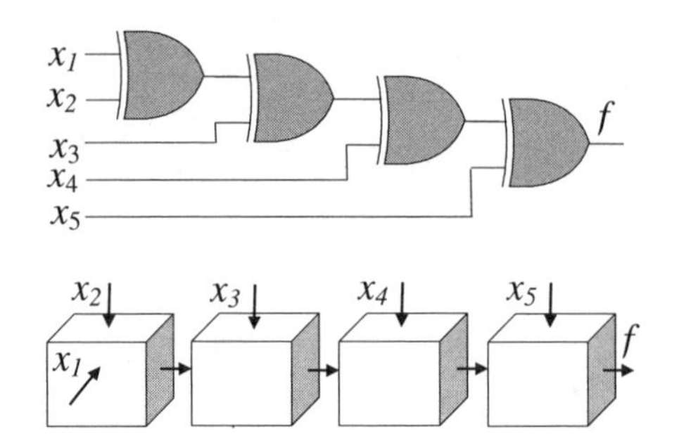

FIGURE 6.8
Linear $\mathcal{N}$-hypercube structure that implements 5-input and 4-level circuits (Example 6.6).

6.5 Numerical evaluation of 3-D structures

In this section, the results of numerical evaluation of $\mathcal{N}$-hypercubes combinational circuits from the ISCAS85 standard database are given. In experiments, circuits from 6 gates to more than 3,500 gates were used.

6.5.1 Experiment on evaluating the $\mathcal{N}$-hypercube

In this experiment, parameters of $\mathcal{N}$-hypercube derived by the decision diagrams were evaluated. We used the parameters acquired from the shared reduced ordered binary decision diagrams (BDDs) built from the circuit netlist.

The results of evaluation are summarized in Table 6.2 in which:

The first three columns named **TEST** include the benchmark type **Name**, the number of inputs and outputs **I/O**, and the number of gates **#G**.

The next four columns named **SPACE SIZE** include the maximum number of dimensions **#Dim** (this resembles the numbers of variables, and it can be less for each separate function represented by a BDD), and the size of the solid in **X, Y** and **Z** coordinates.

The last column **NODES** includes the number of nodes in BDD. Note that the topological characteristics are represented by the total number of nodes in the shared BDD; this number is too large for some circuits and is not shown in the Table.

Consider, for example, circuit c6288. This is a 32 bit multiplier that includes 2,416 NAND gates. Using decision trees of each output, derived from the netlist, $\mathcal{N}$-hypercube structure with 11×11, and 10 primitives connections (lines) in the 3-D topology.

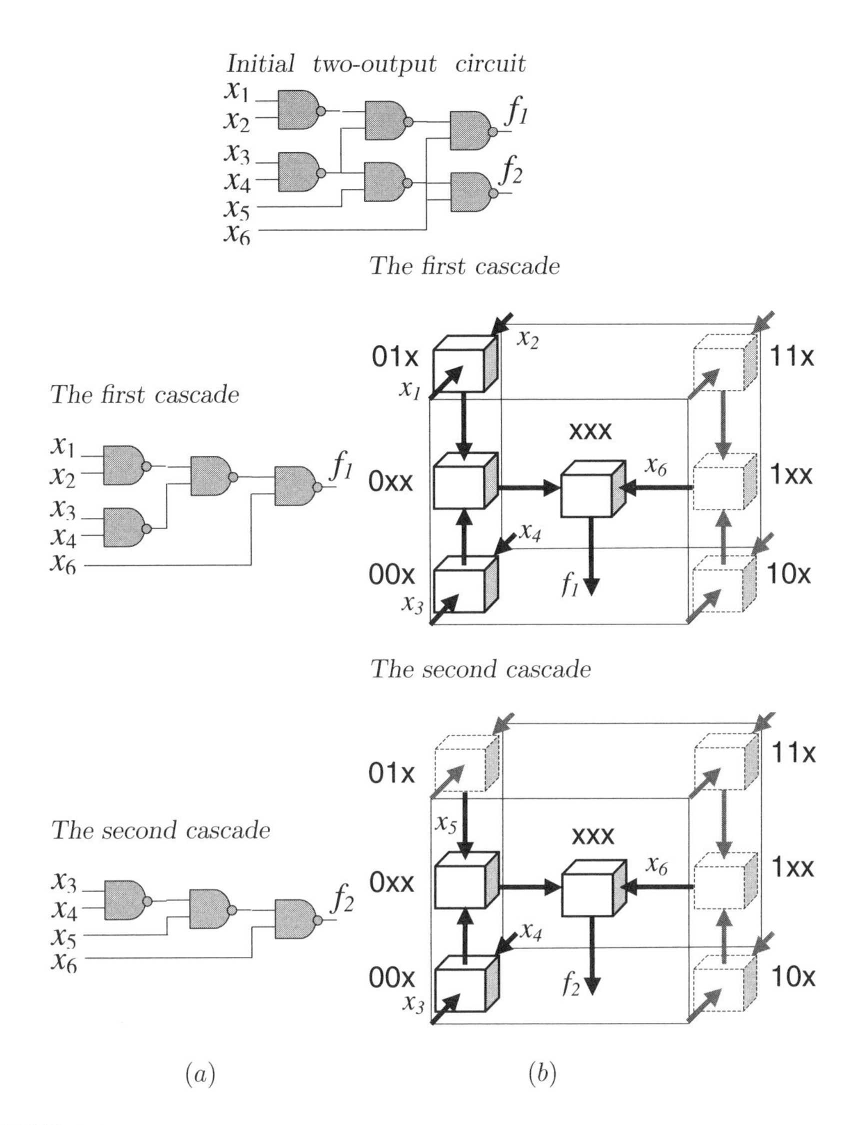

FIGURE 6.9
Cascading of a circuit (a) to represent the outputs f_1 and f_2 by incomplete $\mathcal{N}$-hypercubes (b)(Example 6.7).

Observation. It follows from this experiment that:

- The number of dimensions is equal to the number of variables. It is a reasonable value.
- Spatial parameters X, Y and Z are relevant to the number of levels in the decision tree, i.e., correlated with the number of variables.

TABLE 6.2
Fragment of an experiment on evaluation of $\mathcal{N}$-hypercube parameters derived from a decision diagram.

TEST				SPACE SIZE			NODES
Name	**I/O**	**#G**	**#Dim**	**X**	**Y**	**Z**	**# Node**
c17	5/2	6	5	2	2	1	11
c432	36/7	160	36	88	12	12	1460
c499	41/32	202	41	14	14	13	45922
c880	60/26	383	60	20	20	20	101076
c1355	41/32	546	41	14	14	13	45922
c1908	33/25	889	33	11	11	11	42427
c2670	233/140	1,193	233	78	78	77	-
c3540	50/22	1669	50	17	17	16	422803
c5315	178/123	2307	178	60	59	59	-
c6288	32/32	2,416	32	11	11	10	-
c7552	207/108	3,512	207	69	69	69	-

- ▶ The number of terminal nodes is upperbounded by 2^n for an n-variable function.

6.5.2 Experiment on evaluating the hybrid $\mathcal{N}$-hypercube

In this experiment, the hybrid $\mathcal{N}$-hypercube based approach was evaluated for a selected output in each test circuit. Each gate in the network was replaced by a $\mathcal{N}$-hypercube model. An $\mathcal{N}$-hypercube of each output has been derived from the tree obtained for each output by scanning from output to inputs, and next, by levelization.

The results of evaluation are summarized in Table 6.3 where:

The first three columns named **TEST** include the benchmark **Name**, the number of the selected output, and the number of gates **#G** in the selected output. We have selected the output whose implementation involves the maximum number of gates, and considered the subnetwork that involves the inputs and gates to implement this function.

The next three columns named **SPACE SIZE** include the size of the structure in **X, Y** and **Z**.

The last two columns named **NODES** contain the total number of terminal nodes $\#N_T$ (that is the total number of intermediate nodes of the 2- and 3-D $\mathcal{N}$-hypercubes of the gates), and the total number of intermediate nodes $\#N$ in the $\mathcal{N}$-hypercube.

For example, a 32 bit multiplier c6288 requires $248 \times 248 \times 244$ elementary

TABLE 6.3
Fragment of an experimental study of the DAG based 3-D models.

TEST			SPACE SIZE			NODES	
Name	$\#O_c$	$\#G_c$	**X**	**Y**	**Z**	$\#N_T$	#Node
c17	1	4	8	8	2	16	12
c432	5	126	66	64	66	2022	1896
c499	1	102	28	24	20	468	366
c880	24	130	70	72	70	612	482
c1355	1	322	58	52	54	1346	1024
c1908	25	522	100	104	92	2526	2004
c2670	139	828	82	80	78	3594	2766
c3540	21	1458	132	132	140	9462	8004
c5315	122	937	138	132	126	3750	2813
c6288	32	2327	248	248	244	9246	6916
c7552	107	474	114	112	106	1916	1442

links in the dimension X, Y, Z and 9246 terminal nodes to represent 2327 gates, and 6916 intermediate nodes.

Observation. It follows from the evaluation that:

- ► The hybrid approach does not suffer from exponential complexity of the number of terminal nodes, and demonstrates less values of the terminal nodes than is relevant to the number of gates in the benchmark circuit,
- ► The space sizes X, Y and Z are higher than in the case of shared BDD models (Table 6.2).

6.6 Summary

An arbitrary combinational circuit can be represented by an $\mathcal{N}$-hypercube from the library of $\mathcal{N}$-hypercube gates. The design methodology includes different methods of manipulation of data structures.

1. Designing $\mathcal{N}$-hypercube structure includes:

- ► Representation of a given switching function in the appropriate form (decision tree, decision diagram, logical network);

- Construction of topological structure accordingly to the rules (recursion, connections) and algorithms (embedding, levelization, and cascading).
- Analysis of the topological structure (deriving the flowgraph, verification, simplification).

2. The properties of $\mathcal{N}$-hypercube structure include:

- Topological characteristics (diameter, number of nodes, links, and primitives);
- The number of primitive $\mathcal{N}$-hypercubes;
- Configuration of a set of $\mathcal{N}$-hypercubes, in particular, linear, incomplete or complete $\mathcal{N}$-hypercubes;
- Signal flowgraphs that characterize the distribution of information streams in the $\mathcal{N}$-hypercube.

3. The strategy for designing $\mathcal{N}$-hypercubes involves three basic approaches:

- *Embedding* technique: a switching function must be represented by a tree and then embedded in a $\mathcal{N}$-hypercube. This approach is limited in terms of size when a decision tree or diagram is embedded. Hence, it is useful for representation of small switching functions, for example, elementary switching functions in library of gates design. For large switching functions, DAG of the circuit can be treated as tree, while each gate is represented by a decision tree. The tree derived from DAG is levelized.
- *Levelization* technique: the implemented network is levelized, and each level is considered a level of the tree that is embedded in an $\mathcal{N}$-hypercube; This approach is used in hybrid technique for designing large networks in spatial dimensions.
- *Cascading* technique: the implemented circuit is cascaded, i.e., represented by a logical network of a single output; the number of networks is equal to the number of outputs of a circuit. The logical network is embedded in the $\mathcal{N}$-hypercube. The set of $\mathcal{N}$-hypercubes is represented by an initial circuit. This approach is useful for representation of large circuits in spatial dimensions. Hence, the cascading technique is implemented in the *mixed* strategy, i.e., decision trees are embedded in $\mathcal{N}$-hypercubes to represent the gate, and logical networks are embedded in the $\mathcal{N}$-hypercubes.

Problems

Problem 6.1 Consider a two-level combinational circuit shown in Figure 6.10.

(*a*) Derive a decision tree and a hypercube for each node.

(*b*) Derive a DAG of the circuit and embed it in a hypercube structure. Taking into account the results of (a), evaluate the parameters of the hybrid approach (a DAG-based hypercube in which each node is a hypercube of a gate).

(*c*) Derive two decision trees each of them representing one of two outputs (functions f_1 and f_2) and embed them each in a hypercube; compare the parameters of the decision tree-based hypercubes (total values over both hypercubes) and parameters of part (b).

(*d*) Derive a set of two word-level linear decision diagrams for the levelized circuit. Embed each in a hypercube with a Davio expansion in the nodes.

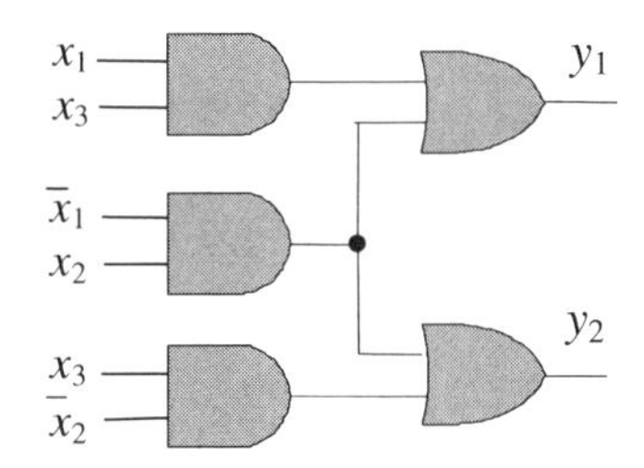

FIGURE 6.10
Two-level logic function for Problem 6.1.

Problem 6.2 Consider two 4-D $\mathcal{N}$-hypercubes, A and B (Figure 6.11a) that implement functions $f_1 = x_1 \vee x_2 \vee x_3 \vee x_4$ and $f_2 = x_1 \vee x_2 \vee x_3 \vee x_5$. Prove that functions f_1 and f_2 can be implemented by 2-D $\mathcal{N}$-hypercubes.

Problem 6.3 Given the $\mathcal{N}$-hypercube derived from a DAG of a switching function, restore the function.

(*a*) The nodes of the $\mathcal{N}$-hypercube in Figure 6.12a implement EXOR function.

(*b*) The nodes of the $\mathcal{N}$-hypercube in Figure 6.12b implement NAND function.

(*c*) In the $\mathcal{N}$-hypercube depicted in Figure 6.12c, the nodes $\{01\mathbf{x}, 0\mathbf{xx}, 1\mathbf{xx}\}$ implement EXOR function, and the rest of nodes implement AND function.

Problem 6.4 Construct a binary decision tree of the switching functions given below and embed it into the $\mathcal{N}$-hypercube

(*a*) 2-input majority circuit

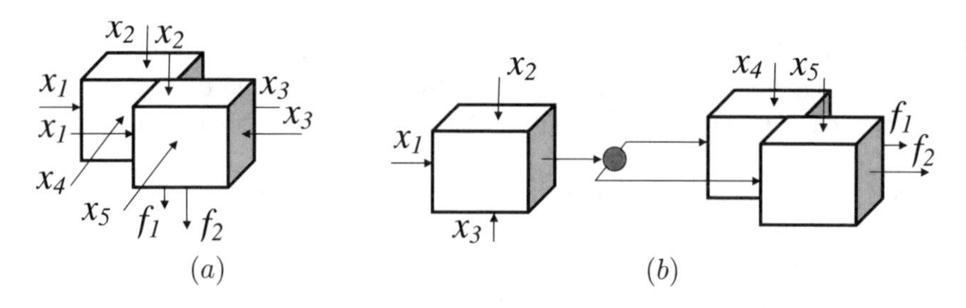

FIGURE 6.11
$\mathcal{N}$-hypercube structure transformation for Problem 6.2.

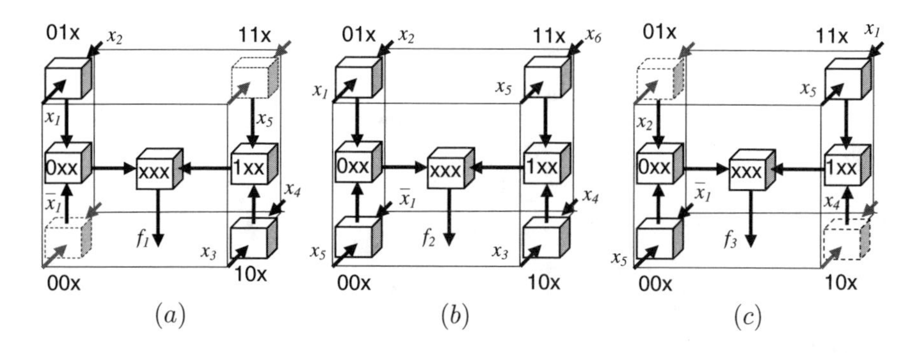

FIGURE 6.12
$\mathcal{N}$-hypercube structures for Problem 6.3.

(*b*) 3-input majority circuit
(*c*) Full adder
(*d*) 2-input adder

Problem 6.5 In Figure 6.13, two $\mathcal{N}$-hypercubes are given.

(*a*) Verify that they represent the same, or different switching functions assuming the NAND function in the nodes.
(*b*) Verify that they represent the same, or different switching functions assuming the EXOR function in the nodes $\{01\mathbf{x}, 0\mathbf{xx}, 1\mathbf{xx}\}$ and AND function in the nodes $\{01\mathbf{x}, 0\mathbf{xx}, 1\mathbf{xx}\}$.

Problem 6.6 Represent by the $\mathcal{N}$-hypercube

(*a*) The NAND circuit given in Figure 6.14a.
(*b*) The circuit given in Figure 6.14b.
(*c*) The circuit given in Figure. 6.14c

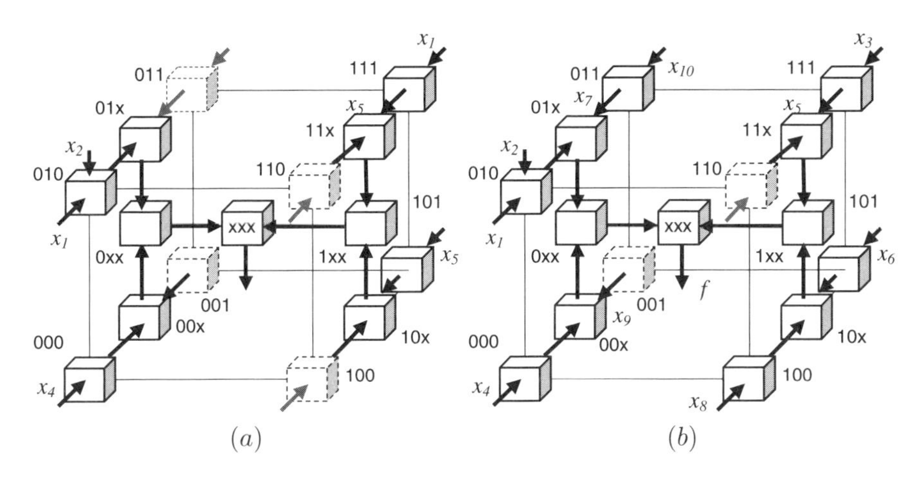

FIGURE 6.13
The $\mathcal{N}$-hypercube for Problem 6.5.

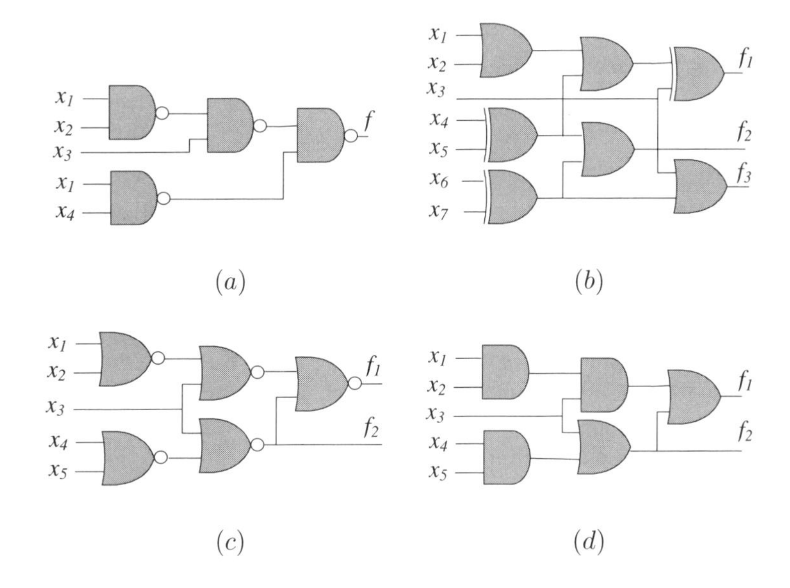

FIGURE 6.14
Circuits for Problem 6.6.

(*d*) The circuit given in Figure 6.14d.

Problem 6.7 For the decision tree given in Figure 6.15a,

(*a*) Derive the switching function,

(*b*) Represent the network (use any number of levels),

(*c*) Design a DAG of the network and embed it into a DAG based $\mathcal{N}$-hypercube.

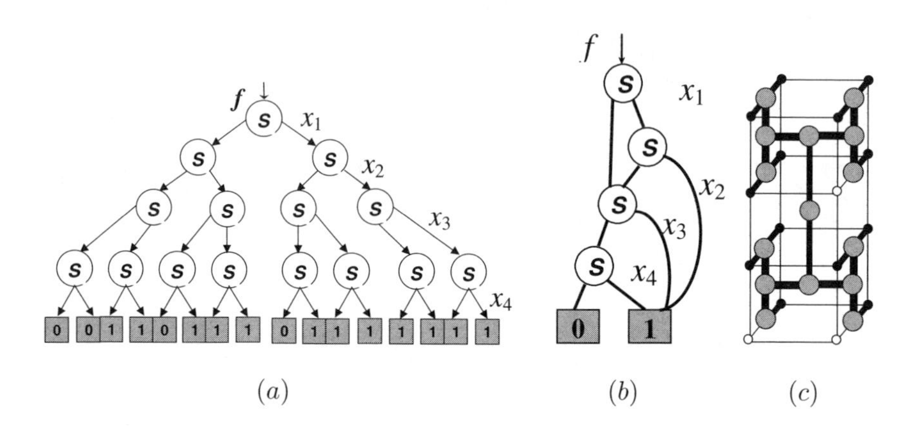

FIGURE 6.15

Decision tree (a), decision diagram (b) and the $\mathcal{N}$-hypercube (c) for Problems 6.8 and 6.9.

Problem 6.8 Justify that the decision tree and decision diagram given in Figure 6.15a,b represent the same switching function.

Problem 6.9 Derive an incomplete $\mathcal{N}$-hypercube from Figure 6.15c.

Problem 6.10 The topology presented in Figure 6.16 implements a DAG based embedded structure consisting of three connected hypercubes. Restore the gate-level circuit.

HINT: to restore the gate functions, analyze the terminal nodes.

Problem 6.11 In logic design techniques, BDD are mapped into a circuit netlist, so that each BDD node is associated with a universal element, e.g., multiplexer. A multiplexer normally consists of two AND and one OR gate, in a gate-level design. Pass-transistor logic may be used to implement a multiplexer with a smaller number of transistors, however, level restoration problem causes the transistor count to increase.

Compare pass-transistor logic that is

(a) A result of mapping a BDD in a network of pass-gate, and

(b) A SET-based model of a decision tree node.

See Chapter 2 for references.

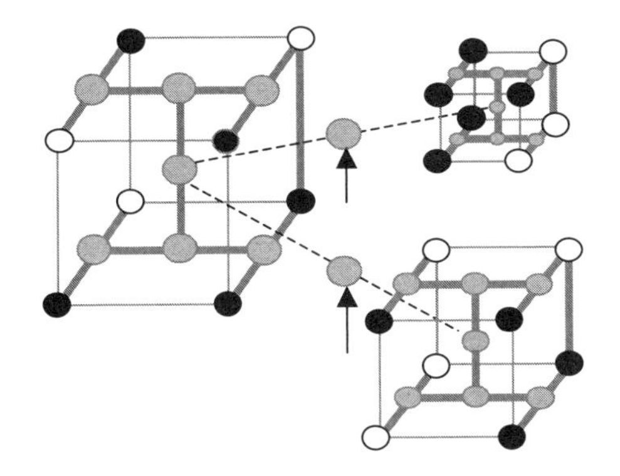

FIGURE 6.16
DAG based hypercube topology for Problem 6.10.

6.7 Further reading

Multilevel circuit design. The best overview of the state-of-the-art multilevel circuit design can be found in [3]. In this and many other references on logic design, the traditional gate-level model of a multilevel switching circuit is a DAG. The major problem of DAG synthesis is factorization of Boolean expressions, since factored form correspond to levelized (cascade) Boolean network.

Decision diagram based models of two-level circuits. References to graph-based models of logic circuits called decision diagrams can be found in Chapter 3. BDDs are derived from two-level AND-OR representations of switching functions, and, thus, require transformation of the gate-level model to the decision diagram. Once the decision diagram is created, a MUX-based or pass-gate based circuit can be designed. These models have been utilized by Asachi, Kasai, Yamada et.al. [1, 2, 4] for single-electron circuit design (see Chapter 2 for details).

6.8 References

[1] Asahi N, Akazawa M, and Amemiya Y. Single-electron logic device based on the binary decision diagram. *IEEE Transactions on Electron Devices*, 44(7):1109–1116, 1997.

[2] Kasai S, and Hasegawa H. A single-electron binary-decision-diagram quantum logic circuit based on Schottky wrap gate control of a GaAs nanowire hexagon. *IEEE Electron Device Letters*, 23(8), 2002.

[3] Sasao T, and Hassoun S., Eds., *Logic Synthesis and Verification.* Kluwer, Dordrecht, 2001.

[4] Yamada T, Kinoshita Y, Kasai S, Hasegawa H, and Amemiya Y. Quantum-dot logic circuits based on shared binary decision diagram. *Journal of Applied Physics*, 40(7):4485–4488, 2001.

7

Linear Word-Level Models of Multilevel Circuits

This chapter extends the word-level technique presented in Chapter 4. In this case, the word-level arithmetic representations are linear. There are several reasons to consider linear models of circuits:

- Linear models are simply embedded in the spatial structures.
- Linear models, being the *boundary* case of more general models, have a number of specific features that can be useful in logic design.
- Design of hypercube structures is very flexible.

One of the major purposes of this chapter is to introduce the methods of linearization of word-level models (arithmetic, sum-of-products, and Reed-Muller) and the former's mapping into spatial dimensions.

In Section 7.1 the basic definitions of linear word-level technique are introduced. They include conditions for linearity, grouping, masking, and methods for computing. The linear model based on arithmetic expression is introduced in Section 7.2. The linear arithmetic models in rigorous mathematical notation for the typical library of gates are designed in Section 7.3. Section 7.4 focuses on designing the linear decision diagrams. The main hypothesis is that a set of linear decision diagrams is a formal model of an arbitrary multilevel circuit. Based on this understanding, a technique of linear models design is introduced in Sections 7.5 and 7.6. The problem of large coefficients is resolved in Section 7.7. In the remaining sections, the linear word-level sum-of-products (Section 7.8) and Reed-Muller (Section 7.9) expressions are introduced. We follow the same strategy of representation: formal definition, specific features of grouping and masking, and linear decision diagram design.

7.1 Linear expressions

In this section, a brief introduction to linearization technique is given. It includes the general algebraic structure, conditions for linearity, and conditions for embedding in the hypercube and hypercube-like topology.

7.1.1 General algebraic structure

The following word-level forms of a switching function are considered (Table 7.1):

- Linear word-level arithmetic expressions,
- Linear word-level sum-of-products expressions, and
- Linear word-level Reed-Muller expressions.

TABLE 7.1
Linear word-level expressions for a switching function and decomposition rules for decision trees and diagram.

Formal description	Decomposition
Linear word-level arithmetic expressions	
$\sum_{i=0}^{n} d_i^* \cdot x_i^{i_j}, \quad x_j^{i_j} = \begin{cases} 1, & i_j = 0; \\ x_j, & i_j = 1. \end{cases}$	Davio arithmetic
Linear word-level sum-of-products	
$\overset{n}{\widehat{\bigvee}}_{i=1} v_i^* \cdot x_i^{i_j}, \quad v_i^* x_i^{i_j} = \begin{cases} v_i' x_i, & i_j = 1; \\ v_i'' \overline{x}_i, & i_j = 0. \end{cases}$	Shannon
Linear word-level Reed-Muller expressions	
$\overset{n}{\widehat{\bigoplus}}_{i=0} w_i^* \cdot x_i^{i_j}, \quad x_j^{i_j} = \begin{cases} 1, & i_j = 0; \\ x_j, & i_j = 1. \end{cases}$	Davio

Inspection of Table 7.1 leads to the conclusions that:

- Various linear word-level expressions have a uniform algebraic structure.
- Multiple decomposition with respect to a given variable can be applied to linear word sum-of-products and Reed-Muller expressions.
- Linear expressions produce linear data structures.

Word-level manipulation of switching functions $f_1, f_2, \ldots, f_r$ using Shannon, Davio and arithmetic analogs of Davio expansion is illustrated in Figure

7.1. Let f be an r-output switching function. A node of a diagram that implemented Shannon or Davio expansion of r switching functions (r-output function) in parallel is called a *word-level* Shannon or Davio expansion, and is denoted as $\widehat{S}$ and $\widehat{pD}$ correspondingly. The arithmetic analog of Davio expansion is applied ones to word-format, i.e. to integer encoded multioutput function.

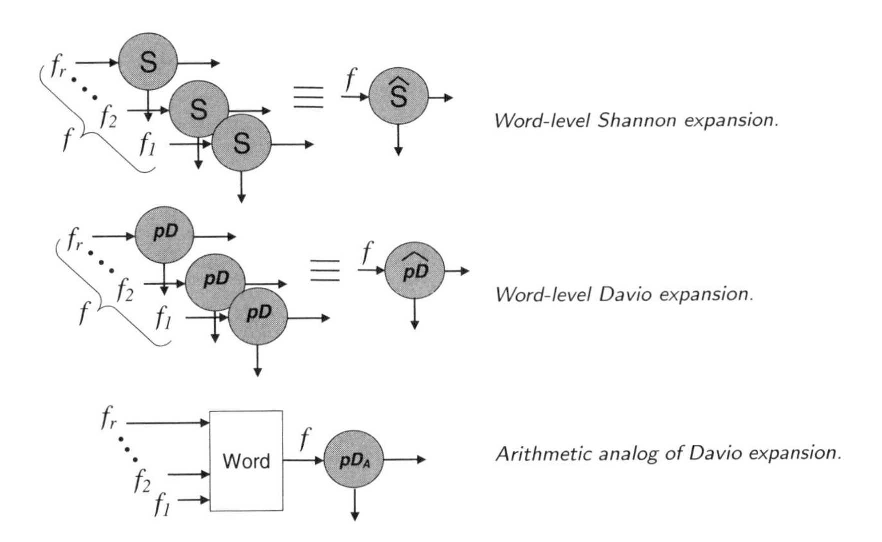

FIGURE 7.1
Diagram-based interpretation of a word-level Shannon, Davio, and arithmetic analog of Davio expansion.

7.1.2 Linearization

The technique of linearization of word-level arithmetic expressions involves algebraic and matrix transforms and comprises:

- ▶ Conditions for linearization,
- ▶ Synthesis methods,
- ▶ Grouping and masking methods,
- ▶ Linear decision diagrams,
- ▶ Methods of computing the coefficients,
- ▶ Methods for representation of circuits, and
- ▶ Methods for embedding the data structures in a hypercube space.

Conditions for linearization. There is a large group of word-level expressions that cannot be linearized by the traditional approaches. To recognize the switching functions that cannot be represented by linear expressions, the conditions for linearization have been developed. These conditions give an understanding of limitations of the word-level format.

Synthesis methods are understood here as the approaches to constructing word-level models of switching functions and circuits under conditions of linearity. In this chapter, we focus on the representation of an arbitrary multilevel circuit by linear expressions and linear decision diagrams level by level. The final result is a set of linear expressions and linear decision diagrams.

Grouping and masking methods. The order of switching functions in a word is dependent on certain criteria. In this chapter, the linearization is achieved through grouping and masking over the standard library of gates. This approach is useful in the hypercube space too.

Linear decision diagram is the result of the direct mapping of linear expressions into a word-level decision diagram. A linear decision diagram can be considered as a boundary case of word-level diagrams and has a number of features useful for classical logic design and nanotechnologies. The most important and promising property for space representation is the simple embedding procedure of linear decision diagrams into 3-D structures (hypercubes, hypercube-like topology, pyramids, etc.).

Methods of coefficients computing. For arithmetic expressions, the main goal is to minimize the effects of large value coefficients in linear arithmetic expressions. The crucial idea is to replace the computation by manipulation of codes of coefficients. This is possible in some cases because the regular structure of coefficients. The problem is simplified significantly for the word-level sum-of-products and Reed-Muller expressions.

Methods for circuit representation are based on a number of a fundamental statements, in particular: (i) the typical library of gates can be represented by the appropriate linear expressions and correspondent linear decision diagrams; (ii) an arbitrary combinational and sequential circuit can be represented by a set of linear models over the library of linear primitives.

Methods for embedding data structures in hypercube space is the final and crucial point of the above technique. It is shown in this chapter, that based on the classical philosophy of logic design (design over the library of cells) it is possible to map a traditional (2-D) circuit solutions into a hypercube structure in 3-D space.

Generalization of the above techniques toward multiple-valued logic is introduced in Chapter 8. Hence, the technique of linearization is universal with respect to data structures which are both binary structures and multivalued structures.

7.2 Linear arithmetic expressions

An arbitrary switching function can be represented by a unique arithmetic expression. For example, $x_1 \vee x_2$ corresponds to $x_1 + x_2 - x_1 x_2$. The remarkable property of the arithmetic expressions is that they can be applied to an r-output function f with outputs $f_1, f_2, ..., f_r$. In this section we focus on the effects of grouping the functions with the goal of representing a word-level expression in linear form.

7.2.1 Grouping

Consider the problem of grouping several switching functions in a word-level format. Let an r-output function f with outputs $f_1, f_2, ..., f_r$ be given. This function is described by the word-level arithmetic expression

$$f = 2^{r-1} f_r + \ldots + 2^1 f_2 + 2^0 f_1, \tag{7.1}$$

and the outputs can be restored in a unique way. Therefore, the outputs of a circuit can be grouped together by using a weighted sum of the outputs. Given the simplest commutator function (Figure 7.2a,b), the direct transmission of input data to outputs is described by

$$f = 2^{r-1} x_1 + 2^{r-2} x_2 + \ldots + x_r.$$

Assume $n = 2$, then $f = 2x_1 + x_2$ (Figure 7.2c). This expression does not include product terms of variables, therefore, it is *linear*.

The linear arithmetic expression of a switching function f of n variables $x_1, \ldots, x_n$ is the expression with $(n+1)$ integer coefficients $d_0^*, d_1^*, \ldots, d_n^*$

$$f = d_0^* + \sum_{i=1}^{n} d_i^* x_i = d_0^* + d_1^* x_1 + \ldots + d_n^* x_n. \tag{7.2}$$

Note that the word-level arithmetic expression

$$f = \sum_{i=0}^{2^n-1} d_i \cdot (x_1^{i_1} \cdots\ x_n^{i_n})$$

can be linear (Equation 7.2) in two cases:

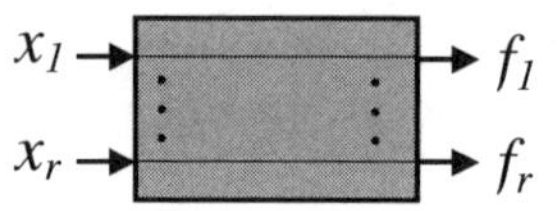

x_1	x_2	f_1	f_2
0	0	0	0
0	1	0	1
1	0	1	0
1	1	1	1

Word-level expression

$f = 2^{r-1}x_1 + 2^{r-2}x_2 + \ldots + x_r$
$\underline{n = 2:}$
$f = 2x_1 + x_2$

(a) (b) (c)

FIGURE 7.2
The direct transmission of input data to outputs (a), truth table (b), and word-level representation (c).

(*a*) Either arithmetic expressions of each f_j is linear, or
(*b*) Neither f_j generates linear expressions separately, but their combination produces a linear arithmetic expression.

Linearization generally means transformation of a nonlinear expression to a linear arithmetic expression (Equation 7.2), with no more than $(n + 1)$ nonzero coefficients. Briefly, the idea of linearization can be explained by a simple example. The function

$$f = x_1 \vee x_2 = x_1 + x_2 - x_1x_2$$

is extended to the 2-output switching function

$$f_1 = 1 \oplus x_1 \oplus x_2,$$
$$f_2 = x_1 \vee x_2,$$

that derives from the linear word-level representation $f = 2^1 f_2 + 2^0 f_1 = x_1+x_2+1$. The position of f_2 (the most significant bit) in this linear expression is indicated by the masking operator

$$\Xi^2\{f\} = \Xi^2\{x_1 + x_2 + 1\}.$$

In other words, to obtain a linear arithmetic expression given the switching function $f_2 = x_1 \vee x_2$, a garbage function $f_1 = 1 \oplus x_1 \oplus x_2$ has to be added. Then, f_2 can be extracted using the masking operator. The problem is how to find this additional function. In the absence of such a technique, i.e., a small amount of multioutput functions can generate linear arithmetic expressions assigning the naive approach. We use such functions to form a fixed library of primitive cells.

Example 7.1 *The half-adder (Figure 7.3) can be represented by the linear arithmetic expression $f = x_1 + x_2$. The permutation of the outputs f_1 and f_2 generates the nonlinear expression:*

$$f = 2x_1 + 2x_2 - 3x_1x_2.$$

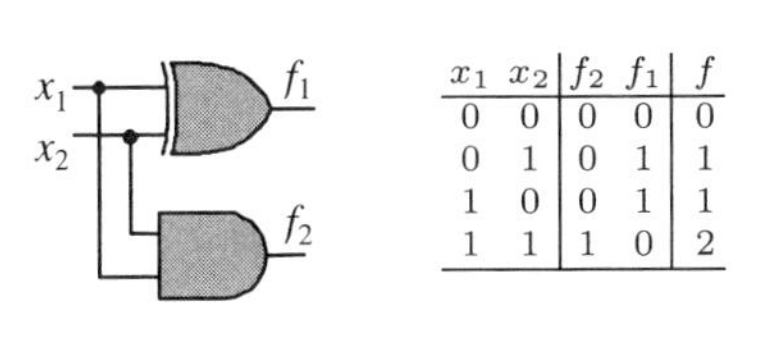

x_1	x_2	f_2	f_1	f
0	0	0	0	0
0	1	0	1	1
1	0	0	1	1
1	1	1	0	2

The outputs of switching function in arithmetic form:

$$f_1 = x_1 \oplus x_2 = x_1 + x_2 - 2x_1x_2$$
$$f_2 = x_1x_2$$

Word-level expression

$$\begin{aligned} f &= 2^1 f_2 + 2^0 f_1 \\ &= 2^1 x_1x_2 + 2^0(x_1 + x_2 - 2x_1x_2) \\ &= x_1 + x_2 \end{aligned}$$

(*a*) (*b*) (*c*)

FIGURE 7.3
Half-adder circuit (a), its truth table (b), and the word-level representation (c) (Examples 7.1 and 7.4).

The above example demonstrates the high sensitivity of linear arithmetic expression (Equation 7.2) to any permutation of outputs in the word-level description. On the other hand, it is a unique representation given the order of the switching function f_1 or f_2.

7.2.2 Computing of the coefficients in the linear expression

Equation 7.2 describes a set of switching functions. Let $n = 1$, then $f = d_0^* + d_1^* x_1$, and the function f is single-output. The coefficients d_0^* and d_1^* can be calculated by the equation

$$\mathbf{D} = \mathbf{P}_{2^1} \cdot \mathbf{F} = \begin{bmatrix} 1 & 0 \\ -1 & 1 \end{bmatrix} \begin{bmatrix} 0 \\ 1 \end{bmatrix} = \begin{bmatrix} 0 \\ 1 \end{bmatrix},$$

i.e., $d_0^* = 0$, $d_1 = 1$. In general, f is the r-output switching function $f_1, \ldots, f_r$. Let f be a 3-output switching function: $f_1 = x_1$, $f_2 = \overline{x}_1$, $f_3 = x_1$, with the truth vector $\mathbf{F} = [2\ 5]^T$. Calculation of the coefficients implies:

$$\mathbf{D} = \mathbf{P}_{2^1} \cdot \mathbf{F} = \begin{bmatrix} 1 & 0 \\ -1 & 1 \end{bmatrix} \begin{bmatrix} 2 \\ 5 \end{bmatrix} = \begin{bmatrix} 2 \\ 3 \end{bmatrix},$$

and $f = 2 + 3x_1$. Assume $n = 2$, Equation 7.2 yields $f = d_0^* + d_1^* x_1 + d_2^* x_2$.

Example 7.2 *The linear word-level arithmetic expression for half-adder function is defined as shown in Figure 7.4.*

7.2.3 Weight assignment

There are two kinds of weight assignments in a linear arithmetic expression in the formulation as follows:

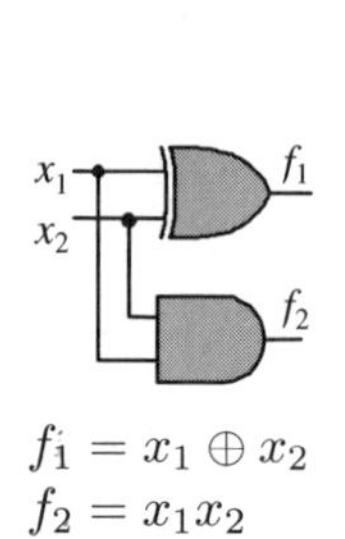

$f_1 = x_1 \oplus x_2$
$f_2 = x_1 x_2$

Truth vector

$$\mathbf{F} = [\ \mathbf{F_2}|\mathbf{F_1}] = \begin{bmatrix} 0\ 0 \\ 0\ 1 \\ 0\ 1 \\ 1\ 0 \end{bmatrix} = \begin{bmatrix} 0 \\ 1 \\ 1 \\ 2 \end{bmatrix}$$

Vector of coefficients

$$\mathbf{D} = \mathbf{P}_{2^2} \cdot \mathbf{F} = \left[\begin{array}{rr|rr} 1 & 0 & 0 & 0 \\ -1 & 1 & 0 & 0 \\ \hline -1 & 0 & 1 & 0 \\ 1 & -1 & -1 & 1 \end{array}\right] \begin{bmatrix} 0 \\ 1 \\ 1 \\ 2 \end{bmatrix} = \begin{bmatrix} 0 \\ 1 \\ 1 \\ 0 \end{bmatrix}$$

Word-level linear arithmetic expression $f = x_1 + x_2$

FIGURE 7.4
Constructing the linear word-level expression for a half-adder by matrix method (Example 7.2).

- ► The weight assignment to each of functions in a set of switching functions, and
- ► The weight assignment to each linear arithmetic expression in a set of expressions.

The weight assignment to the set of switching functions is defined by Equation 7.1. The weight assignment to the set of linear arithmetic expressions is defined as follows. Let f_i be the i-th, $i = 1, 2, \ldots, r$, linear word-level arithmetic expression of n_i variables. A linear expression of elementary switching function is represented by

$$t_i = \lceil \log_2 n_i \rceil + 1 \ \ \textit{bits}. \tag{7.3}$$

where $\lceil x \rceil$ is a ceiling function (the least integer greater than or equal to x). Suppose that f_i is the description of some primitive. In this formulation, the problem is relevant to the representation of an arbitrary level of a combinational circuit by linear arithmetic expression as an n-input r-output switching function. To construct a word-level expression of r linear arithmetic expressions, the weight assignment must be made appropriately. This means that applying any pattern to the inputs of f, an output of each expression f_i cannot affect the outputs of others.

Figure 7.5a illustrates the problem. Formally, the weight assignments without overlapping of functions are determined by the equation

$$f = \sum_{i=0}^{r-1} 2^{T_i} f_{i+1}, \tag{7.4}$$

where

$$T_i = \begin{cases} 0 & \text{for } i = 0 \\ T_{i-1} + t_i & \text{for } i > 0 \end{cases}.$$

and t_i is calculated by Equation 7.3.

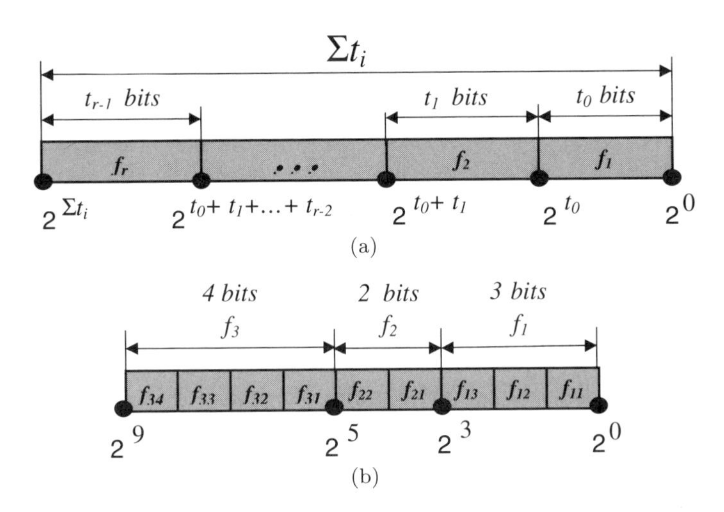

FIGURE 7.5
Word-level format for the set of r linear arithmetic expressions f_i (a) and example for f_1, f_2, and f_3 that are respectively 3rd, 2nd and 4th subfunctions (b).

Example 7.3 *Let the word-level arithmetic expression f consist of three linear arithmetic expressions f_1, f_2 and f_3 (Figure 7.5b). The expressions are constructed as follows*

$$f_1 = 2^0 f_{11} + 2^1 f_{12} + 2^2 f_{13},$$
$$f_2 = 2^3 f_{21} + 2^4 f_{22},$$
$$f_3 = 2^5 f_{31} + 2^6 f_{32} + 2^7 f_{33} + 2^8 f_{34}.$$

by Equation 7.4, the weight assignment of linear expressions f_1, f_2 and f_3 is resulted in

$$f = 2^0 f_1 + 2^3 f_2 + 2^5 f_3.$$

7.2.4 Masking

The masking operator $\Xi^{\tau}\{f\}$ is discussed in Chapter 3. It indicates the position $t \in (1, \ldots, r)$ of a given function f in the word-level expression that represents a set of r switching functions.

Example 7.4 *The arithmetical expression of half adder $f = x_1 + x_2$ is a two-output switching function: $f_1 = x_1 \oplus x_2$, $f_2 = x_1 x_2$ (Figure 7.3). The most significant bit (f_2) can be extracted by the masking operator*

$$\Xi^2\{x_1 + x_2\}$$

whereas function f_1 is encoded by the least significant bit which can be recovered by the masking operator

$$\Xi^1\{x_1 + x_2\}.$$

7.3 Linear arithmetic expressions of elementary functions

As was shown earlier, the majority of functions cannot be converted to a linear arithmetic expression, since their arithmetic equivalent includes nonlinear products. In this section, we focus on linearization of elementary switching functions.

7.3.1 Functions of two and three variables

Elementary switching functions of two variables from the typical library of cells can be represented by a linear arithmetic expressions.

Theorem 7.1 *An arbitrary switching function of two variables can be described, in a unique way, by a linear arithmetic expression.*

PROOF There are 16^2 pairs of 16 possible switching functions of two variables, each pair forming a two-output switching function that generates an arithmetic expression. This set includes linear arithmetic expressions (Equation 7.2). Indeed, two functions, represented by their four values, $f = (f_0, f_1, f_2, f_3)$ and $f' = (f'_0, f'_1, f'_2, f'_3)$, form 256 possible 2-output functions with outputs f, f', or integer-valued functions with values $2f_0 + f'_0, 2f_1 + f'_1, 2f_2 + f'_2, 2f_3 + f'_3$. Such a function generates the arithmetic expression $2^1 f + 2^0 f'$. This satisfies (7.2) iff

$$2f_3 + f'_3 = -(2f_0 + f'_0) + (2f_1 + f'_1) + (2f_2 + f'_2).$$

The last equation is valid for only 44 two-output functions among those 256. However, each of the possible 16 functions of two variables can be found at least once in the first or second place in these two outputs. It means that any single-output function of two variables can be extended to a function of two variables, being its first or second output, such that this extension generates a linear arithmetic expression. ∎

The conditions of linearity can be formulated in a rigorous way (see "Further Reading" Section). It follows from the Theorem that in order to represent an arbitrary switching function of two variables by the linear arithmetic expression, at least one garbage function must be added.

In Table 7.2 the linear expressions for the two-input primitive functions are shown. For example, function AND is the second function in the word described by $x_1 + x_2$.

The above method of proof can be extended to switching functions of many variables. The linear expressions for the three-input primitives are given in Table 7.2.

Example 7.5 *A switching function $x_1 \oplus x_2$ is represented by a nonlinear arithmetic expression $x_1 + x_2 - 2x_1x_2$ because of the product term $2x_1x_2$. To linearize, this function is expanded to the 2-input function $f_1 = x_1 \oplus x_2$ and $f_2 = x_1x_2$. The switching function f_1 is extracted from the linear expression by the masking operator $f = \Xi^1\{x_1 + x_2\}$.*

7.3.2 AND, OR, and EXOR functions of n variables

Bearing in mind that linear combination of linear expressions for some elementary switching functions produces linear expressions, we look now at the method of linearization of elementary switching functions. An approach to design linear expressions for two- and three-input elementary switching functions is discussed in the previous section. An elegant method to design linear arithmetic expressions for many-input elementary switching functions has been developed by *Malyugin* (see "Further Reading" Section). We introduce *Malyugin's theorem's* theorem without proof.

Let the input variable of a primitive gate can be x_j or $\overline{x}_j$. Denote the j-th input, $j = 1, 2, \ldots, n$, as

$$x_j^{i_j} = \begin{cases} x_j \text{ if } j_i = 0, \\ \overline{x}_j \text{ if } i_j = 1. \end{cases}$$

Theorem 7.2 *The n-variable AND function $x_1^{i_1} \ldots x_n^{i_n}$ can be represented by the linear arithmetic expression*

$$f = 2^{t-1} - n + \sum_{j=1}^{n} (i_j + (-1)^{i_j} x_j), \tag{7.5}$$

generated by an r-output function, in which the function AND is the most significant bit, as indicated by the masking operator $\Xi^r\{f\}$.

TABLE 7.2
Linear arithmetic expressions for 2– and 3–input gates.

Function	2-input	3-input
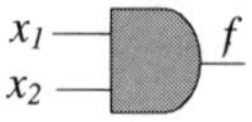 $f = x_1 x_2$	$\Xi^2\{x_1 + x_2\}$	$\Xi^3\{x_1 + x_2 + x_3\}$
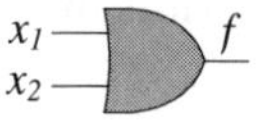 $f = x_1 \vee x_2$	$\Xi^2\{1 + x_1 + x_2\}$	$\Xi^3\{3 + x_1 + x_2 + x_3\}$
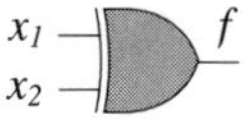 $f = x_1 \oplus x_2$	$\Xi^1\{x_1 + x_2\}$	$\Xi^1\{x_1 + x_2 + x_3\}$
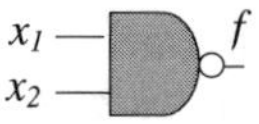 $f = \overline{x_1 x_2}$	$\Xi^2\{3 - x_1 - x_2\}$	$\Xi^3\{6 - x_1 - x_2 - x_3\}$
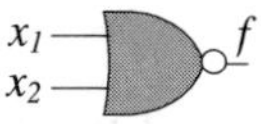 $f = \overline{x_1 \vee x_2}$	$\Xi^2\{2 - x_1 - x_2\}$	$\Xi^3\{4 - x_1 - x_2 - x_3\}$

Theorem 7.3 *The n-variable OR function $x_1^{i_1} \vee \ldots \vee x_n^{i_n}$ can be represented by the linear arithmetic expression*

$$f = 2^{t-1} - 1 + \sum_{j=1}^{n} (i_j + (-1)^{i_j} x_j) \tag{7.6}$$

of an r-output function, so that the function OR is the most significant bit $f = \Xi^r\{f\}$.

Theorem 7.4 *The n-variable EXOR function $x_1^{i_1} \oplus \ldots \oplus x_n^{i_n}$ can be represented by the linear arithmetic expression*

$$f = \sum_{j=1}^{n} (i_j + (-1)^{i_j} x_j) \tag{7.7}$$

of an r-output function, in which the function EXOR is in the least significant bit, $\Xi^1\{f\}$.

In the above statements, the parameter t (the number of bits in a linear word-level representation of a given switching function) is defined by Equation 7.4.

Note that the expression $1 \oplus x_j^{i_j}$ must be avoided in Equation 7.7. Before applying Equation 7.7, we have to replace $\overline{x}_j$ with $x_j \oplus 1$, or replace $x_j \oplus \overline{x}_j$ with 1 in order to cancel 1's.

TABLE 7.3
Linear arithmetic expressions for the AND, OR, and EXOR functions of n-input.

Function	Linear arithmetic expression
$f = x_1^{i_1} \ldots x_n^{i_n}$	$\Xi^t\{2^{t-1} - n + \sum_{j=1}^{n}(i_j + (-1)^{i_j} x_j)\}$
$f = x_1^{i_1} \vee \ldots \vee x_n^{i_n}$	$\Xi^t\{2^{t-1} - 1 + \sum_{j=1}^{n}(i_j + (-1)^{i_j} x_j)\}$
$f = x_1^{i_1} \oplus \ldots \oplus x_n^{i_n}$	$\Xi^t\{\sum_{j=1}^{n}(i_j + (-1)^{i_j} x_j)\}$

Table 7.3 contains three n-input primitives and corresponding linear expressions. For NOT function, the corresponding linear arithmetic expression equals $\Xi^1\{1 - x\}$. Based on expressions from Table 7.3, it is possible to describe modified gates, for example,

$$\begin{aligned}
\overline{x}_1 \vee x_2 &= \Xi^2\{1 + (1 - x_1) + x_2\} \\
&= \Xi^2\{2 - x_1 + x_2\}, \\
\overline{x}_1 \oplus x_2 &= \Xi^1\{(1 - x_1) + x_2\} \\
&= \Xi^1\{1 - x_1 + x_2\}.
\end{aligned}$$

7.3.3 "Garbage" functions

Linear arithmetic expressions are word-level arithmetic expressions that possess specific properties. First, the linear expression involves extra functions called *garbage functions*. The number of garbage functions G increases with the number of variables in a function that has been linearized:

$$G = t - 1 = \lceil \log_2 n \rceil. \tag{7.8}$$

Example 7.6 *Consider the linear expression for two-input AND function (Table 7.2) $f = x_1 + x_2$. To derive this linear form, the garbage function f_1 has been added, so that the given function AND is the most significant bit in the word-level description, f_2 (Figure 7.6a). To derive a linear representation of the three-input AND function, two garbage functions have been added through the two least significant bits of the 3-bit word (Figure 7.6b).*

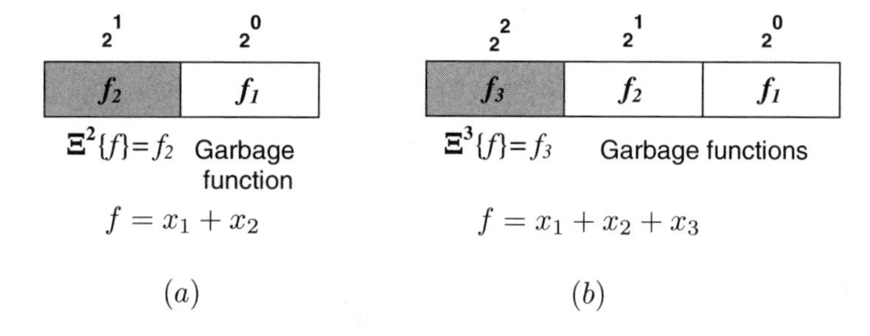

FIGURE 7.6
Garbage functions in linear arithmetic expression two-input (a) and three-input AND function (Example 7.6).

7.4 Linear decision diagrams

A *linear* decision diagram used to represent a multioutput switching function is a word-level arithmetic decision diagram upon the condition of linearity. A set of the linear diagrams is a formal model used to represent a multilevel circuit. While related as particular (linear) case, the linear diagram based model has a number of attractive properties discussed in this chapter.

In linear word-level diagram, a node realizes the *arithmetic analog of positive Davio expansion*

$$pD_A : f = \underbrace{f_{x_i=0}}_{left\ term} + \underbrace{x_i(-f_{x_i=0} + f_{x_i=1})}_{right\ term}$$

and terminal nodes correspond to integer-valued coefficients of switching function f.

A linear decision diagram is a decision diagram to represent an arbitrary network described by a linear word-level arithmetic expression; the nodes correspond to pD_A expansion and the terminal nodes are assigned the coefficients of the linear expression. Linear decision diagram for n-input linear arithmetic expression includes n nonterminal and $n+1$ terminal nodes.

Example 7.7 *Figure 7.7a,b shows a linear decision diagram for AND function. This diagram includes three nodes. Lexigraphic order of variables is used:* x_1, x_2.

Step 1 : Compute $f_{x_1=0}$ *and* $f_{x_1=1}$*:*

$$f_{x_1=0} = 0 \cdot x_2 = 0, \text{ and } f_{x_1=1} = 1 \cdot x_2 = x_2.$$

Step 2 : A terminal node with the value 1 *is generated since the left term of the* pD_A *expansion is the constant* $f_{x_1=0} = 0$. *The right term is equal to*

$$f_{x_1=1} - f_{x_1=0} = x_2 - 0 = x_2$$

and requires further decomposition.

Step 3 : Compute

$$f\Big|_{\substack{x_1=0\\x_2=0}} = 0, \qquad f\Big|_{\substack{x_1=0\\x_2=1}} = 0.$$

The terminal node is equal to

$$x_2(-f_{x_2=0} + f_{x_2=1}) = -f\Big|_{\substack{x_1=0\\x_2=0}} + f\Big|_{\substack{x_1=0\\x_2=1}} = -0 + 0 = 0.$$

The embedded linear decision diagram in 2-D $\mathcal{N}$-hypercube is given in Figure 7.7c.

In Table 7.4, the linear decision diagrams for two-input gates from the typical gate library are given.

Summarizing,

- Elementary switching functions can be represented by linear arithmetic expressions as shown in Table 7.2 and Table 7.3,
- Linear composition of linear expressions (gates in the level of a circuit) produce linear arithmetic expression,
- Linear expression directly maps into linear decision diagram (Table 7.4).

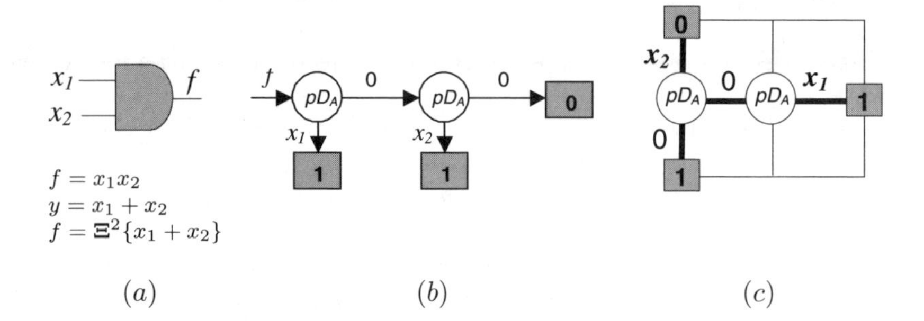

FIGURE 7.7
The linear word-level arithmetic expression of AND function (a), linear decision diagram (b), and embedding of a linear decision diagram in a 2-D $\mathcal{N}$-hypercube (c) (Example 7.7).

7.5 Representation of a circuit level by linear expression

Suppose a multilevel circuit over the typical library of gates is given. The problem is formulated as follows: representation of an arbitrary level of this circuit by linear word-level models, algebraic equation and decision diagram. The solution is based on the following theorem.

Theorem 7.5 *A circuit level with n inputs $x_1, ..., x_n$ and r gates (r outputs) is modeled by linear diagram with n nodes, assigned input variables and $n+1$ terminal nodes, assigned coefficients of the linear expression.*

PROOF The proof follows immediately from the fact that an arbitrary n-input r-output function can be represented by a weighted arithmetic expression. ∎

Example 7.8 *A level of a circuit is shown in Figure 7.8. The linear arithmetic expressions describing the first, the second, and the third gate are given in Table 7.2. Combining these expressions, we compile f where parameters T_0, T_1, T_2 are calculated by (7.4). The final result is*

$$\begin{aligned} f &= 2^0(x_1 + x_2) + 2^2(\overline{x}_1 + x_2) + 2^4(\overline{x}_2 + x_3) \\ &= 2^0(x_1 + x_2) + 2^2(1 - x_1 + x_2) + 2^4(1 - x_2 + x_3) \\ &= -3x_1 - 12x_2 + 17x_3 + 20. \end{aligned}$$

Let us design a set of linear decision diagrams for the circuit from Example 7.8. Note that the order of variables in the diagram can be arbitrary. Let us choose the lexigraphical order, i.e., x_1, x_2.

TABLE 7.4
Linear decision diagrams derived from linear word-level arithmetic expressions for two-variable functions.

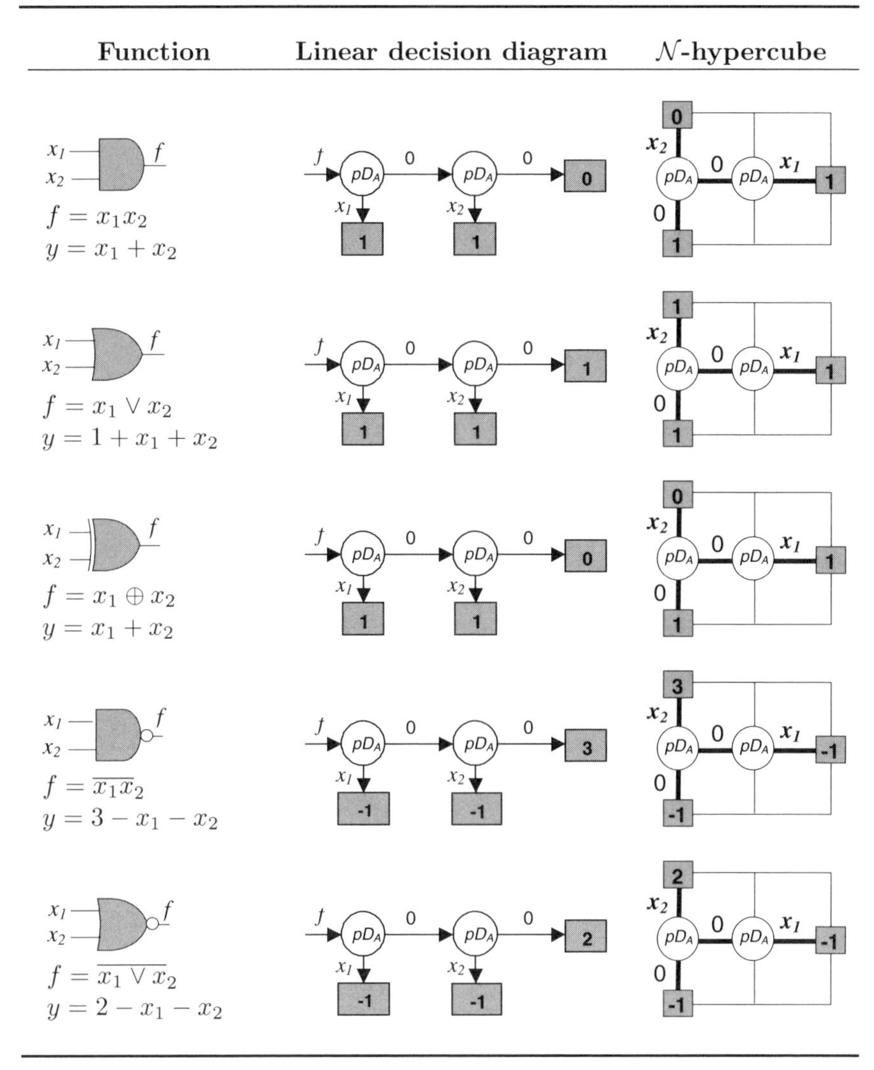

Example 7.9 *The linear decision diagram for the expression* $-3x_1 - 12x_2 + 17x_3 + 20$ *consists of three nodes. There are three steps to design the diagram:*

Step 1 : Compute

$$f_{x_1=0} = -3 \cdot 0 - 12x_2 + 17x_3 + 20,$$
$$f_{x_1=1} = -3 \cdot 1 - 12x_2 + 17x_3 + 20.$$

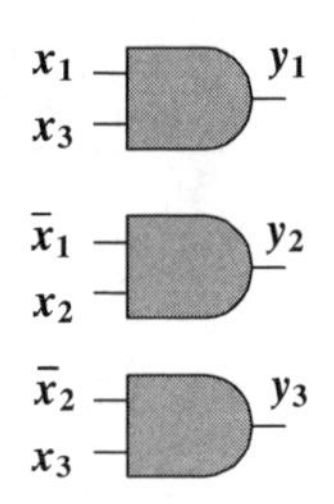

$f_1 = x_1 + x_3$
$f_2 = 1 - x_1 + x_2$
$f_3 = 1 - x_2 + x_3$
Outputs
$y_1 = \Xi^2\{x_1 + x_2\}$
$y_2 = \Xi^2\{1 - x_1 + x_2\}$
$y_3 = \Xi^2\{1 - x_2 + x_3\}$

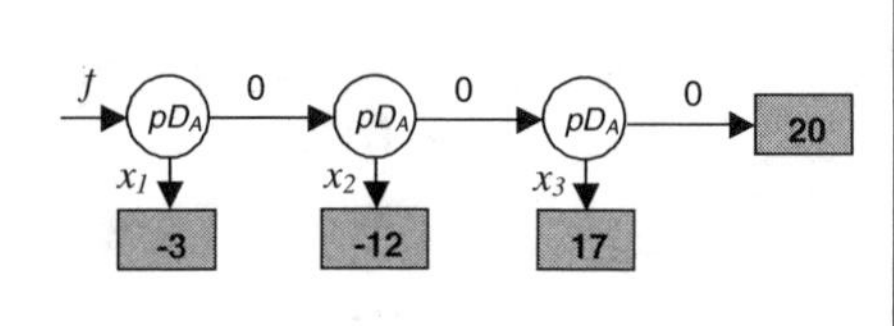

Level description

$f = 2^{T_2} f_3 + 2^{T_1} f_2 + 2^{T_0} f_1$

Masking parameters

$t_0 = \lceil log_2 2 \rceil + 1 = 2$ bits
$t_1 = \lceil log_2 2 \rceil + 1 = 2$ bits
$t_2 = \lceil log_2 2 \rceil + 1 = 2$ bits
$T_0 = 0$
$T_1 = 2$
$T_2 = 4$

Linear word-level arithmetic expression

$f = -3x_1 - 12x_2 + 17x_3 + 20$

Arithmetic positive Davio expansion

$pD_A : f = f_{x_i=0} + x_i(-f_{x_i=0} + f_{x_i=1})$

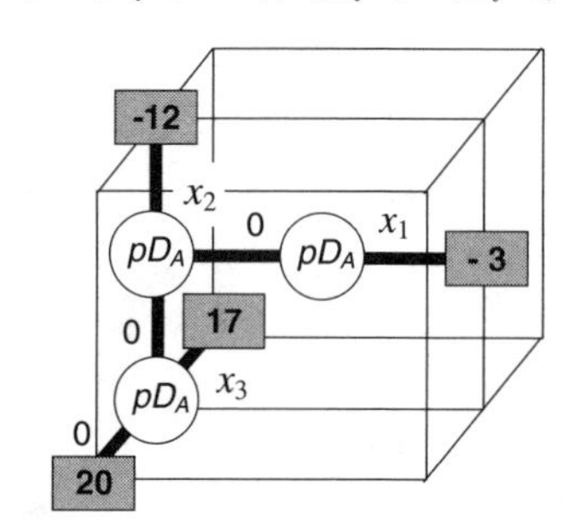

FIGURE 7.8
Technique of the representation of a level of circuit by linear word-level arithmetic expression and decision diagram (Examples 7.8 and 7.9).

The terminal node is equal to -3 *because the right product is a constant* $f_{x_1=1} - f_{x_1=0} = -3$. *The left product needs further decomposition.*

<u>*Step 2*</u>: *Compute*

$$f\Big|_{\substack{x_1=0\\x_2=0}} = 17x_3 + 20, \qquad f\Big|_{\substack{x_1=0\\x_2=1}} = -12 + 17x_3 + 20.$$

The terminal node equals

$$f\Big|_{\substack{x_1=0\\x_2=1}} - f\Big|_{\substack{x_1=1\\x_2=0}} = -12.$$

<u>*Step 3*</u>: *By analogy, the terminal node for variable* x_3 *is* 17, *and the free terminal node is* 20.

From Examples 7.8 and 7.9, one can observe that the coefficients in linear expressions are quite large even for the small circuits. Therefore, a special technique is needed to alleviate this effect.

7.6 Linear decision diagrams for circuit representation

In this section, an arbitrary r-level combinational circuits is represented by r linear decision diagrams, i.e. for each level of a circuit a linear diagram is designed. The complexity of this representation is $O(G)$, where G is the number of gates in the circuit. The outputs of this model are calculated by transmission data through this set of diagrams. This approach is the basis for representation circuits in spatial dimensions:

< 2-D circuit> ⇒ < A set of linear diagrams>⇒ <Hypercube-like topology>.

7.6.1 The basic statement

Theorem 7.6 *An arbitrary m-level switching network can be uniquely described by a set of m linear decision diagrams, and vice versa, this set of linear decision diagrams corresponds to a unique network.*

PROOF One of m levels with r n-input gates from a fixed library is described by one linear arithmetic expression. Fixing order of the gates in the level, i.e. keeping unambiguity of the structure, we can derive the unique linear decision diagram for this level, as well as for other $m-1$ levels of the network. ▮

From this statement follows:

- The order of gates in a level of circuit must be fixed.
- The complexity of linear decision diagram does not depend on the order of variables.
- Data transmission through linear decision diagrams must be provided.

7.6.2 Examples

Three examples below focus on various details of computing by linear models.

Example 7.10 *Representation of a three-level (L_1, L_2 and L_3) circuit by linear arithmetic expressions is shown in Figure 7.9.*

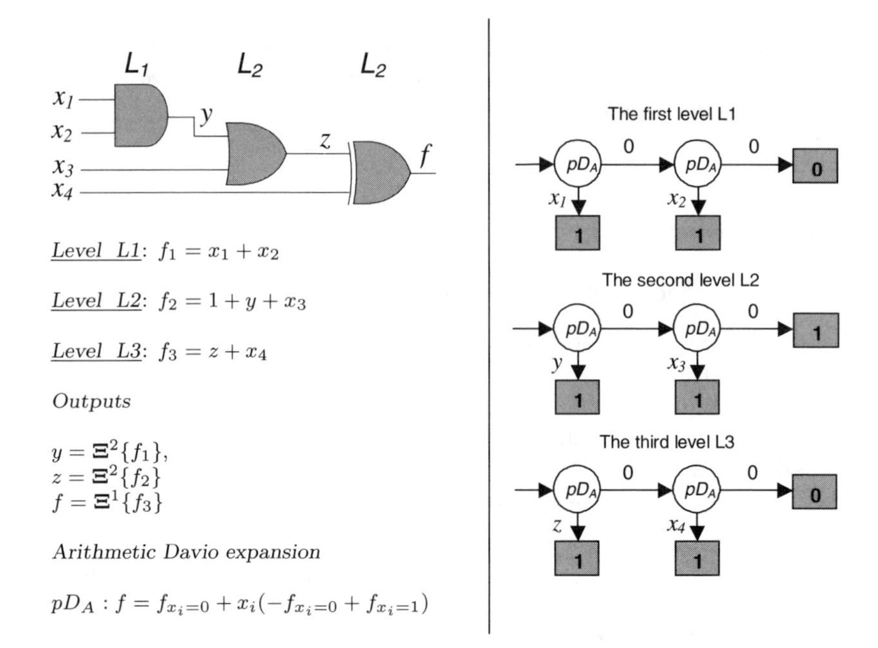

FIGURE 7.9
Representation of a three-level circuit by a set of linear decision diagrams (Example 7.10).

The next example illustrates the technique of a linear model design for a circuit with two gates at the first level.

Example 7.11 *A circuit in Figure 7.10 consists of three AND gates G_1, G_2 and G_3 with known linear expressions for gates. Output f is defined by double masking operator.*

The example below illustrates the technique of a linear model design for the data permutation, the typical procedure in spatial dimension processing.

Example 7.12 *A two-level communication network shown in Figure 7.11a can be described by two linear arithmetic expressions below:*

$$f_1 = 2^0x_1 + 2^1x_2 + 2^2x_3 + 2^3x_4,$$
$$f_2 = 2^0y_1 + 2^1y_2 + 2^2y_3 + 2^3y_4.$$

Details of calculations, the final diagram and the $\mathcal{N}$-hypercube representation are given in Figure 7.11b,c,d.

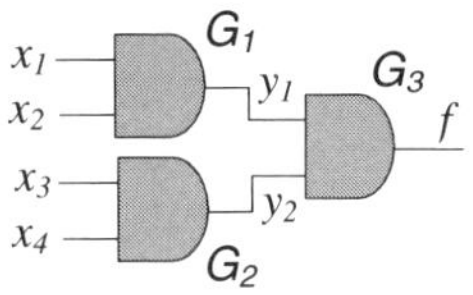

Gate G_1: $f_1 = x_1 + x_2$

Gate G_2: $2 = x_3 + x_4$

Gate G_3: $f_3 = y_1 + y_2$

Output:

$$
\begin{aligned}
f &= \Xi^2\{f_3\} = \Xi^2\{y_1 + y_2\} \\
&= \Xi^2\{\Xi^2\{f_1\} + \Xi^2\{f_2\}\}
\end{aligned}
$$

FIGURE 7.10
Deriving the linear models design for a two-level circuit (Example 7.11).

7.7 Technique for manipulating the coefficients

The problem of large coefficients (weights) seems to be a drawback of linear arithmetic expression. Direct computation is impossible for circuits with hundreds of gates in the level. Fortunately, the large coefficients can be encoded in order to avoid calculation of exponents and to manipulate these codes easy.

7.7.1 The structure of coefficients

The purpose of this section is to determine the structure of the coefficients and to apply the appropriate encoding thereof. The following theorem specifies the structure of coefficients assigned to the terminal nodes in the linear decision diagram.

Theorem 7.7 *The circuit level with n inputs $x_1, ..., x_n$ and r gates is described by the linear expression:*

$$
f_l = W_0 + W_{x_1}x_1 + ... + W_{x_n}x_n, \tag{7.9}
$$

where terminal weights W_0 *and* W_{x_i}, $i = 1, ..., n$, *are formed as below*

$$
\begin{aligned}
W_0 &= a_{0,1}2^{T_1} + \ldots + a_{0,r}2^{T_r}, \\
W_{x_i} &= a_{i,1}2^{T_1} + \ldots + a_{i,r}2^{T_r},
\end{aligned} \tag{7.10}
$$

$a_{i,k} \in \{0, +1, -1\}$ *and* $a_{0,k}$ *are positive integer numbers, and* T_i *is defined by Equation 7.4.*

PROOF Let the output of the first gate in the level be described by $f(y_1) = 2^1 f_2 + 2^0 f_1$. Since

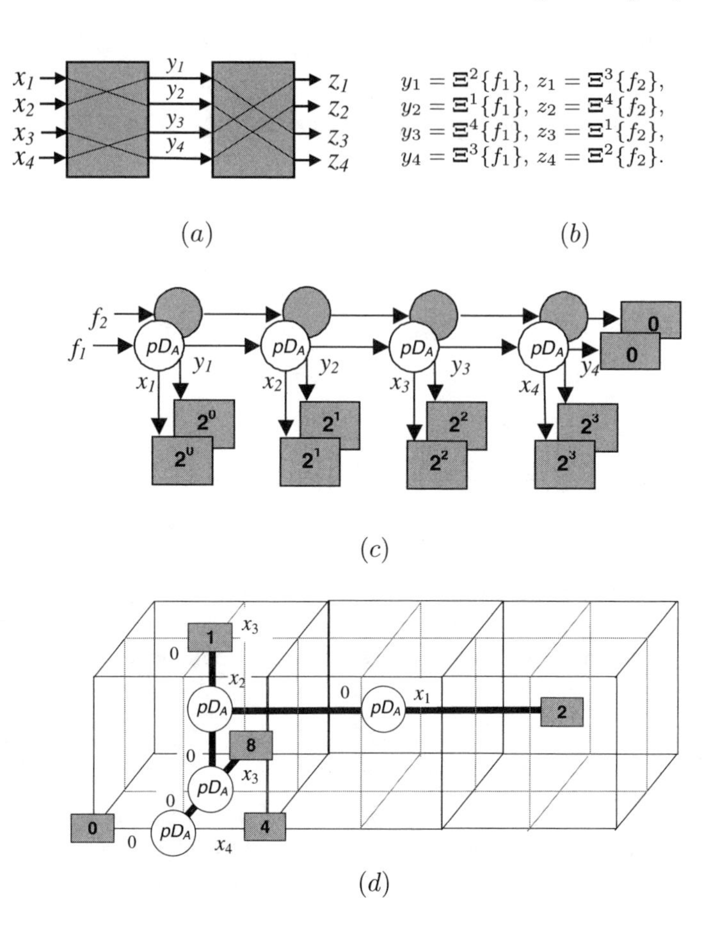

FIGURE 7.11
Communication network (a), formal description by the linear arithmetic expressions (b), a set of linear decision diagrams (c), and 4-D $\mathcal{N}$-hypercube representation of function f_1 (d) (Example 7.12).

$$f_1 = a_{0,1} + a_{1,1}x_2 + a_{2,1}x_1 + a_{3,1}x_1x_2$$
$$f_2 = a_{0,2} + a_{1,2}x_2 + a_{2,2}x_1 + a_{3,2}x_1x_2$$

then

$$\begin{aligned} f(y_1) &= 2^1 f_2 + 2^0 f_1 \\ &= (2a_{0,2} + a_{0,1}) + (2a_{1,2} + a_{1,1})x_2 \\ &\quad + (2a_{2,2} + a_{2,1})x_1 + (2a_{3,2} + a_{3,1})x_1x_2. \end{aligned}$$

In the linear expression, $2a_{3,2} + a_{3,1} = 0$. Therefore,

$$f(y_1) = (2a_{0,2} + a_{0,1}) + (2a_{1,2} + a_{1,1})x_2 + (2a_{2,2} + a_{2,1})x_1.$$

Let

$$\begin{aligned} W_0 &= 2^1 a_{0,2} + 2^0 a_{0,1}, \\ W_{x_1} &= 2^1 a_{1,2} + 2^0 a_{1,1}, \\ W_{x_2} &= 2^1 a_{2,2} + 2^0 a_{2,1}, \end{aligned}$$

then the output of the first gate is described by the linear expression

$$f(y_1) = W_0 + W_{x_1} x_1 + W_{x_2} x_2.$$

Applying the same expansion to other gates, we obtain a linear arithmetic expression (Equation 7.9) whose coefficients are structured by Equation 7.10.
▮

It follows from the above that:

- ▶ The weights are linear expression over $a_{i,j}$,
- ▶ The co-factors $a_{i,j}$ in W_{x_i} are always to the power of 2.

7.7.2 Encoding

Based on the above statement that properties of the coefficients are suitable for effective encoding, the technique of manipulation of the coefficients has been developed. This technique is introduced by two examples below.

Example 7.13 *The level of a circuit is depicted in Figure 7.8. There are a number of steps to encode the terminal nodes of the corresponding decision diagram.*

Step 1: *Determine the outputs $f(y_i)$, $i = 1, 2, 3$, in the form of liner expression (Table 7.4) with respect to all functions; because there are two input gates, two functions are involved in a linear form of each gate.*

Step 2: *Define f_1, f_2 and f_3 by Equation 7.9 and Equation 7.10.*

Step 3: *Define masking parameters T_i.*

Step 4: *Define the weights $W_0, W_{x_1}, W_{x_2}, W_{x_3}$.*

Step 5: *Define the f_l.*

Details are given in Figure 7.12.

It is useful to compare the above example to Example 7.8 and Example 7.9.

Step 1

$f_1 = 2^0(x_1 + x_3)$
$f_2 = 2^2(1 - x_1 + x_2)$
$f_3 = 2^4(1 - x_1 + x_3)$

Step 2: *General equation*

$$f = f(y_1) + f(y_2) + f(y_3) = W_0 + W_{x_1}x_1 + W_{x_2}x_2 + W_{x_3}x_3$$

Step 3: *Masking parameters*

$t_1 = \lceil \log_2 2 \rceil + 1 = 2$
$t_2 = \lceil \log_2 2 \rceil + 1 = 2$
$T_1 = t_0 = 0$
$T_2 = t_0 + t_1 = 0 + 2 = 2$
$T_3 = t_0 + t_1 + t_2 = 4$

Step 4: *Computing the weights*

$$W_0 = 0 \cdot 2^{T_1} + 1 \cdot 2^{T_2} + 1 \cdot 2^{T_3} = 2^2 + 2^4,$$
$$W_{x_1} = 1 \cdot 2^{T_1} - 1 \cdot 2^{T_2} + 0 \cdot 2^{T_3} = 2^0 - 2^2,$$
$$W_{x_2} = 0 \cdot 2^{T_1} + 1 \cdot 2^{T_2} - 1 \cdot 2^{T_3} = 2^2 - 2^4,$$
$$W_{x_3} = 1 \cdot 2^{T_1} + 0 \cdot 2^{T_2} + 1 \cdot 2^{T_3} = 2^0 + 2^4$$

Step 5: *Equations with encoding values of* W_{x_i}

$$f = 2^2 + 2^4 + (2^0 - 2^2)x_1 + (2^2 - 2^4)x_2 + (2^0 + 2^4)x_3$$

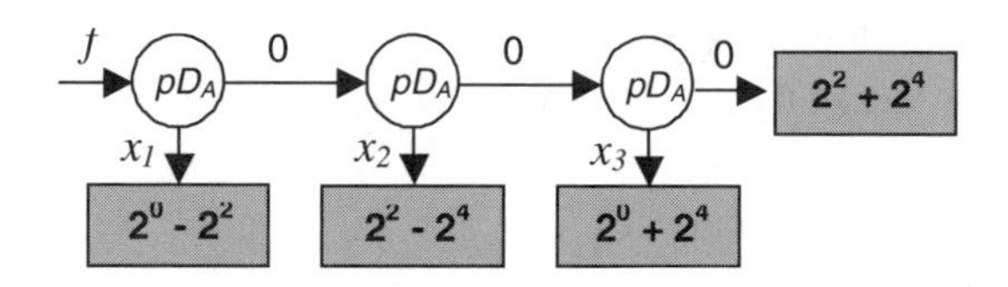

FIGURE 7.12
Technique of encoding the weight coefficients in linear arithmetic expression (Example 7.13).

Example 7.14 *Let a terminal node be assigned with the weight*

$$W_{x_i} = 0 + 2^3 - 2^6 - 2^9 - 2^{12} + 0 + 0 + 2^{21} + 2^{24}.$$

We apply the following encoding rules for $0 \to 00,\ 1 \to 01,\ -1 \to 11$ *as shown in Figure 7.13.*

The below example illustrates restoration (decoding) of coefficients $a_{ij}2^T$.

Example 7.15 *Given the terminal node* $W_{x_1} = -2^4 + 2^6 - 2^9$. *The encoding of the weights is given in Figure 7.14.*

It is observed from Example 7.15 that the computing of $a_{i,j} = 0$ is a recurrent procedure. This can be accomplished by using manipulations on a spectral type of tree.

2^{24}	2^{21}	2^{18}	2^{15}	2^{12}	2^{9}	2^{6}	2^{3}	2^{0}
01	01	00	00	11	11	11	01	00

FIGURE 7.13
Encoding the weights of a terminal node (Example 7.14).

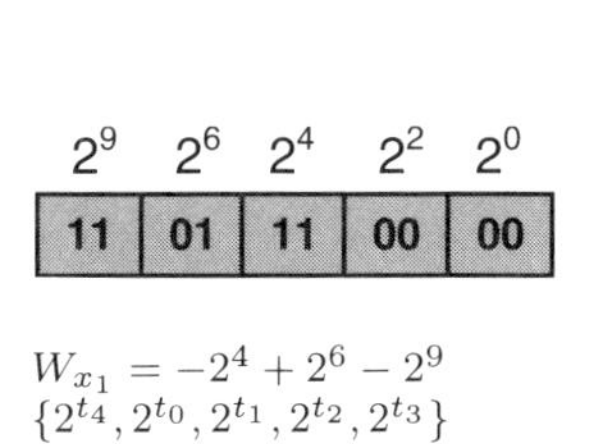

Encoding coefficients
$a_{1,0} = 0,\ a_{1,1} = 0,$
$a_{1,2} = -1,\ a_{1,3} = 1,\ a_{1,4} = -1$

$$\begin{aligned}
a_{1,0}2^{T_0} &= a_{1,0}2^{t_0} = 0,\\
a_{1,1}2^{T_1} &= a_{1,1}2^{t_0+t_1} = 0,\\
a_{1,2}2^{T_2} &= a_{1,2}2^{t_0+t_1+t_2} = -2^4,\\
a_{1,3}2^{T_3} &= a_{1,3}2^{t_0+t_1+t_2+t_3} = 2^6\\
a_{1,4}2^{T_4} &= a_{1,4}2^{t_0+t_1+t_2+t_3+t_4} = -2^9
\end{aligned}$$

FIGURE 7.14
Encoding the weights of a terminal node (Example 7.15).

7.7.3 W-trees

There are two types of terminal nodes in a linear decision diagram: W_{x_i} and W_0 formed by Equation 7.10. The W-tree to represent a terminal node corresponding to a variable x_i, $i \in (1, ..., n)$, is a linear tree with r nodes assigned the coefficients 2^{t_j}, r 1-terminal nodes and one 0-terminal node, and the successors assigned (multiplied) with the coefficients $a_{i,j} \in \{0, 1, -1\}$. In a W-tree, we encode the coefficients $a_{i,j}$ which take the values 0, 1 or -1, by the codes 00, 01 and 11 correspondingly.

The weight W_{x_i} assigned to the i-th terminal node includes the exponents such that each upper one is a sum of the previous ones: $T_j = t_0 + t_1 + \ldots + jt_{j-1}$. Let the nodes in a W-tree be assigned with the coefficients 2^{t_1}, 2^{t_2}, ..., $2^{t_{j-1}}$. Then the paths from the root to a terminal node involve the components

$$\begin{aligned}
2^{T_1} &= 2^{t_0},\\
2^{T_2} &= 2^{t_0} \cdot 2^{j_1} = 2^{t_0+t_1},\\
2^{T_r} &= 2^{t_0} \cdot \ldots \cdot 2^{t_{r-1}} = 2^{t_0+\ldots+t_{r-1}}
\end{aligned}$$

of W_{x_i} that the weights in the nodes along the path are multiplied by the coefficients $a_{i,j}$, so that the j-th terminal node corresponds to

$$a_{i,j}2^{T_j} = a_{i,j}2^{t_0+\ldots+t_{j-1}}.$$

These values are summarized, so that the linear W-tree describes the weight W_{x_i} (Equation 7.10).

Example 7.16 *(Continuation of Example 7.15). A W-tree corresponding to the terminal node, $W_{x_1} = -2^4 + 2^6 - 2^9$ is given in Figure 7.15. This value can be described by the W-tree with values 2^{t_0}, 2^{t_1}, 2^{t_2}, 2^{t_3} and 2^{t_4} assigned to the nodes, where $t_1 = 2$, $t_2 = 2$, $t_3 = 2$ and $t_4 = 3$. The successors of these nodes are assigned with*

$$a_{1,0} = 0,\ a_{1,1} = 0,\ a_{1,2} = -1,\ a_{1,3} = 1,\ a_{1,4} = 1,$$

so that the paths from the root to the terminal nodes correspond to the exponents $-2^4,\ 2^6,\ -2^9$.

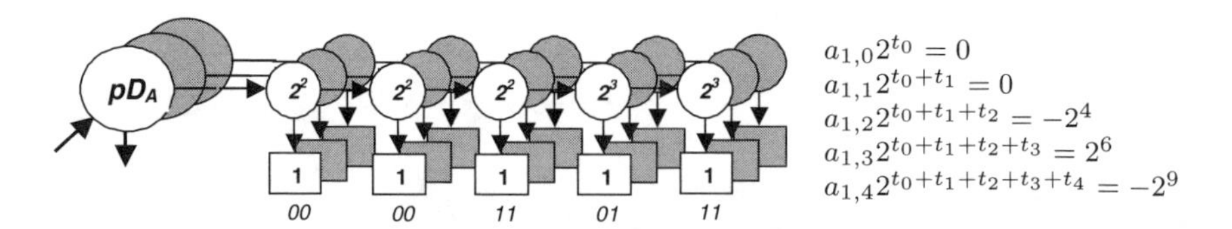

FIGURE 7.15
W-trees represent terminal weights; the first terminal node $W_{x_1} = -2^4 + 2^6 - 2^9$ is represented by the front tree consisting of five nodes (Example 7.16).

7.8 Linear word-level sum-of-products expressions

The most important drawback of the linear word-level arithmetic decision diagrams is the fact that even using the weight encoding technique, the problem of large coefficients is still a difficult challenge to tackle. This is a motivation to define the linear word-level sum-of-products expressions and study their properties.

7.8.1 Definition

In Chapter 4, a word-level sum-of-products expression for an n-input r-output switching function f is defined as the bitwise of sum-of-products expressions of f_j, $j = 1, \ldots, r$,

$$f = 2^{r-1} f_r \,\widehat{\vee}\, \cdots \,\widehat{\vee}\, 2^1 f_2 \,\widehat{\vee}\, 2^0 f_1 \;=\; \widehat{\bigvee}_{i=1}^{2^n} v_i \cdot \underbrace{(x_1^{i_1} \cdots\; x_n^{i_n},)}_{i-th\ product}$$

where $x_i^{i_j}$ is equal to $\overline{x}_i$ if $i_j = 0$, and $x_i^{i_j}$ is equal to x_i if $i_j = 1$.

A *linear* word-level sum-of-products of a switching function f of n variables $x_1, \ldots, x_n$ is the expression with $2n$ integer coefficients $v_1^*, \ldots, v_n^*$

$$f = v_1^* x_1^{i_1} \,\widehat{\vee}\, \ldots \,\widehat{\vee}\, v_n^* x_n^{i_n} \;=\; \widehat{\bigvee}_{i=1}^{n} v_i^* \cdot x_i^{i_j}, \tag{7.11}$$

where the product $v_i^* \cdot x_i^{i_j}$ is a pair

$$v_i^* x_i^{i_j} = \begin{cases} v_i^{'} x_i, \; i_j = 1; \\ v_i^{''} \overline{x}_i, \; i_j = 0. \end{cases}$$

Definition of a linear word-level sum-of-products (Equation 7.11) implies that:

- ▶ Linear word-level sum-of-products expression can be recognized by the products. For example, for given $n = 2$, the linear expression includes no more than two literals. Expression $2x_1 \,\widehat{\vee}\, 3x_2$ is linear, whereas $x_1 \,\widehat{\vee}\, 2\overline{x}_1 x_3 \,\widehat{\vee}\, 3x_2$ is not.
- ▶ The same literal $x_i^{i_j}$ can be included in the expression complemented and non-complemented, with the corresponding coefficients $v_i^{'}$ and $v_i^{''}$ (Equation 7.8.1). For example, $f = x_1 \,\widehat{\vee}\, 5\overline{x}_1$.
- ▶ A coefficient v_i^* carries information about the distribution of terms over the switching functions: 1's in its binary representation indicates the index of function that consists of variable $x_i^{i_j}$. For example, $v_2^* = 6 = 110$ carries information about the variable $x_2^{i_j}$: this variable is present in the functions f_3 (f_3) and f_2 (f_2), and does present in the function f_1 (f_2).
- ▶ A product $v_i^* x_i^{i_j}$ generates, in general, two products: $v_i^{'} x_i$ and $v_i^{''} \overline{x}_i$ (Figure 7.16).
- ▶ To represent the product by Equation 7.11, DeMorgan's rule might be applied. For example, $f = x_1 x_2 = \overline{\overline{\overline{x}_1 \overline{x}_2}} = \overline{\overline{x}_1 \vee \overline{x}_2}$.

7.8.2 Grouping, weight assignment, and masking

In contrast to linear arithmetic expression, the linear word-level sum-of-products model is not sensitive to any permutation of outputs. Each output in this model needs only one bit. Consequently, the parameter t is always equal to 1, so the masking operators are simplified.

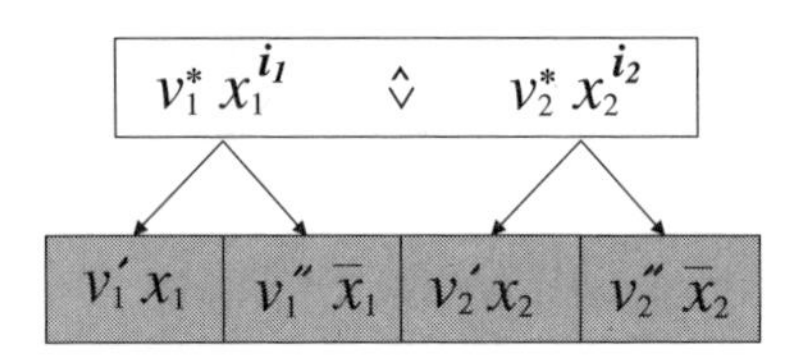

- *A coefficient v_i^* carries information about the distribution of terms over the switching functions.*
- *Distribution: 1's in its binary representation indicates the index of function that depends on variable $x_i^{i_j}$.*
- *A product $v_i^* x_i^{i_j}$ generates, in general, two products: $v_i^{'} x_i$ and $v_i^{''} \overline{x}_i$.*

FIGURE 7.16
Structure of products $v_i^* x_i^{i_j}$, $i = 1, 2$, in the linear sum-of-products expression.

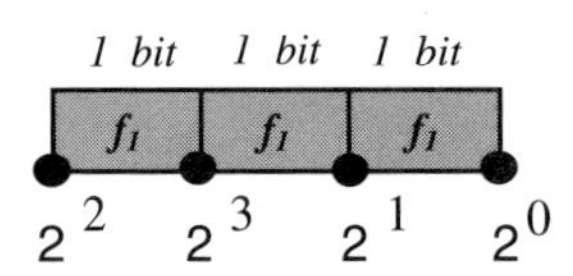

Bits of word-level format:

$$\begin{cases} f_1 = \Xi^1\{f\} = x_1 \vee x_3 \\ f_2 = \Xi^2\{f\} = x_3 \vee x_4 \\ f_3 = \Xi^3\{f\} = \overline{x}_2 \vee x_4 \end{cases}$$

Word-level representation:

$$f = 2^2 f_3 \,\widehat{\vee}\, 2^1 f_2 \,\widehat{\vee}\, 2^0 f_1 = x_1 \,\widehat{\vee}\, 4\overline{x}_2 \,\widehat{\vee}\, 3x_3 \,\widehat{\vee}\, 6x_4$$

FIGURE 7.17
Word-level format for the set of three linear sum-of-products expressions f_1, f_2, and f_3 (Example 7.17).

Example 7.17 *The three-output function, $f_1 = x_1 \vee x_3$, $f_2 = x_3 \vee x_4$, $f_3 = \overline{x}_2 \vee x_4$ is represented by the word-level sum-of-products as shown in Figure 7.17. Let us change the order of functions to $f_2 = x_3 \vee x_4$, $f_1 = x_1 \vee x_3$, $f_3 = \overline{x}_2 \vee x_4$. Then, $f = 2x_1 \,\widehat{\vee}\, 4\overline{x}_2 \,\widehat{\vee}\, 3x_3 \,\widehat{\vee}\, 5x_4$. Hence, linearity is not relevant to the order of the functions in word-level sum-of-products expressions.*

7.8.3 Linear expressions of elementary functions

In Table 7.5, two word-level models of elementary switching functions are shown: linear arithmetic and sum-of-products. Linear sum-of-products expressions for OR, AND, NOR and NOT functions can be determined directly from Equation 7.11. For example, an OR function requires two bits for representation by arithmetic expression (the first function is a garbage function). In a word-level sum-of-products, the OR function needs one bit only. The remarkable feature is that these expressions are determined with respect to two different definitions of linearity: arithmetic operations and OR bitwise operation respectively.

TABLE 7.5
Comparison of linear word-level arithmetic and sum-of-products models of two-input elementary functions.

Gate	Linear word-level arithmetic model	Linear word-level sum-of-products model
OR	$\Xi^2\{1+x+y\}$	$\Xi^1\{x \vee y\}$
AND	$\Xi^2\{x+y\}$	$\Xi^1\{\overline{\overline{x} \vee \overline{y}}\}$
NOR	$\Xi^2\{2-x-y\}$	$\Xi^1\{\overline{x \vee y}\}$
NAND	$\Xi^2\{3-x-y\}$	$\Xi^1\{\overline{x} \vee \overline{y}\}$

7.8.4 Linear decision diagrams

A linear decision diagram derived from the linear word-level sum-of-products Equation 7.11 is a linear tree with nodes realizing the Shannon expansion of r switching functions f_j, $j = 1, 2, \ldots, r$, in parallel

$$f_j = f_j(x_i = 0) \vee x_i f_j(x_i = 1), \tag{7.12}$$

where $j = 1, 2, \ldots, r$. The term $v_i^* x_i^{i_j}$ of the word-level sum-of-products Equation 7.11 carries information about:

- ▶ The number of times Shannon expansions with respect to variables x_i were applied, and
- ▶ The functions f_j to which the Shannon expansion was applied.

Example 7.18 *Consider product $5x_1$ in a word-level sum-of-products. We observe that the Shannon expansion with respect to variable x_1 is used twice: the number of 1s in the coefficient $5 = 101$ is equal to 2 ($1+0+1=2$). This coefficient also carries information about the values of functions f_j ($f_1 = 1, f_2 = 0, f_3 = 1$) given $x_1 = 1$. (Figure 7.18).*

In general, a linear decision diagram includes $2n$ nonterminal and $2n+1$ terminal nodes as follows from Equation 7.11.

Example 7.19 *Linear word-level decision diagram using Shannon expansion (Equation 7.12) for the OR and AND functions are shown in Figure 7.19a and b respectively:*

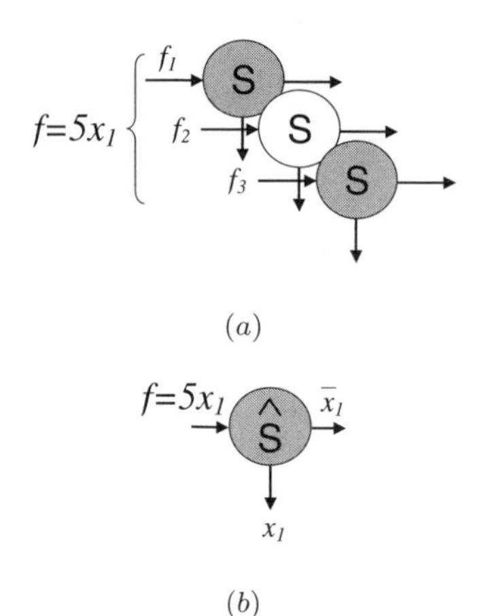

- *Bitwise Shannon expansion is implemented in parallel.*
- *The coefficient* $5 = 101$ *indicates to which functions Shannon expansion has been applied (*f_1 *and* f_3*).*

$$f_1 = f_1(x_1 = 0) \vee x_1 f_1(x_1 = 1)$$
$$f_3 = f_3(x_1 = 0) \vee x_1 f_3(x_1 = 1)$$

- *Equivalent denotation of the word-level Shannon expansion requires one node.*

FIGURE 7.18
Shannon expansion for the product $5x_1$ in the word-level format of sum-of-products (a) and its equivalent denotation (b) (Example 7.18).

(*a*) *The diagram for OR function is derived from the linear expression* $x_1 \vee x_2$ *using Shannon expansion* $f = \overline{x}_i f_0 \vee x_i f_1 = f_0 \overset{\frown}{\vee} x_i f_1$ *that is denoted by* $\overset{\frown}{S}$;

(*b*) *The diagram for AND function is derived from the linear expression* $\overline{x}_1 \vee \overline{x}_2$ *using Shannon expansion* $f = \overline{x}_i f_0 \vee x_i f_1 = \overline{x}_i f_0 \overset{\frown}{\vee} f_1$ *also denoted by* $\overset{\frown}{S}$.

From Example 7.19, we observe that the word-level Shannon expansions (Equation 7.12) for elementary functions means the Shannon expansion applied to each participation function, i.e., bitwise.

7.8.5 Technique of computation

Here, the details of computing the linear sum-of-products (expressions and diagrams) for a circuit are introduced by three examples.

Representation of a circuit level. An arbitrary level of a multilevel circuit can be represented by a linear word-level sum-of-products over the library OR, AND, NOR, NAND, and NOT gates.

Example 7.20 *The level of a circuit is presented in Figure 7.20. The linear sum-of-products describing three gate outputs* f_1, f_2, *and* f_3 *are given in Table*

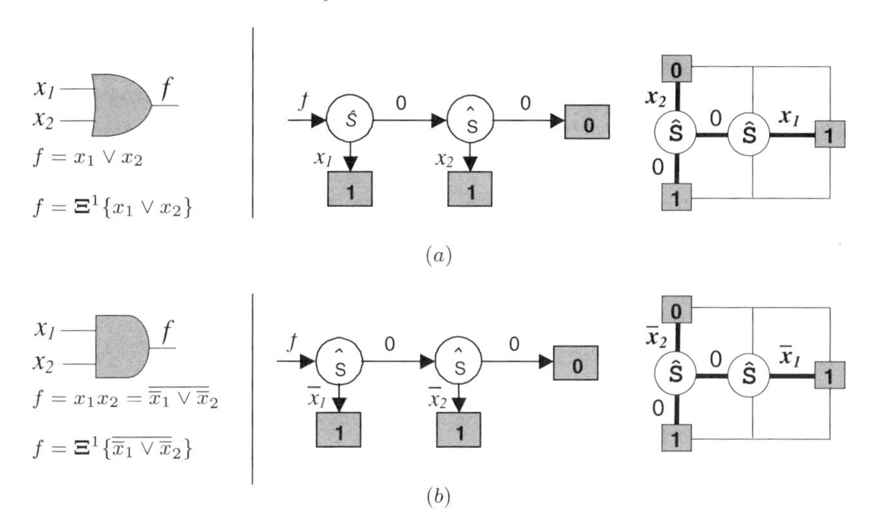

FIGURE 7.19
Linear decision diagram and $\mathcal{N}$-hypercube for the OR (a) and AND (b) gate (Example 7.19).

7.5. The final result is $f = x_1 \widehat{\vee}\ 2\overline{x}_1 \widehat{\vee}\ 2x_2 \widehat{\vee}\ 4\overline{x}_2 \widehat{\vee}\ 5x_3$. *We observe, that literal* x_1 *is generated by* f_1 *because the coefficient* $v_1^{'} = 1 = 001$. *Literal* $\overline{x}_1$ *is generated by* f_2 *because* $v_1^{''} = 1 = 010$. *The coefficient* $v_2^{'} = 2 = 010$ *means that* $x_2 \in f_2$; $v_2^{''} = 4 = 100$, *that is* $\overline{x}_2 \in f_3$ *and* $x_3 \in f_3$, $f_3 = \overline{x}_2 \vee x_3$.

Application of the word-level Shannon expansion. The two examples below demonstrate details of computation.

Let us design a linear decision diagram given the result of Example 7.20. Note that the order of variables in the diagram can be arbitrary. Let us choose the lexicographical order, i.e., x_1, x_2.

Example 7.21 *(Continuation of Example 7.20). The linear decision diagram for the expression* $f = x_1 \widehat{\vee}\ 2\overline{x}_1 \widehat{\vee}\ 2x_2 \widehat{\vee}\ 4\overline{x}_2 \widehat{\vee}\ 5x_3$ *consists of five nodes, each node performing a word-level Shannon expansion. Details are given in Table 7.6 and Figure 7.21.*

Some comments for this example are useful. There are five steps in the diagram design with respect to variables $x_1, \overline{x}_1, x_2, \overline{x}_2$ and x_3. Each step in the Shannon expansion is computed as follows.

With respect to x_1 : The literal $x_1 \in f_1$, $f_1 = x_1 \vee x_3$ ($v_1^{'} = 1 = 001$). The first terminal node is equal to 1 (right branch). The left branch needs further decomposition.

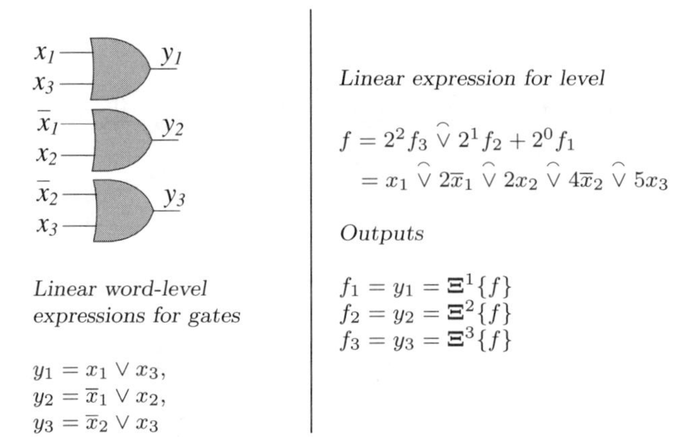

FIGURE 7.20
A circuit level and the corresponding linear word-level sum-of-products (Example 7.20).

TABLE 7.6
Linear decision diagram design (Example 7.21).

		Right branch	**Left branch**
x_1	$f_1(x_1 = 0) = x_3$ $f_1(x_1 = 1) = 1$	$x_1 f_1(x_1 = 1) = 1$	$x_1 \,\widehat{\vee}\, 2\overline{x}_1 \,\widehat{\vee}\, 2x_2 \,\widehat{\vee}\, 4\overline{x}_2 \,\widehat{\vee}\, 5x_3$
$\overline{x}_1$	$f_2(\overline{x}_1 = 0) = x_2$ $f_2(\overline{x}_1 = 1) = 1$	$\overline{x}_1 f_2(\overline{x}_1 = 1) = 1$	$x_1 \,\widehat{\vee}\, 2 \,\widehat{\vee}\, 2x_2 \,\widehat{\vee}\, 4\overline{x}_2 \,\widehat{\vee}\, 5x_3$
x_2	$f_2(x_2 = 0) = 1$ $f_2(x_2 = 1) = 1$	$x_2 f_2(x_2 = 1) = 1$	$x_1 \,\widehat{\vee}\, 2 \,\widehat{\vee}\, 2 \,\widehat{\vee}\, 4\overline{x}_2 \,\widehat{\vee}\, 5x_3$
$\overline{x}_2$	$f_3(\overline{x}_2 = 0) = x_3$ $f_3(\overline{x}_2 = 1) = 1$	$\overline{x}_2 f_3(\overline{x}_2 = 1) = 1$	$x_1 \,\widehat{\vee}\, 2 \,\widehat{\vee}\, 2 \,\widehat{\vee}\, 4 \,\widehat{\vee}\, 5x_3$
x_3	$f_1(x_3 = 0) = 1$ $f_1(x_3 = 1) = 1$ $f_3(x_3 = 1) = 1$ $f_3(x_3 = 1) = 1$	$x_3 f_1(x_3 = 1) = 1$ $x_3 f_3(x_3 = 1) = 1$	$1 \,\widehat{\vee}\, 2 \,\widehat{\vee}\, 2 \,\widehat{\vee}\, 4 \,\widehat{\vee}\, 5 = 7$

With respect to $\overline{x}_1$: The literal $\overline{x}_1 \in f_2$, $f_2 = \overline{x}_1 \vee x_2$ ($v_1^{''} = 2 = 010$). The second terminal node is equal to 1 (right branch), thus $010 = 2$ in the notation of a word-level Shannon expansion. The left branch needs further decomposition.

With respect to x_2 : The literal $x_2 \in f_2$, $f_2 = \overline{x}_1 \vee x_2$ ($v_2^{'} = 2 = 010$).

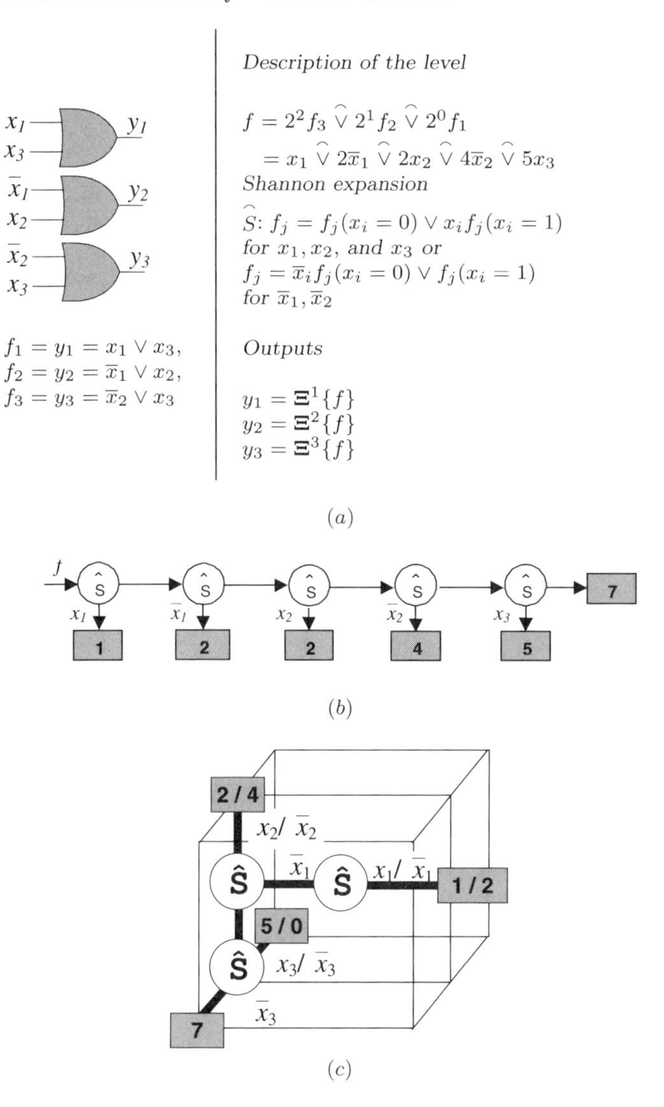

FIGURE 7.21
Technique of the representation of a level of circuit by a linear word-level sum-of-products expression (a) linear decision diagram (b), and $\mathcal{N}$-hypercube (c) (Example 7.21).

The third terminal node is equal to 1 (right branch) that is 010 = 2 in the notation of a word-levelShannon expansion. The left branch needs further decomposition.

With respect to $\overline{x}_2$: The literal $\overline{x}_2 \in f_3$, $f_3 = \overline{x}_2 \vee x_3$ ($v_2^{''} = 100$). The fourth terminal node is equal to 1 (right branch) that is 100 = 4 in the notation of a word-level Shannon expansion. The left branch needs further decomposition.

With respect to x_3 : The literal $x_3 \in f_1$, $f_1 = x_1 \vee x_3$ ($v_1^{'} = 1 = 001$) and $f_3 = \overline{x}_2 \vee x_3$ ($v_3^{'} = 4 = 100$). The fifth terminal node is equal to 1 (right branch) that is $001 \; \widehat{\vee} \; 100 = 101 = 5$ in the notation of a word-level Shannon expansion. The left branch (sixth terminal node) is equal to seven.

The example below demonstrates two aspects of linearization technique by the word-level sum-of-products: representation of circuit level and computing by a set of linear decision diagrams given assignment of inputs.

Example 7.22 *A three-level circuit is shown in Figure 7.22. For the assignment* $x_1x_2x_3x_4 = 0\ 1\ 1\ 1$*, the value of the function* f *is calculated as follows. The outputs* $y_1 = 1$ *and* $y_2 = 0$*, since* $f_1 = \overline{0} \vee 3 \cdot \overline{1} \vee 2 \cdot \overline{1} = 01_2$*. Similarly,* $z = 1$ *because* $f_2 = \overline{1} \vee \overline{1} = 1_2$*. Finally,* $f = 0$*, since* $f_3 = \overline{1} \vee \overline{1} = 0$*.*

7.9 Linear word-level Reed-Muller expressions

The linear word-level sum-of-products decision diagrams are not efficient for EXOR circuits. However, in some technologies the cost of EXOR logic is acceptable compared to NAND and NOR logic. The modification of a linear sum-of-products model can be done to avoid this drawback.

7.9.1 Definition

In Chapter 3, the word-level Reed-Muller expression for an n-input r-output switching function f is defined as the bitwise of sum-of-products expressions of f_j, $j = 1, \ldots, r$,

$$f = 2^{r-1} f_r \,\widehat{\oplus}\, \cdots \,\widehat{\oplus}\, 2^1 f_2 \,\widehat{\oplus}\, 2^0 f_1 \;=\; \widehat{\bigoplus_{i=0}^{2^n-1}} \; w_i \cdot \underbrace{(x_1^{i_1} \cdots \; x_n^{i_n \cdot})}_{i-th\ product}$$

where $x_i^{i_j}$ is equal to 1 if $i_j = 0$, and $x_i^{i_j}$ is equal to x_i if $i_j = 1$.

The *linear* word-level Reed-Muller expression of a switching function f of n variables $x_1, \ldots, x_n$ is the expression with integer coefficients $w_1^*, \ldots, w_n^*$

$$f = w_1^* x_1^{i_1} \,\widehat{\oplus}\, \ldots \,\widehat{\oplus}\, w_n^* x_n^{i_n} \;=\; \widehat{\bigoplus_{i=0}^{n}} \; w_i^* \cdot x_i^{i_j}. \tag{7.13}$$

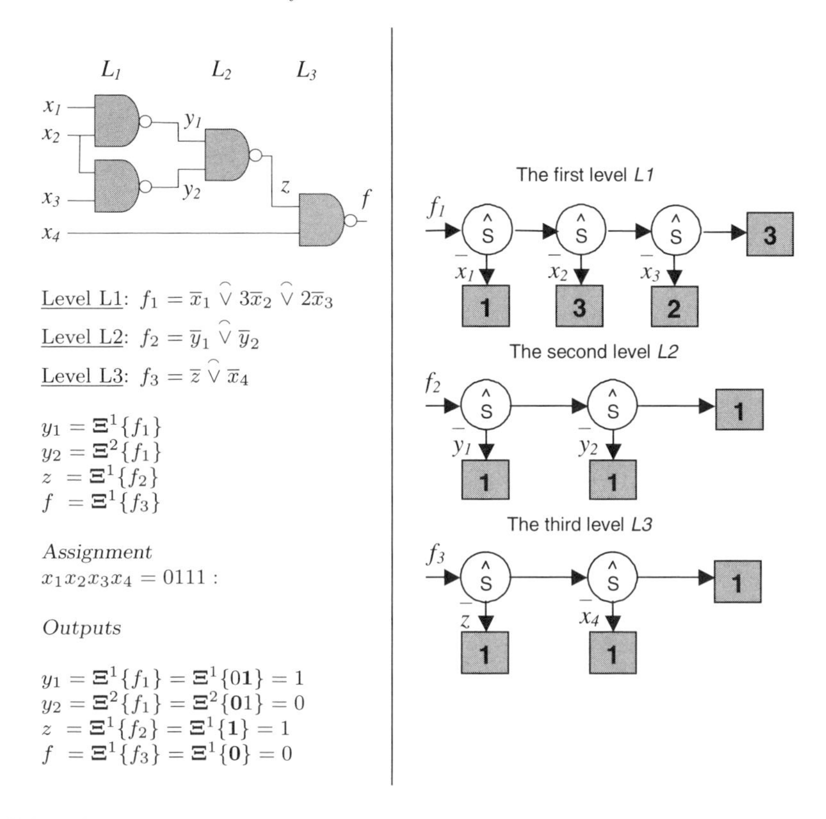

FIGURE 7.22
Representation of a three-level circuit by a set of linear decision diagrams (Example 7.22).

where

$$w_i^* x_i^{i_j} = \begin{cases} w_i^{'} x_i, & i_j = 1; \\ w_i^{''} \overline{x}_i, & i_j = 0. \end{cases}$$

The above properties are similar to ones of linear word-level sum-of-products expressions.

7.9.2 Grouping, weight assignment, and masking

Grouping, weight assignment and masking in the format of linear word Reed-Muller expressions are similar to the linear word-level sum-of-products model. For instance, the linearity property does not depend on the order of the functions in word-level expression.

7.9.3 Linear Reed-Muller expressions of primitives

In Table 7.7, two word-level models of two-input elementary functions are given: linear word-level arithmetic and linear word-level Reed-Muller expressions. Linear Reed-Muller expressions for two-input EXOR function can be determined directly from Equation 7.13 for $n = 2$.

TABLE 7.7
Comparison of linear word-level arithmetic and Reed-Muller models of two variables elementary functions.

Gate	Linear word-level arithmetic model	Linear word-level sum-of-products model
EXOR	$\Xi^1\{x + y\}$	$\Xi^1\{x \oplus y\}$
EXNOR	$\Xi^1\{1 + x + y\}$	$\Xi^1\{\overline{x \oplus y}\}$
NOT	$\Xi^1\{1 + x\}$	$\Xi^1\{x \oplus 1\}$

7.9.4 Linear decision diagrams

A linear decision diagram derived from the linear word-level Reed-Muller expression (Equation 7.13) is a linear tree with nodes in which the *multiple* Davio expansion is implemented:

$$f = f_j(x_i = 0) \oplus x_i f_j(x_i = 1), \tag{7.14}$$

where $j = 1, 2, \ldots, r$. The term $w_i^* x_i^{i_j}$ of a word-level sum-of-products (7.13) carries information about:

- ▶ The number of required Davio expansions with respect to variables x_i,
- ▶ The functions f_j to which Davio expansion has been applied.

Example 7.23 *Consider the term $5x_1$ in a word-level Reed-Muller expression. We observe that the Davio expansion with respect to variable x_1 is used twice as indicated by coefficient 5: the number of 1s in $5 = 101$ is equal to 2, $1 + 0 + 1 = 2$. This coefficient also carries information about the function of action: Davio expansion is applied to f_1 and f_3 (Figure 7.23).*

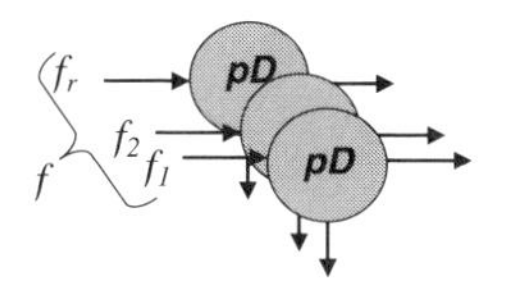

The coefficient $w^* = 5 = 101$ carries information about the functions to which Davio expansion has been applied to (f_1 and f_3)

$$f_1 = f_1(x_1 = 0) \oplus x_1 f_1(x_1 = 1)$$
$$f_3 = f_3(x_1 = 0) \oplus x_1 f_3(x_1 = 1)$$

FIGURE 7.23
Davio expansion is applied to f_1 and f_3 (Example 7.23).

In general, a linear decision diagram includes a $2n$ nonterminal and $2n + 1$ terminal nodes as follows from Equation 7.13.

Example 7.24 *Apply Davio expansion (Equation 7.14) to the EXOR expression with respect to x_1: $f_0 = f_{x_1=0} \oplus x_1 f_{x_1=1}$. The right branch (first terminal node) is a constant $x_1 f_{x_1=1} = 1$ (Figure 7.24). The left branch leads to x_2. Next, expand with respect to x_2: the right branch is $x_2 f_1 = 1$ (second terminal node) and the left branch is $f_0 = 0$ (third terminal node).*

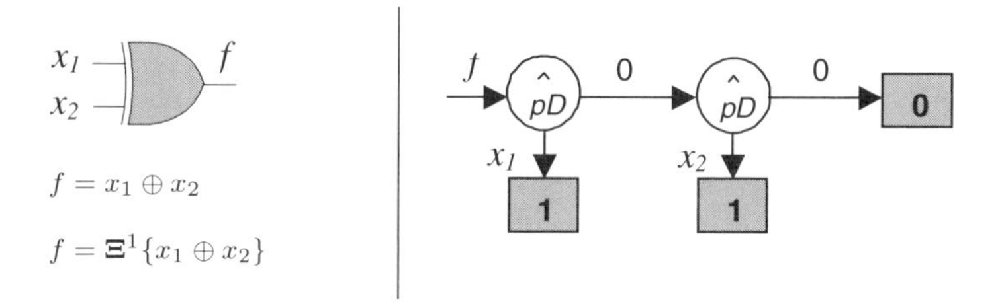

FIGURE 7.24
Linear decision diagram for the EXOR function (Example 7.24).

For circuits, different linear word-level logics and diagrams can be used. Especially simple is the application of linear word-level sum-of-products and Reed-Muller expressions. This "mixed" technique can be used, that leads to linear decision diagrams with Shannon and Davio expansion nodes.

7.10 Summary

Linearization is a technique aimed at conversion of an arbitrary multiinput multioutput switching function into a linear word-level expression and, eventually, in a linear decision diagram. The linear decision diagram is then embedded in a spatial structure, $\mathcal{N}$-hypercube.

1. The linearization technique

- Is based on grouping several switching functions into a word. The grouping can be done on various algebraic basis (arithmetic, sum-of-products, Reed-Muller) that provides linearity over corresponding arithmetic or logic operations.
- Can be implemented for any unary function based on Theorems 7.2, 7.3, and 7.4, on this basis, for multioutput functions.
- Provides the significant simplification of embedding decision diagrams in hypercubes.
- Provides the additional resources to massive parallel computation on words instead of bits.
- Enhance possibilities for verification of spatial structures.

2. There are three levels of parallelism in linear word-level expressions in spatial dimensions:

- Parallelism of information flows on a hypercube,
- Parallelism of word-level computation of functions, and
- Parallelism of computing the assignments of variables.

The hypercube structure and word-level linear expressions are inherently parallel in implementation and computation. This property is integrated into the nodes of decision diagrams and hypercube structures.

3. The advantages and disadvantages of the linearization technique are as follows:

Arithmetic word-level linear expressions are linear polynomial expansions. While the class of single-output functions that can generate linear expressions is very limited, the word-level combination of several functions can produce linear expression even if involved function does not have linear representations. To represent an arbitrary switching function by linear word-level expression, a special technique based on adding extra bits, or garbage functions, is applied. Moreover, additional resources are required for masking operators and encoding of coefficients to alleviate the effect of large values. The node of linear word-level diagram implements the arithmetic analog of Davio expansion with arithmetic operations. The number of intermediate nodes (processing elements) and terminal nodes in linear diagram does not exceed n and $n+1$ correspondingly, where n is a number of variables.

Sum-of-products word-level linear expressions provides the linearity of word-level representation over OR operation. The node of linear word-level diagram implements the word-level Shannon expansion that is a multiple-input multiplexor. The problem of large coefficients is not present here. However, masking operator is applied to recover the initial functions (bits) from the word. The number of intermediate nodes (processing elements) and terminal nodes in linear diagram not exceed $2n$ and $2n+1$ correspondingly.

Reed-Muller word-level linear expressions are utilized for representation of Reed-Muller expressions (EXOR circuits). The number of intermediate nodes (processing elements) and terminal nodes in linear diagram does not exceed $2n$ and $2n+1$ correspondingly.

4. The above techniques for word-level linearization utilized the properties of linearization by Theorems 7.2, 7.3, and 7.4.

5. Criteria to choose the method of linearization are as follows:

- ▶ Function of the node in the computing structure,
- ▶ Preferences on the type of operation and masking operator, and
- ▶ Library of available elements or logic gates.

In conclusion, the linear models of word-level computation provide the necessary level of parallelism and are perfect candidates for massive parallel computations on 2-D and 3-D structures as it will be shown in Chapter 11.

7.11 Problems

Problem 7.1 In the circuit presented in Figure 7.25a, a output y_1 is equal to 1 if variable x_2 is greater than x_1, and output y_2 is set if both of variables x_1 and x_2 are equal. Construct:

(*a*) An arithmetic linear expression for each gate,
(*b*) An arithmetic linear expression for the whole circuit.

Use Example 7.2 and Example 7.3.

Problem 7.2 Given the linear arithmetic expression $5x_1 + x_2 + 4x_3 + 5$, find two masking operators which extract the switching functions $1 \oplus x_1 \oplus x_2$ and $x_1 \vee x_3$. You may use Example 7.4 for reference.

Problem 7.3 Construct a linear decision diagram for the level of a circuit presented in Figure 7.25b.

Problem 7.4 Restore ta circuit level containing of two gates, f_1 and f_2, represented by a linear decision diagram in Figure 7.25c, *if* $f_1 = \Xi^3\{f\}$, *and* $f_1 = \Xi^1\{f\}$.

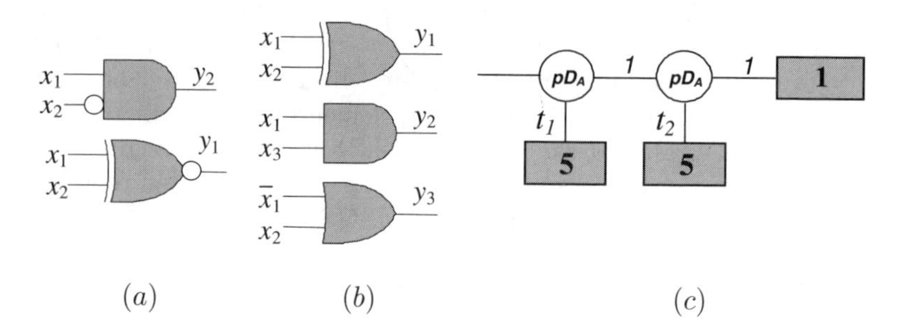

FIGURE 7.25
Circuit for Problem 7.1 (a), Problem 7.3 (b), and a linear decision diagram for Problem 7.4 (c).

Problem 7.5 For the linear decision diagram given in Figure 7.25c, evaluate the output value given the input assignment $x_1x_2 = 10$.
HINT: Use the method from Example 7.7.

Problem 7.6 Given the two-input multiplexer realizing the switching function $f = \overline{s}x_1 \vee sx_2$, construct:

(*a*) Word-level sum-of-products expression and,
(*b*) Propose a simple BDD-based implementation,
(*c*) A $\mathcal{N}$-hypercube model.

Problem 7.7 Consider a full-adder 3-D presented in Figure 7.26a. Linear decision diagram of linear arithmetic expressions and $\mathcal{N}$-hypercube for each level of the circuit are given in Figure 7.26b,c.

(*a*) Construct linear logic expressions (use the appropriate logic operation)
(*b*) Restore sum-of-products expression.

Problem 7.8 Consider linear decision diagrams for the three-input majority circuit presented in Figure 7.27. The three-input majority function, $x_1+x_2+x_3 \geq 2$, is implemented by the circuit, whose level is described by the linear arithmetic expressions:

$$L_i = \begin{cases} (2^0 + 2^2)x_1 + (2^0 + 2^4)x_2 + (2^2 + 2^4)x_3, & i = 1, \\ 2^0 y_1 + 2^0 y_2 + 2^0 y_3, & i = 2. \end{cases}$$

(*a*) Find the masking operators for the corresponding logic functions of the first and the second level.
(*b*) Evaluate the value of f given assignments 001, 101, 111 of x_1, x_2 and x_3.

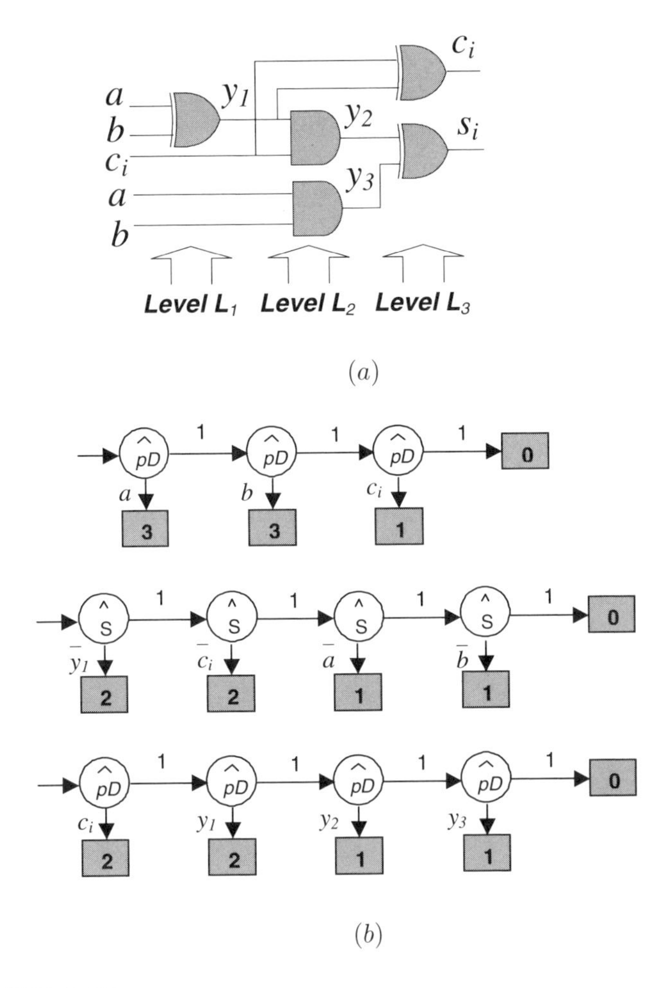

FIGURE 7.26
Full adder circuit (a) and its linear decision diagrams (b) for Problem 7.7.

Problem 7.9 Given a set of linear word-level decision diagrams (Figure 7.28a) and $\mathcal{N}$-hypercubes(Figure 7.28b), restore the word-level logic expressions (the type of logic is shown in the nodes) and the initial functions for:

(*a*) $L1$, assuming that logic operation in the nodes is $\widehat{\oplus}$

(*b*) $L2$, assuming that logic operation in the nodes is $\widehat{\vee}$

(*c*) $L3$, assuming that logic operation in the nodes is $\widehat{\oplus}$

(*d*) $L4$, assuming that logic operation in the nodes is $\widehat{\vee}$

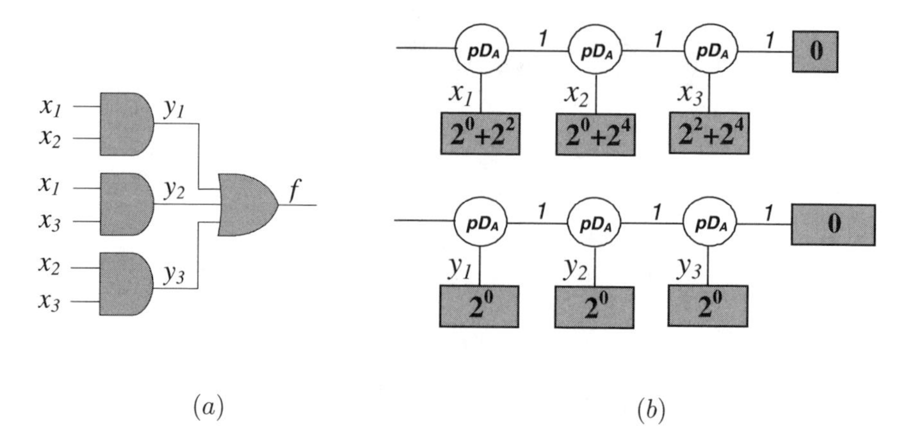

FIGURE 7.27
Three-input majority function (a) and corresponding linear decision diagrams (b) for Problem 7.8.

Problem 7.10 The 2-level AND/OR implementation of 2-input multiplexer and a set of linear decision diagrams are given in Figure 7.29. Represent the multiplexer using binary decision diagram technique and compare the results.

7.12 Further reading

Linearization technique. An elegant method for linearization of AND, OR, EXOR functions of an arbitrary number of variables was introduced by Malyugin [2]. The method is based on the so-called *the algebra of corteges.* Different aspects of linearization technique can be found in [5]. Additional references can be found in Chapter 9.

Linear transformation of variables is a method for optimization of the representation of a switching function. In terms of spectral technique, linear transformation of variables is a method to reduce the number of nonzero coefficients in the spectrum of a switching function. This approach is developed in [1, 3, 4].

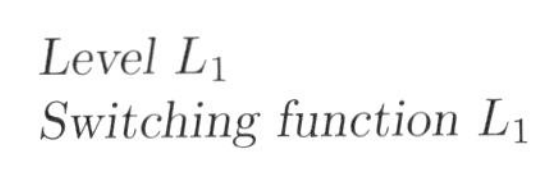

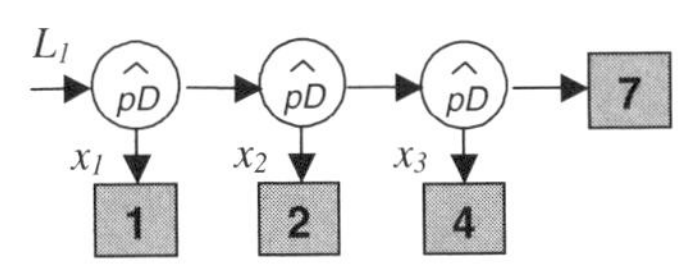

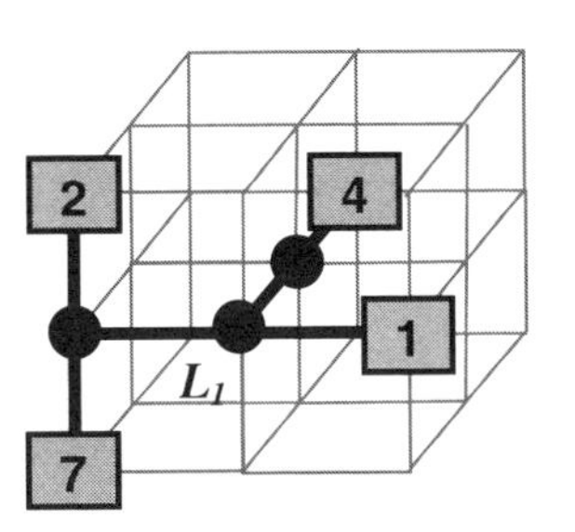

Level L_2
Switching function L_2

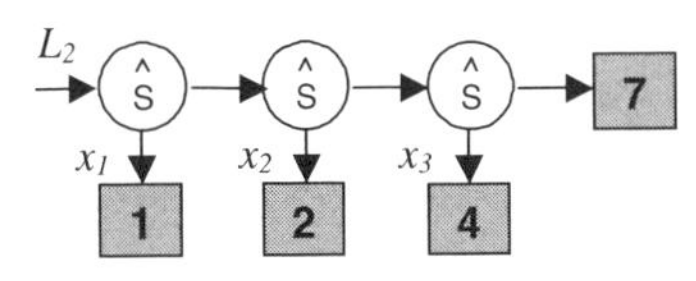

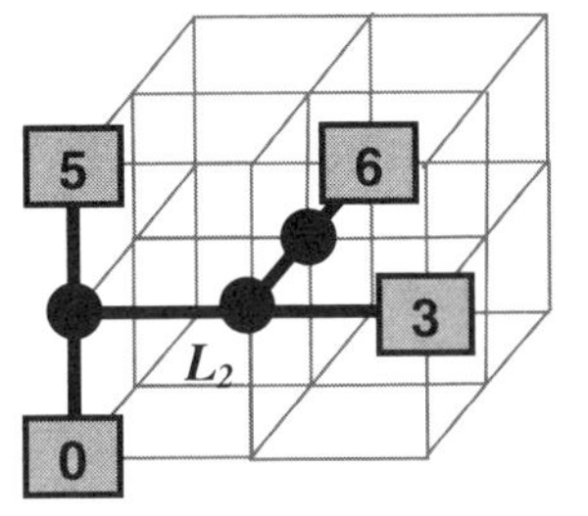

Level L_3
Switching function L_3

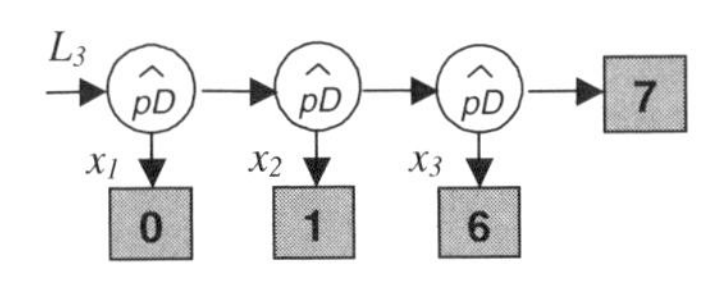

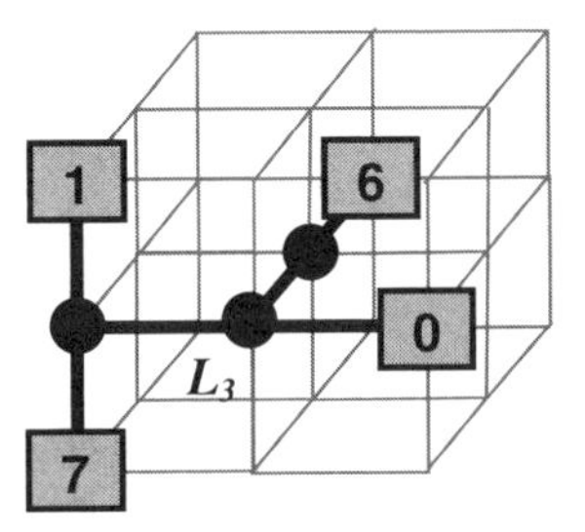

Level L_4
Switching function L_4

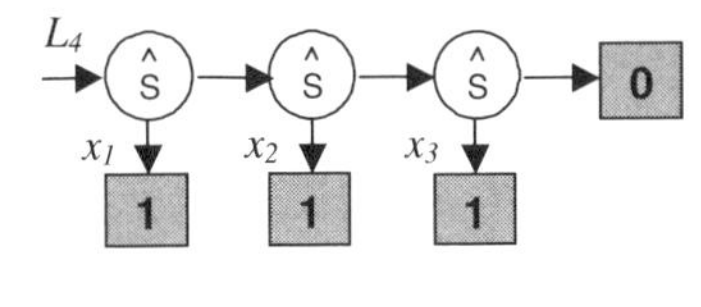

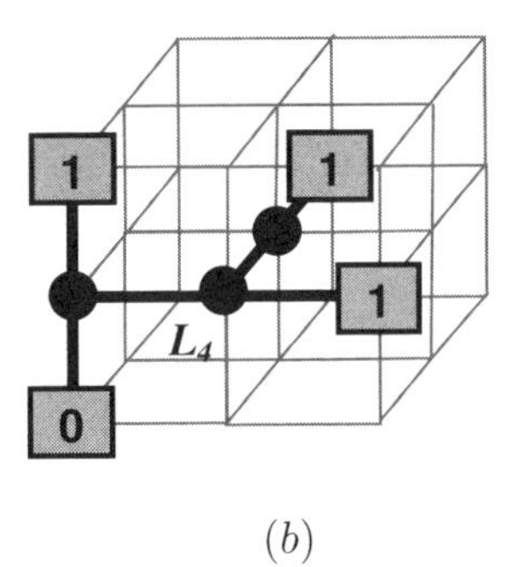

(a) (b)

FIGURE 7.28
Three-input circuits: word-level linear decision diagrams (a) and $\mathcal{N}$-hypercubes (b) for Problem 7.9.

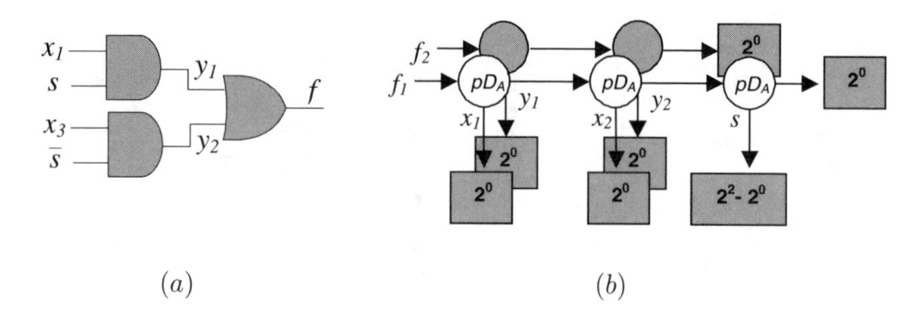

(a) (b)

FIGURE 7.29
AND/OR circuit to implement two-input multiplexer (a) and corresponding set of linear decision diagrams (b) for Problem 7.10.

7.13 References

[1] Karpovsky MG. *Finite Orthogonal Series in the Design of Digital Devices.* John Wiley & Sons, New York, 1976.

[2] Malyugin VD. Realization of Boolean function's corteges by means of linear arithmetical polynomial. *Automation and Remote Control* (Kluwer/Plenum Publishers), 45(2):239–245, 1984.

[3] Moraga C. On some applications of the Chrestenson functions in logic design and data processing. *Mathematics and Computers in Simulation,* 27:431–439, 1985.

[4] Stanković RS, and Astola JT. Some remarks on linear transform of variables in representation of adders by word-level expressions and spectral transform decision diagrams. In *Proceedings IEEE 32nd International Symposium on Multiple-Valued Logic*, pp. 116–122, 2002.

[5] Yanushkevich SN, Shmerko VP, and Dziurzanski P. Linearity of word-level models: new understanding. In *Proceedings IEEE/ACM 11th International Workshop on Logic and Synthesis,* pp. 67–72, New Orleans, LA, 2002.

8

Event-Driven Analysis of Hypercube-Like Topology

This chapter contributes to the field of logic design of nanoICs that analyzes behavior of computing structures in terms of change. The notation of elementary change in a system is useful at some phases of analysis and synthesis. This model is serviceable in the study of "static" and "dynamic" changes in a circuit, caused by an "event" (e.g., a fault on the line of a circuit). This analysis is event-driven, and the mathematical tool for it is logic differential calculus. Differential operators, of which the basic one is Boolean difference, provide an opportunity for analysis of circuit properties, such as flexibility (ability to be modified without compromising functionality), symmetry, monotony, i.e., detection of properties that are prerequisite for optimization. It provides an additional opportunity for analysis of circuit behavior (consequences of "events", sensitivity analysis, testability, etc.) In this chapter, we discuss:

- The definition of a mathematical tool for detection of an "event" (change) in a binary system, a Boolean difference of a switching function;
- Differential operators for analysis of the properties and behavior of switching functions; and
- Data structures and technique for computing differential operators.

Event-driven analysis is applied to

- Sensitivity and observability analysis;
- Testing;
- Symmetry recognition;
- Power dissipation estimation; and
- Verification.

The Boolean difference is a certain analog of the Taylor cofactor of an algebraic function. The analog of Taylor expansion on switching theory is Reed-Muller expansion, as well as arithmetic and Walsh forms. Thus, Boolean difference can be utilized to calculate Reed-Muller, arithmetical and Walsh coefficients. On the other hand, Taylor expansion gives a useful interpretation of spectral technique because of the structure of each spectral coefficient in terms of change.

In this chapter, we briefly introduce various aspects of applied event-driven analysis. The "Further Reading" Section provides references for extended reading on the subject.

In Section 8.1, the formal model of change, a Boolean difference, is given. The basics of technique for computing Boolean differences are the focus of Section 8.2. The models for independent and dependent changes are introduced in Section 8.3. In Section 8.4, the computing of change in matrix form is presented. To detect the direction of change, special operators are utilized (Section 8.5). The technique of local computing via Boolean differential operators is given in Section 8.6. Taylor expansion, a polynomial model for representation of various forms of a switching function relevant to concept of change, is the subject of Section 8.7 (Reed-Muller expansion) and Section 8.8 (arithmetic expression). After the Summary (Section 8.9), a set of problems is provided (Section 8.10). Finally, additional information on state-of-the-art logic differential calculus is referred to in Section 8.11.

8.1 Formal definition of change in a binary system

To model a binary system in terms of change, we need to:

(*a*) Specify a formal notation of change;

(*b*) Develop the rules to manipulate this model to describe the system behavior; and

(*c*) Apply this technique to solving the problems of logic design.

In this section, we give the formal definition of change, and introduce a technique of detection of change using various data structures.

8.1.1 Detection of change

Detection of a change in a binary system. A signal in a binary system is represented by two logical levels, 0 and 1. Let us formulate the task as detection of the change in this signal. The simplest solution is to deploy an EXOR operation, modulo 2 sum of the signal s_{i-i} before an "event" and the signal s_i after the "event" (e.g., a faulty signal), i.e., $s_{i-i} \oplus s_i$.

Example 8.1 *For the signal depicted in Figure 8.1, four possible combinations of the logical values or signals 0 and 1 are analyzed.*

It follows from this example that if not change itself but direction of change is the matter, then two logical values 0 and 1 can characterize the behavior of the logic signal $s_i \in \{0, 1\}$ in terms of change, where 0 means any change of a signal, and 1 indicates that one of two possible changes has occurred $0 \rightarrow 1$ or $1 \rightarrow 0$.

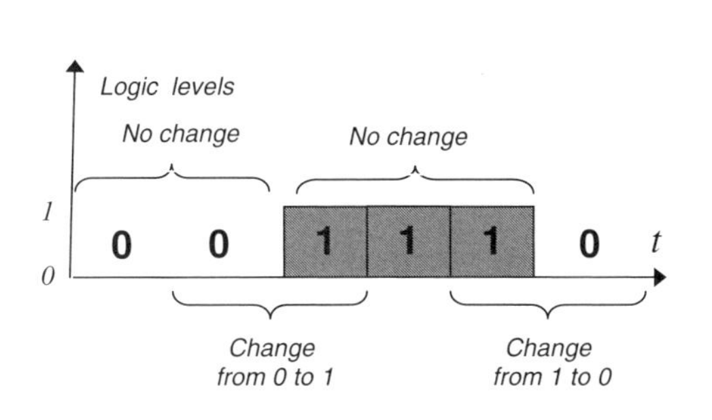

FIGURE 8.1
The change of a binary signal and its detection (Example 8.1).

Detection of change in a switching function. Let the i-th input of a switching function have been changed from the value x_i to the opposite value, $\overline{x}_i$. This causes the circuit output to be changed from the initial value. Note that values $f(x_i)$ and $f(\overline{x}_i)$ are not necessarily different. The simplest way to recognize whether or not they are different is to try to find a difference between $f(x_i)$ and $f(\overline{x}_i)$.

Model of single change: Boolean difference. The Boolean difference of a switching function f of n variables with respect to a variable x_i is defined by equation

$$\frac{\partial f}{\partial x_i} = \underbrace{f(x_1, \ldots, x_i, \ldots, x_n)}_{Initial\ function} \oplus \underbrace{f(x_1, \ldots, \overline{x}_i, \ldots, x_n)}_{Function\ with\ x_i\ complemented} \tag{8.1}$$

It follows from the definition of Boolean difference that

$$\begin{aligned}\frac{\partial f}{\partial x_i} &= \underbrace{f(x_1, \ldots, 0, \ldots, x_n)}_{x_i\ is\ replaced\ with\ 0} \oplus \underbrace{f(x_1, \ldots, 1, ..., x_n)}_{x_i\ is\ replaced\ with\ 1} \\ &= f_{x_i=0} \oplus f_{x_i=1}.\end{aligned} \tag{8.2}$$

Therefore, the simplest (but optimal) algorithm to calculate the Boolean difference of a switching function with respect to a variable x_i includes two steps:

(a) Replace x_i in the switching function with 0 to get a cofactor $f_{x_i=0}$; similarly, replacement of x_i with 1 yields $f_{x_i=1}$, and (*b*) Find modulo 2 sum of the two cofactors.

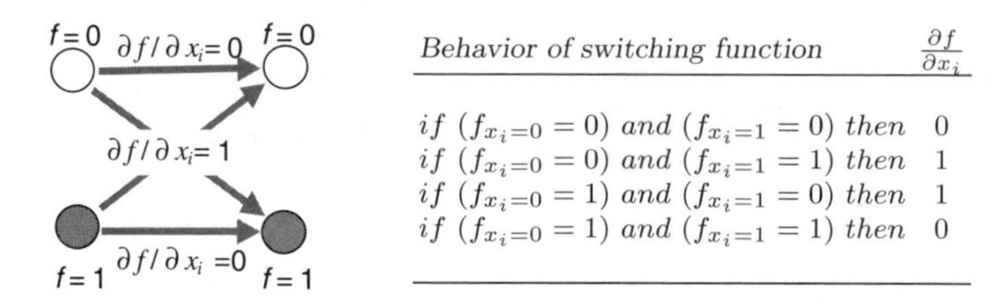

Behavior of switching function	$\frac{\partial f}{\partial x_i}$
if $(f_{x_i=0} = 0)$ *and* $(f_{x_i=1} = 0)$ *then*	0
if $(f_{x_i=0} = 0)$ *and* $(f_{x_i=1} = 1)$ *then*	1
if $(f_{x_i=0} = 1)$ *and* $(f_{x_i=1} = 0)$ *then*	1
if $(f_{x_i=0} = 1)$ *and* $(f_{x_i=1} = 1)$ *then*	0

FIGURE 8.2
The formal description of change by Boolean difference.

Figure 8.2 gives an interpretation of Boolean difference (Equation 8.1).

Example 8.2 *There are four combinations of possible changes of the output function* $f = x_1 \vee x_2$ *with respect to input* x_1 *(*x_2*). The Boolean differences of a switching function* f *with respect to* x_1 *and* x_2 *are calculated by Equation 8.2 in Figure 8.3.*

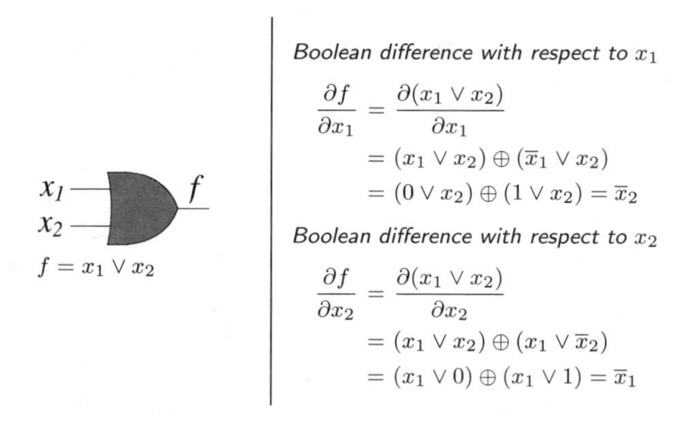

Boolean difference with respect to x_1

$$\frac{\partial f}{\partial x_1} = \frac{\partial(x_1 \vee x_2)}{\partial x_1} = (x_1 \vee x_2) \oplus (\overline{x}_1 \vee x_2) = (0 \vee x_2) \oplus (1 \vee x_2) = \overline{x}_2$$

Boolean difference with respect to x_2

$$\frac{\partial f}{\partial x_2} = \frac{\partial(x_1 \vee x_2)}{\partial x_2} = (x_1 \vee x_2) \oplus (x_1 \vee \overline{x}_2) = (x_1 \vee 0) \oplus (x_1 \vee 1) = \overline{x}_1$$

FIGURE 8.3
Computing Boolean differences for a two-input OR gate (Example 8.2).

The Boolean difference (Equation 8.1) possesses the following properties:

- The Boolean difference is a switching function calculated by the Exclusive OR operation of the primary function and the function derived by complementing variable x_i; otherwise, it can also be calculated as EXOR of co-factors $f_{x_i=0}$ and $f_{x_i=1}$.

- ▶ The Boolean difference is a switching function of $n-1$ variables $x_1, x_2, \ldots, x_{i-1}, x_{i+1}, \ldots, x_n$, i.e., it does not depend on variable x_i.
- ▶ The value of the Boolean difference reflects the fact of local change of the switching function f with respect to changing the i-th variable x_i: the Boolean difference is equal to 0 when such change occurs, and it is equal to 1 otherwise.

The Boolean difference (Equation 8.1) has a number of limitations, in particular: it cannot recognize the direction of change and cannot recognize the change in a function while changing a group of variables. This is the reason to extend the class of differential operators.

Model for simultaneous change: Boolean difference with respect to vector of variables. Consider the model of change with respect to simultaneously changed values of input signals. This model is called Boolean difference with respect to *vector of variables.* For a switching function f Boolean difference of n variables $x_1 \ldots x_n$ with respect to the vector of k variables $x_{i_1}, \ldots, x_{i_k}$, $i_1, \ldots, i_n \in \{1, \ldots, n\}$, is defined as follows

$$\frac{\partial f}{\partial(x_{i_1}, x_{i_2}, \ldots, x_{i_k})} = \overbrace{f(x_1, \ldots, x_{i_1}, x_{i_2}, \ldots, x_{i_k}, \ldots, x_n)}^{Initial\ function} \oplus \underbrace{f(x_1, \ldots, \overline{x}_{i_1}, \overline{x}_{i_2}, \ldots, \overline{x}_{i_k}, \ldots, x_n)}_{Function\ while\ \overline{x}_{i_1}, \overline{x}_{i_2}, \ldots, \overline{x}_{i_k}} \tag{8.3}$$

Given $k = 2$, it follows from Equation 8.3 that

$$\frac{\partial f}{\partial(x_i, x_j)} = f(x_1, \ldots, x_i, x_j, \ldots, x_n) \oplus f(x_1, \ldots, \overline{x}_i, \overline{x}_j, \ldots, x_n) \tag{8.4}$$

$$\tag{8.5}$$

Example 8.3 *Calculate Boolean difference of the switching function $f = x_1x_2 \vee x_3$ with respect to a vector of variables using Equation 8.4:*

$$\begin{aligned}\frac{\partial f}{\partial(x_1, x_2)} &= f(x_1, x_2, x_3) \oplus f(\overline{x}_1\overline{x}_2x_3 = (x_1x_2 \vee x_3) \oplus (\overline{x}_1\overline{x}_2 \vee x_3) \\ &= x_1x_2 \oplus x_3 \oplus x_1x_2x_3 \oplus \overline{x}_1\overline{x}_2 \oplus x_3 \oplus \overline{x}_1\overline{x}_2x_3 = (x_1x_2 \oplus \overline{x}_1\overline{x}_2)\overline{x}_3 \\ &= x_1x_2\overline{x}_3 \vee \overline{x}_1\overline{x}_2\overline{x}_3\end{aligned}$$

Figure 8.4 illustrates the Boolean differences with respect to vectors of variables of the function considered above using $\mathcal{N}$-hypercube.

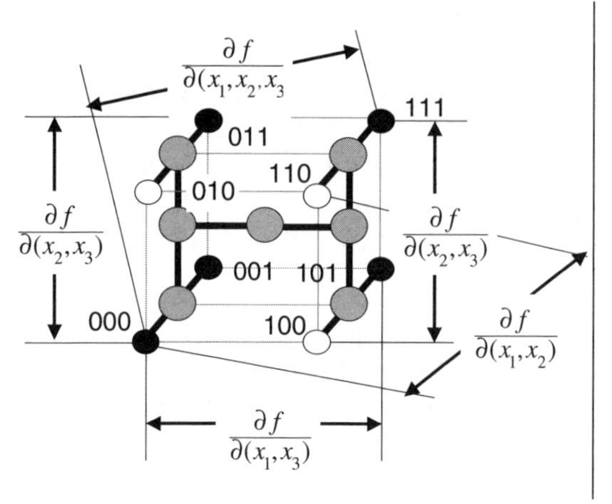

Boolean difference $\frac{\partial f}{\partial(x_1,x_2)}$ *with respect to vector of variables* (x_1, x_2)

$$\frac{\partial f}{\partial x_1} \oplus \frac{\partial f}{\partial x_2} \oplus \frac{\partial^2 f}{\partial x_1 \partial x_2}$$
$$= x_2\overline{x}_3 \oplus x_1\overline{x}_3 \oplus \overline{x}_3$$
$$= \overline{(x_1 \oplus x_2)}\overline{x}_3$$
$$= x_1 x_2 \overline{x}_3 \vee \overline{x}_1 \overline{x}_2 \overline{x}_3$$

FIGURE 8.4
Interpretation of a Boolean difference with respect to a vector of variables by $\mathcal{N}$-hypercube (Example 8.4).

Model of multiple change: k-ordered Boolean differences. Multiple, or k-ordered, Boolean difference is defined as

$$\frac{\partial^k f}{\partial x_{i_1} \partial x_{i_2} \ldots \partial x_{i_k}} = \underbrace{\frac{\partial}{\partial x_{i_1}} \left(\frac{\partial}{\partial x_{i_2}} \left(\cdots \frac{\partial f}{\partial x_{i_k}} \right) \cdots \right)}_{Either\ way}. \tag{8.6}$$

It follows from Equation 8.6 that

- ▶ High-order differences can be obtained from single-order differences;
- ▶ The order of calculation of the Boolean differences does not influence the result.

Let $k = 2$, then the second order Boolean difference with respect to variables x_i and x_j will be

$$\frac{\partial^2 f}{\partial x_i \partial x_j} = \underbrace{\frac{\partial}{\partial x_i} \left(\frac{\partial}{\partial x_j} \right) = \frac{\partial}{\partial x_j} \left(\frac{\partial}{\partial x_i} \right)}_{Either\ way}. \tag{8.7}$$

Relationship of a Boolean difference with respect to vector of variables and multiple Boolean differences. There is a relationship between the second order Boolean difference (Equation 8.7) and Boolean difference

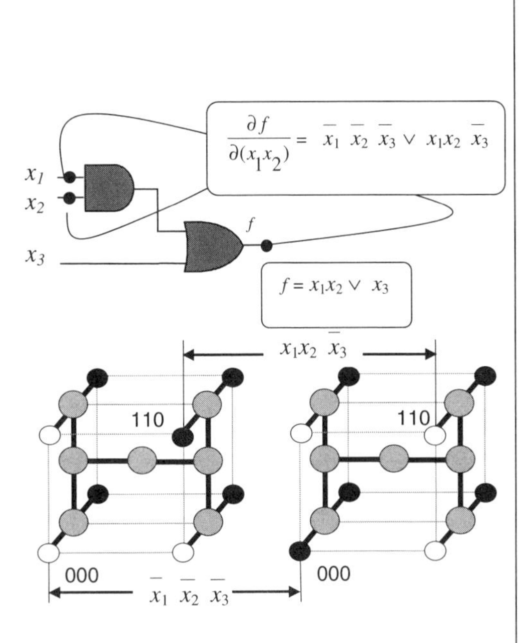

Boolean difference with respect to x_1 and x_2

$$\frac{\partial f}{\partial x_1} = x_2 \oplus x_2 x_3 = x_2 \overline{x}_3$$
$$\frac{\partial f}{\partial x_2} = x_1 \oplus x_1 x_3 = x_1 \overline{x}_3$$

Second order Boolean difference with respect to x_1 and x_2

$$\begin{aligned}\frac{\partial^2 f}{\partial x_1 \partial x_2} &= \frac{\partial}{\partial x_1}\left(\frac{\partial f}{\partial x_2}\right)\\ &= \frac{\partial}{\partial x_1}(x_1 \oplus x_1 x_3)\\ &= (0 \oplus 0x_3) \oplus (1 \oplus 1x_3)\\ &= 1 \oplus x_3 = \overline{x}_3\end{aligned}$$

Boolean difference with respect to vector of variables (x_1, x_2)

$$\begin{aligned}\frac{\partial f}{\partial(x_1, x_2)} &= \frac{\partial f}{\partial x_1} \oplus \frac{\partial f}{\partial x_2} \oplus \frac{\partial^2 f}{\partial x_1 \partial x_2}\\ &= x_2\overline{x}_3 \oplus x_1\overline{x}_3 \oplus \overline{x}_3\\ &= \overline{(x_1 \oplus x_2)}\,\overline{x}_3\\ &= x_1 x_2 \overline{x}_3 \vee \overline{x}_1 \overline{x}_2 \overline{x}_3\end{aligned}$$

FIGURE 8.5
Measuring the input sensitivity of a circuit in the case of simultaneous change of input signals (Example 8.3).

with respect to vector of two variables (Equation 8.4)

$$\begin{cases} \frac{\partial f}{\partial(x_i, x_j)} = \frac{\partial f}{\partial x_i} \oplus \frac{\partial f}{\partial x_j} \oplus \frac{\partial^2 f}{\partial x_i \partial x_j} \\ \\ \frac{\partial^2 f}{\partial x_i \partial x_j} = \frac{\partial f}{\partial x_i} \oplus \frac{\partial f}{\partial x_j} \oplus \frac{\partial f}{\partial(x_i, x_j)} \end{cases} \tag{8.8}$$

This relationship for two variables can be generalized for $k \leq n$ variables, i.e., between multiple or k-ordered Boolean difference (Equation 8.6) and Boolean difference with respect to vector of k variables (Equation 8.3).

Example 8.4 *Calculation of 2-ordered Boolean difference of the switching function $f = x_1 x_2 \vee x_3$ with respect to variables x_1, x_2 and the vector of variables $(x_, x_2)$, is shown in Figure 8.4. To calculate Boolean difference $\frac{\partial f}{\partial(x_1, x_2)}$, Equation 8.8 was used.*

8.1.2 Symmetric properties of Boolean difference

By inspection of Equation 8.1, one can observe the symmetry in the computation:

$$\frac{\partial f_{x_i=0}}{\partial x_i} = \frac{\partial f_{x_i=1}}{\partial x_i}.$$

The signal graph of the computation has a symmetrical structure well-known as "butterfly" (in signal processing) The graph input is the truth vector **F** of the given switching function f, and the result is the truth vector of the Boolean difference.

Example 8.5 *Figure 8.6 illustrates the flowgraphs whose input is the truth column vector* **F** *of an initial function* f*, output is truth column vector of Boolean differences* $\frac{\partial \mathbf{F}}{\partial x_i}$*,* i*=1,2,3, and* $\frac{\partial^3 \mathbf{F}}{\partial x_1 \partial x_2 \partial x_3}$*, and EXOR operation is implemented in the nodes.*

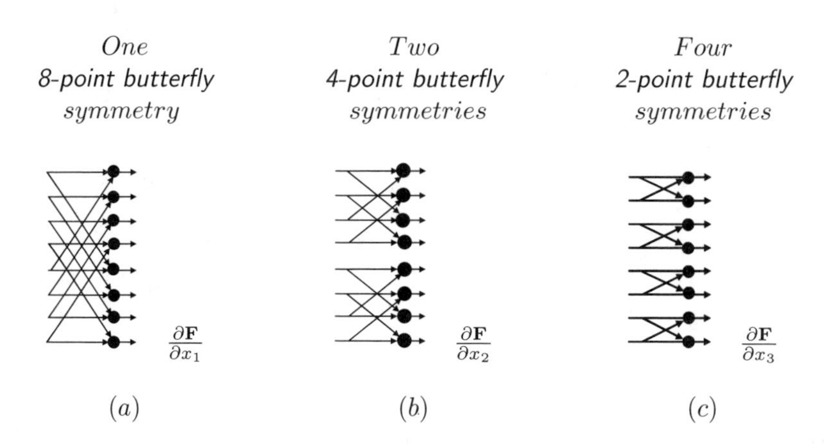

FIGURE 8.6
Illustration of symmetric properties of Boolean difference by flowgraphs for switching function of three variables: with respect to the first variable (a), the second variable (b), and the third variable (c) (Example 8.5).

We observe from the above example that

- The Boolean difference is symmetric with respect to x_i;
- Symmetries are represented by the "butterfly" configuration of the signal graphs.

Note that symmetries of the multiple Boolean differences as shown in Figure 8.7 are composed of the Boolean differences with respect to variables.

8.2 Computing Boolean differences

In this section, the technique for computation of Boolean differences using a decision tree and an $\mathcal{N}$-hypercube is introduced. There are two approaches:

The first approach is based on interpretation of decision tree and $\mathcal{N}$-hypercube which nodes implement Shannon expansion. This attractive technique allows us to get values of Boolean differences without extra manipulation of the data structure (tree or hypercube);

The second approach is oriented to the Davio tree and a corresponding $\mathcal{N}$-hypercube structure.

Both approaches include two phases: (a) computing of Boolean differences, and (b) analysis of behavior of the switching function in terms of change.

8.2.1 Boolean difference and $\mathcal{N}$-hypercube

The problem of computation of a decision tree or $\mathcal{N}$-hypercube is formulated as the analysis of the behavior of data structure in terms of change. The example below introduces this technique.

Example 8.6 *The $\mathcal{N}$-hypercube in Figure 8.7 represents the switching function $x_1x_2 \vee x_3$. To analyze the behavior of this function, let us detect the changes as follows.*

- *Boolean difference with respect to variable x_1 is $\frac{\partial f}{\partial x_1} = x_2\overline{x}_3$. The logic equation $x_2\overline{x}_3 = 1$ yields the solution $x_2x_3 = 10$. This specifies the conditions to detect the changes at x_1: when $x_2x_3 = 10$, a change at x_1 cause a change at f. This can be seen on the decision tree and on the $\mathcal{N}$-hypercube (Figure 8.7a).*
- *Boolean difference with respect to variable x_2 is $\frac{\partial f}{\partial x_1} = x_1\overline{x}_3$. The logic equation $x_2\overline{x}_3 = 1$ specifies the condition of observation as a change at f while changing x_i: $x_2x_3 = 10$(Figure 8.7b).*
- *Boolean difference with respect to variable x_3 is $\frac{\partial f}{\partial x_1} = \overline{x_1x_2}$. The logic equation $\overline{x_1x_2} = 1$ determines the condition: $x_1x_2x_3 = \{00, 01, 10\}$ (Figure 8.7c).*

8.2.2 Boolean difference, Davio tree, and $\mathcal{N}$-hypercube

Here, we show how to use

- The Davio decision tree, and
- The $\mathcal{N}$-hypercube which implements positive Davio expansion in the nodes

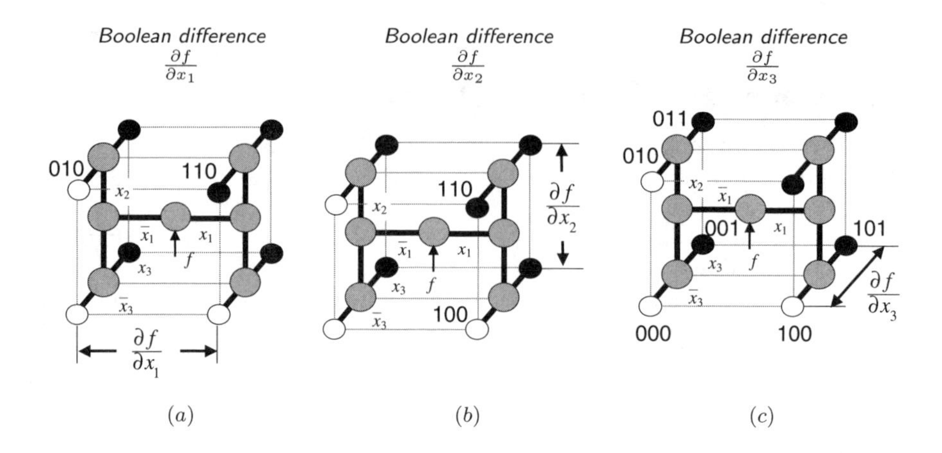

FIGURE 8.7
Interpretation of Boolean differences by $\mathcal{N}$-hypercube: Boolean difference with respect to x_1 (a), x_2 (b), and x_3 (c) (Example 8.6).

to compute Boolean differences. Let us rewrite positive Davio expansion in the form

$$\begin{aligned} f &= f|_{x_i=0} \oplus x_i \; (f|_{x_i=0} \oplus f|_{x_i=1}) \\ &= \underbrace{f|_{x_i=0}}_{Left\ branch} \oplus \underbrace{x_i \frac{\partial f}{\partial x_i}}_{Right\ branch} \end{aligned}$$

It follows from this form that:

- ▶ Branches of the Davio decision tree carry information about Boolean differences;
- ▶ Terminal nodes are the values of Boolean differences for corresponding variable assignments.
- ▶ Computing of Reed-Muller coefficients can be implemented on the Davio decision tree as a data structure;
- ▶ The Davio tree includes values of all single and multiple Boolean differences given a variable assignment $x_1 x_2 \ldots x_n = 00\ldots0$. This assignment corresponds to the calculation of Reed-Muller expansion of polarity 0, so in the Davio tree, positive Davio expansion is implemented at each node. We do not consider other polarities with respect to the $\mathcal{N}$-hypercube representation, though it should be noted that any polarity can be represented by the corresponding Davio tree (with positive and negative expansion at the nodes).
- ▶ Representation of a switching function in terms of change is a unique representation; it means that the corresponding decision diagram is canonical;

► The values of terminal nodes correspond to coefficients of logic Taylor expansion.

The Davio tree can be embedded in an $\mathcal{N}$-hypercube, and the above-mentioned properties are valid for that data structure as well. In addition, the $\mathcal{N}$-hypercube enables computing of the Reed-Muller coefficients/Boolean differences, assuming that the processing is organized using parallel-pipelined, or systolic processing (see Chapter 10 for details).

Example 8.7 *Figure 8.8 shows a Davio decision tree and corresponding $\mathcal{N}$-hypercube for an arbitrary switching function of two and three variables.*

Example 8.8 *Let $f = x_1 \vee x_2$. The values of Boolean differences given assignments $x_1x_2 = \{00, 01, 10, 11\}$ are: $f(00) = 0$, $\frac{\partial f(01)}{\partial x_1} = x_2 = 1$, $\frac{\partial f(10)}{\partial x_2} = x_1 = 1$, and $\frac{\partial^2 f(11)}{\partial x_1 \partial x_2} = 1$. They correspond to the terminal nodes of the Davio tree and $\mathcal{N}$-hypercube (Figure 8.9).*

One can conclude from Example 8.8 that data structure in the form of a Davio decision tree carries information about

► Reed-Muller representation of switching functions, and
► Representation of switching functions in terms of change.

The edges and values in terminal nodes of a Davio decision tree and $\mathcal{N}$-hypercube carry information about the behavior of a switching function. For example, let us compare the decision trees in Figure 8.7 and Figure 8.9. They demonstrate the relationship of Boolean differences and logic Taylor expansion (Section 8.7). Moreover, manipulation of a decision tree can be interpreted in terms of change: reduction of the decision tree to a decision diagram leads to minimization of Reed-Muller expression and can be used as a behavioral model of this function in terms of change.

8.3 Models of logic networks in terms of change

In this section, we consider the simplest behavior models of combinational circuits in terms of change. The problem is formulated as follows: given a multiinput multioutput combinational circuit, analyze its behavior in terms of change.

8.3.1 Event-driven analysis of switching function properties: dependence, sensitivity, and fault detection

Consider a binary system with n inputs and, for simplicity's sake, with one output. Suppose that input signals are changed *independently*. The problem

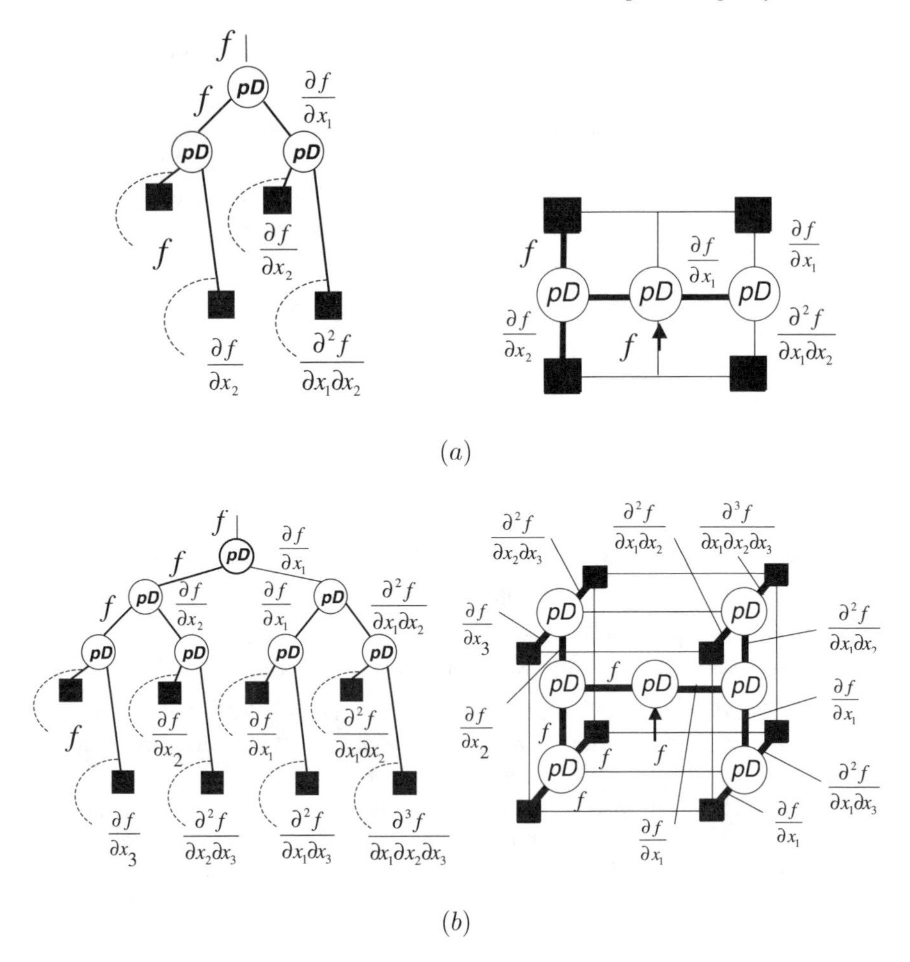

FIGURE 8.8
Computing Boolean differences by Davio decision tree and $\mathcal{N}$-hypercube for a switching function of two (a) and three (b) variables (Example 8.7).

is to analyze the behavior of this system in terms of change. Boolean difference (Equation 8.1) has the ability to detect dependence of a function f on a variable x_i, i.e., the sensitivity of switching function f to change at x_i. Formally, the unconditional independence/dependence of output on the input x_i can be detected as follows:

- Switching function f is *unconditionally independent* of i-th variable x_i if $\frac{\partial f}{\partial x_i} \equiv 0$. This is because

$$if \quad \frac{\partial f}{\partial x_i} = 0, \quad then \quad f(x_1, \ldots, x_i, \ldots, x_n) = f(x_1, \ldots, \overline{x}_i, \ldots, x_n).$$

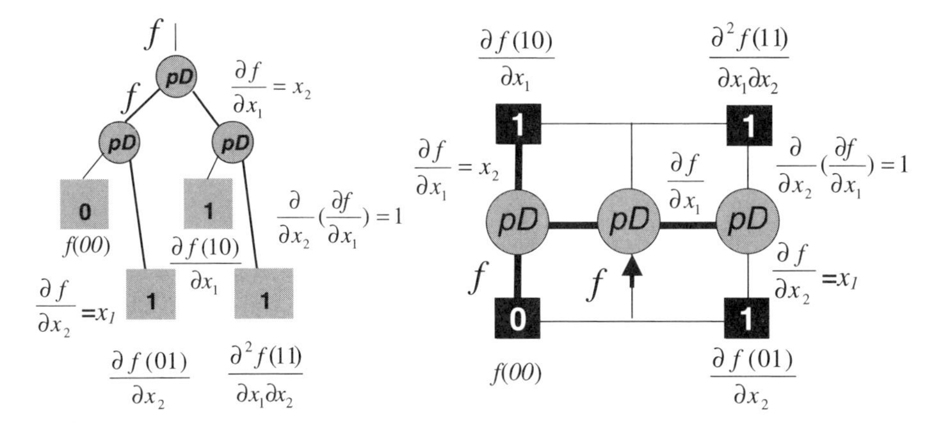

FIGURE 8.9
Computing Boolean differences of the switching function $f = x_1 \vee x_2$ (Example 8.8).

► Switching function f is *unconditionally dependent* on i-th variable x_i if $\frac{\partial f}{\partial x_i} \equiv 1$. This is because

$$if \quad \frac{\partial f}{\partial x_i} = 1, \quad then \quad f(x_1, \ldots, x_i, \ldots, x_n) \neq f(x_1, \ldots, \overline{x}_i, \ldots, x_n).$$

Therefore, given a switching function,

(a) $\frac{\partial f}{\partial x_i} = 0$ specify the conditions (variables assignments) under which f is independent on x_i; and

(b) $\frac{\partial f}{\partial x_i} = 1$ generates the conditions under which f is dependent on x_i.

Example 8.9 *The technique of calculation of Boolean differences and analysis of sensitivity of a switching function to changes of variable values is given in Figure 8.10.*

Behavior of elementary functions. Here we show that the differential operators considered above allow us to effectively extract information about the behavior of a circuit. For simplification, we consider a typical library of gates.

Table 8.1 summarizes results of analysis of change in elementary gates using three types of Boolean differential operators: Boolean difference with respect to a variable, vector of variables and multiple Boolean difference. Consider four cases:

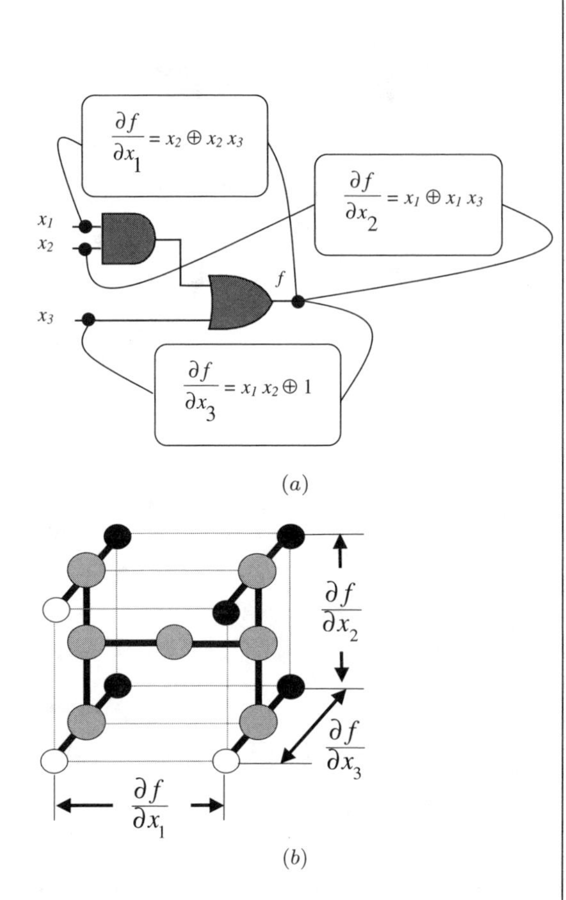

Boolean difference with respect to x_1

$$
\begin{aligned}
\frac{\partial f}{\partial x_1} &= f(0, x_2, x_3) \oplus f(1, x_2, x_3) \\
&= (0x_2 \vee x_3) \oplus (1x_2 \vee x_3) \\
&= x_3 \oplus \overline{\overline{x_2 \vee x_3}} \\
&= x_3 \oplus (x_2 \oplus 1)(x_3 \oplus 1) \oplus 1 \\
&= x_2 \oplus x_2 x_3 = x_2 \overline{x}_3
\end{aligned}
$$

By analogy, Boolean difference with respect to x_2 and x_3

$$
\begin{aligned}
\frac{\partial f}{\partial x_2} &= x_1 \oplus x_1 x_3 = x_1 \overline{x}_3 \\
\frac{\partial f}{\partial x_3} &= x_1 x_2 \oplus 1 = \overline{x_1 x_2}
\end{aligned}
$$

Conditions of dependence

- *on x_1:*

$x_2 \oplus x_2 x_3 = 1,\ x_2 = 1,\ x_3 = 0$

- *on x_2:*

$x_1 \oplus x_1 x_3 = 1,\ x_1 = 1,\ x_3 = 0$

- *on x_3:*

$x_1 x_2 \oplus 1 = 1,\ x_1 = 0,\ x_2 = 0$

Conditions of independence

- *on x_1:*

$x_2 \oplus x_2 x_3 = 0$, $\{x_2 = 0,\ x_3 = 0\}$, $\{x_2 = 0,\ x_3 = 1\}$, $\{x_2 = 1,\ x_3 = 1\}$

- *on x_2:*

$x_1 \oplus x_1 x_3 = 0$, $\{x_1 = 0,\ x_3 = 0\}$, $\{x_1 = 0,\ x_3 = 1\}$, $\{x_1 = 1,\ x_3 = 1\}$,

- *on x_3:*

$x_1 x_2 \oplus 1 = 0,\ x_1 = 1,\ x_2 = 1.$

FIGURE 8.10
Measuring of sensitivity of the output to changes at the inputs of a circuit (a) and $\mathcal{N}$-hypercube (b) in terms of change (Example 8.9).

Case 1: The dependence of switching function f on either variable x_i or x_j that is $\frac{\partial f}{\partial x_i} = 1$ and $\frac{\partial f}{\partial x_j} = 1$.

Case 2: The dependence of switching function f on either variable x_i or x_j but not both. This case is formalized as

$$\frac{\partial f}{\partial x_i} \vee \frac{\partial f}{\partial x_j} = 1. \tag{8.9}$$

TABLE 8.1
Boolean differences of two-input switching function functions.

	$\frac{\partial f}{\partial x_1}$	$\frac{\partial f}{\partial x_2}$	$\frac{\partial f}{\partial (x_1,x_2)}$	$\frac{\partial^2 f}{\partial x_1 \partial x_2}$
AND	x_2	x_1	$x_1 \sim x_2$	1
OR	$\overline{x}_2$	$\overline{x}_1$	$x_1 \sim x_2$	1
EXOR	1	1	0	0
NOR	$\overline{x}_2$	$\overline{x}_1$	$x_1 \sim x_2$	1
NAND	x_2	x_1	$x_1 \sim x_2$	1

Case 3: The dependence of switching function f on either variable x_i or x_j or both x_i and x_j simultaneously. This case is described as

$$\frac{\partial f}{\partial x_i \partial x_j} \vee \frac{\partial f}{\partial x_i} \vee \frac{\partial f}{\partial x_j} = 1. \tag{8.10}$$

Case 4: The dependence of switching function f on both variables x_i and x_j, i.e., these variables changing simultaneously. In this case, the following formula is used:

$$\frac{\partial f}{\partial (x_i, x_j)} = \frac{\partial f}{\partial x_i} \oplus \frac{\partial f}{\partial x_j} \oplus \frac{\partial^2 f}{\partial x_i \partial x_j} = 1 \tag{8.11}$$

Example 8.10 *The OR function analysis based on Table 8.1 is given in Figure 8.11. Note that Equations 8.9, 8.10, and 8.11 generate conditions of independence when the right part is equal to 0.*

Fault detection. Consider the simplest case of application of Boolean differences to fault detection.

Let us analyze what happens if a fault has occurred in a line (connection) that transmits binary signals in a circuit. *Stuck-at-0* or *stuck-at-1* is a fault type that causes a wire to be stuck-at-zero or one respectively. The conditions to observe the fault at input x_i and its transportation to output are described by the Boolean equation $\frac{\partial f}{\partial x_i} = 1$. Solutions to the equations $x_i \frac{\partial f}{\partial x_i} = 1$ and $\overline{x}_i \frac{\partial f}{\partial x_i} = 1$ specify the tests for detecting both *stuck-at-0* and *stuck-at-1* faults.

Example 8.11 *Figure 8.12 shows the switching function in the form of truth vector* **F** *and conditions to detect stuck-at-0 and stuck-at-1 faults. The tests to detect the fault are shown as well.*

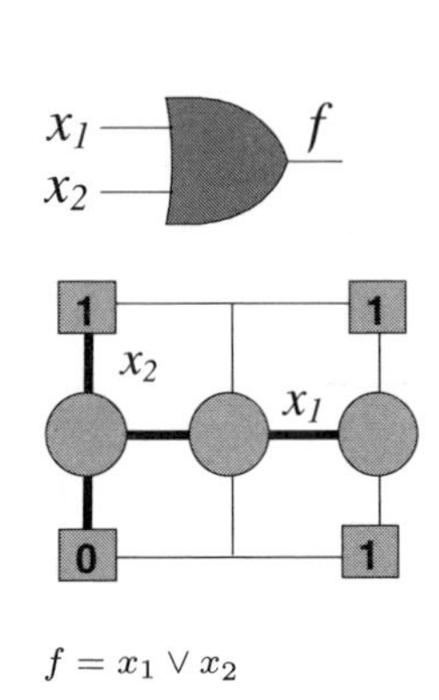

Case 1
The dependence of switching function f on either variables x_i or x_j.

- *Input* x_1*:* $\overline{x}_2 = 1$,
  $\{(x_1 = 0, x_2 = 0), (x_1 = 1, x_2 = 0)\}$, *and*
- *Input* x_2*:* $\overline{x}_1 = 1$,
  $\{(x_1 = 0, x_2 = 0), (x_1 = 0, x_2 = 1)\}$

Case 2
The dependence of switching function f on either x_1 or x_2.
Input x_1 *or* x_2*:* $\overline{x}_1 \vee \overline{x}_2 = 1$,
$\{(x_1 = 0, x_2 = 0), (x_1 = 0, x_2 = 1), (x_1 = 1, x_2 = 0)\}$

Case 3
The dependence of switching function f on either x_1 or x_2 or both x_1 and x_2 simultaneously.
Input x_1 *or* x_2 *or both* x_1 *and* x_2*:* $1 \vee \overline{x}_1 \vee \overline{x}_2 = 1$,
$\{(x_1 = 0, x_2 = 0), (x_1 = 0, x_2 = 1), (x_1 = 1, x_2 = 0),$
$(x_1 = 1, x_2 = 1)\}$

Case 4
The dependence of switching function f on both variables x_1 and x_2
Inputs x_1 *and* x_2*:* $x_1 \sim x_2 = 1$,
$\{x_1 = 0, x_2 = 0\}$

FIGURE 8.11
Event-driven analysis of OR gate (Example 8.10).

8.3.2 Useful rules

Rule 1. A complement of a switching function f does not change the Boolean difference with respect to variable x_i:

$$\frac{\partial \overline{f}}{\partial x_i} = \frac{\partial f}{\partial x_i}.$$

For instance, the tests for AND and NAND gates are identical.

Rule 2. Given a constant function c, $\partial c/\partial x_i = 0$.

Rule 3. (Operations with a constant.) Let c be a constant and f be a switching function. Then

$$\frac{\partial (cf)}{\partial x_i} = c\frac{\partial f}{\partial x_i}, \quad \frac{\partial (c \vee f)}{\partial x_i} = \overline{c}\frac{\partial f}{\partial x_i}, \; and \; \frac{\partial (c \oplus f)}{\partial x_i} = \frac{\partial f}{\partial x_i}$$

These formulas describe situations when the constant value feeds one of the inputs of AND, OR, EXOR gate (Figure 8.13).

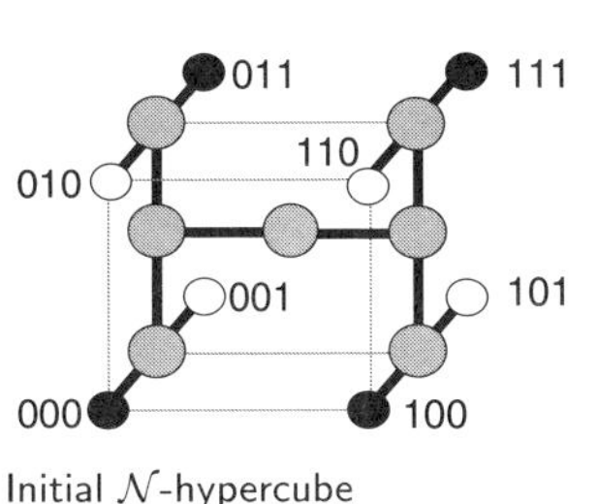

Initial $\mathcal{N}$-hypercube
$f = \overline{x}_1\overline{x}_2x_3 \vee \overline{x}_1x_2\overline{x}_3 \vee x_1\overline{x}_2\overline{x}_3 \vee x_1x_2x_3$

$x_1x_2x_3$	F	$x_3\frac{\partial \mathbf{F}}{\partial x_3}$	$\overline{x}_3\frac{\partial \mathbf{F}}{\partial x_3}$
0 0 0	0	0	1
0 0 1	1	1	0
0 1 0	1	0	1
0 1 1	0	1	0
1 0 0	1	0	1
1 0 1	0	1	0
1 1 0	0	0	1
1 1 1	1	1	0

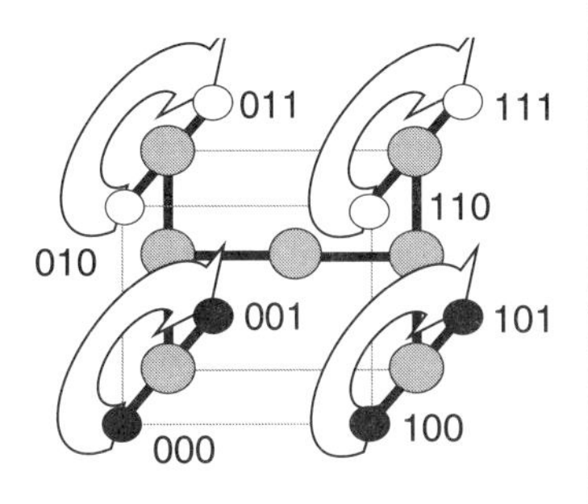

Stuck-at-0
Stuck-at-0 at x_3 *causes each value* $f|_{x_3=1}$ *to be changed to* $f|_{x_3=0}$
Equation to find test to detect stuck-at-0 faults:

$$x_3\frac{\partial \mathbf{F}}{\partial x_3} = 1$$

Solution: tests $\{001, 011, 101, 111\}$

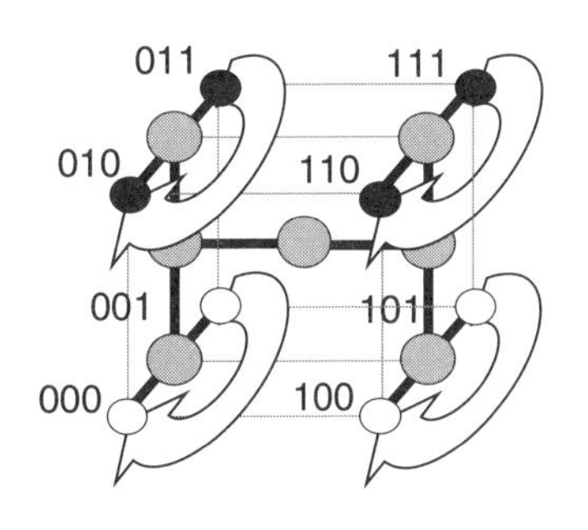

Stuck-at-1
Stuck-at-1 at x_3 *causes each value* $f|_{x_3=0}$ *to be changed to* $f|_{x_3=1}$
Equation to find test to detect stuck-at-1 faults:

$$\overline{x}_3\frac{\partial \mathbf{F}}{\partial x_3} = 1$$

Solution: tests $\{000, 010, 100, 110\}$

FIGURE 8.12
Deriving the tests to detect *stuck-at-0* and *stuck-at-1* faults (Example 8.11).

Rule 4. Let f and g be switching functions that depend on x_i. Then (Figure 8.14):

$$\frac{\partial(f \oplus g)}{\partial x_i} = \frac{\partial f}{\partial x_i} \oplus \frac{\partial g}{\partial x_i},$$

$$\frac{\partial(f \wedge g)}{\partial x_i} = f\frac{\partial g}{\partial x_i} \oplus g\frac{\partial f}{\partial x_i} \oplus \frac{\partial f}{\partial x_i}\frac{\partial g}{\partial x_i},$$

$$\frac{\partial(f \vee g)}{\partial x_i} = \overline{f}\frac{\partial g}{\partial x_i} \oplus \overline{g}\frac{\partial f}{\partial x_i} \oplus \frac{\partial f}{\partial x_i}\frac{\partial g}{\partial x_i}$$

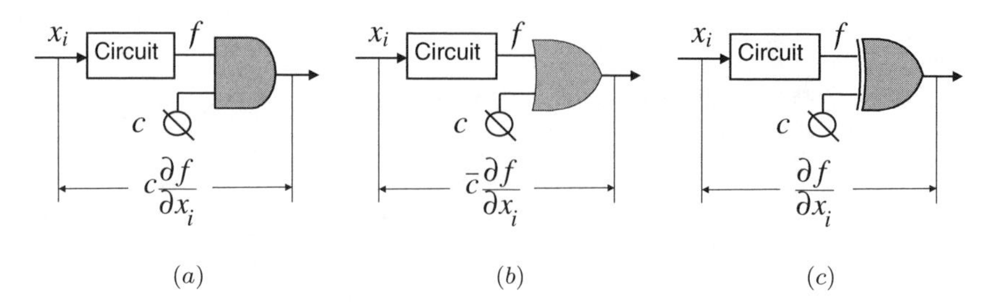

FIGURE 8.13
Boolean difference with respect to variable x_i of AND (a), OR (b), and EXOR operation (c) while the second input is a constant.

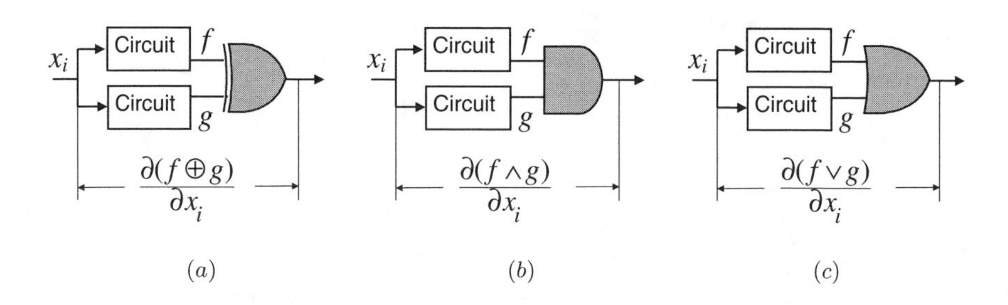

FIGURE 8.14
Boolean difference with respect to variable x_i on logical operations of switching f of n variables x_i, $i = 1, 2, \ldots, n$: EXOR operation (a), AND operation (b), and OR operation (c).

Rule 5. Multilevel circuit analysis includes computing of Boolean difference with respect to a function z. There are three cases:

- Let $z = x_1 \wedge x_2$, then

$$\frac{\partial f}{\partial(x_i \wedge x_j)} = \frac{\partial f}{\partial x_i} \oplus \frac{\partial f}{\partial x_j} \oplus \frac{\partial^2 f}{\partial x_i \partial x_j}. \tag{8.12}$$

- Let $z = x_1 \vee x_2$, then

$$\frac{\partial f}{\partial(x_i \vee x_j)} = \frac{\partial f}{\partial x_i} \vee \frac{\partial f}{\partial x_j} \vee \frac{\partial f}{\partial(x_i, x_j)}. \tag{8.13}$$

- Let $z = x_1 \oplus x_2$ then

$$\frac{\partial f}{\partial(x_i \oplus x_j)} = \frac{\partial f}{\partial x_i} \vee \frac{\partial f}{\partial x_j}. \tag{8.14}$$

In Figure 8.15, these three situations are interpreted. Note that the graphical interpretation supposes that Boolean differences are calculated for inputs x_i and $\overline{x}_i$.

Example 8.12 *Equations 8.12, 8.13, and 8.14 are illustrated in Figure 8.15.*

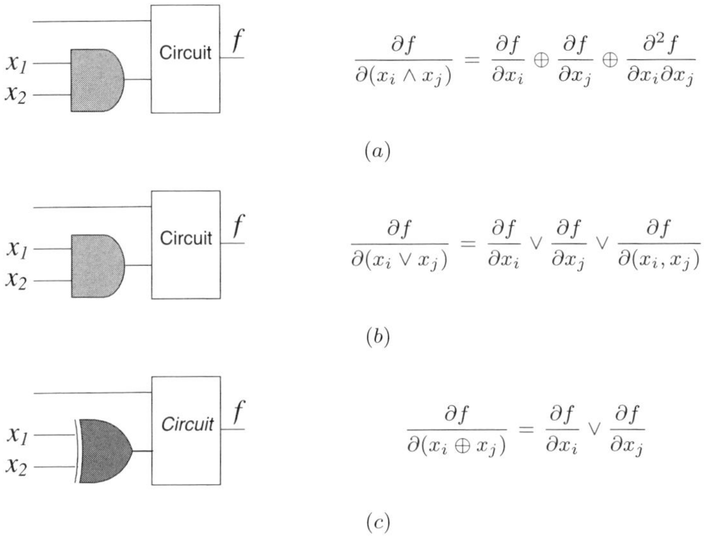
Circuit
f
x_1
x_2
$\frac{\partial f}{\partial (x_i \wedge x_j)} = \frac{\partial f}{\partial x_i} \oplus \frac{\partial f}{\partial x_j} \oplus \frac{\partial^2 f}{\partial x_i \partial x_j}$
(a)
Circuit
f
x_1
x_2
$\frac{\partial f}{\partial (x_i \vee x_j)} = \frac{\partial f}{\partial x_i} \vee \frac{\partial f}{\partial x_j} \vee \frac{\partial f}{\partial (x_i, x_j)}$
(b)
Circuit
f
x_1
x_2
$\frac{\partial f}{\partial (x_i \oplus x_j)} = \frac{\partial f}{\partial x_i} \vee \frac{\partial f}{\partial x_j}$
(c)

FIGURE 8.15
Boolean differences for AND (a), OR (b), and EXOR (c) gates with simultaneously changed inputs x_1 and x_2 (Example 8.12).

8.3.3 Probabilistic model

The probabilistic model of implementations of switching functions has been considered as twofold:

- Probabilistic models utilize pseudo-Boolean expression, e.g., arithmetic polynomials, to evaluate the probabilities of logic signals in the network of gates; and
- Probabilistic, or random, Boolean networks, also called *Kauffman nets*; the latter share some properties of cellular automata (see Chapter 10), the simplest 1-D type of which are called *stochastic cellular automata* (see "Further Reading" Section).

The influence of the variable x_i on the switching function f is the expectation of the Boolean difference with respect to x_i:

$$I_i(f) = E\left[\frac{\partial f(x)}{\partial x_i}\right].$$

The expectation is equal to the probability of Boolean difference being at value 1: $P\left\{\frac{\partial f(x)}{\partial x_i} = 1\right\}$. In this context, $\partial f(x)/\partial x_i$ is also called *distribution* $D(x_i)$.

Example 8.13 *If the vector of Boolean difference contains four ones in the truth table given* $n = 3$*, then* $P\left\{\frac{\partial f(x)}{\partial x_i} = 1\right\} = 0.5$.

Calculation of distribution is relevant to evaluation of switching activity or *transition* density of signals in the circuit, i.e., calculated as transition density of each cube in sum-of-products form of the switching function f. The signal probability is defined as the probability of a signal being at value 1. The signal probabilities can be found by BDDs for each x_i with respect to the primary inputs of the circuits.

Given a circuit with n inputs x_i, and transition densities $D(x_i)$. Transition densities of all cubes are calculated by equation

$$D(c_j) = \sum_{i=1}^{n} E\left(\frac{\partial f(x)}{\partial x_i}\right) D(x_i),$$

where $D(x_i) = p(x_i)$ is the probability of x_i being at value 1.

Example 8.14 *Figure 8.16 illustrates the computing of transition densities* $D(x_i)$ *for a switching function of four variables (the independence of events is assumed). For the simplification, the transition densities of the inputs are are equal to signal probabilities of corresponded inputs.*

Probabilistic properties of Boolean differences can be easy interpreted by decision trees and $\mathcal{N}$-hypercubes.

8.4 Matrix models of change

Symbolic manipulations using the rules above are costly in terms of time complexity. Hence, efficient algorithms are needed to compute models based on differential operators. In this section, matrix methods of computing are introduced. These methods are useful in different design styles, in particular, in massive parallel computing:

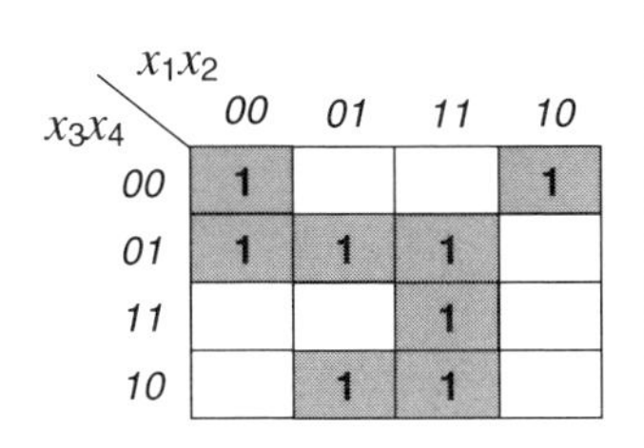

Given:
Signal probabilities:
$p(x_1) = 0.9$
$p(x_2) = 0.3$
$p(x_3) = 0.5$
$p(x_4) = 0.8$

Transition densities:
$D(x_1) = 0.9$
$D(x_2) = 0.3$
$D(x_3) = 0.5$
$D(x_4) = 0.8$

Transition densities of the cube $c = \overline{x}_2 x_3 x_4$:

$$\begin{aligned} D_{c_j} &= p(x_3)p(x_4)D(x_2) \\ &+ p(\overline{x}_2)p(x_4)D(x_3) \\ &+ p(\overline{x}_2)p(x_3)D(x_1) \\ &= 0.5 \cdot 0.8 \cdot 0.3 \\ &+ 0.7 \cdot 0.8 \cdot 0.5 \\ &+ 0.7 \cdot 0.5 \cdot 0.9 \\ &= 0.715 \end{aligned}$$

FIGURE 8.16
Computing the transition densities for switching function (Example 8.14).

- ▶ Decision diagram technique,
- ▶ Cellular arrays, and
- ▶ Systolic arrays.

8.4.1 Boolean difference with respect to a variable in matrix form

Boolean difference of a switching function f with respect to a variable x_i is defined by Equation 8.1. Let 2×2 matrix $\tilde{D}_2$ be

$$\tilde{D}_2 = \begin{bmatrix} 0 & 1 \\ 1 & 0 \end{bmatrix}$$

Let us form the 2×2 matrix D_2 by the rule

$$D_2 = I_2 \oplus \tilde{D}_2 = \begin{bmatrix} 1 & 0 \\ 0 & 1 \end{bmatrix} \oplus \begin{bmatrix} 0 & 1 \\ 1 & 0 \end{bmatrix} = \begin{bmatrix} 1 \oplus 0 & 0 \oplus 1 \\ 0 \oplus 1 & 1 \oplus 1 \end{bmatrix} = \begin{bmatrix} 1 & 1 \\ 1 & 1 \end{bmatrix}$$

where I_2 is the identity matrix.

The matrix form of a Boolean difference (Equation 8.1) with respect to the i-th variable x_i of a switching function f of n variable given by truth vector $\mathbf{F}$ is defined as

$$\frac{\partial \mathbf{F}}{\partial x_i} = D^i_{2^n}\mathbf{F}, \qquad (8.15)$$

where $2^n \times 2^n$ matrix $D_{2^n}^i$ is called a *Boolean differential matrix* generated by the rule

$$D_{2^n}^i = I_{2^{i-1}} \otimes \begin{bmatrix} 1 & 1 \\ 1 & 1 \end{bmatrix} \otimes I_{2^{n-i}}, \tag{8.16}$$

$I_{2^{i-1}}, \quad I_{2^{n-i}}$ are the identity matrices.

Example 8.15 *$D_{2^2}^1$ and $D_{2^3}^1$ are constructed by Equation 8.16 as follows:*

$$\begin{aligned}
D_{2^2}^{(1)} &= I_{2^{1-1}} \otimes D_2 \otimes I_{2^{2-1}} \\
&= 1 \otimes \begin{bmatrix} 1 & 1 \\ 1 & 1 \end{bmatrix} \otimes \begin{bmatrix} 1 & 0 \\ 0 & 1 \end{bmatrix} = \begin{bmatrix} I_2 & I_2 \\ I_2 & I_2 \end{bmatrix} = \begin{bmatrix} 1 & & 1 & \\ & 1 & & 1 \\ 1 & & 1 & \\ & 1 & & 1 \end{bmatrix}; \\
D_{2^3}^{(1)} &= I_{2^{1-1}} \otimes D_2 \otimes I_{2^{3-1}} \\
&= 1 \otimes \begin{bmatrix} 1 & 1 \\ 1 & 1 \end{bmatrix} \otimes \begin{bmatrix} 1 & & & \\ & 1 & & \\ & & 1 & \\ & & & 1 \end{bmatrix} = \begin{bmatrix} I_{2^2} & I_{2^2} \\ I_{2^2} & I_{2^2} \end{bmatrix} = \begin{bmatrix} 1 & & & & 1 & & & \\ & 1 & & & & 1 & & \\ & & 1 & & & & 1 & \\ & & & 1 & & & & 1 \\ 1 & & & & 1 & & & \\ & 1 & & & & 1 & & \\ & & 1 & & & & 1 & \\ & & & 1 & & & & 1 \end{bmatrix}
\end{aligned}$$

Example 8.16 *Figure 8.17 illustrates computing the Boolean differences with respect to variable x_1, x_2 and x_3 by multiplication of the truth vector $\mathbf{F} = [f(0)\ \ f(1) \ldots\ f(7)]^T$ of the switching function function and the corresponding matrix $D_{2^n}^i$ for $n = 3$ and $i = 1, 2, 3$. Let $f = x_1 x_2 \vee x_3$ and $\mathbf{F} = [01010111]^T$. Then*

$$\begin{aligned}
\frac{\partial \mathbf{X}}{\partial x_1} &= [00100010]^T, \quad \frac{\partial f}{\partial x_1} = \overline{x}_3(\overline{x}_1 x_2 \vee x_1 x_2), \\
\frac{\partial \mathbf{X}}{\partial x_2} &= [00001010]^T, \quad \frac{\partial f}{\partial x_2} = \overline{x}_3(x_1 \overline{x}_2 \vee x_1 x_2).
\end{aligned}$$

8.4.2 Boolean difference with respect to a vector of variables in matrix form

Matrix technique can be used for computing of Boolean differences with respect to a vector of k variables, and thus, multiple Boolean differences.

Matrix form of Boolean difference with respect to a vector of k variables $x_{i_1}, ..., x_{i_k}$, $i_1, ..., i_n \in \{1, ..., n\}$ (Equation 8.3) is defined as

$$\frac{\partial f}{\partial x_1} = (I_{2^{1-1}} \otimes \begin{bmatrix} 1 & 1 \\ 1 & 1 \end{bmatrix} \otimes I_{2^{3-1}})\mathbf{F} = \begin{bmatrix} 1 & & & & 1 & & & \\ & 1 & & & & 1 & & \\ & & 1 & & & & 1 & \\ & & & 1 & & & & 1 \\ 1 & & & & 1 & & & \\ & 1 & & & & 1 & & \\ & & 1 & & & & 1 & \\ & & & 1 & & & & 1 \end{bmatrix} \begin{bmatrix} f \\ f \\ f \\ f \\ f \\ f \\ f \\ f \end{bmatrix}$$

$$\frac{\partial f}{\partial x_2} = (I_{2^{2-1}} \otimes \begin{bmatrix} 1 & 1 \\ 1 & 1 \end{bmatrix} \otimes I_{2^{3-2}})\mathbf{F} = \begin{bmatrix} 1 & & 1 & & & & & \\ & 1 & & 1 & & & & \\ 1 & & 1 & & & & & \\ & 1 & & 1 & & & & \\ & & & & 1 & & 1 & \\ & & & & & 1 & & 1 \\ & & & & 1 & & 1 & \\ & & & & & 1 & & 1 \end{bmatrix} \begin{bmatrix} f(0) \\ f(1) \\ f(2) \\ f(3) \\ f(4) \\ f(5) \\ f(6) \\ f(7) \end{bmatrix}.$$

$$\frac{\partial f}{\partial x_3} = (I_{2^{3-1}} \otimes \begin{bmatrix} 1 & 1 \\ 1 & 1 \end{bmatrix} \otimes I_{2^{3-3}})\mathbf{F} = \begin{bmatrix} 1 & 1 & & & & & & \\ 1 & 1 & & & & & & \\ & & 1 & 1 & & & & \\ & & 1 & 1 & & & & \\ & & & & 1 & 1 & & \\ & & & & 1 & 1 & & \\ & & & & & & 1 & 1 \\ & & & & & & 1 & 1 \end{bmatrix}$$

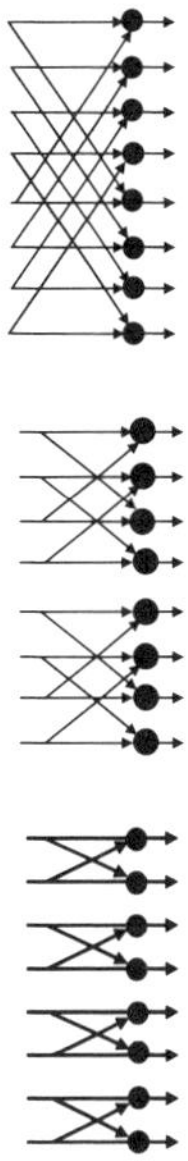

FIGURE 8.17
Matrix based computing of the Boolean differences and corresponding flow-graphs of the algorithm; a node implements EXOR operation (Example 8.16).

$$\frac{\partial \mathbf{F}}{\partial(x_{i_1}, x_{i_2}, ..., x_{i_k})} = \underbrace{\mathbf{F}(x_{i_1}, x_{i_2}, ..., x_{i_k})}_{Initial\ variables} \oplus \underbrace{\mathbf{F}(\overline{x}_{i_1}, \overline{x}_{i_2}, ..., \overline{x}_{i_k})}_{Complemented\ variables}. \quad (8.17)$$

It follows from Equation 8.17 that

$$\begin{aligned} \frac{\partial \mathbf{F}}{\partial(x_i, x_j)} &= \mathbf{F}(x_1, ..., x_i, ..., x_j, ..., x_n) \oplus \mathbf{F}(x_1, ..., \overline{x}_i, ..., \overline{x}_j, ..., x_n) \\ &= \underbrace{\mathbf{F}(x_1, ..., 0, ..., 0, ..., x_n)}_{x_i\ is replaced\ with\ 0} \oplus \underbrace{\mathbf{F}(x_1, ..., 1, ..., 1, ..., x_n)}_{x_j\ is replaced\ with\ 1} \end{aligned} \quad (8.18)$$

Calculation of the Boolean difference with respect to two variables can be simplified using the equation:

$$\frac{\partial \mathbf{F}}{\partial(x_i, x_j)} = \frac{\partial \mathbf{F}}{\partial x_i} \oplus \frac{\partial \mathbf{F}}{\partial x_j} \oplus \frac{\partial^2 \mathbf{F}}{\partial x_i \partial x_j} \quad (8.19)$$

Matrix form of a multiple Boolean difference Given the truth vector $\mathbf{F}$ of a switching function f of n variables $x_1...x_n$, its Boolean difference with respect to k variables $x_{i_1},...,x_{i_k}$ is defined by

$$\frac{\partial^k \mathbf{F}}{\partial x_{i_1} \partial x_{i_2} ... \partial x_{i_k}} = \prod_{p \in \{i_1 ... i_k\}} D_{2^n}^{(p)} \mathbf{F} \quad over\ GF(2) \tag{8.20}$$

Example 8.17 *The flowgraph of the calculation of the second-order Boolean difference is given in Figure 8.18. Here, Equation 8.20 is used.*

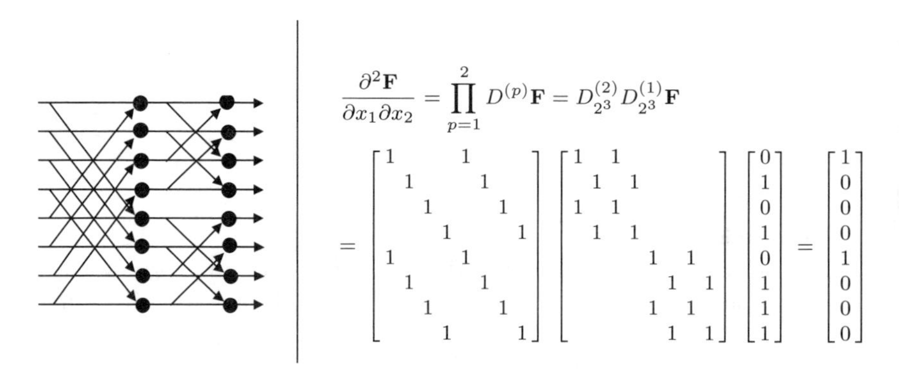

FIGURE 8.18
Flowgraphs for computing multiple Boolean difference $\frac{\partial^2 \mathbf{F}}{\partial x_1 \partial x_2}$ (Example 8.17).

8.5 Models of directed changes in algebraic form

While the Boolean difference indicates output (function) change with respect to input (variables) changes, a directed Boolean difference investigates the direction of the changes. There are *direct* and *inverse* Boolean differences with respect to a variable for switching functions, in accordance with the direction of changes of the values of the function and variables.

8.5.1 Model for direct change

The direct Boolean difference of a switching function f with respect to a variable x_i specifies the conditions of change of f and x_i along the same directions:

$$\frac{\partial f}{\partial_+ x_i} = \frac{\partial f(0 \to 1)}{\partial x_i(0 \to 1)} = \frac{f(1 \to 0)}{\partial x_i(1 \to 0)}. \tag{8.21}$$

Here $\partial_+ x_i$ denotes change $0 \longrightarrow 1$ of x_i. Table 8.2 contains all possible changes of f while changing x to $\overline{x}$.

TABLE 8.2
Truth table of the direct Boolean difference with respect to a variable.

Changing	**Changing switching function f**			
$x \to \overline{x}$	$f(0 \to 0)$	$f(0 \to 1)$	$f(1 \to 0)$	$f(1 \to 1)$
$x(0 \to 1)$	0	1	0	0
$x(1 \to 0)$	0	0	1	0

$$\begin{aligned} \frac{\partial f}{\partial_+ x_i} &= \frac{\partial f(0 \to 1)}{\partial x_i(0 \to 1)} = \frac{\partial f(1 \to 0)}{\partial x_i(1 \to 0)} \\ &= (\overline{x}_i \oplus f)\frac{\partial f}{\partial x_i} = \overline{f}|_{x_i=0} f|_{x_i=1} \end{aligned} \tag{8.22}$$

Example 8.18 *Equation 8.22 is illustrated in Figure 8.19 for AND gate. The directed Boolean difference is equal to 1 if and only if the following changes take place:*

(a) $f : 0 \longrightarrow 1$ *while* $x_i : 0 \longrightarrow 1$, *and*
(b) $f : 1 \longrightarrow 0$ *while* $x_i : 1 \longrightarrow 0$.

Thus, if the changes in the function and variable occur in unison with each other, this fact is shown by the directed Boolean difference.

8.5.2 Model for inverse change

The inverse Boolean difference of a switching function f with respect to a variable x_i defines conditions of change of f and x_i in the opposite direction

$$\frac{\partial f}{\partial x_i} = \frac{\partial f(0 \to 1)}{\partial x_i(1 \to 0)} = \frac{f(1 \to 0)}{\partial x_i(0 \to 1)}. \tag{8.23}$$

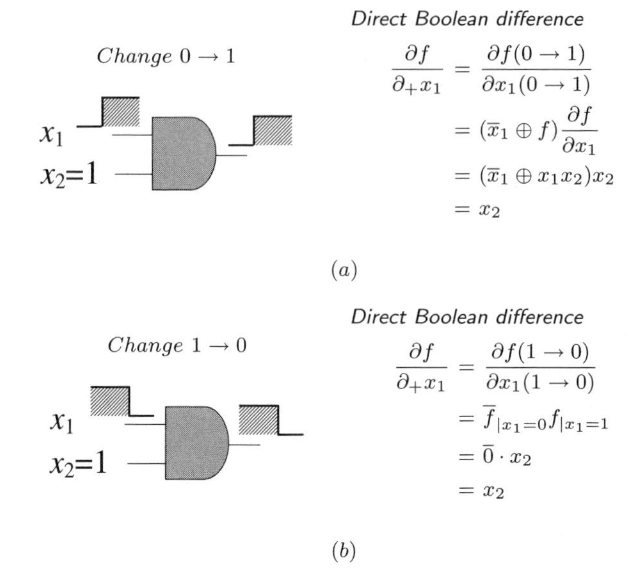

FIGURE 8.19
AND gate: conditions of change of output f and input x_1 of gate AND in the same direction (Example 8.18).

TABLE 8.3
Truth table of the inverse Boolean difference with respect to a variable.

Changing	**Changing switching function f**			
$x \to \overline{x}$	$f(0 \to 0)$	$f(0 \to 1)$	$f(1 \to 0)$	$f(1 \to 1)$
$x(0 \to 1)$	0	0	1	0
$x(1 \to 0)$	0	1	0	0

To calculate the inverse Boolean difference the following formulas are used:

$$\frac{\partial f}{\partial_- x_i} = \frac{\partial f(0 \to 1)}{\partial x_i(1 \to 0)} = (x_i \oplus f)\frac{\partial f}{\partial x_i} = \overline{f}|_{x_i=1} f|_{x_i=0}$$

Example 8.19 *Figure 8.20 illustrates calculation of inverse Boolean difference for AND gate.*

Table 8.4 gives more details via the primitive gates. There are 4 types of

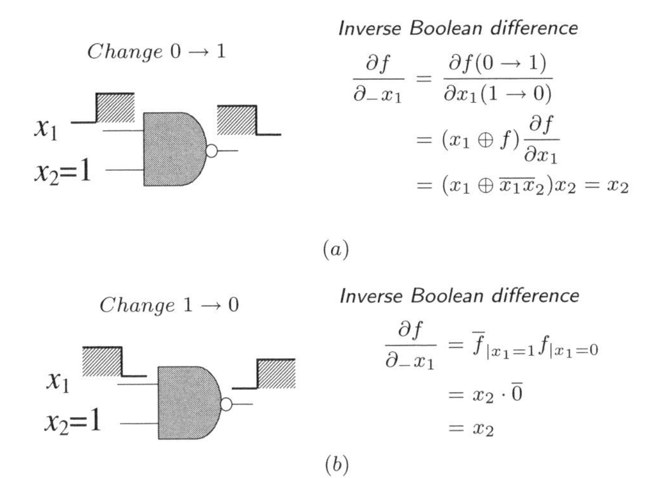

FIGURE 8.20
NAND gate: conditions of change of output f and input x_1 of gate NAND in the opposite direction (Example 8.19).

direct and four types of inverse Boolean differences for switching functions with respect to variable x_i.

TABLE 8.4
Direct and inverse Boolean differences for the gates.

Gate	$\frac{\partial f}{\partial_+ x_1}$	$\frac{\partial f}{\partial_+ x_2}$	$\frac{\partial f}{\partial_- x_1}$	$\frac{\partial f}{\partial_- x_2}$
AND	x_2	x_1	0	0
OR	$\overline{x}_2$	$\overline{x}_1$	0	0
EXOR	$\overline{x}_2$	$\overline{x}_1$	0	0
NAND	0	0	x_2	x_1
NOR	0	0	$\overline{x}_2$	$\overline{x}_1$

Notice that the relationship between generic and directed/inverse Boolean differences is expressed by the relation:

$$\frac{\partial f}{\partial x_i} = \frac{\partial f}{\partial_+ x_i} \vee \frac{\partial f}{\partial_- x_i}.$$

The direct and inverse Boolean differences do not depend on the variable of derivation, but they generally involve more literals or products in their expression.

Example 8.20 *Figure 8.21 shows the circuits used to implement the switching function $f = x_1x_2 \oplus x_3$, and the calculation of the direct and inverse Boolean differences. This also illustrates the relationship between Boolean differences of different types.*

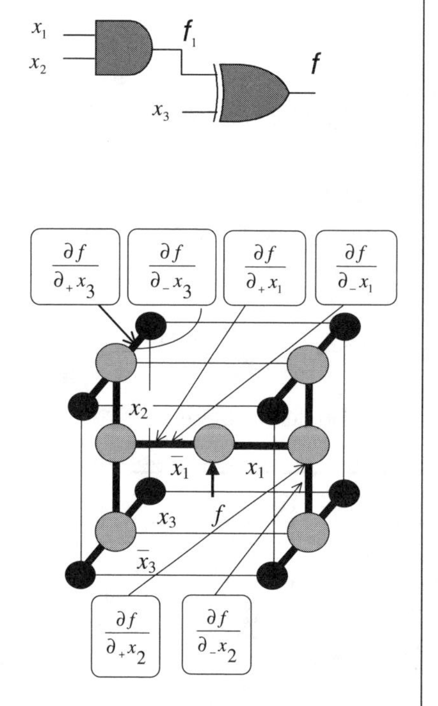

Direct Boolean difference

$$\frac{\partial f}{\partial_+ x_1} = (\overline{x}_1 \oplus f)\,\frac{\partial f}{\partial x_1} = (\overline{x}_1 \oplus (x_1x_2 \oplus x_3)) \cdot x_2 = x_2\overline{x}_3$$

Inverse Boolean difference

$$\frac{\partial f}{\partial_- x_1} = (x_1 \oplus f)\,\frac{\partial f}{\partial x_1} = (x_1 \oplus (x_1x_2 \oplus x_3)) \cdot x_2 = x_2x_3$$

Direct Boolean difference

$$\frac{\partial f}{\partial_+ x_2} = (\overline{x}_2 \oplus f)\,\frac{\partial f}{\partial x_2} = (\overline{x}_2 \oplus (x_1x_2 \oplus x_3)) \cdot x_1 = x_1\overline{x}_3$$

Inverse Boolean difference

$$\frac{\partial f}{\partial_- x_2} = (x_2 \oplus f)\,\frac{\partial f}{\partial x_2} = (x_2 \oplus (x_1x_2 \oplus x_3)) \cdot x_1 = x_1x_3$$

Direct Boolean difference

$$\frac{\partial f}{\partial_+ x_3} = (\overline{x}_3 \oplus f)\,\frac{\partial f}{\partial x_3} = (\overline{x}_3 \oplus (x_1x_2 \oplus x_3)) \cdot 1 = \overline{x_1x_2}$$

Inverse Boolean difference

$$\frac{\partial f}{\partial_- x_3} = (x_3 \oplus f)\,\frac{\partial f}{\partial x_3} = (x_3 \oplus x_1x_2 \oplus x_3)) \cdot 1 = x_1x_2$$

$$\frac{\partial f}{\partial x_1} = \frac{\partial f}{\partial_+ x_1} \vee \frac{\partial f}{\partial_- x_1} = x_2$$

$$\frac{\partial f}{\partial x_2} = \frac{\partial f}{\partial_+ x_2} \vee \frac{\partial f}{\partial_- x_2} = x_1$$

$$\frac{\partial f}{\partial x_3} = \frac{\partial f}{\partial_+ x_3} \vee \frac{\partial f}{\partial x_3} = 1$$

FIGURE 8.21
The circuit, computing direct and inverse Boolean differences, and interpretation by $\mathcal{N}$-hypercube (Example 8.20).

8.6 Local computation via partial Boolean difference

The basic property of a partial Boolean difference is formulated as follows. Given a switching functions $f = f(f_1)$ and $f_1 = f_1(x_i)$, the partial Boolean difference of function f with respect to variable x_i via the internal function f_k, is equal to

$$\frac{\partial f}{\partial(x_i/f_1)} = \frac{\partial f}{\partial f_1} \cdot \frac{\partial f_1}{\partial x_i} \tag{8.24}$$

The above relation is important for efficient calculation of Boolean differences on the functional paths from the primary input nodes to the output nodes in the logic network. For instance, the path is "$f_{i_0} - f_{i_1} - \ldots - f_{i_r}$", thus, the partial Boolean difference is locally calculated provided that we scan the path through the whole network. The extensions to Equation 8.24 enable local calculations for multilevel networks as demonstrated in Figures 8.22 and 8.23.

Example 8.21 *Let us consider the third relation from Figure 8.22, and let the three cascades of the circuit have inputs x_1, x_2 and x_3. The partial Boolean differences imply:*

$$\frac{\partial f}{\partial f_2} = \overline{x}_3, \quad \frac{\partial f_2}{\partial f_1} = x_2, \quad \frac{\partial f_1}{\partial x_i} = \overline{x}_1,$$

and thus

$$\frac{\partial f}{\partial x_i} = \overline{x}_1 x_2 \overline{x}_3.$$

On the other hand, Boolean difference calculated "globally," yields:

$$\frac{\partial f}{\partial x_i} = f|_{x_i=0} \oplus f|_{x_i=1} = \overline{\overline{x}_1 x_2 \vee x_3} \oplus \overline{x}_3 = \frac{\partial f}{\partial x_i} = \overline{x}_1 x_2 \overline{x}_3.$$

8.7 Generating Reed-Muller expressions by logic Taylor series

The Boolean difference is relevant to Reed-Muller expansion of a switching function. The *l*ogic Taylor series for a switching function f of n variables at the point $c \in 0, 1, \ldots, 2^n - 1$ is defined as

$$f = \bigoplus_{i=0}^{2^n-1} f_i^{(c)} \underbrace{(x_1 \oplus c_1)^{i_1} \ldots (x_n \oplus c_n)^{i_n}}_{i-th\ product},$$

1. *Dependence of f on x_i via function f_1*

$$\frac{\partial f}{\partial(x_i/f_1)} = \frac{\partial f}{\partial f_1}\frac{\partial f_1}{\partial x_1}$$

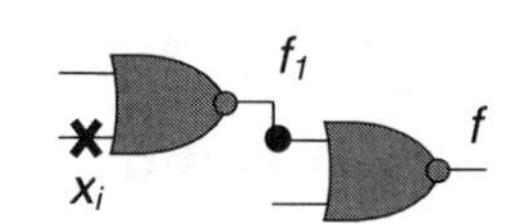

2. *Reduction of calculations in the path $\{x_i \rightarrow f_1 \rightarrow f_1\}$*

$$\frac{\partial f}{\partial(x_i/f_1/f_1)} = \frac{\partial f}{\partial(x_i/f_1)}$$

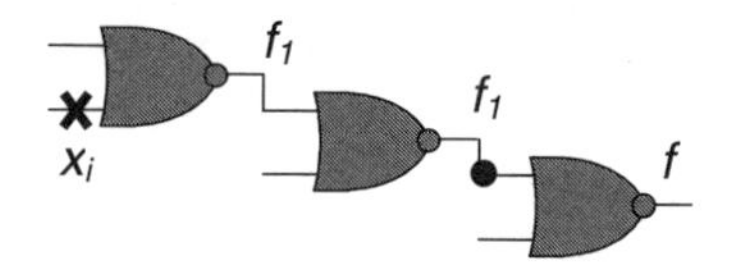

3. *Dependence of f on x_i via functions f_1, f_2*

$$\frac{\partial f}{\partial(x_i/f_1/f_2)} = \frac{\partial f}{\partial f_2}\frac{\partial f_2}{\partial f_1}\frac{\partial f_1}{\partial x_i}$$

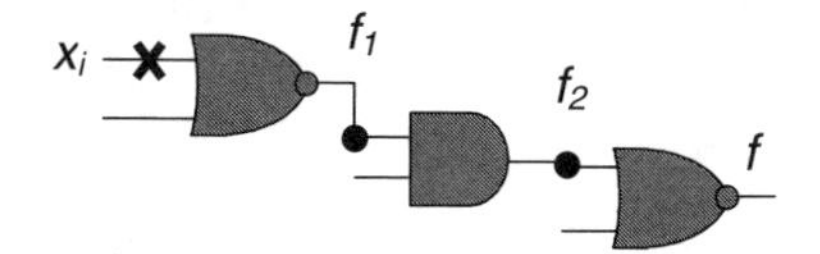

4. *Dependence of f on both x_i and x_j simultaneously via function f_1*

$$\frac{\partial f}{\partial(x_i,x_j/f_1)} = \frac{\partial f}{\partial f_1}\frac{\partial f_1}{\partial(x_i,x_j)}$$

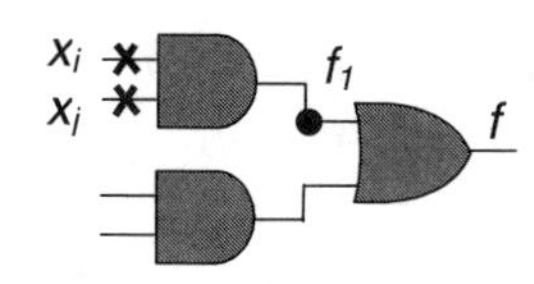

FIGURE 8.22
Technique of application of partial Boolean differences.

where $c_1, c_2, \ldots, c_n$ and $i_1, i_2, \ldots, i_n$ are the binary representations of c and i respectively, and the i-th coefficient is defined as

$$f_i^{(c)}(d) = \left.\frac{\partial^n f(c)}{\partial x_1^{i_1}\partial x_2^{i_2}\ldots\partial x_n^{i_n}}\right|_{d=c} \quad \text{and} \quad \partial x_i^{i_j} = \begin{cases} 1, & i_j = 0 \\ \partial x_j, & i_j = 1 \end{cases}$$

that is a value of the n-ordered Boolean difference of f where $x_1 = c_1, ..., x_n = c_n$. Note that c is called a *polarity* of an expansion, i.e., it is an expansion of a function at the point c.

It follows from this definition that

- The logic Taylor expansion generates 2^n Reed-Muller expressions corresponding to 2^n polarities;
- In terms of spectral interpretation, this means that expressions are a spectrum of a Boolean function in one of 2^n polarities. A variable x_j

5. *Dependence of f on both x_i and x_j simultaneously via functions f_1, f_2*

$$\frac{\partial f}{\partial(x_i,x_j/f_1/f_2)} = \frac{\partial f}{\partial f_2}\frac{\partial f_2}{\partial f_1}\frac{\partial f_1}{\partial(x_i,x_j)}$$

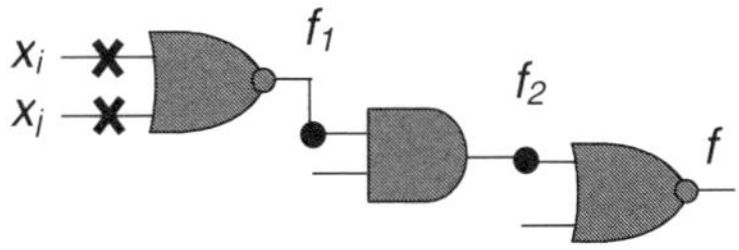

6. *Dependence of f on either x_i or x_j or both x_i and x_j via function f_1*

$$\frac{\partial f}{\partial(x_i \vee x_j/f_1)} = \frac{\partial f}{\partial f_1}\frac{\partial f_1}{\partial(x_i \vee x_j)} = \frac{\partial f}{\partial f_1}\left(\frac{\partial f_1}{\partial(x_i,x_j)} \vee \frac{\partial f_1}{\partial x_i} \vee \frac{\partial f_1}{\partial x_j}\right)$$

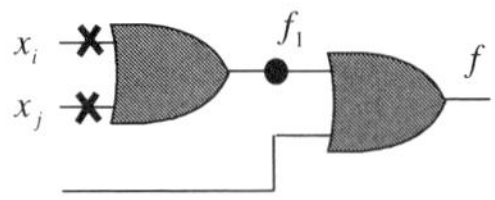

7. *Dependence of f on both x_i and x_j simultaneously via functions f_1, f_2*

$$\frac{\partial f}{\partial(x_i,x_j/f_1,f_2)} = \frac{\partial f}{\partial(f_1,f_2)}\frac{\partial f_1}{\partial(x_i,x_j)}\frac{\partial f_2}{\partial(x_i,x_j)}$$

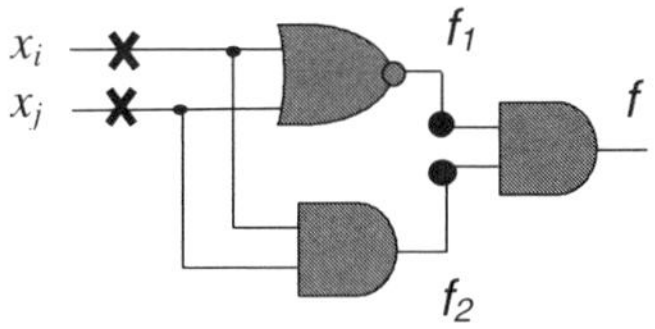

FIGURE 8.23
Technique of application of partial Boolean differences (Continuation of Figure 8.22).

is 0-polarized if it enters into the expansion uncomplemented, and 1-polarized otherwise. The components of the logic Taylor series are Boolean differences.

While the i-th spectral coefficient is described by a Boolean expression, it can be calculated in different ways, for example, matrix transformations, cube-based technique, decision diagram technique, and probabilistic methods.

Example 8.22 *The Reed-Muller spectrum of an arbitrary switching function of two variables ($n = 2$) and the polarity $c = c_1, c_2 = 0, 1$ (x_1 is uncomplemented and x_2 is complemented) is defined as a logic Taylor expansion of this* *function*

$$f = \bigoplus_{i=0}^{7} f_i^{(1)} (x_1 \oplus 0)^{i_1} (x_2 \oplus 1)^{i_2}$$

$$= f(1) \oplus \frac{\partial f(1)}{\partial x_2}\overline{x}_2 \oplus \frac{\partial f(1)}{\partial x_1} x_1 \oplus \frac{\partial^2 f(1)}{\partial x_1 \partial x_2} x_1 \overline{x}_2.$$

Example 8.23 *Let $f = x_1 \vee x_2$, then its Reed-Muller expansion of polarity $c = 01$ is $f = 1 \oplus \overline{x}_2 \oplus x_1\overline{x}_2$. The Reed-Muller spectral coefficients $f^{(i)}$ for all four polarities and their relationship to the Boolean differences are shown in Table 8.5. For example, the OR gate in polarity $c = 3$ ($c_1c_2 = 11$) is represented by two nonzero spectral components $f^{(0)}(3)$ and $f^{(3)}(3)$. This is an optimal spectral representation of the OR function (optimal polarity).*

TABLE 8.5
Reed-Muller as a Taylor expansion of elementary switching functions.

$\mathcal{N}$-hypercube Function	F	$\frac{\partial \mathbf{F}}{\partial x_2}$	$\frac{\partial \mathbf{F}}{\partial x_1}$	$\frac{\partial^2 \mathbf{F}}{\partial x_1 \partial x_2}$	Reed-Muller expression
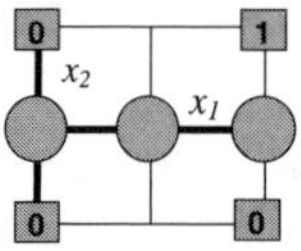	0	0	0	1	x_1x_2
	0	0	1	1	$x_1 \oplus x_1\overline{x}_2$
	0	1	0	1	$x_2 \oplus \overline{x}_1x_2$
$f = x_1 \wedge x_2$	1	1	1	1	$1 \oplus \overline{x}_2 \oplus \overline{x}_1 \oplus \overline{x}_1\overline{x}_2$
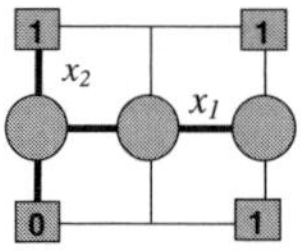	0	1	1	1	$x_2 \oplus x_1 \oplus x_1x_2$
	1	1	0	1	$1 \oplus \overline{x}_2 \oplus x_1\overline{x}_2$
	1	0	1	1	$1 \oplus \overline{x}_1 \oplus \overline{x}_1x_2$
$f = x_1 \vee x_2$	1	0	0	1	$1 \oplus \overline{x}_1\overline{x}_2$
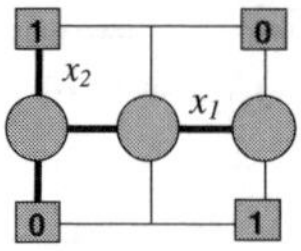	0	1	1	0	$x_2 \oplus x_1$
	1	1	1	0	$1 \oplus \overline{x}_2 \oplus x_1$
	1	1	1	0	$1 \oplus x_2 \oplus \overline{x}_1$
$f = x_1 \oplus x_2$	0	1	1	0	$\overline{x}_2 \oplus \overline{x}_1$

It follows from this example that it is possible to calculate separate spectral coefficients of a logic Taylor expansion (Equation 8.7). Moreover, the logic

Taylor expansion generates a family of 2^n Reed-Muller spectra of a Boolean function. In terms of signal processing theory, this means implementation of a transform in one of 2^n bases.

8.8 Arithmetic analogs of Boolean differences and logic Taylor expansion

In this section,

- Arithmetic analogs of Boolean differences, and
- Arithmetic analogs of logic Taylor expansion

are considered. An arithmetic analog of Boolean difference is called the *arithmetic difference* of a switching function. It is utilized to derive a representation of the function in arithmetic form.

8.8.1 Arithmetic analog of Boolean difference

arithmetic analog of Boolean difference in algebraic form is defined by equation

$$\frac{\widetilde{\partial} f}{\widetilde{\partial} x_i} = -f_{\overline{x}_i} + f_{x_i}. \tag{8.25}$$

The matrix form of the arithmetic difference (Equation 8.25) with respect to the i-th variable x_i of a switching function f of n variables given by truth vector $\mathbf{F}$ is defined as

$$\frac{\widetilde{\partial} f}{\widetilde{\partial} x_i} = \widetilde{D}_{2^{n-i}}\mathbf{F} \tag{8.26}$$

where the $2^{n-i} \times 2^{n-i}$ matrix $\widetilde{D}_{2^{n-i}}$ is generated by the rule

$$\widetilde{D}_{2^{n-i}} = I_{2^{i-1}} \otimes \begin{bmatrix} -1 & 1 \\ 1 & -1 \end{bmatrix} \otimes I_{2^{n-i}} \tag{8.27}$$

and $I_{2^{n-i}}$ is the identity matrix.

A k-th-order, $k = 1, ..., n$, arithmetical difference with respect to a subset of k variables $x_{i_1}, \ldots, x_{i_k}$, is defined in algebraic and matrix form as follows

$$\frac{\widetilde{\partial}^k f}{\widetilde{\partial} x_1 \cdots \widetilde{\partial} x_k} = \frac{\widetilde{\partial}}{\widetilde{\partial} x_1}\left(\frac{\widetilde{\partial}}{\widetilde{\partial} x_2}\left(\cdots \frac{\widetilde{\partial} f}{\widetilde{\partial} x_k}\right)\cdots\right), \tag{8.28}$$

$$\frac{\widetilde{\partial}^k \mathbf{F}}{\widetilde{\partial} x_1 \cdots \widetilde{\partial} x_k} = \widetilde{D}_{2^n}^{(1)} \widetilde{D}_{2^n}^{(2)} \ldots \widetilde{D}_{2^n}^{(n)}. \tag{8.29}$$

Example 8.24 *The structural properties of the flowgraphs of algorithms for computing arithmetic differences with respect to all the variables are explained in Figure 8.24.*

Boolean differences

$$\frac{\partial \mathbf{F}}{\partial x_1 \partial x_2 \partial x_3} = \overbrace{\begin{bmatrix} 1 & & & & 1 & & & \\ & 1 & & & & 1 & & \\ & & 1 & & & & 1 & \\ & & & 1 & & & & 1 \\ 1 & & & & 1 & & & \\ & 1 & & & & 1 & & \\ & & 1 & & & & 1 & \\ & & & 1 & & & & 1 \end{bmatrix}}^{D_{2^3}^{(1)}} \overbrace{\begin{bmatrix} 1 & & 1 & & & & & \\ & 1 & & 1 & & & & \\ 1 & & 1 & & & & & \\ & 1 & & 1 & & & & \\ & & & & 1 & & 1 & \\ & & & & & 1 & & 1 \\ & & & & 1 & & 1 & \\ & & & & & 1 & & 1 \end{bmatrix}}^{D_{2^3}^{(2)}} \overbrace{\begin{bmatrix} 1 & 1 & & & & & & \\ 1 & 1 & & & & & & \\ & & 1 & 1 & & & & \\ & & 1 & 1 & & & & \\ & & & & 1 & 1 & & \\ & & & & 1 & 1 & & \\ & & & & & & 1 & 1 \\ & & & & & & 1 & 1 \end{bmatrix}}^{D_{2^3}^{(3)}} \mathbf{F}$$

Arithmetic differences

$$\frac{\widetilde{\partial} \mathbf{F}}{\widetilde{\partial} x_1 \widetilde{\partial} x_2 \widetilde{\partial} x_3} = \overbrace{\begin{bmatrix} 1 & & & & -1 & & & \\ & 1 & & & & -1 & & \\ & & 1 & & & & -1 & \\ & & & 1 & & & & -1 \\ -1 & & & & 1 & & & \\ & -1 & & & & 1 & & \\ & & -1 & & & & 1 & \\ & & & -1 & & & & 1 \end{bmatrix}}^{\widetilde{D}_{2^3}^{(1)}} \overbrace{\begin{bmatrix} 1 & & 1 & & & & & \\ & 1 & & 1 & & & & \\ 1 & & 1 & & & & & \\ & 1 & & 1 & & & & \\ & & & & 1 & & 1 & \\ & & & & & 1 & & 1 \\ & & & & 1 & & 1 & \\ & & & & & 1 & & 1 \end{bmatrix}}^{\widetilde{D}_{2^3}^{(2)}} \overbrace{\begin{bmatrix} 1 & 1 & & & & & & \\ 1 & 1 & & & & & & \\ & & 1 & 1 & & & & \\ & & 1 & 1 & & & & \\ & & & & 1 & 1 & & \\ & & & & 1 & 1 & & \\ & & & & & & 1 & 1 \\ & & & & & & 1 & 1 \end{bmatrix}}^{\widetilde{D}_{2^3}^{(3)}} \mathbf{F}$$

FIGURE 8.24
Matrix based computing of third-order Boolean and arithmetical differences and flowgraphs (the nodes realize EXOR and arithmetic sum respectively (Example 8.24).

8.8.2 Arithmetic analog of logic Taylor expansion

An analog of the logic Taylor series for a switching function called the *arithmetical Taylor expansion* is expressed by the equation

$$P_c = \sum_{j=0}^{2^n-1} p_c^{(j)} (x_1 \oplus c_1)^{j_1} (x_2 \oplus c_2)^{j_2} \ldots (x_n \oplus c_n)^{j_n}, \tag{8.30}$$

where $c_1 c_2 \ldots c_n$ and $j_1 j_2 \ldots j_n$ are the binary representations of c (polarity) and j respectively, and the j-th coefficient is defined as

$$p_c^{(j)} = \frac{\widetilde{\partial}^n f(c)}{\widetilde{\partial} x_1^{j_1} \widetilde{\partial} x_2^{j_2} \ldots \widetilde{\partial} x_n^{j_n}} \tag{8.31}$$

that is, a value of the arithmetical analog of an n-ordered Boolean difference of f given c, i.e., $x_1 = c_1$, $x_2 = c_2$, ..., $x_n = c_n$. The coefficients $p_c^{(j)}$ (Equation

8.31) are also called the *arithmetic spectrum* of the switching function f.

Hence, the arithmetic Taylor expansion produces 2^n arithmetic expressions corresponding to 2^n polarities. Similarly to multiple Boolean differences, one can draw the flowgraph for any subset of variables to calculate multiple arithmetic differences.

Example 8.25 *Arithmetic spectrum of polarity c of an arbitrary three-variable ($n = 3$) switching function can be represented by the arithmetic Taylor series (Equation 8.30). In particular, given polarity $c = 3$ ($c_1c_2c_3 = 011$),*

$$\begin{aligned}P_3 &= \sum_{j=0}^{7} p_3^{(j)} (x_1 \oplus 0)^{j_1} (x_2 \oplus 1)^{j_2} (x_3 \oplus 1)^{j_3} \\ &= f(3) + \frac{\widetilde{\partial} f(3)}{\widetilde{\partial} x_3}\overline{x}_3 + \frac{\widetilde{\partial} f(3)}{\widetilde{\partial} x_2}\overline{x}_2 + \frac{\widetilde{\partial}^2 f(3)}{\widetilde{\partial} x_2 \widetilde{\partial} x_3}\overline{x}_2\overline{x}_3 + \frac{\widetilde{\partial} f(3)}{\widetilde{\partial} x_1}x_1 + \frac{\widetilde{\partial}^2 f(3)}{\widetilde{\partial} x_1 \widetilde{\partial} x_3}x_1\overline{x}_3 \\ &+ \frac{\widetilde{\partial}^2 f(3)}{\widetilde{\partial} x_1 \widetilde{\partial} x_2}x_1\overline{x}_2 + \frac{\widetilde{\partial}^3 f(3)}{\widetilde{\partial} x_1 \widetilde{\partial} x_2 \widetilde{\partial} x_3}x_1\overline{x}_2\overline{x}_3.\end{aligned}$$

The arithmetic differences of elementary switching functions represented by $\mathcal{N}$-hypercubes and corresponding arithmetic expressions as Taylor expansion in all polarities are given in Table 8.6.

8.9 Summary

Change is the basic concept for analyzing discrete systems, and can be useful in analyzing a system in spatial dimensions (fault detection, verification, symmetry detection, etc.).

1. The essence of this concept is an event that occurs in a system if a single variable changes its value. Formally it corresponds to the problem of detecting changes in the value of a switching function f in response to change of x_i. The function used to detect it is called *Boolean difference* or *Boolean derivative.*
2. The technique based on the Boolean difference is aimed at analysis of the behavior of a switching function represented by $\mathcal{N}$-hypercube:
   - Sensitivity of a function to the change of inputs;
   - Detection of local changes in a circuit;
   - Detection of directions of changes.
3. The technique of computing in terms of change utilizes the following properties of differential operators:

TABLE 8.6
Arithmetic differences and arithmetic expressions of elementary switching functions.

$\mathcal{N}$-hypercube Function	F	$\frac{\tilde{\partial}\mathbf{F}}{\tilde{\partial}x_2}$	$\frac{\tilde{\partial}\mathbf{F}}{\tilde{\partial}x_1}$	$\frac{\tilde{\partial}^2\mathbf{F}}{\tilde{\partial}x_1\tilde{\partial}x_2}$	Arithmetic expression
$f = x_1 \wedge x_2$	0	0	0	1	x_1x_2
	0	0	1	-1	$x_1 - x_1\overline{x}_2$
	0	1	0	-1	$x_2 - \overline{x}_1x_2$
	1	-1	-1	1	$1 - \overline{x}_2 - \overline{x}_1 + \overline{x}_1\overline{x}_2$
$f = x_1 \vee x_2$	0	1	1	-1	$x_2 + x_1 - x_1x_2$
	1	-1	0	1	$1 - \overline{x}_2 + x_1\overline{x}_2$
	1	0	-1	1	$1 - \overline{x}_1 + \overline{x}_1x_2$
	1	0	0	-1	$1 - \overline{x}_1\overline{x}_2$
$f = x_1 \oplus x_2$	0	1	1	-2	$x_2 + x_1 - 2x_1x_2$
	1	-1	-1	2	$1 - \overline{x}_2 - x_1 + 2x_1\overline{x}_2$
	1	-1	-1	2	$1 - x_2 - \overline{x}_1 + 2\overline{x}_1x_2$
	0	1	1	-2	$\overline{x}_2 + \overline{x}_1 - 2\overline{x}_1\overline{x}_2$

- Symmetry properties ("butterfly" configuration of flowgraphs of basic operators);
- Homogeneity and regularity;
- Fast iterative algorithm for multiorder differences.

4. The coefficients of logic Taylor expansion are values of Boolean differences. By analogy, arithmetic differences are coefficients of arithmetic Taylor expansion. The properties of logic and arithmetic Taylor expansion include:

- Generation of the Reed-Muller or arithmetic expression in a given polarity;
- Mutual relation of spectral coefficients (Reed-Muller and arithmetic).

  It should be noted that extension to other forms of switching function representation, for example, Walsh forms, will deploy the same approach and possesses the same properties. For this, so-called Walsh differences must be defined (see "Further Reading" Section).
- Generalization towards multiple-valued functions.

8.10 Problems

Problem 8.1 Calculate Boolean differences $\partial f/\partial x_1$, $\partial f/\partial x_2$, $\partial f/\partial x_3$, $\partial f/\partial(x_1, x_2)$, and $\partial f/\partial f_1$ for the circuits given in:

(*a*) Figure 8.25a
(*b*) Figure 8.25b
(*c*) Figure 8.25c
(*d*) Figure 8.25d

You may use generic Boolean difference or partial difference in symbolic form (Equation 8.1 or Equation 8.24).

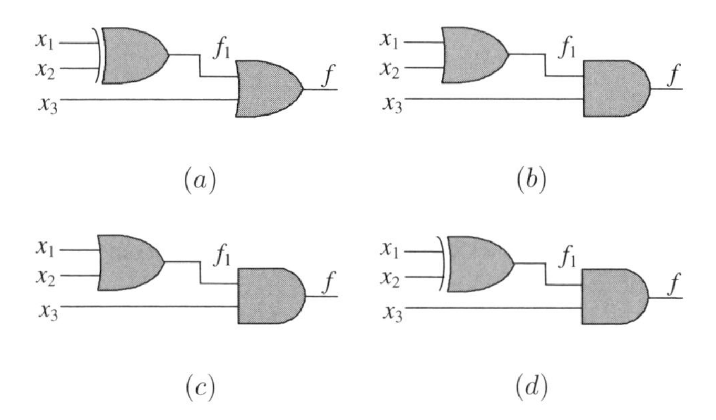

FIGURE 8.25
Combinational circuits (Problem 8.1).

Problem 8.2 Given the circuit shown in Figure 8.25a, calculate Boolean differences $\partial f/\partial x_1$, $\partial f/\partial x_2$, $\partial f/\partial x_3$, and $\partial f/\partial(x_1, x_2)$:

(*a*) In matrix form (Equation 8.15)
(*b*) On hypercube (see, for example, Equation 8.7)
(*c*) On Davio tree embedded in hypercube (use polarity $c = 0$ to calculate Boolean difference at the point c)

Problem 8.3 Calculate Boolean differences for the circuit depicted in Figure 8.15. Follow Example 8.12.

(a) Let $< Circuit1 >= OR$ and $< Circuit2 >= EXOR$ (Figure 8.15a)
(b) Let $< Circuit1 >= NOR$ and $< Circuit2 >= EXOR$ (Figure 8.15b)
(c) Let $< Circuit1 >= EXOR$ and $< Circuit2 >= AND$ (Figure 8.15c)

Problem 8.4 Justify the result given in Figure 8.22 and Figure 8.23 by comparison of generic Boolean difference techniques with partial Boolean difference. Consider:

(a) Relation 2
(b) Relation 3
(c) Relation 4
(d) Relation 7

Problem 8.5 Justify the equations given in Figure 8.26 by comparison of the resulting generic Boolean difference with partial Boolean difference. Consider:

(a) Relation 2
(b) Relation 3
(c) Relation 4
(d) Relation 7

Problem 8.6 Table 8.7 contains arithmetic differences (column vectors) for the switching function $f = x_1\overline{x}_2 \vee \overline{x}_3$. These differences are coefficients in the arithmetic Taylor expansion (row vectors).

(a) Find the optimal polarity, i.e., the polarity that contains the minimal number of literals.
(b) Calculate the coefficients of the Taylor expansion of polarity $c = 001$ via symbolic calculation of Boolean differences.
(c) Draw the arithmetic Davio tree for polarity $c = 001$.
(d) Draw an $\mathcal{N}$-hypercube of the function and show how to calculate arithmetic difference with respect to variable x_3 on the hypercube.

Problem 8.7 Prove the following equation given in Table 8.8:

(a) Rule (c)
(b) Rule (d)
(c) Rule (e)
(d) Rule (f)
(d) Rule (h)

Problem 8.8 A parametric Boolean difference of a switching function f with respect to variable X with integer (positive or negative) parameter (distance) t is defined by Bochmann and Posthoff [5]

$$\frac{df}{d(tX)} = X \oplus f(X + t)$$

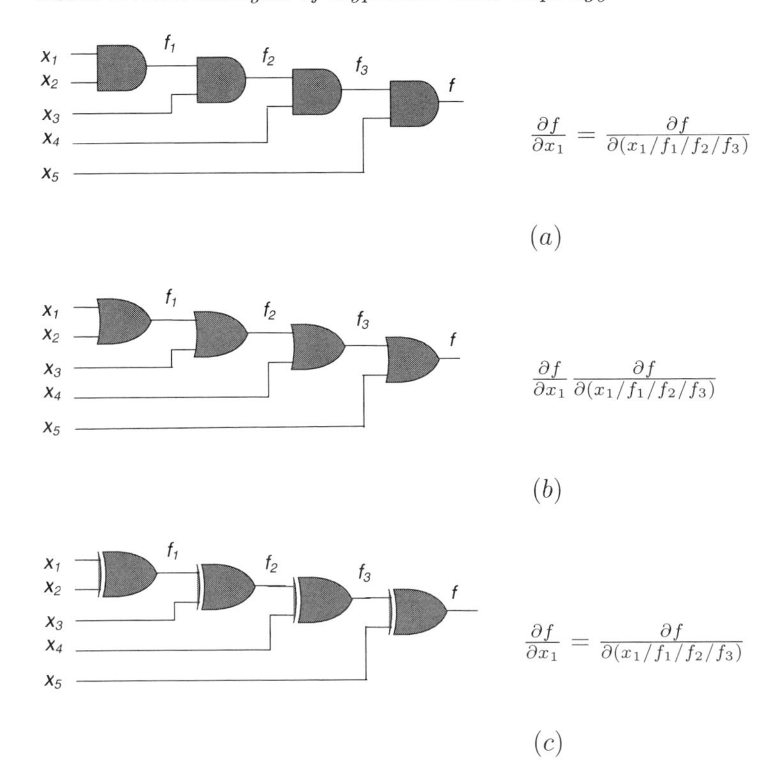

FIGURE 8.26
Local analysis in multilevel circuits by partial Boolean difference (Problem 8.5).

TABLE 8.7
Calculation of eight arithmetic expressions by Taylor expansion for the switching function $f = x_1\overline{x}_2 \vee \overline{x}_3$ (Problem 8.6).

		Coefficients of arithmetic expression/Arithmetic differences							
		000	001	010	011	100	101	110	111
c	$\mathbf{P}_c$	**F**	$\frac{\tilde{\partial}\mathbf{F}}{\tilde{\partial}x_3}$	$\frac{\tilde{\partial}\mathbf{F}}{\tilde{\partial}x_2}$	$\frac{\tilde{\partial}^2\mathbf{F}}{\tilde{\partial}x_2\tilde{\partial}x_3}$	$\frac{\tilde{\partial}\mathbf{F}}{\tilde{\partial}x_1}$	$\frac{\tilde{\partial}^2\mathbf{F}}{\tilde{\partial}x_1\tilde{\partial}x_3}$	$\frac{\tilde{\partial}^2\mathbf{F}}{\tilde{\partial}x_1\tilde{\partial}x_2}$	$\frac{\tilde{\partial}^3\mathbf{F}}{\tilde{\partial}x_1\tilde{\partial}x_2\tilde{\partial}x_3}$
0	$\mathbf{P}_0$	1	-1	0	0	0	1	0	-1
1	$\mathbf{P}_1$	0	1	0	0	1	-1	-1	1
2	$\mathbf{P}_2$	1	-1	0	0	0	0	0	1
3	$\mathbf{P}_3$	0	1	0	0	0	0	1	-1
4	$\mathbf{P}_4$	1	0	0	-1	0	-1	0	1
5	$\mathbf{P}_5$	1	0	-1	1	-1	1	1	-1
6	$\mathbf{P}_6$	1	-1	0	1	0	0	0	-1
7	$\mathbf{P}_7$	0	1	1	-1	0	0	-1	1

TABLE 8.8
The rules for simplification of manipulation of Boolean differences (Problem 8.7)

	Rule	Hint
(a)	$\frac{\partial c}{\partial x_i} = 0$	Proof: $\frac{\partial c}{\partial x_i} = a \oplus a = 0$
(b)	$\frac{\partial \overline{f}}{\partial x_i} = \frac{\partial f}{\partial x_i}$	Proof: $\frac{\partial \overline{f}}{\partial x_i} = \overline{f}_{\vert x_1=0} \oplus \overline{f}_{\vert x_1=1} = 1 \oplus f_{\vert x_1=0} \oplus 1 \oplus f_{\vert x_1=1} = f_{\vert x_1=1} \oplus f_{\vert x_1=0} = \frac{\partial f}{\partial x_i}$
(c)	$\frac{\partial (cf)}{\partial x_i} = c\frac{\partial f}{\partial x_i}$	
(d)	$\frac{\partial (c \vee f)}{\partial x_i} = \overline{c}\frac{\partial f}{\partial x_i}$	
(e)	$\frac{\partial (c \oplus f)}{\partial x_i} = \frac{\partial f}{\partial x_i}$	
(f)	$\frac{\partial (f \oplus g)}{\partial x_i} = \frac{\partial f}{\partial x_i} \oplus \frac{\partial g}{\partial x_i}$	
(g)	$\frac{\partial (f \wedge g)}{\partial x_i} = f\frac{\partial g}{\partial x_i} \oplus g\frac{\partial f}{\partial x_i} \oplus \frac{\partial f}{\partial x_i}\frac{\partial g}{\partial x_i}$	Hint: start with the right part: $(\overline{x}_i f_{\vert x_i=0} \oplus x_i f_{\vert x_i=1})(g_{\vert x_i=0} \oplus g_{\vert x_i=1}) \oplus (\overline{x}_i g_{\vert x_i=0} \oplus \overline{x}_i g_{\vert x_i=1})(f_{\vert x_i=0} \oplus f_{\vert x_i=1}) \oplus (f_{\vert x_i=0} \oplus f_{\vert x_i=1})(f g_{\vert x_i=0} \oplus f_{\vert x_i=1})$. Further simplification yields the left part of (g).
(h)	$\frac{\partial (f \vee g)}{\partial x_i} = \overline{f}\frac{\partial g}{\partial x_i} \oplus \overline{g}\frac{\partial f}{\partial x_i} \oplus \frac{\partial f}{\partial x_i}\frac{\partial g}{\partial x_i}$	

where + denotes arithmetic sum. Parametric Boolean difference verifies the fact of change of the switching function value when the argument X is changed to $X+t$. Note that in this particular case (special value of t and separate parts of truth vector of the function) the parametric Boolean difference becomes the Boolean difference with respect to variables.

In Figure 8.27 parametric Boolean difference of a switching function $f(X)$ given by its truth vector $F = [\ \ldots 10001010111 \ldots\ ]$ for $t = 3$ is calculated.

Parametric Boolean difference of a switching function $f(X, Y)$ with respect to its variables X and Y with integer positive parameters t_1 and t_2 accordingly is defined by the following equation

$$\frac{df(X,Y)}{d(t_1 X, t_2 Y)} = f(X,Y) \oplus f(X + t_1, Y + t_2).$$

This allows us to perform processing (computing Boolean difference, minimum, maximum or other operators) with respect to any direction,

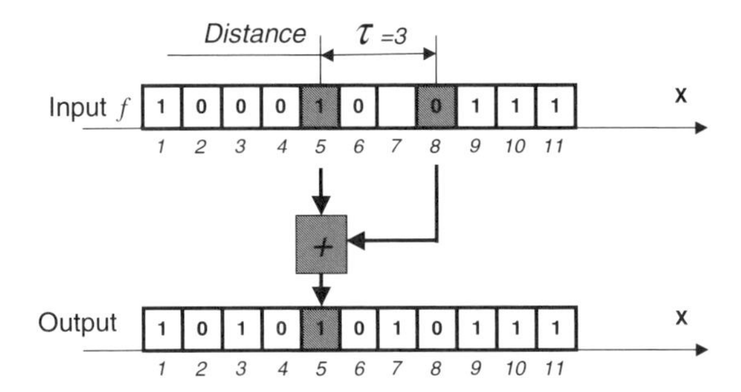

FIGURE 8.27
Parametric Boolean difference for distance $t = 3$ (Problem 8.8).

Problem 8.9 Figure 8.28 demonstrate the meaning of the terminal nodes in the positive Davio tree: these are values of the Boolean differences at point 0. Correspondence of the levels of the tree to calculation of the appropriate Boolean differences is shown as well.

(a) Find the function implemented by the tree.
(b) Derive the Davio tree that implements negative Davio expansion on the upper level, and positive on the upper level
(c) Formulate the algorithm of forming spectral (Reed-Muller) transform based on truth table (or decision diagram).

Problem 8.10 Consider the design possibilities for implementing positive or negative Davio expansion on a single-electron device and evaluate the characteristics of the devices.

8.11 Further reading

Boolean differential calculus. Akers introduced the concept of Boolean difference in order to derive and define various formal properties of switching functions: dependent and independent variables, series expansion as analogous to the classical MacLauren and Taylor series, functional decomposition, and solution of Boolean equations [1]. Fundamentals of Boolean differential calculus are developed in [5, 7, 21, 23].

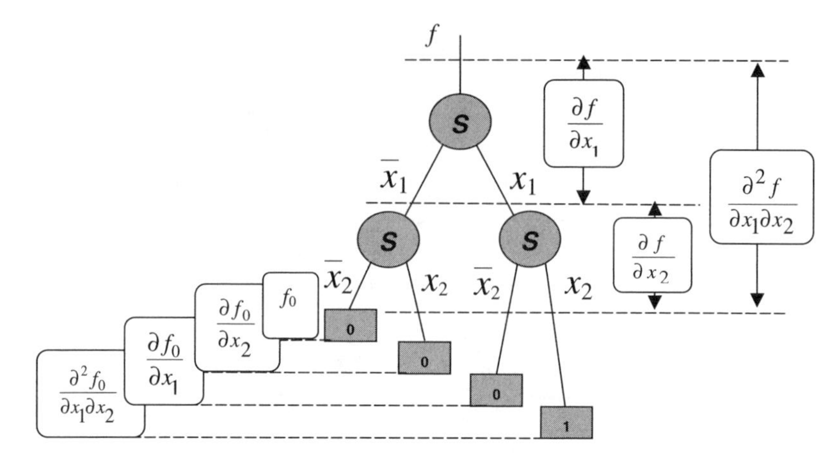

FIGURE 8.28
Positive Davio tree (Problem 8.9).

Gibbs differences. In 1969, Gibbs and Millard introduced so-called *dyadic derivative* for a complex function on finite dyadic groups [9]. During the 1970s this direction was called harmonic differential calculus in Galois fields. Work in this direction has been continued by Stankovic who has developed the theoretical and computational aspects of Gibbs's dyadic differentiator [6, 20].

Applications. It has been proposed by Sellers and et al. [18] to use Boolean differences to find tests for switching circuits. In a number of publications, the usefulness of differential models has been shown [8, 13, 15, 16]. Boolean differences are useful for decomposition of logic functions [7]. A comprehensive guide to application of Boolean differences in switching theory and digital design can be found in [7], [17], and also in selected chapters of [14]. A review of applications of logic differential calculus is given by Bochmann et al. [3].

Arithmetic analogs of Boolean differences. Tosic introduced arithmetical difference of switching functions [22]. The arithmetical difference was used as an analog of Taylor-Maclaurin series by Davio et al. [7]. Theoretical and applied aspects have been studied in [28].

Matrix notation of Boolean and arithmetic differences has been proposed by Yanushkevich [26, 28]. This technique has been developed towards multivalued functions in [25]. Matrix-based approach to solution of Boolean and multivalued differential equations can be found in [27, 29].

Probabilistic models. The probabilistic, or random, Boolean networks have been introduced by Kauffman [10, 11, 12]. The simplest, 1-D ones are called stochastic cellular automata and have been studied in [2]. Boolean difference has been utilized by Vichniac [24] to characterize phase transition of the random Boolean network. It has been also used by Shmulevich et al. [19] to evaluate the expectation of the Boolean difference with respect to distribution.

8.12 References

[1] Akers SB. On a theory of Boolean functions. *Society for Industrial and Applied Mathematics*, 7(4):487–498, 1959.

[2] Bagnoli F, and Rechtman R. Synchronization and maximum Lyapunov exponents of cellular automata. *Physical Review*, E59(2):R1307-R1310, 1999.

[3] Bochmann D, Yanushkevich S, Stanković R, Tosic Z, and Shmerko V. Logic differential calculus: progress, tendencies and applications. *Automation and Remote Control* (Kluwer/Plenum Publishers), 61(1):1033–1047, 2000.

[4] Bochmann D. Modelle fur ereignisdiskrete systeme im Booleschen differentialkalkul. *Automatisierungstechnic*, 3:99–106, 1997.

[5] Bochmann D, and Posthoff Ch. *Binäre Dynamishe Systeme.* Akademieverlag, Berlin, 1981.

[6] Butzer PL, and Stanković RS., Eds., *Theory and Applications of Gibbs Derivatives.* Mathematical Institute, Belgrade, 1990.

[7] Davio MJ, Deschamps P, and Thayse A. *Discrete and Switching Functions.* McGraw-Hill, New York, 1978.

[8] Edwards CR. The Gibbs dyadic differentiator and its relationship to the Boolean difference. *Computers and Electronic Engineering*, 5(4):335–344, 1978.

[9] Gibbs JE, and Millard MS. Walsh functions as solutions of a logical differential equation. *DES Report No.1 National Physical Laboratory* Middlesex, England, 1969.

[10] Kauffman SA. Metabolic stability and epigenesis in randomly constructed genetic nets. *Journal of Theoretical Biology* 22:437–467, 1969.

[11] Kauffman SA. Emergent properties in random complex automata. *Physica D*, 10:145–156, 1984.

[12] Kauffman SA. *The Origins of Order, Self-Organization and Selection in Evolution.* Oxford University Press, Oxford, 1993.

[13] Larrabee T. Test pattern generation using Boolean satisfiability. *IEEE Transactions on Computer-Aided Design of Integrated Circuits and Systems*, 11(1):4–15, 1992.

[14] Lee SC. *Modern Switching Theory and Digital Desing.* Prentice-Hall, New Jersey, 1978.

[15] Marinos P. Derivation of minimal complete sets of test-input sequences using Boolean differences. *IEEE Transactions on Computers*, 20(1):25–32, 1981.

[16] Najm FN. A survey of power estimation techniques in VLSI circuits. *IEEE Transactions on VLSI*, 2(4):446–455, Dec., 1994.

[17] Posthoff Ch, and Steinbach B. *Logic Functions and Equations.* Springer-Verlag, Heidelberg, 2004.

[18] Sellers FF, Hsiao MY, and Bearson LW. Analyzing errors with the Boolean difference. *IEEE Transactions on Computers*, 1:676–683, 1968.

[19] Shmulevich I, Dougherty ER, Kim S, and Zhang W. Probabilistic Boolean networks: a rule-based uncertainty model for gene regulatory networks. *Bioinformatics* 18:274–277, 2002.

[20] Stanković RS. Fast algorithm for calculation of Gibbs derivative on finite group. *Approximation Theory and its Applications,* 7(2):1–19, 1991.

[21] Thayse A, and Davio M. Boolean differential calculus and its application to switching theory. *IEEE Transactions on Computers*, 22:409–420, 1973.

[22] Tosic Z. Arithmetical representation of logic functions. In *Discrete Automatics and Networks,* USSR Academy of Sciences/Nauka, Moscow, 1970, pp. 131–136.

[23] Tucker JH, Tapia MA, and Bennet AW. Boolean integral calculus for digital systems. *IEEE Transactions on Computers,* 34:78–81, 1985.

[24] Vichniac G. Simulating physics with cellular automata. *Physica D*, 10:96–115, 1984.

[25] Yanushkevich SN. Systolic Synthesis algorithms for arithmetic polynomial forms of k-valued functions of Boolean algebra. *Automation and Remote Control*, Kluwer/Plenum Publishers, 55(12):1812–1823, 1994.

[26] Yanushkevich SN. Development of methods of Boolean differential calculus for arithmetic logic. *Automation and Remote Control*, Kluwer/Plenum Publishers, 55(5):715–729, 1994.

[27] Yanushkevich SN. Matrix method to solve logic differential equations. *IEE Proceedings, Pt.E, Computers and Digital Technique*, 144(5):267–272, 1997.

[28] Yanushkevich SN. *Logic Differential Caluclus in Multi-Valued Logic Design.* Technical University of Szczecin Academic Publishers, Poland, 1998.

[29] Yanushkevich SN. Matrix and combinatorics solution of Boolean differential equations. *Discrete Applied Mathematics*, 117:279–292, 2001.

9

Nanodimensional Multivalued Circuits

The primary advantage of a multivalued system is the ability to encode more information per variable than a binary system is capable of doing. Hence, less area is used for interconnections since each interconnection carries more information. This chapter generalizes the design paradigms introduced in previous chapters toward multivalued logic systems. These generalizations are made in the following directions:

- Technique of computing the sum-of-products, Reed-Muller, and arithmetic representations,
- Event-driven analysis based on logic differential operations,
- Word-level representation of logic functions,
- Linearization of word-level expressions, and
- Linear decision diagrams.

The main contribution of this chapter to logic design of multivalued circuits in nanodimensions includes:

- Representation of multivalued functions (circuits) by hypercube-like structures,
- Concept of change in spatial multivalued circuit analysis, and
- Linear word-level models of computing using hypercube-like structures.

The type of data structure is as critical in nanocomputing of multivalued logic functions as switching ones. The data structure must carry information in a form suitable for extraction of this information. Matrix (spectral) technique is very flexible and satisfies this requirement: spectral coefficients carry information about the form of representation, and the structure of these coefficients carry information about the behavior of the logic function in terms of change (logic difference, by analogy with Boolean differences in a binary system). Spectral representation is closely related to Taylor expansion and is a useful model for representation of multivalued functions. It can be applied to generate different forms of multivalued functions: logic and arithmetic. Following Chapter 4, where the word-level representations of switching functions have been given, this chapter also discusses generalizations of multivalued functions:

- ▶ Linearization technique based on arithmetic expressions,
- ▶ Linear word-level sum-of-products, and
- ▶ Linear word-level Reed-Muller expressions.

The motivation for this is that linear expression is mapped in linear word-level decision diagrams, and its embedding in spatial topologies becomes very simple.

The material is introduced as follows. In Section 9.1 the basics of multiple-valued logic are introduced. Sections 9.2 and 9.3 offer a brief introduction to spectral and decision diagram technique for multivalued functions respectively. The focus of Section 9.4 is the concept of change in multivalued systems. After formal definition of a model of change, a logic derivative, generalized logic Taylor expansion is considered in Section 9.5. The approach to linearization of word-level expressions (Chapter 8) is developed toward multivalued logic functions in Section 9.6. Also, in Section 9.7 development of linearization technique (Chapter 8) applied to multiplevalued functions in the word-level of nonarithmetic expressions is introduced. Section 9.10 is a brief overview of results in multivalued logic theory and its applications. A set of problems on the techniques of multivalued logic circuit design is given in Section 9.9.

9.1 Introduction to multivalued logic

Boolean algebra is the mathematical foundation of binary systems. Boolean algebra is defined on a set of two elements, $M = \{0, 1\}$. The operations of Boolean algebra must adhere to certain properties, called laws, or axioms, for the two binary operations $\wedge, \vee$ and complement. These are AND, OR, and NOT operations, respectively. The axioms can be used to prove more general laws about Boolean expressions, for example, to simplify expressions, factorize them, etc. Multivalued algebra is a generalization of Boolean algebra towards a set of m elements $M = \{0, 1, 2, \ldots, m\}$ and corresponding operations. The focus of this section is

- ▶ Operators of m-valued logic; the set of elementary functions is more large compared to Boolean algebra,
- ▶ Algebras that are specified on a universal set of operations, and finally,
- ▶ Data structures for representation and manipulation of multivalued functions.

9.1.1 Operations of multivalued logic

A multivalued logic function $f = f(x_1, x_2, \ldots, x_n)$ of n variables $x_1, x_2, \ldots, x_n$ is a logic function defined on the set $M = \{0, 1, \ldots, m-1\}$ and satisfying the mapping: $\{0, 1, \ldots, m-1\} \times \{0, 1, \ldots, m-1\}$. This means that multivalued logic circuits operate with multivalued logic signals (Figure 9.1).

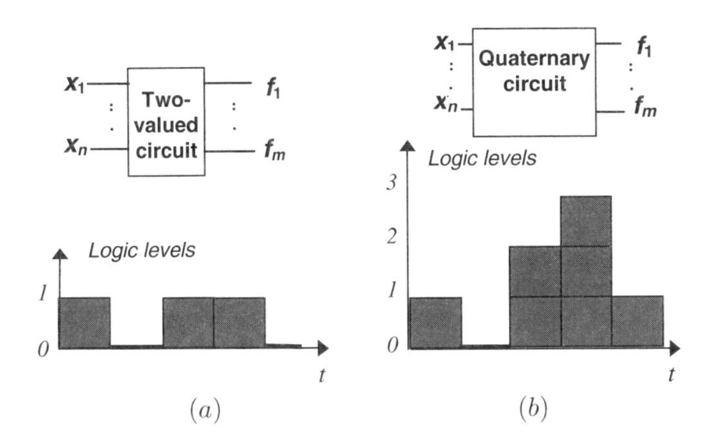

FIGURE 9.1
A switching circuit operates with two-level logic signals (a), and a quaternary circuit operates with four-valued logic signals (b).

Each of the logic operations has a corresponding logic gate. Multivalued logic gates are closely linked to hardware implementation given in Table 9.1 and Table 9.2.

Below we list the implementation-oriented operations of m-valued logic.

MAX operation is defined as

$$MAX(x_1, x_2) = \begin{cases} x_1 & \text{if } x_1 \geq x_2 \\ x_2 & \text{otherwise.} \end{cases}$$

When $m = 2$, this operation turns to an OR operation. The properties of MAX operations in ternary logic resemble those of Boolean algebra, i.e., $MAX(x, x) = x$ that is $x \vee x = x$ in binary circuit, $x \vee 0 = x$, and $x \vee 2 = 2$ that is $x \vee 0 = x$ and $x \vee 1 = 1$ in binary case. Hence, to propagate a signal x through MIN gate we must apply logical "2" to the second input (this is "1" in binary circuit) (Figure 9.2). MAX function of n variables is

$$MAX(x_1, x_2, \ldots, x_n) = x_1 + x_2 + \ldots + x_n.$$

TABLE 9.1
Library of ternary ($m = 3$) two-variable elementary functions.

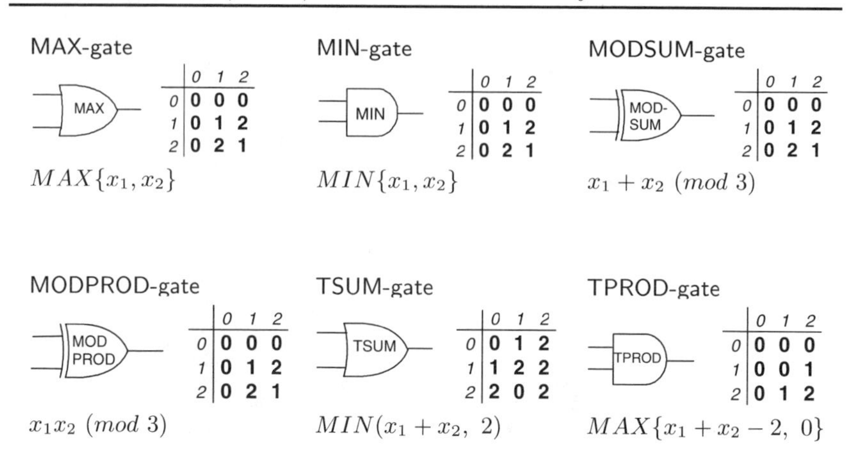

MAX-gate: $MAX\{x_1, x_2\}$

	0	1	2
0	0	0	0
1	0	1	2
2	0	2	1

MIN-gate: $MIN\{x_1, x_2\}$

	0	1	2
0	0	0	0
1	0	1	2
2	0	2	1

MODSUM-gate: $x_1 + x_2 \ (mod\ 3)$

	0	1	2
0	0	0	0
1	0	1	2
2	0	2	1

MODPROD-gate: $x_1 x_2 \ (mod\ 3)$

	0	1	2
0	0	0	0
1	0	1	2
2	0	2	1

TSUM-gate: $MIN(x_1 + x_2,\ 2)$

	0	1	2
0	0	1	2
1	1	2	2
2	2	0	2

TPROD-gate: $MAX\{x_1 + x_2 - 2,\ 0\}$

	0	1	2
0	0	0	0
1	0	0	1
2	0	1	2

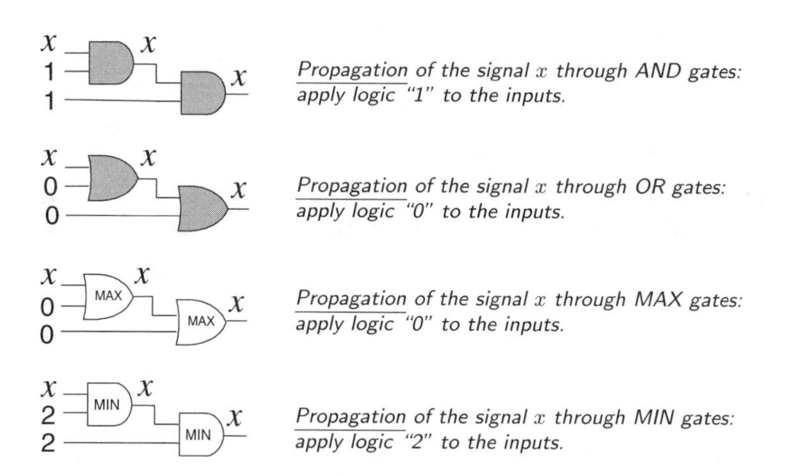

FIGURE 9.2
Propagation properties of the binary AND, OR, and ternary MAX and MIN gates.

MIN operation is defined as

$$MIN(x_1, x_2) = \begin{cases} x_2 & \text{if } x_1 \geq x_2 \\ x_1 & \text{otherwise.} \end{cases}$$

Example 9.1 *Propagation properties of the ternary MIN operation are illus-*

trated in Figure 9.2: signal x is propagated to the output when the second input is "2" $MIN(x, 2) = x$. This corresponds to $x \wedge 1 = x$ in binary circuit. Also, $MIN(x, x) = x$ in a ternary circuit and $x \vee x = x$ in a binary one. MAX operation for n variables is

$$MIN(x_1, x_2, \ldots, x_n) = x_1 x_2 \ldots x_n.$$

Modulo m product operation (MODPROD) is defined by

$$MODPROD(x_1, x_2, \ldots, x_n) = x_1 x_2 \ldots x_n \quad mod\ (m).$$

Example 9.2 *Let $m = 2$, then $MODPROD(x_1, x_2, \ldots, x_n) = x_1 \cdot x_2 \cdot, \ldots, \cdot x_n$, i.e., a unary AND function of n variables.*

Modulo sum operation (MODSUM) is defined below as

$$MODSUM(x_1, x_2, \ldots, x_n) = x_1 + x_2 + \ldots + x_n \quad mod\ (m).$$

It has been proven that modulo m sum operation $MODSUM$, modulo m product operation $MODPROD$, and constant "1" constitute a universal set of operations, and defined as

Example 9.3 *Let $m = 2$, then $MODSUM(x_1, x_2, \ldots, x_n) = x_1 \oplus x_2 \oplus \ldots \oplus x_n$, i.e., it is the EXOR function.*

Truncated sum operation (TSUM) of n variables is specified by

$$TSUM(x_1, x_2, \ldots, x_n) = MIN(x_1 + x_2 + \ldots + x_n, m - 1).$$

Example 9.4 *Let $m = 2$, $n = 2$, then $TSUM(x_1, x_2) = MIN(x_1 + x_2, 1) = x_1 \vee x_2$.*

Truncated product operation (TPROD) is defined by

$$TPROD(x_1, x_2, \ldots, x_n) = MIN(x_1 x_2 \ldots x_n,\ (m - 1)).$$

Example 9.5 *Given a switching function ($m = 2$, $n = 2$), $TPROD(x_1, x_2) = MIN(x_1 x_2,\ 1) = x_1 x_2$.*

Webb function is defined below as

$$x_1 \uparrow x_2 = MAX(x_1, x_2) + 1.$$

The unique property of this operation is that it represents a universal set itself, i.e., one can use Webb-gate to design an arbitrary multivalued logic network. The well known Pierce operation is a binary analog of Webb operation.

Complement operation is specified by

$$\overline{x} = (m-1) - x,$$

where $x \in M$ is a unary operation. For example, in ternary logic, $\overline{x} = 2 - x$. Notice, the property $\overline{\overline{x}} = x$ can be used in multivalued logic. This is because $(m-1) - \overline{x} = (m-1) - ((m-1) - x) = x$.

Example 9.6 *Let $m = 2$, then $\overline{x} = (2-1) - x = 1 - x$.*

TABLE 9.2
Library of ternary ($m = 3$) logic functions.

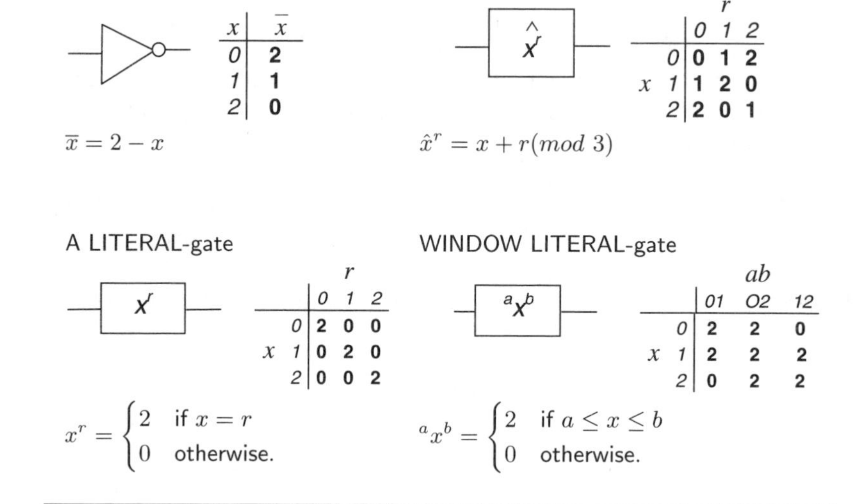

COMPLEMENT-gate

x	$\overline{x}$
0	**2**
1	**1**
2	**0**

$\overline{x} = 2 - x$

CYCLIC COMPLEMENT-gate

x \ r	0	1	2
0	**0**	**1**	**2**
1	**1**	**2**	**0**
2	**2**	**0**	**1**

$\hat{x}^r = x + r(mod\ 3)$

A LITERAL-gate

x \ r	0	1	2
0	**2**	**0**	**0**
1	**0**	**2**	**0**
2	**0**	**0**	**2**

$$x^r = \begin{cases} 2 & \text{if } x = r \\ 0 & \text{otherwise.} \end{cases}$$

WINDOW LITERAL-gate

x \ ab	01	02	12
0	**2**	**2**	**0**
1	**2**	**2**	**2**
2	**0**	**2**	**2**

$${}^a x^b = \begin{cases} 2 & \text{if } a \le x \le b \\ 0 & \text{otherwise.} \end{cases}$$

Clockwise cycle operation (r-order cyclic complement) is defined by

$$\hat{x}^r = x + r \quad (mod\ m).$$

This implies that

$$\hat{x}^0 = x \quad (mod\ m),$$
$$\hat{x}^m = x + m = x \quad (mod\ m).$$

The operation MIN and the clockwise cycle operation form a complete system as well.

Example 9.7 *Let $m = 2$, then the system is $\{AND, NOT\}$, i.e., NAND that is known to be complete.*

Literal operation (LITERAL) is specified below

$$x^y = y^x = \begin{cases} m-1 & \text{if } x = y, \\ 0 & \text{otherwise.} \end{cases}$$

A particular case of a literal is a multiple-valued input binary output function

$$x^y = \begin{cases} 1 & \text{if } x = y, \\ 0 & \text{otherwise.} \end{cases}$$

Window literal operation (WINDOW LITERAL) is defined as

$${}^a x^b = \begin{cases} m-1 & \text{if } a \le x \le b, \\ 0 & \text{otherwise.} \end{cases}$$

Any m-valued single-output system can be described by a functionally complete set of primitive operations. Various algebras exist to provide functional completeness for $m > 2$.

9.1.2 Multivalued algebras

The universal set of operations for each multivalued algebra below is specified.

Post algebra is based on two operations: 1-cycle inversion $(x+1)_{mod\ m}$, and MAX operation $x \vee y = MAX(x, y)$. Using these operations, one can describe any m-valued logic function. Analogs of Post operations in Boolean algebra are NOT and OR operations that also constitute a universal system.

Webb algebra is based on one operation – the Sheffer-Stroke operation (a functionally complete operation) that is specified by $x|y = MAX(x, y) + 1(mod\ m\ )$.

Bernstein algebra or modulo-sum and modulo-product algebra, includes modulo m sum $x_1 + x_2\ (mod\ m)$ and modulo m product $xy\ (mod\ m)$.

Allen and Givone algebras A universal set consists of MIN operation $MIN(x_1, x_2)$, $MAX(x_1, x_2)$, and WINDOW LITERAL operation

$${}^a x^b = \begin{cases} m-1 & \text{if } a \le x \le b \\ 0 & \text{otherwise.} \end{cases}$$

There are a lot of other algebras oriented mostly toward circuit implementations. For example, there are algebras based on MIN, MAX, and CYCLIC LITERAL operations. Other examples are MIN, TSUM, and WINDOW LITERAL operations.

9.1.3 Data structures

We will use the following representations of logic functions:

- Symbolic (algebraic) notations including sum-of-products and polynomial forms,
- Vector notations, i.e., truth column vector and coefficients column vectors,
- Matrix (two dimensional) notation relevant to word-level representation,
- Graph-based representation, or direct acyclic graph (DAG) and decision diagram technique,
- Embedded graph-based 3-D data structures.

Multivalued network. Graph-based representation of a network of logic gates (netlist), that is, the DAG, is relevant to gate-level design. It aims to make a library of logic gates available.

Figure 9.3 shows the library of gates for design of binary and multivalued circuits.

Truth table and truth column vector. The simplest way to represent a multivalued logic function is the truth table. The truth table of a logic function is the representation that tabulates all possible input combinations with their associated output values.

A truth column vector of a multivalued logic function f of n m-valued variables $x_1, x_2, ..., x_n$ is defined as $\mathbf{F}= [f(0), f(1), ..., f(m^n - 1)]^T$. The index i of the element $x^{(i)}$ corresponds to the assignments $i_1 i_2 ... i_n$ of variables $x_1, x_2, ..., x_n$ ($i_1, i_2, ..., i_n$ is binary representation of $i, i = 0, ..., m^n - 1$). For example, the truth column vector $\mathbf{F}$ of a ternary MIN function of two variables is $\mathbf{F}= [000011012]^T$ (Table 9.3).

Example 9.8 *A ternary function $MAX(x_1, x_2)$ with truth table given in Table 9.3 can be represented in algebraic (sum-of-products) form as follows:*

$$MAX(x_1, x_2) = 0 \cdot x_1^0 x_2^0 + 1 \cdot x_1^0 x_2^1 + 2 \cdot x_1^0 x_2^2 + 1 \cdot x_1^1 x_2^0 + 1 \cdot x_1^1 x_2^1 + 2 \cdot x_1^1 x_2^2 + 2 \cdot x_1^2 x_2^0 + 2 \cdot x_1^2 x_2^1 + 2 \cdot x_1^2 x_2^2$$

Decision diagrams technique. Any m-valued function can be given by a multivalued decision tree and decision diagram, a canonical graph-based data structure. A multivalued tree and diagram is specified in the same way as binary decision tree and binary decision diagram (BDD), except the nodes

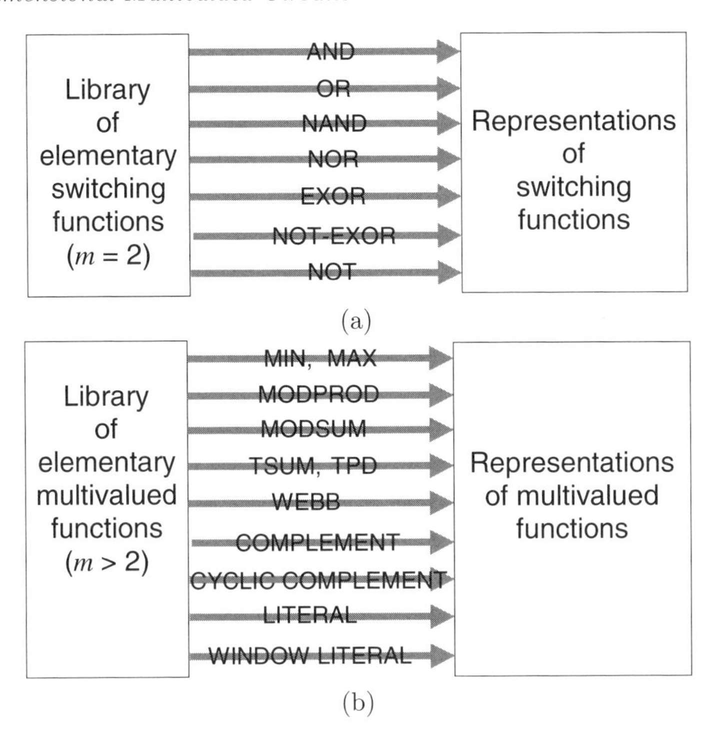

FIGURE 9.3
Representation of switching (a) and multivalued (b) functions.

TABLE 9.3
Truth table for elementary three-valued functions.

x_1	x_2	**MAX**	**MIN**	**MODSUM**	**MODPROD**	**TSUM**	**TPROD**
0	0	0	0	0	0	0	0
0	1	1	0	1	0	1	0
0	2	2	0	2	0	2	0
1	0	1	0	1	0	1	0
1	1	1	1	2	1	2	1
1	2	2	1	0	2	2	2
2	0	2	0	2	0	2	0
2	1	2	1	0	2	2	2
2	2	2	2	1	1	2	2

become more complex due to usage of Shannon and Davio expansion for multivalued functions.

Example 9.9 *A function of two ternary variables is represented by its truth* *table* $\mathbf{F}=[111210221]^T$. *In algebraic form the function is expressed by*

$$f = 0 \cdot x_1^1 x_2^2 + 1 \cdot (x_1^0 x_2^0 + x_1^0 x_2^1 + x_1^0 x_2^2 + x_1^1 x_2^1 + x_1^2 x_2^2)$$
$$+ 2 \cdot (x_1^1 x_2^0 + x_1^2 x_2^0 + x_1^2 x_2^1).$$

The map and decision diagram of the function is given in Figure 9.4 (the nodes implement Shannon expansion for ternary logic function).

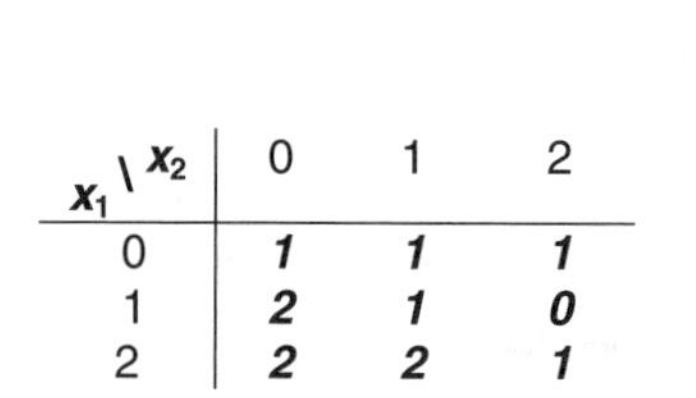

$x_1 \backslash x_2$	0	1	2
0	*1*	*1*	*1*
1	*2*	*1*	*0*
2	*2*	*2*	*1*

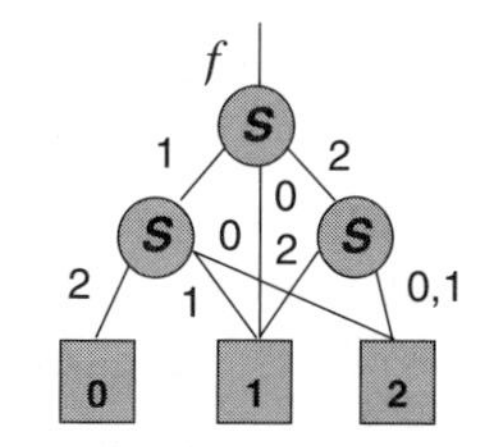

FIGURE 9.4
The map of the ternary function and its graph representation by decision diagram (Example 9.9).

9.2 Spectral technique

In Chapter 3, spectral technique has been used to represent switching functions in various forms. In this section, the generalization of spectral methods for multivalued functions is introduced.

9.2.1 Terminology

Because of interdisciplinary study of logic functions, different interpretations of the same concepts are often used, and new terminology bears clarification.

Spectrum of logic function. Below we draw parallels between the terminology of spectral technique and classical logic design.

Reed-Muller spectrum is the set of coefficients of a Reed-Muller expression of a switching function as the result of the Galois field transform. In some cases the term *generalized* is added to emphasize that the Reed-Muller transform is applied to a multivalued function.

Arithmetic spectrum is the set of coefficients of an arithmetic expression of a switching function or coefficients of a word-level representation of switching functions in arithmetic form. The additional term *generalized* is used to emphasize its representation of multivalued logic functions.

Walsh spectrum is the coefficients of a Walsh expression of switching function or a word-level representation of switching functions in Walsh form.

Spectral transforms of logic functions usually utilize matrix technique.

Reed-Muller spectral transform is a transformation of a logic function to a Reed-Muller expression.

Arithmetic spectral transform is a transformation of a logic function to an arithmetic expression.

Walsh transform is the transformation of a logic function to a Walsh expression.

Families of spectral transforms currently used in the practice of logic design:

- ▶ Reed-Muller transform,
- ▶ Arithmetic transform,
- ▶ Walsh transform, and
- ▶ Galois field transforms for multivalued functions.

Classification of spectral transforms used in logic design:

Walsh transform is defined as a Fourier transform on the diadic group. The arithmetic transform is derived from the Walsh transform by recoding 1 and -1 by 0 and 1 respective to the values of Boolean variables.

Vilenkin-Chrestenson transform is a generalization of the Walsh transform for $m > 2$.

Galois field transform is a generalization of the Reed-Muller transform over $GF(m)$.

9.2.2 Generalized Reed-Muller transform

In Chapter 3, a pair of Reed-Muller transforms for switching functions was defined. Generalization for multivalued functions is quite straightforward.

Direct and inverse transform generalized Reed-Muller transforms over $GF(m)$ are defined by matrix equations

$$\begin{aligned} \mathbf{R} &= \mathbf{R}_{m^n}^{(c)} \mathbf{F} \quad && \textit{over } GF(m) \\ \mathbf{F} &= \mathbf{R}_{m^n}^{-1(c)} \mathbf{R} \quad && \textit{over } GF(m) \end{aligned} \tag{9.1}$$

where $c = c_1c_2 \ldots c_n$ is a m-valued representation of $c = 1, 2, \ldots, m^n$. The pair of matrices $\mathbf{R}_{m^n}^{(c)}$ and $\mathbf{R}_{m^n}^{-1(c)}$ in Equation 9.1 are calculated as

$$\begin{array}{ll} \mathbf{R}_{m^n}^{(c)} &= \mathbf{R}_m^{(c_1)} \otimes \mathbf{R}_m^{(c_2)} \otimes \cdots \otimes \mathbf{R}_m^{(c_n)} \\ \mathbf{R}_{m^n}^{-1(c)} &= \mathbf{R}_m^{-1(c_1)} \otimes \mathbf{R}_m^{-1(c_2)} \otimes \cdots \otimes \mathbf{R}_m^{-1(c_n)} \end{array} \quad (9.2)$$

Elements of matrix $\mathbf{R}_m^{(c_j)}$ and $\mathbf{R}_m^{-1(c_j)}$, $j = 1, 2, \ldots, n$, are obtained as the solution to the logic equation

$$\mathbf{R}_m^{-1(c_j)}\mathbf{R}_m^{(c_j)} = \mathbf{I}_m \quad over\ GF(m) \quad (9.3)$$

where $\mathbf{I}_m$ is $m \times m$ identity matrix. This solution is represented by matrices. Observation of solutions results in useful formal representation.

Polarity. The pair of equations in Equation 9.2 is useful in technique manipulation of logic functions for the following reasons:

- They are a formal justification of the statement that an arbitrary logic function can be represented by m^n different generalized Reed-Muller expressions or *polarities.*
- They are a formal notation of the problem of optimal representation of multivalued functions by generalized Reed-Muller expressions (see Example 9.11). This is because it is possible to find optimal (minimal literals) representation among the m^n different generalized Reed-Muller expressions.
- They are a formal description of the behavior of multivalued function in terms of change (see Section 9.4).

It is worthwhile to note that in signal processing, the matrix representations in the form of Equation 9.2 are known as *factorized* representations of transform matrices. These equations play the central role in synthesis of so-called *fast algorithms.* However, in advanced logic design we observe another valuable feature of this classic result.

$$R^{(c)} = \sum_{j=0}^{m^n-1} r_j (x_1 + c_1)^{i_1} (x_2 + c_2)^{i_2} \cdots (x_n + c_n)^{i_n} \quad over\ GF(m) \quad (9.4)$$

where

$$(x_j + c_j)^{i_j} = \begin{cases} x_j + c_j = \hat{x}_j^{c_j}, & i_j \neq 0; \\ 1, & i_j = 0. \end{cases} \quad (9.5)$$

TABLE 9.4
Basic transform matrices, polarity $c = 0, 1, 2$, of generalized Reed-Muller and arithmetic transforms of ternary logic functions.

c	Reed-Muller transform Direct	Reed-Muller transform Inverse	Arithmetic transform Direct	Arithmetic transform Inverse
0	$\mathbf{R}_3^{(0)} = \begin{bmatrix} 1 & 0 & 0 \\ 0 & 2 & 1 \\ 2 & 2 & 2 \end{bmatrix}$	$\mathbf{R}_3^{-1(0)} = \begin{bmatrix} 1 & 0 & 0 \\ 1 & 1 & 1 \\ 1 & 2 & 1 \end{bmatrix}$	$\mathbf{P}_3^{(0)} = \begin{bmatrix} 2 & 0 & 0 \\ -3 & 4 & -1 \\ 1 & -2 & 1 \end{bmatrix}$	$\mathbf{P}_3^{-1(0)} = \begin{bmatrix} 1 & 0 & 0 \\ 1 & 1 & 1 \\ 1 & 2 & 1 \end{bmatrix}$
1	$\mathbf{R}_3^{(1)} = \begin{bmatrix} 0 & 0 & 1 \\ 2 & 1 & 0 \\ 2 & 2 & 2 \end{bmatrix}$	$\mathbf{R}_3^{-1(1)} = \begin{bmatrix} 1 & 1 & 1 \\ 1 & 2 & 1 \\ 1 & 0 & 0 \end{bmatrix}$	$\mathbf{P}_3^{(1)} = \begin{bmatrix} 0 & 0 & 2 \\ 4 & -1 & -3 \\ -2 & 1 & 1 \end{bmatrix}$	$\mathbf{P}_3^{-1(1)} = \begin{bmatrix} 1 & 1 & 1 \\ 1 & 2 & 1 \\ 1 & 0 & 0 \end{bmatrix}$
2	$\mathbf{R}_3^{(2)} = \begin{bmatrix} 0 & 1 & 0 \\ 1 & 0 & 2 \\ 2 & 2 & 2 \end{bmatrix}$	$\mathbf{R}_3^{-1(2)} = \begin{bmatrix} 1 & 2 & 1 \\ 1 & 0 & 0 \\ 1 & 1 & 1 \end{bmatrix}$	$\mathbf{P}_3^{(2)} = \begin{bmatrix} 0 & 2 & 0 \\ -1 & -3 & 4 \\ 1 & 1 & -2 \end{bmatrix}$	$\mathbf{P}_3^{-1(2)} = \begin{bmatrix} 1 & 2 & 1 \\ 1 & 0 & 0 \\ 1 & 1 & 1 \end{bmatrix}$

Example 9.10 *Given truth vector* $\mathbf{F} = [201000102]^T$ *of a ternary logic function of two variables. There are nine Reed-Muller expressions to represent this function that correspond to polarities of* $c_1c_2 = \{00, 01, 02, \ldots, 22\}$. *In Figure 9.5, the computing of Reed-Muller expression for* $c_1c_2 = 01$ *is given.*

Example 9.11 *It is possible to manipulate multivalued expressions in different polarities, for example, to choose the optimal form. Table 9.5 contains all 9 forms for the function given by truth vector* $[020120000]^T$. *The optimal (in terms of number of literals) polarity is* $c = 8$.

9.2.3 Generalized arithmetic transform

Arithmetic expressions of multivalued functions are an alternative approach to description of logic circuits sharing useful properties of generalized Reed-Muller expressions and permitting, at the same time, simplified representations of multioutput functions. In many applications where switching functions need to be analyzed, arithmetic expressions provide a better insight into related problems and offer efficient solutions. Examples: satisfiability, tautology, equivalence checking, etc.

Direct and inverse transform. Direct and inverse arithmetic transforms are defined as follows:

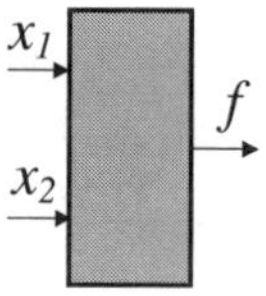

x_1, x_2 → f

x_1	x_2	**F**
0	0	2
0	1	0
0	2	1
1	0	0
1	1	0
1	2	0
2	0	1
2	1	0
2	2	2

$$\mathbf{R}^{(2)} = \mathbf{R}_{3^2}^{(2)}\,\mathbf{F} = \left(\mathbf{R}_3^{(0)} \otimes \mathbf{R}_3^{(1)}\right)\mathbf{F}$$

$$= \left(\begin{bmatrix} 1 & 0 & 0 \\ 0 & 2 & 1 \\ 2 & 2 & 2 \end{bmatrix} \otimes \begin{bmatrix} 0 & 0 & 1 \\ 2 & 1 & 0 \\ 2 & 2 & 2 \end{bmatrix}\right)\mathbf{F}$$

$$= \left[\begin{array}{ccc|ccc|ccc} 0&0&1&&&&&& \\ 2&1&0&&&&&& \\ 2&2&2&&&&&& \\ \hline &&&0&0&2&0&0&1 \\ &&&1&2&0&2&1&0 \\ &&&1&1&1&2&2&2 \\ \hline 0&0&2&0&0&2&0&0&2 \\ 1&2&0&1&2&0&1&2&0 \\ 1&1&1&1&1&1&1&1&1 \end{array}\right] \begin{bmatrix} 2\\0\\1\\0\\0\\0\\1\\0\\2 \end{bmatrix} = \begin{bmatrix} 1\\1\\0\\2\\2\\0\\0\\0\\0 \end{bmatrix} \quad over\ GF(3)$$

$$R^{(1)} = \sum_{j=0}^{8} r_j x_1^{j_1} \hat{x}_2^{j_2}$$

$$= 1 + \hat{x}_2 + 2x_1 + 2x_1\hat{x}_2 \quad over\ GF(3)$$

FIGURE 9.5
Representation of a ternary logic function of two variables by Reed-Muller expression of polarity $c = 1$ (Example 9.10).

TABLE 9.5
Generalized Reed-Muller expression in polarities $c = 0, 1, \ldots 8$ of ternary logic functions of two variables.

Polarity	**Reed-Muller expression over $GF(3)$**
$c = 0\ (c_1c_2 = 00)$	$R^{(0)} = 2x_1 + x_1^2 + x_1x_2 + 2x_1^2x_2$
$c = 1\ (c_1c_2 = 01)$	$R^{(1)} = 2 + 2x_1 + 2x_1^2 + 2\hat{x}_2 + 2x_1\hat{x}_2 + 2x_1^2\hat{x}_2$
$c = 2\ (c_1c_2 = 02)$	$R^{(2)} = 2x_1\hat{\hat{x}}_2 + 2x_1^2\hat{\hat{x}}_2 + 2\hat{\hat{x}}_2$
$c = 3\ (c_1c_2 = 10)$	$R^{(3)} = 1 + 2\hat{x}_1 + \hat{x}_1^2 + 2x_2 + \hat{x}_1x_2 + 2\hat{x}_1^2x_2$
$c = 4\ (c_1c_2 = 11)$	$R^{(4)} = 2 + \hat{x}_1 + 2\hat{x}_1^2 + 2\hat{x}_2 + \hat{x}_1\hat{x}_2 + 2\hat{x}_1^2\hat{x}_2$
$c = 5\ (c_1c_2 = 12)$	$R^{(5)} = 2\hat{x}_1\hat{\hat{x}}_2 + \hat{x}_1^2\hat{\hat{x}}_2 + 2\hat{\hat{x}}_2$
$c = 6\ (c_1c_2 = 20)$	$R^{(6)} = 2\hat{\hat{x}}_1^2x_2 + \hat{\hat{x}}_1$
$c = 7\ (c_1c_2 = 21)$	$R^{(7)} = 2\hat{\hat{x}}_1^2\hat{x}_2 + 2\hat{\hat{x}}_1$
$c = 8\ (c_1c_2 = 22)$	$R^{(8)} = 2\hat{\hat{x}}_1^2\hat{\hat{x}}_1$

$$\mathbf{P} = \frac{1}{(m-1)^n}\,\mathbf{P}_{m^n}^{(c)}\mathbf{F},$$
$$\mathbf{F} = \mathbf{P}_{m^n}^{-1(c)}\mathbf{P}. \tag{9.6}$$

The matrices $\mathbf{P}_{m^n}^{(c)}$ and $\mathbf{P}_{m^n}^{-1(c)}$ in pair of arithmetic transforms (Equation

9.6) are calculated by

$$
\begin{aligned}
\mathbf{P}_{m^n}^{(c)} &= \mathbf{P}_m^{(c_1)} \otimes \mathbf{P}_m^{(c_2)} \otimes \cdots \otimes \mathbf{P}_m^{(c_n)}, \\
\mathbf{P}_{m^n}^{-1(c)} &= \mathbf{R}_m^{-1(c_1)} \otimes \mathbf{P}_m^{-1(c_2)} \otimes \cdots \otimes \mathbf{P}_m^{-1(c_n)}.
\end{aligned}
\tag{9.7}
$$

Elements of the matrices $\mathbf{P}_m^{(c_j)}$ and $\mathbf{P}_m^{-1(c_j)}$, $j = 1, 2, \ldots n$, are obtained as solutions of the equation

$$
\mathbf{P}_m^{-1(c_j)}\mathbf{P}_m^{(c_j)} = \mathbf{I}_m. \tag{9.8}
$$

Polarity. Equation 9.7 is useful as a technique for manipulation of logic functions for the following reasons:

- It is a formal justification of the statement that an arbitrary m-valued logic function can be represented by m^n different generalized arithmetic expressions, or *polarities*.
- It is a formal notation of the problem of optimal arithmetic expressions derived from multivalued functions. This is because it is possible to find optimal (in terms of the literals) representation among m^n different generalized arithmetic expressions.
- A formal description of the behavior of a multivalued function in terms of change is derived by the equation

$$
P^{(c)} = \frac{1}{(m-1)^n} \sum_{j=0}^{m^n-1} p_j (x_1 + c_1)^{i_1} (x_2 + c_2)^{i_2} \cdots (x_n + c_n)^{i_n} \tag{9.9}
$$

where

$$
(x_j + c_j)^{i_j} = \begin{cases} x_j + c_j, & i_j \neq 0 \ \ (mod\ m); \\ 1, & i_j = 0. \end{cases} \tag{9.10}
$$

Note that the coefficients p_j are cofactors in the Taylor expansion.

Example 9.12 *In Figure 9.6, the calculation of an arithmetic expression of polarity $c_1c_2 = 02$ for a ternary function of two variables given truth vector $\mathbf{F} = [010311202]^T$ is given. There are nine arithmetic expressions to represent this function that correspond to the polarities $c_1c_2 = \{00, 01, 02, \ldots, 22\}$.*

Word-level representation. Similar to word-level representation of switching functions, the properties of linearity and superposition are utilized in computing arithmetic expressions of multioutput logic functions.

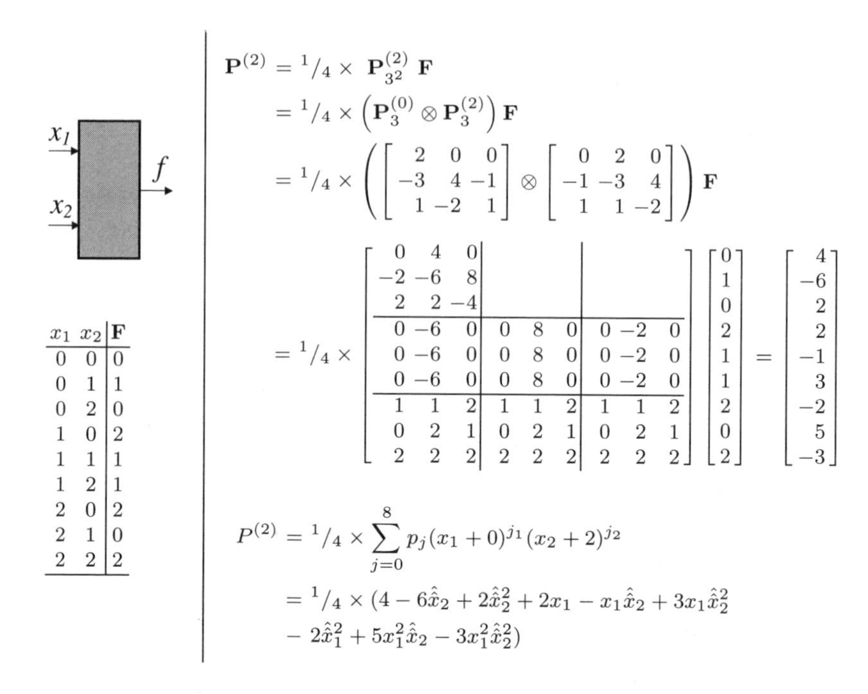

FIGURE 9.6
Representation of a ternary logic function of two variables by arithmetic expression of polarity $c = 1$ (Example 9.12).

Example 9.13 *Consider a three-output ternary logic function of two variables given by truth vectors (Figure 9.7). Accordingly to the properties of linearity and superposition, the first method is based on three iterations of a direct arithmetic transform (Equation 9.6) of truth-vectors* $\mathbf{F}_0, \mathbf{F}_1$ *and* $\mathbf{F}_2$. *The resulting vectors of coefficients* $\mathbf{P}_0, \mathbf{P}_1$ *and* $\mathbf{P}_2$ *form the vector* $\mathbf{D}$ *calculated as a weighted sum*

$$3^0\mathbf{P}_0 + 3^1\mathbf{P}_1 + 3^2\mathbf{P}_2.$$

Alternatively, the direct arithmetic transform (Equation 9.6) is applied to the vector $\mathbf{F}$ *calculated as a weighted sum*

$$3^0\mathbf{F}_0 + 3^1\mathbf{F}_1 + 3^2\mathbf{F}_2.$$

9.2.4 Relation of spectral representations to data structures, behavior models, and massive parallel computing

Here, we focus on the conceptual value of a spectral representation of logic functions, starting from the trivial role of transformation and emphasizing

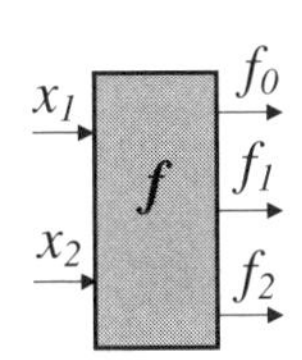

x_1	x_2	$\mathbf{F}_2$	$\mathbf{F}_1$	$\mathbf{F}_0$
0	0	0	1	2
0	1	1	1	1
0	2	0	2	1
1	0	0	2	1
1	1	2	2	2
1	2	0	0	0
2	0	2	2	1
2	1	2	1	0
2	2	1	1	1

Method 1

$\mathbf{P}_0 = {}^1/_4 \times \mathbf{P}_{3^2}^{(0)} \mathbf{F}_0 = {}^1/_4 \times [\,8 - 6\ 2 - 6\ 3\ 3 - 17\ 2 - 17\ 9\,]^T,$

$\mathbf{P}_1 = {}^1/_4 \times \mathbf{P}_{3^2}^{(0)} \mathbf{F}_1 = {}^1/_4 \times [\,4\ 0\ 0\ 6\ 11 - 9\ \ - 2 - 7\ 5\,]^T,$

$\mathbf{P}_2 = {}^1/_4 \times \mathbf{P}_{3^2}^{(0)} \mathbf{F}_2 = {}^1/_4 \times [\,0\ 8 - 4 - 4\ 19 - 9\ 4 - 11\ 5\,]^T.$

$\mathbf{D} = 3^0\mathbf{P}_0 + 3^1\mathbf{P}_1 + 3^2\mathbf{P}_2$

$= {}^1/_4 \times [20\ 66 - 34 - 24\ 237 - 125\ 32 - 137\ 69]^T$

Method 2

$\mathbf{F}_D = [\mathbf{F}_2|\mathbf{F}_1|\mathbf{F}_0] = 3^2\mathbf{F}_2 + 3^1\mathbf{F}_1 + 3^0\mathbf{F}_0$

$$= 3^2 \begin{bmatrix} 0\\1\\0\\0\\2\\0\\2\\2\\1 \end{bmatrix} + 3^1 \begin{bmatrix} 1\\1\\2\\2\\2\\0\\2\\1\\1 \end{bmatrix} + 3^0 \begin{bmatrix} 2\\1\\1\\1\\2\\0\\1\\0\\1 \end{bmatrix} = \begin{bmatrix} 5\\13\\4\\7\\26\\0\\25\\21\\13 \end{bmatrix}$$

$\mathbf{D} = {}^1/_4 \times\ \mathbf{P}_{3^2}^{(0)} \mathbf{F}_D$

$= [20\ 66\ - 34\ - 24\ 237\ - 125\ 32\ - 137\ 69]^T$

$$\mathbf{P}_{3^2}^{(0)} = \frac{1}{4}\left(\begin{bmatrix} 2 & 0 & 0\\ -3 & 4 & -1\\ 1 & -2 & 1 \end{bmatrix} \otimes \begin{bmatrix} 2 & 0 & 0\\ -3 & 4 & -1\\ 1 & -2 & 1 \end{bmatrix}\right)$$

$$= \frac{1}{4}\left[\begin{array}{ccc|ccc|ccc} 4 & 0 & 0 & & & & & & \\ -6 & 8 & -2 & & & & & & \\ 2 & -4 & 2 & & & & & & \\ \hline -6 & 0 & 0 & 8 & 0 & 0 & -2 & 0 & 0\\ 9 & -12 & 3 & -12 & 16 & -4 & 3 & -4 & 1\\ -3 & 6 & -3 & 4 & -8 & 4 & -1 & 2 & -1\\ \hline 2 & 0 & 0 & -4 & 0 & 0 & 2 & 0 & 0\\ -3 & 4 & -1 & 6 & -8 & 2 & -3 & 4 & -1\\ 1 & -2 & 1 & -2 & 4 & -2 & 1 & -2 & 1 \end{array}\right]$$

$D = {}^1/_4 \times (20 + 66x_2 - 34x_2^2 - 24x_1 + 237x_1x_2$
$- 125x_1x_2^2 + 32x_1^2 - 137x_1^2x_2 + 69x_1^2x_2^2)$

FIGURE 9.7
Representation of a three-output ternary function of two variables by a word-level arithmetic expression of polarity $c = 0$ (Example 9.13).

hidden information about the behavior of logic functions, its testing and verification.

The essence of the spectral technique in advanced logic design is formalized by the following statements:

- The direct and inverse matrix transforms of logic functions.
- Factorization of transform matrices.

The latter property leads directly to the fast Fourier-like transform algorithms and, thus, to implementation of the spectral transforms of signal processors, including parallel-pipelined processors.

The goal of using spectral transforms is to "extract" the information about the function, interpret this information, and utilize it in computing. It is utilized in many practical applications:

- Technology-dependent gate-level implementation, for example, Reed-Muller forms are implemented using AND, EXOR, and NOT gates,
- Optimization of representations in the chosen domain, i.e., between a variety of bases and transforms,
- Determination of functional properties, and
- Event-driven analysis via Taylor expansion.

Families of spectral transforms. Families of spectral transforms that are used in logic design can be divided into two classes: *logic* and *arithmetic* transforms.

Class of logic transforms. Utilizing various basic matrices allows us to generate m^n Reed-Muller expressions over $GF(m)$ for a given m-valued function.

Class of arithmetic transforms. The first family (m^n arithmetic expressions for a given m-valued function) is generated by changing the basic matrix. The next family is known as Vilenkin-Chrestenson transforms. From these transforms various forms can be derived, including a complex representation. In complex representation, the coefficients are represented by a complex number that is reasonable for large m because the additional resource of parallel computing can be developed. The particular cases: the Walsh and Walsh-like transform are known as global transforms in logic design. The usefulness of many other transforms has not been proven; however, it is clear that they can be used in certain areas of logic design. For example, Haar and Haar-like transforms are suitable for "catching" group behavior of logic functions.

Information about the behavior of logic functions in terms of change. Matrices of factorized transforms carry information about the behavior of a logic function in notation of change.

Elementary change of a logic function is formally described by logic or arithmetic difference. Each iteration of spectrum computing carries information about the behavior of a logic function with respect to a variable. Note that it is essential to calculate observability and sensitivity functions.

Taylor expansion. The factorized matrix transform can be viewed as a Taylor expansion if the result of each iteration is interpreted in terms of logic differences.

Testibility properties of a logic function can be analyzed using spectral coefficients.

Massive parallel processing. The intrinsic nature of massive parallelism of matrix transforms can be revised in the light of data structures and topology used for logic calculations. This involves:

Parallelism of word-level representations based on replacing bitwise parallelism with word-wide parallelism.

Parallelism of multidimensional processing of a logic function, defined as a parallel processing in many dimensions based on embedding a decision tree or decision diagram into an appropriate spatial structure. In this book, the latter is a hypercube-like structure, which inherits parallel properties of decision trees.

9.3 Multivalued decision trees and decision diagrams

Multivalued decision trees and diagrams are the result of straightforward generalization of binary decision trees and diagrams. In this section, we briefly review designing these multivalued decision data structures.

9.3.1 Operations in *GF(m)*

In the Galois field p complements of $p-$valued variable are considered. They are defined by

$$^{i-}x = x + i,$$

$i = 1, \dots, p-1$, where $+$ denotes the addition in the considered Galois field. For example, in $GF(4)$, four literals are considered: x, ^{1-}x, ^{2-}x, and ^{3-}x for each variable x. Two elementary operations, addition and multiplication in $GF(4)$, are given in Table 9.6.

TABLE 9.6
Addition and multiplication in $GF(4)$.

+	*0*	*1*	*2*	*3*
0	0	1	2	3
1	1	0	3	2
2	2	3	0	1
3	3	2	1	0

·	*0*	*1*	*2*	*3*
0	0	0	0	0
1	0	1	2	3
2	0	2	3	1
3	0	3	1	2

9.3.2 Shannon trees for ternary functions

The type of tree is characterized by the Shannon expansion applied in its nodes. Ternary Shannon expansion is defined by

$$f = x_i^0 f_0 \vee x_i^1 f_1 \vee x_i^2 f_2, \tag{9.11}$$

where $f_0 = f(x_i = 0)$, $f_1 = f(x_i = 1)$, and $f_2 = f(x_i = 2)$.

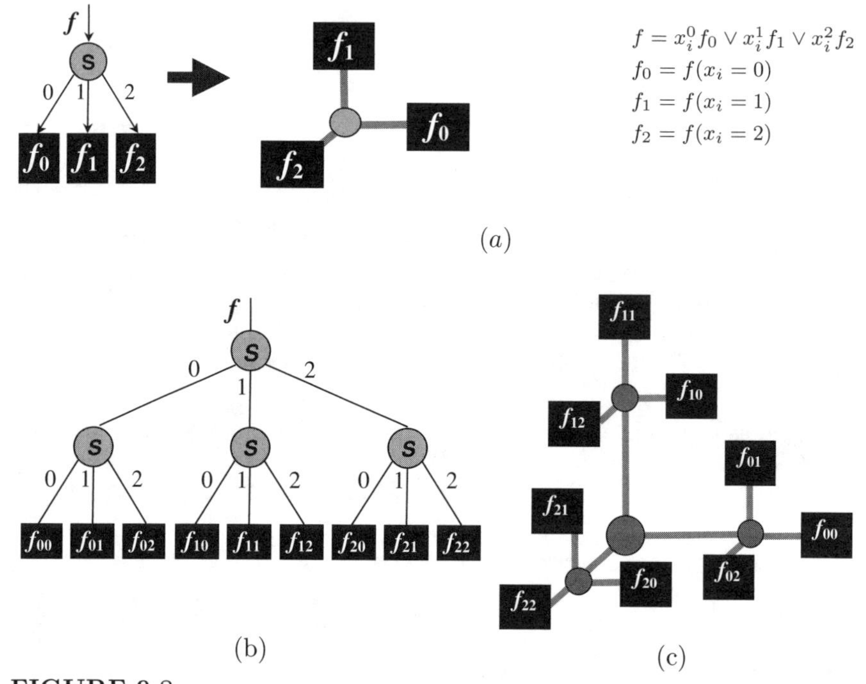

FIGURE 9.8
The node of a Shannon tree (a), decision tree for ternary function of two variables (b), and embeddings in hypercube-like structures (c) (Example 9.14).

Example 9.14 *An arbitrary ternary function f of two variables can be represented by the decision tree shown in Figure 9.8. Here, at the upper level, the function is expanded as*

$$f = x_1^0 f_0 \vee x_1^1 f_1 \vee x_1^2 f_2.$$

At the lower level, further expansion is implemented

$$f_0 = x_2^0 f_{00} \vee x_2^1 f_{01} \vee x_2^2 f_{02},$$
$$f_1 = x_2^0 f_{10} \vee x_2^1 f_{11} \vee x_2^2 f_{12},$$
$$f_2 = x_2^0 f_{20} \vee x_2^1 f_{21} \vee x_2^2 f_{22}.$$

It follows from the above example that this decision tree represents a ternary function f of two variables in sum-of-products form.

$$f = x_1^0 x_2^0 x f_{00} \vee x_1^0 x_2^1 f_{01} \vee x_1^0 x_2^2 f_{02} \vee x_1^1 x_2^0 f_{10} \vee x_1^1 x_2^1 f_{11} \vee x_1^1 x_2^2 f_{12} \vee x_1^2 x_2^0 f_{20} \vee x_1^2 x_2^1 f_{21} \vee x_1^2 x_2^2 f_{22}.$$

9.3.3 Shannon and Davio trees for quaternary functions

The decision tree design for four-valued functions is based on the following types of expansion (types of nodes): S, pD, and nD ($nD^{'}$, $nD^{''}$, $nD^{'''}$). In Table 9.7, $J_i(x)$, $i = 0, \ldots, k-1$, are the characteristic functions, denoted by $J_i(x) = 1$, if $x = i$ and $J_i(x) = 0$, otherwise.

9.3.4 Embedding decision tree in hypercube-like structure

Figure 9.9a illustrates the embedding of a node of a quaternary decision tree in a 2-D hypercube-like structure. In this topology

- The node is assigned with a Shannon expansion as given in Table 9.7.
- The i-th terminal node corresponds to the characteristic function $J_i(x)$.
- The outgoing branches are assigned with x^0, x^1, x^2, x^3.

In Figure 9.9b, a quaternary decision tree of two variables is embedded into a 3-D hypercube-like structure that is recursively generated from 2-D structure.

Let one node together with outgoing branches and connected nodes (terminal or intermediate nodes) be a primitive structure, or primitive. In Table 9.8, some parameters of ternary and quaternary two-inputs MIN, MAX and inverter gates represented in hypercube-like structure are given. Here,

- The number of primitives is equal to $m + n - 1$ for m-valued n-input functions,
- The number of active nodes corresponds to the number of nonzero elements in the truth table of the function,
- The number of terminal nodes is m^n, where n is a number of variables,
- The connectivity is calculated by $\sum_{i=1}^{n} m^i$,
- The number of intermediate nodes is $\sum_{i=1}^{n-1} m^i$.

TABLE 9.7
Analogues of Shannon and Davio expansions in $GF(4)$.

Type	Rule of expansion
f → S → 1, 2, 3, 4	$f = \overbrace{J_0(x) \cdot f_{\vert x=0}}^{Leaf\ 1} + \overbrace{J_1(x) \cdot f_{\vert x=1}}^{Leaf\ 2} + \underbrace{J_2(x) \cdot f_{\vert x=2}}_{Leaf\ 3} + \underbrace{J_3(x) \cdot f_{\vert x=3}}_{Leaf\ 4}$
f → pD → 1, 2, 3, 4	$f = f_{\vert x=0} + x \cdot (f_{\vert x=1} + 3f_{\vert x=2} + 2f_{\vert x=3})$ $+x^2 \cdot (f_{\vert x=1} + 2f_{\vert x=2} + 3f_{\vert x=3})$ $+x^3 \cdot (f_{\vert x=0} + f_{\vert x=1} + f_{\vert x=2} + f_{\vert x=3})$
f → nD′ → 1, 2, 3, 4	$f = f_{\vert x=1} + {}^{1-}x \cdot (f_{\vert x=0} + 2f_{\vert x=2} + 3f_{\vert x=3})$ $+{}^{1-}x^2 \cdot (f_{\vert x=0} + 3f_{\vert x=2} + 2f_{\vert x=3})$ $+{}^{1-}x^3 \cdot (f_{\vert x=0} + f_{\vert x=1} + f_{\vert x=2} + f_{\vert x=3})$
f → nD″ → 1, 2, 3, 4	$f = f_{\vert x=2} + {}^{2-}x \cdot (3f_{\vert x=0} + 2f_{\vert x=1} + f_{\vert x=3})$ $+{}^{2-}x^2 \cdot (2f_{\vert x=0} + 3f_{\vert x=1} + f_{\vert x=3})$ $+{}^{2-}x^3 \cdot (f_{\vert x=0} + f_{\vert x=1} + f_{\vert x=2} + f_{\vert x=3})$
f → nD‴ → 1, 2, 3, 4	$f = f_{\vert x=3} + {}^{3-}x \cdot (2f_{\vert x=0} + 3f_{\vert x=1} + f_{\vert x=2})$ $+{}^{3-}x^2 \cdot (3f_{\vert x=0} + 2f_{\vert x=1} + f_{\vert x=2})$ $+{}^{3-}x^3 \cdot (f_{\vert x=0} + f_{\vert x=1} + f_{\vert x=2} + f_{\vert x=3})$

9.4 Concept of change in multivalued circuits

This section introduces the basics of event-driven technique, the development of the binary case represented in Chapter 8. The focus of this section is:

- Formal definition of change for multivalued functions,
- Computing of change, and
- Generalization of logic Taylor expansion for multivalued functions.

9.4.1 Formal definition of change for multivalued functions

Changing a signal. Consider a three-valued signal with three logic values 0, 2, and 3 (Figure 9.10). There are four possible situations (for simplification, the direction of change is not considered):

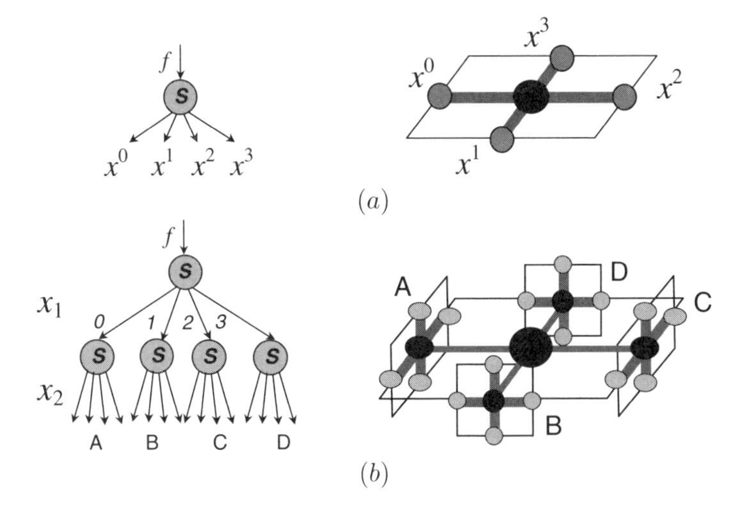

FIGURE 9.9
Hypercube-like structures for single- (a) and two-variable (b) quaternary functions.

- Change $0 \leftrightarrow 1$,
- Change $0 \leftrightarrow 2$,
- Change $1 \leftrightarrow 2$,
- No change $(0 \leftrightarrow 0,\ 1 \leftrightarrow 1,\ 2 \leftrightarrow 2)$.

The problem is formulated as detection of changes in a ternary function f if the ternary variable x_i is changed.

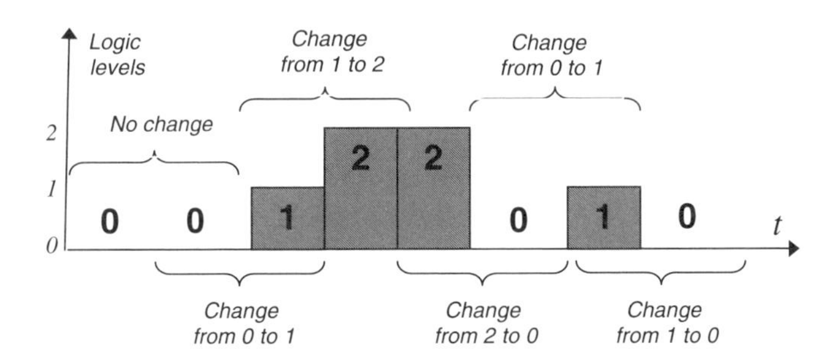

FIGURE 9.10
Change of three-valued signals.

TABLE 9.8
Parameters of a hypercube-like representation of ternary and quaternary single-input and two-input gates.

Parameters	Ternary gates			Quaternary gates		
	MIN	MAX	INV	MIN	MAX	INV
Total number of primitives	4	4	1	5	5	1
Number of active nodes	4	8	2	9	15	3
Number of terminal nodes	9	9	3	16	16	4
Number of intermediate nodes	3	3	0	4	4	0
Total number of nodes	13	13	4	21	21	5
Connectivity	12	12	3	20	20	4

Formal model of change. In contrast to formal notation of Boolean difference, where the complement of binary variable x_i is defined as $\overline{x}_i$, in multivalued logic *the cyclic complement* to a multivalued variable x_i is used. Let f be an m-valued (m is prime) logic function of n variables. The t_i-th order cyclic complement to a variable x_i, $i = 1, 2 \dots, n$, is

$$\overset{t_i}{\hat{x}_i} = x_i + t_i \mod (m),$$

where $t_i \in \{0, 1, 2, \dots, m-1\}$. The logic difference of a function f with respect to the t_i-order cyclic complement of the variable x_i is defined as:

$$\partial f / \partial \overset{t_i}{\hat{x}_i} = \sum_{p=0}^{m-1} r_{m-t_i,p}\ f(x_1, ..., \overset{p}{\hat{x}_i}, ..., x_n) \quad \text{over GF}(m), \tag{9.12}$$

where $r_{m-t_i,p}$ is the $(m - t_i, p)$-th element of the multivalued Reed-Muller transform matrix $R_m^{(0)}$ (Equation 9.2). It follows from Equation 9.12 that

- Logic difference reflects the change of the value of the multivalued function f with respect to t_i-th cyclic complement of the multivalued variable x_i,
- There exist $m-1$ different logic differences with respect to a given variable x_i for an m-valued logic function because there exist $m-1$ complements to x_i.
- In contrast to Boolean difference, the Equation 9.12 involves m cofactors in the sum over $GF(m)$.

Given a switching function ($m = 2$), Equation 9.12 turns to a Boolean difference

$$\partial f/\partial \overset{t_i}{\hat{x}_i} = 1 \cdot f(x_1, ..., \overset{0}{\hat{x}_i}, ..., x_n) + 1 \cdot f(x_1, ..., \overset{1}{\hat{x}_i}, ..., x_n)$$
$$= \underbrace{f(x_1, ..., x_i, ..., x_n)}_{Initial\ function} \oplus \underbrace{f(x_1, ..., \overline{x}_i, ..., x_n)}_{x_i\ replaced\ by\ \overline{x}_i} = \frac{\partial f}{\partial x_i},$$

since $\overset{1}{\hat{x}_i} = \overline{x}_i = x_i \oplus 1$, and the coefficients $r_{2-1,0} = r_{2-1,1} = 1$ are taken from the matrix

$$R_2^{(0)} = \begin{bmatrix} r_{00}\ r_{01} \\ r_{10}\ r_{11} \end{bmatrix} = \begin{bmatrix} 1\ 0 \\ 1\ 1 \end{bmatrix}.$$

Example 9.15 *Figure 9.11 illustrates changes in switching and ternary functions described by Equation 9.12. The logic differences*

$$\partial f/\partial \hat{x}_i, \quad \partial f/\partial \hat{\hat{x}}_i, \quad \partial f/\partial \hat{\hat{\hat{x}}}_i$$

correspond to the behavior of a quaternary function

$$f(\hat{x}_i), \quad f(\hat{\hat{x}}_i), \quad f(\hat{\hat{\hat{x}}}_i)$$

for $x_i \rightarrow \{\hat{x}_i,\ \hat{\hat{x}}_i,\ \hat{\hat{\hat{x}}}_i\}$.

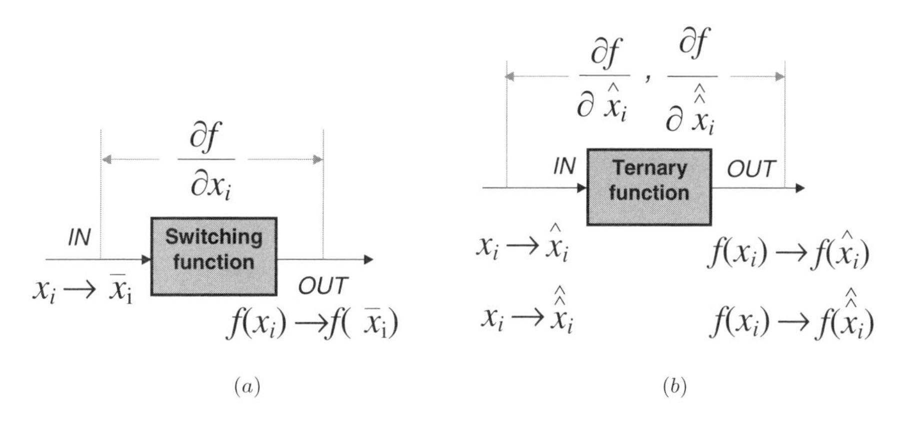

FIGURE 9.11
Logic differences of switching (a) and ternary (b) functions (Example 9.15).

Change of a switching function f (a change in the value of f) caused by a change of the variable x_i to $\overline{x}_i$ is detected by the Boolean difference. In the

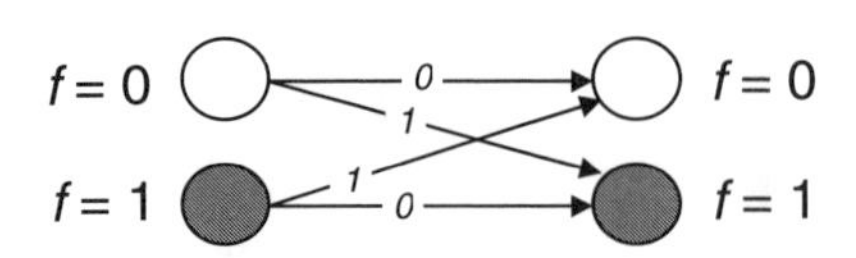

Switching function, Boolean difference

$\frac{\partial f}{\partial x_i} = f_{|x_i=0} + f_{|x_i=1}$

Change of x_i : $x_i \rightarrow \overline{x}_i$

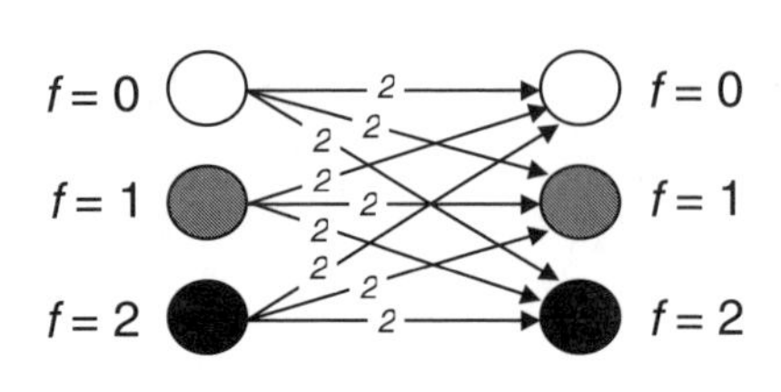

Ternary function, first-order difference

$\frac{\partial f}{\partial \hat{x}_i} = 2f_{|x_i} + 2f_{|\hat{x}_i} + 2f_{|\hat{\hat{x}}_i}$

Change of x_i : $x_i \rightarrow \hat{x}_i$

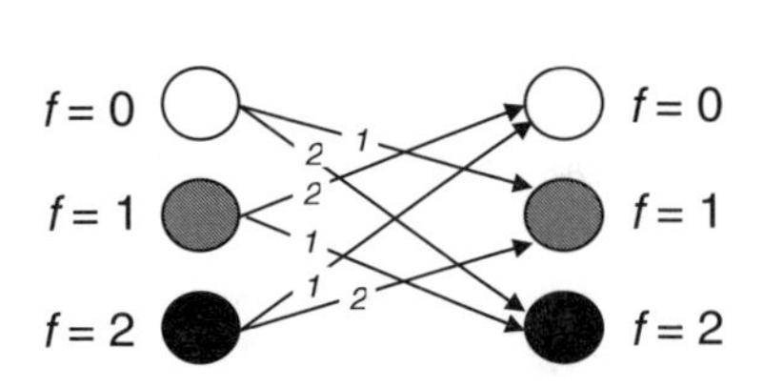

Ternary function, second-order difference

$\frac{\partial f}{\partial \hat{\hat{x}}_i} = 2f_{|\hat{x}_i} + f_{|\hat{\hat{x}}_i}$

Change of x_i : $x_i \rightarrow \hat{\hat{x}}_i$

FIGURE 9.12
Change of switching and ternary functions with respect to a variable (Example 9.16) and logic differences.

ternary logic function f, a combination difference $\partial f/\partial \hat{x}_i = 2$ and $\partial f/\partial \hat{\hat{x}}_i$ recognizes the type of change.

Example 9.16 *Two logic differences with respect to a variable x_i for a ternary system are calculated by the Equation 9.12:*

$$\partial f/\partial \hat{x}_i = \sum_{p=0}^{2-1} r_{3-1,p}\ f(x_1, ..., \overset{p}{\hat{x}}_i, ..., x_n) \ \ over \ \ GF(3),$$

$$\partial f/\partial \hat{\hat{x}}_i = \sum_{p=0}^{2-1} r_{3-2,p}\ f(x_1, ..., \overset{p}{\hat{x}}_i, ..., x_n) \ \ over \ \ GF(3).$$

Since

$$R_3^{(0)} = \begin{bmatrix} r_{00} & r_{01} & r_{02} \\ r_{10} & r_{10} & r_{10} \\ r_{20} & r_{21} & r_{22} \end{bmatrix} = \begin{bmatrix} 1 & 0 & 0 \\ 0 & 2 & 1 \\ 2 & 2 & 2 \end{bmatrix},$$

the coefficients $r_{m-t_i,p}$ *are derived as follows:*

1-order cyclic complement of a variable x_i*:* $t_i = 1$*, and we take coefficients from the last row of* $R_3^{(0)}$ $r_{3-1,0} = r_{3-1,1} = r_{3-1,2} = 2$*;*

2-order cyclic complement of a variable x_i*:* $t_i = 2$*, and we take coefficients from the middle row of* $R_3^{(0)}$ $r_{3-2,0} = 0,\ r_{3-2,1} = 2,\ r_{3-2,2} = 1$.

Figure 9.12 illustrates the changes of x_i *and* f *that are involved in calculation of the logic differences.*
Note that

$$\hat{x}_i = x_i + 1 \quad (mod\ 3)$$
$$\hat{\hat{x}}_i = x_i + 2 \quad (mod\ 3)$$

9.4.2 Computing logic difference

The matrix interpretation of the logic difference of an m-valued function f of n-variables with respect to a variable x_i with the t_i-order cyclic complement, $i = 1, 2, \ldots, n$, is given below:

$$\frac{\partial \mathbf{F}}{\partial \overset{t_i}{\hat{x}_i}} = \overset{t_i}{\hat{D}}{}^{(i)}_{m^n} \mathbf{F}, \tag{9.13}$$

where the matrix $\overset{t_i}{\hat{D}}{}^{(i)}_{m^n}$ is formed by the Kronecker product

$$\overset{t_i}{\hat{D}}{}^{(i)}_{m^n} = (m-1) I_{m^{i-1}} \otimes \Big(\sum_{p=0}^{m-1} r_{m-t_i,p}\ I_m^{(p\rightarrow)}\Big) \otimes I_{m^{n-i}}, \tag{9.14}$$

and $I_m^{(p\rightarrow)}$ is the $m \times m$ matrix generated by p-th right cyclic shift of elements of the identity matrix I_m.

Note that the denotation of matrix $\overset{t_i}{\hat{D}}{}^{(i)}_{m^n}$ carries information about

- The size of the matrix (m^n),
- The number of variables (n),
- The order of cyclic complement (t_i), and
- The variable with respect to which the difference is calculated (x_i).

Example 9.17 *Let $m = 3,\ t_i = 2$, then*

$$\sum_{p=0}^{m-1} r_{m-t_i,p}\ I_m^{(p\rightarrow)} = \sum_{p=0}^{2} r_{1,p}\ I_3^{(p\rightarrow)}$$
$$= 0 \cdot I_3^{(0\rightarrow)} + 2 \cdot I_3^{(1\rightarrow)} + 1 \cdot I_3^{(2\rightarrow)}$$
$$= 2 \cdot \begin{bmatrix} 0\,1\,0 \\ 0\,0\,1 \\ 1\,0\,0 \end{bmatrix} + 1 \cdot \begin{bmatrix} 0\,0\,1 \\ 1\,0\,0 \\ 0\,1\,0 \end{bmatrix} = \begin{bmatrix} 0\,2\,1 \\ 1\,0\,2 \\ 2\,1\,0 \end{bmatrix}.$$

Given a switching function ($m = 2$), Equation 9.13 is the Boolean differences in matrix form

$$\frac{\partial \mathbf{F}}{\partial x_i} = D_{2^n}^{(i)} \mathbf{F}, \qquad (9.15)$$

where matrix $D_{2^n}^{(i)}$ is formed by Equation 9.14

$$D_{2^n}^{(i)} = I_{2^{i-1}} \otimes D_2 \otimes I_{2^{n-i}}, \qquad D_2 = \begin{bmatrix} 1\,1 \\ 1\,1 \end{bmatrix}.$$

Example 9.18 *The structure of matrix $\overset{t_i}{\hat{D}}{}_{m^n}^{(i)}$ for the parameters below is illustrated in Figure 9.13:*

(a) *Switching function of two variables, $m = 2,\ n = 2$, logic difference with respect to the variable x_2, $i = 2$.*
(b) *Ternary function of two variables, $m = 3,\ n = 2$, one- and two-complement logic differences with respect to the variable x_2, $i = 2$.*
(c) *Quaternary function of two variables, $m = 4,\ n = 2$, one-, two-, and three-complement logic differences with respect to the variable x_2, $i = 2$.*

Example 9.19 *Given the truth-vector $\mathbf{F} = [0123112322233333]^T$ of a quaternary ($m = 4$) logic function of two variables ($n = 2$), the logic difference $\partial \mathbf{F}/\partial \hat{\hat{x}}_1$ is calculated by Equation 9.13 and Equation 9.14 as follows:*

$$\frac{\partial \mathbf{F}}{\partial \hat{\hat{x}}_1} = \hat{\hat{D}}{}_{4^2}^{(1)} \mathbf{F}$$
$$= \left(\begin{bmatrix} 0\,1\,2\,3 \\ 1\,0\,3\,2 \\ 2\,3\,0\,1 \\ 3\,2\,1\,0 \end{bmatrix} \otimes \begin{bmatrix} 1 & & & \\ & 1 & & \\ & & 1 & \\ & & & 1 \end{bmatrix} \right) \mathbf{F}$$

BINARY MATRIX

$$D_{2^2}^{(1)} = \begin{bmatrix} 1 & 1 & & \\ 1 & 1 & & \\ & & 1 & 1 \\ & & 1 & 1 \end{bmatrix}$$

TERNARY MATRICES

$$\hat{D}_{3^2}^{(1)} = \begin{bmatrix} 1&1&1&&&&&& \\ 1&1&1&&&&&& \\ 1&1&1&&&&&& \\ &&&1&1&1&&& \\ &&&1&1&1&&& \\ &&&1&1&1&&& \\ &&&&&&1&1&1 \\ &&&&&&1&1&1 \\ &&&&&&1&1&1 \end{bmatrix} \qquad \hat{\hat{D}}_{3^2}^{(1)} = \begin{bmatrix} 1&1&1&&&&&& \\ 1&1&1&&&&&& \\ 1&1&1&&&&&& \\ &&&1&1&1&&& \\ &&&1&1&1&&& \\ &&&1&1&1&&& \\ &&&&&&1&1&1 \\ &&&&&&1&1&1 \\ &&&&&&1&1&1 \end{bmatrix}$$

QUATERNARY MATRICES

$$\hat{D}_{4^2}^{(2)} = \begin{bmatrix} 1&1&1&1&&&&&&&&&&&& \\ 1&1&1&1&&&&&&&&&&&& \\ 1&1&1&1&&&&&&&&&&&& \\ 1&1&1&1&&&&&&&&&&&& \\ &&&&1&1&1&1&&&&&&&& \\ &&&&1&1&1&1&&&&&&&& \\ &&&&1&1&1&1&&&&&&&& \\ &&&&1&1&1&1&&&&&&&& \\ &&&&&&&&1&1&1&1&&&& \\ &&&&&&&&1&1&1&1&&&& \\ &&&&&&&&1&1&1&1&&&& \\ &&&&&&&&1&1&1&1&&&& \\ &&&&&&&&&&&&1&1&1&1 \\ &&&&&&&&&&&&1&1&1&1 \\ &&&&&&&&&&&&1&1&1&1 \\ &&&&&&&&&&&&1&1&1&1 \end{bmatrix} \qquad \hat{\hat{D}}_{4^2}^{(2)} = \begin{bmatrix} 0&1&2&3&&&&&&&&&&&& \\ 1&0&3&2&&&&&&&&&&&& \\ 2&3&0&1&&&&&&&&&&&& \\ 3&2&1&0&&&&&&&&&&&& \\ &&&&0&1&2&3&&&&&&&& \\ &&&&1&0&3&2&&&&&&&& \\ &&&&2&3&0&1&&&&&&&& \\ &&&&3&2&1&0&&&&&&&& \\ &&&&&&&&0&1&2&3&&&& \\ &&&&&&&&1&0&3&2&&&& \\ &&&&&&&&2&3&0&1&&&& \\ &&&&&&&&3&2&1&0&&&& \\ &&&&&&&&&&&&0&1&2&3 \\ &&&&&&&&&&&&1&0&3&2 \\ &&&&&&&&&&&&2&3&0&1 \\ &&&&&&&&&&&&3&2&1&0 \end{bmatrix}$$

$$\hat{\hat{\hat{D}}}_{4^2}^{(2)} = \begin{bmatrix} 0&1&3&2&&&&&&&&&&&& \\ 1&0&2&3&&&&&&&&&&&& \\ 3&2&0&1&&&&&&&&&&&& \\ 2&3&1&0&&&&&&&&&&&& \\ &&&&0&1&3&2&&&&&&&& \\ &&&&1&0&2&3&&&&&&&& \\ &&&&3&2&0&1&&&&&&&& \\ &&&&2&3&1&0&&&&&&&& \\ &&&&&&&&0&1&3&2&&&& \\ &&&&&&&&1&0&2&3&&&& \\ &&&&&&&&3&2&0&1&&&& \\ &&&&&&&&2&3&1&0&&&& \\ &&&&&&&&&&&&0&1&3&2 \\ &&&&&&&&&&&&1&0&2&3 \\ &&&&&&&&&&&&3&2&0&1 \\ &&&&&&&&&&&&2&3&1&0 \end{bmatrix}$$

FIGURE 9.13
Logic difference matrices with respect to variable x_2 for switching, ternary and quaternary functions of two variables (Example 9.18).

$$= \begin{bmatrix} &&&&1&&&&2&&&&3&&& \\ &&&&&1&&&&2&&&&3&& \\ &&&&&&1&&&&2&&&&3& \\ &&&&&&&1&&&&2&&&&3 \\ 1&&&&&&&&3&&&&2&&& \\ &1&&&&&&&&3&&&&2&& \\ &&1&&&&&&&&3&&&&2& \\ &&&1&&&&&&&&3&&&&2 \\ 2&&&&3&&&&&&&&1&&& \\ &2&&&&3&&&&&&&&1&& \\ &&2&&&&3&&&&&&&&1& \\ &&&2&&&&3&&&&&&&&1 \\ 3&&&&2&&&&1&&&&&&& \\ &3&&&&2&&&&1&&&&&& \\ &&3&&&&2&&&&1&&&&& \\ &&&3&&&&2&&&&1&&&& \end{bmatrix} \begin{bmatrix} 0\\1\\2\\3\\1\\1\\2\\3\\2\\2\\2\\3\\3\\3\\3\\3 \end{bmatrix} = \begin{bmatrix} 0\\0\\3\\0\\0\\1\\2\\0\\0\\2\\1\\0\\0\\3\\0\\0 \end{bmatrix}$$

9.5 Generation of Reed-Muller expressions

Reed-Muller representations of switching functions possess the following virtues:

- Reed-Muller expressions are associated with analysis of switching functions in terms of change through logic Taylor expansion,
- The corresponding Reed-Muller decision tree and diagram provide a useful opportunity for detailed analysis of switching functions, including switching activity,
- The decision tree embedded into hypercube-like structure allows word-wise computation and manipulation of Reed-Muller expressions of various polarities,
- The cost of implementation using Reed-Muller expression is often less compared to sum-of-products expressions, and
- Reed-Muller expression can be efficiently computed using matrix transforms and, thus, the calculations are mapped onto massive parallel tools.

These attractive features of Reed-Muller expression apply to multivalued functions. The relationship between Reed-Muller expressions and generalized logic Taylor expansion, which is important for analysis of multivalued functions, is presented below.

9.5.1 Logic Taylor expansion of a multivalued function

The logic analog of the Taylor series for an m-valued function f of n variables at the point $c \in 0, 1, \ldots, m^n - 1$, is defined as

$$f = \sum_{i=0}^{m^n-1} r_i^{(c)} \underbrace{(x_1 \oplus c_1)^{i_1} \ldots (x_n \oplus c_n)^{i_n}}_{i-th\ term} \quad \bmod (m). \tag{9.16}$$

In this expression:

- m is a prime number;
- $c_1 c_2 \ldots c_n$ (polarity) and $i_1 i_2 \ldots i_n$ is the m-valued representation of c and i correspondingly;
- $r_i^{(c)}$ is the i-th coefficient, the value of the multiple (n-ordered) logic difference at the point $d = m - c$

$$r_i^{(c)} = \left. \frac{\partial^n f(d)}{\partial \overset{m-i_1}{\hat{x}_1} \partial \overset{m-i_2}{\hat{x}_2} \ldots \partial \overset{m-i_n}{\hat{x}_n}} \right|_{d=m-c} \tag{9.17}$$

▶ $\partial \overset{m-i_j}{\hat{x}_i}$ indicates with respect to which variables the multiple logic difference is calculated, and is defined by

$$\partial \overset{m-i_j}{\hat{x}_i} = \begin{cases} 1, & m = i_j, \\ \partial \overset{m-i_j}{\hat{x}_j}, & m \neq i_j \end{cases} \tag{9.18}$$

9.5.2 Computing Reed-Muller expressions

It follows from Equation 9.16 that:

- ▶ Logic Taylor expansion produces m^n Reed-Muller expressions that correspond to m^n polarities. In spectral interpretation this means that each of these expressions is a spectrum of the m-valued function at one of m^n polarities.
- ▶ A variable x_j is 0-polarized if it enters into the expansion uncomplemented, and c_j-polarized otherwise.
- ▶ The coefficients in the logic Taylor series are logic differences.

While the i-th coefficient r_i is described by a logical expression, it can be calculated in different ways, for example, by matrix transformations, cube-based technique, decision diagram technique, and probabilistic methods. It is possible to calculate separate coefficients or their arbitrary sets using logic differences.

Example 9.20 *By Equation 9.16, the Reed-Muller expression of an arbitrary ternary ($m = 3$) function of two variables ($n = 2$) and the 7-th polarity $c = 7$, $c_1, c_2 = 2, 1$, is defined as a logic Taylor expansion of this function (Figure 9.14).*

Example 9.21 *(Continuation of Example 9.20) Consider the function $f = MAX(x_1, x_2)$. The values of the first three coefficients at the point $c = 7$ are given in Figure 9.15. The other differences can be calculated in a similar way. Finally, the vector of coefficient is $\mathbf{R} = [200012201]^T$, that yields*

$$f = 2 + \hat{\hat{x}}_1\hat{x}_2 + 2\hat{\hat{x}}_1\hat{x}_2{}^2 + 2\hat{\hat{x}}_2{}^2 + \hat{\hat{x}}_2{}^2\hat{x}_2{}^2.$$

9.5.3 Computing Reed-Muller expressions in matrix form

The logic Taylor expansion consists of n logic differences with respect to each variable and $m^n - n - 1$ multiple logic differences.

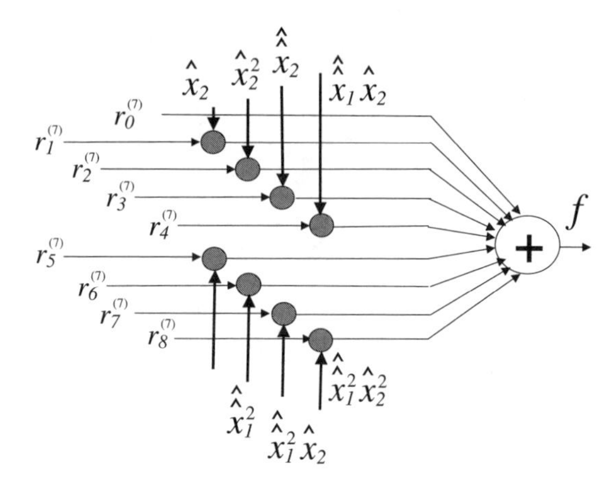

Step 1. *Apply Equation 9.16 for* $m = 2,\ n = 2$:

$$f = \sum_{i=0}^{3^2-1} r_1^{(7)} \, (x_1 \oplus 2)^{i_1} \, (x_2 \oplus 1)^{i_2} = \sum_{i=0}^{8} r_i^{(7)} \, \hat{x}_1^{i_1} \, \hat{x}_1^{i_2}$$

Step 2. *Reed-Muller expression:*

$$f = r_0^{(7)} + r_1^{(7)} \, \hat{x}_2 + r_2^{(7)} \, \hat{x}_2^2 + r3^{(7)} \, \hat{\hat{x}}_2 + r_4^{(7)} \, \hat{\hat{x}}_1 \hat{x}_2 + r_5^{(7)} \, \hat{x}_1 \hat{x}_2^2 + r_6^{(7)} \, \hat{\hat{x}}_1{}^2$$
$$+ r_7^{(7)} \, \hat{\hat{x}}_1{}^2 \hat{x}_2 + r_8^{(7)} \, \hat{\hat{x}}_1{}^2 \hat{x}_2^2$$

Step 3. *Logic derivatives*

$$r_i^{(c)} = \frac{\partial^2 f(7)}{\overset{3-i_1}{\partial \, \hat{x}_1} \; \overset{3-i_2}{\partial \, \hat{x}_2}}$$

$$\overset{3-i_j}{\partial \, \hat{x}_i} = \begin{cases} 1, & 3 = i_j \\ \overset{3-i_j}{\partial \, \hat{x}_j}, & 3 \neq i_j \end{cases}$$

$r_1 = \partial f(7)/\partial \hat{\hat{x}}_2$	$r_5 = \partial^2 f(7)/\partial \hat{\hat{x}}_1 \partial \hat{x}_2$
$r_2 = \partial f(7)/\partial \hat{x}_2$	$r_6 = \partial f(7)/\partial \hat{x}_2$
$r_3 = \partial^2 f(7)/\partial \hat{x}_1 \partial \hat{x}_2$	$r_7 = \partial^2 r(7)/\partial \hat{x}_1 \partial \hat{\hat{x}}_2$
$r_4 = \partial^2 f(7)/\partial \hat{x}_1 \partial \hat{\hat{x}}_2$	$r_8 = \partial^2 r(7)/\partial \hat{x}_1 \partial \hat{x}_2$

FIGURE 9.14
Constructing the logic difference for a logic Taylor expansion of an arbitrary ternary ($m = 3$) function of two ($n = 2$) variables for polarity $c = 7$ (Example 9.20).

9.5.4 $\mathcal{N}$-hypercube representation

Let us utilize Davio decision tree and hypercube-like structure, which implements positive Davio expansion in the nodes (Table 9.7), to compute Boolean differences. The positive Davio expansion is given in the form

$$f = f_{|x=0} + x \cdot (f_{|x=1} + 3f_{|x=2} + 2f_{|x=3})$$
$$+ x^2 \cdot (f_{|x=1} + 2f_{|x=2} + 3f_{|x=3}) + x^3 \cdot (f_{|x=0} + f_{|x=1} + f_{|x=2} + f_{|x=3}).$$

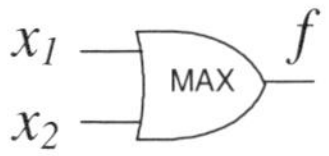

x_1, x_2 → MAX → f

	0	1	2
0	0	0	0
1	0	1	2
2	0	2	1

$f = \mathrm{MAX}(x_1, x_2)$

Reed-Muller coefficients (logic differences)

$$
\begin{aligned}
r_0 &= f(7) = f(2,1) = 2,\\
r_1 &= \frac{\partial f(7)}{\partial \hat{\hat{x}}_2}\\
&= 2f(x_1, \hat{x}_2) + f(x_1, \hat{\hat{x}}_2)\\
&= 2f(2, \hat{1}) + f(2, \hat{\hat{1}})\\
&= 2f(2,2) + f(2,0) = 2 \cdot 2 + 2 = 0 \pmod 3,\\
r_2 &= \frac{\partial f(7)}{\partial \hat{x}_2}\\
&= 2f(x_1, x_2) + 2f(x_1, \hat{x}_2) + 2f(x_1, \hat{\hat{x}}_2)\\
&= 2f(2,1) + 2f(2, \hat{1}) + 2f(2, \hat{\hat{1}})\\
&= 2f(2,1) + 2f(2,2) + 2f(2,0)\\
&= 2 \cdot 2 + 2 \cdot 2 + 2 \cdot 2 = 0 \ \ (mod\ 3)
\end{aligned}
$$

FIGURE 9.15
Taylor expansion of the ternary ($m = 3$) function MAX of two ($n = 2$) variables for polarity $c = 7$ (Example 9.21).

Figure 9.16 illustrates the computing of logic differences by different data structures: decision tree and hypercube-like structure.

It follows from this form that:

- ▶ Branches of the Davio decision tree carry information about logic differences;
- ▶ Terminal nodes are the values of logic differences for corresponding variable assignments;
- ▶ Computing of Reed-Muller coefficients can be implemented on the Davio tree as a data structure;
- ▶ The Davio decision tree includes values of all single and multiple logic differences given a variable assignment $x_1 x_2 ... x_n = 00 \ldots 0$. This assignment corresponds to calculation of Reed-Muller expansion of polarity 0, so in the Davio tree, positive Davio expansion is implemented at each node;
- ▶ Representation of a logic function in terms of change is a unique representation; it means that the corresponding decision diagram is canonical;
- ▶ The values of terminal nodes correspond to coefficients of logic Taylor expansion.

The Davio tree can be embedded into hypercube-like structure, and the above mentioned properties are valid for that data structure as well.

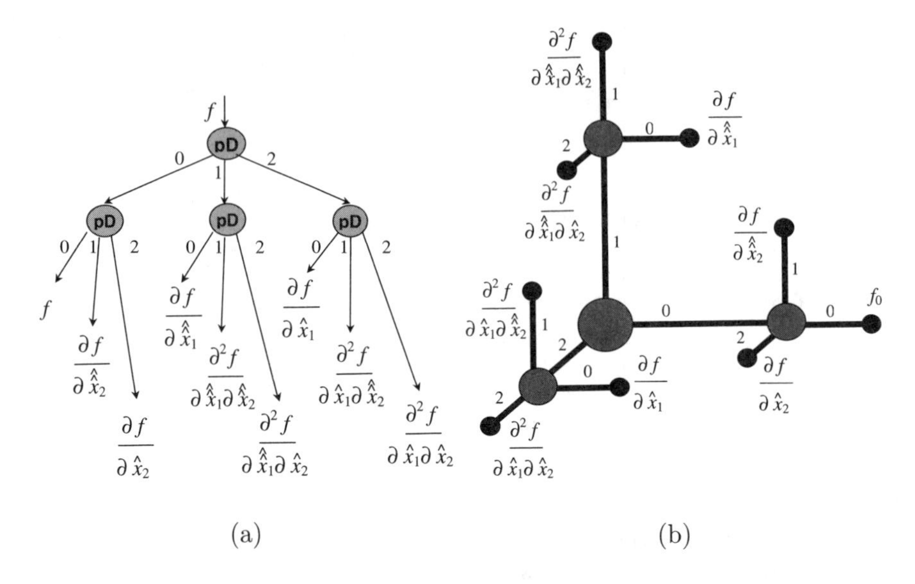

FIGURE 9.16
Computing logic differences by Davio decision tree (a) and hypercube-like structure for ternary logic function of two variables.

9.6 Linear word-level expressions of multivalued functions

In this section, generalization of linear word-level data structures toward multivalued functions is considered. In a similar manner to binary functions, linear word-level expressions and decision diagrams are distinguished by their type of decomposition (expansion). There exist three linear word-level forms for multivalued valued functions:

- Linear word-level arithmetic expressions;
- Linear word-level Reed-Muller (modulo m) expressions; and
- Linear word-level sum-of-products expressions.

The last two forms are considered in the next sections. The focus of this section is an approach to representation of m-valued function of n variables by the linear word-level expression

$$f = d_0 + d_1 x_1^{\circ} + d_2 x_2^{\circ} + \cdots + d_n x_n^{\circ}. \tag{9.19}$$

An arbitrary multivalued function can be represented in linear form (Equation 9.19). However, in this section, the library of linear models (linear expressions, decision diagrams, and hypercube-like structures) includes elementary functions only. Then, the different techniques can be used to design an arbitrary multivalued circuit over this library of gates.

The main reason for introducing this approach to linearization of multivalued functions is that linear word-level expressions can be represented by linear word-level diagrams that are

- ▶ Easy to embed in hypercube-like structures, and
- ▶ Are intrinsically parallel and can be calculated by massive parallel arrays.

9.6.1 Approach to linearization

The approach to linearization includes the following phases (Figure 9.17):

Phase 1. Partitioning of the truth vector $\mathbf{F}$ of the m-valued function f of n m-valued variables to a set of subvectors $\mathbf{F}_j^{\circ}$,

Phase 2. Encoding the multivalued variables x_i. The new, binary variables x_i° are called *pseudo-variables*, and

Phase 3. Representation of the multivalued function f by the linear word-level arithmetic expression that depends on binary pseudo-variables x_i°.

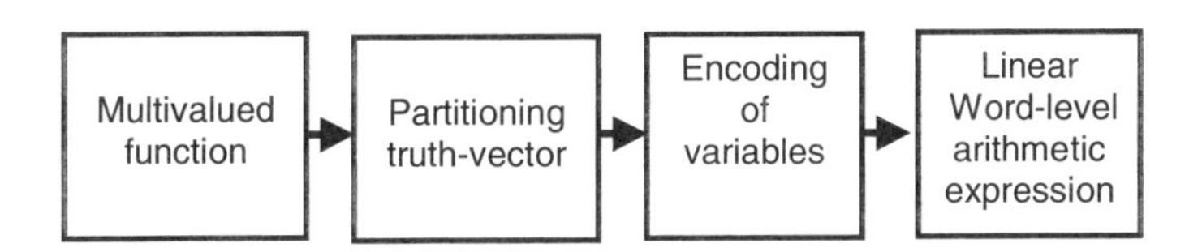

FIGURE 9.17
The main phases of an algorithm to represent a multivalued function (circuit) by a linear word-level expression.

9.6.2 Algorithm for linearization of multivalued functions

Phase 1: Partition.

Given the truth vector $\mathbf{F}$ of an m-valued n-variable logic function f. Let us partition this vector into τ subvectors,

$$\tau = \left\lceil \frac{m^n}{n+1} \right\rceil, \tag{9.20}$$

where $\lceil a \rceil$ denotes the least integer greater than or equal to a. The order of the partition is fixed (with respect to assignments of variables). The index

μ of subvector $\mathbf{F}_\mu$ that contains the i-th element of the initial truth table is equal to

$$\mu = \left\lfloor \frac{i}{n+1} \right\rfloor, \tag{9.21}$$

where $\lfloor a \rfloor$ is the greatest integer less than or equal to a.

Example 9.22 *Partitioning the truth vector* $\mathbf{F}$ *of lengh* $3^3 = 27$ *of a ternary* $(m = 3)$ *function of three variables* $(n = 3)$ *is illustrated in Figure 9.18. The location of the 20-th element of the truth vector* $\mathbf{F}$ *is determined by the index* $\mu = \left\lfloor \frac{20}{3+1} \right\rfloor = 5$ *of subvector* $\mathbf{F}_\mu$*. This element belongs to subvector* $\mathbf{F}_5$*.*

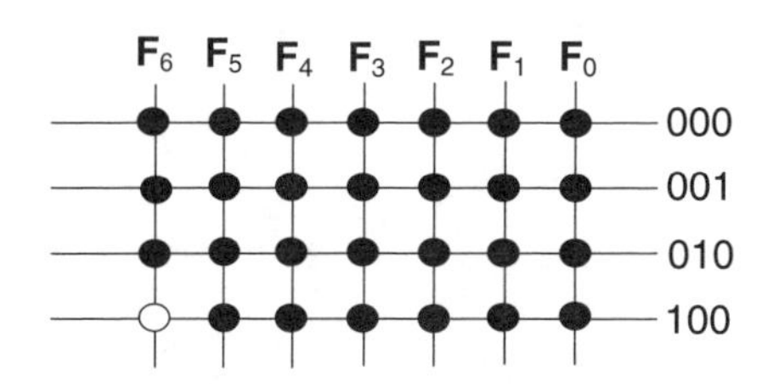

The vector $\mathbf{F}$ *is partitioned to*

$$\tau = \left\lceil \frac{3^3}{3+1} \right\rceil = \left\lceil \frac{27}{4} \right\rceil = \lceil 6.7 \rceil = 7$$

subvectors $\mathbf{F}_0, \mathbf{F}_1, \ldots, \mathbf{F}_6$
The 21-th element is located in the truth-vector $\mathbf{F}_\mu$*,*

$$\mu = \left\lfloor \frac{21}{3+1} \right\rfloor = 5$$

FIGURE 9.18
Representation of truth-vector of a multivalued function by 2-D data structure (Example 9.22).

Phase 2: Encoding.

Consider the μ-th subvector $\mathbf{F}_\mu$, $\mu = 0, 1, \ldots, \tau - 1$. The length of the subvector $\mathbf{F}_\mu$ is $n + 1$. Hence, the i-th element is allocated in the subvector $\mathbf{F}_\mu$. Its position inside μ is specified by the index $j = Res\left(\frac{i}{n+1}\right) = 5$. The assignments of n variables $x_1^\circ, x_2^\circ, \ldots, x_n^\circ$ in $\mathbf{F}_\mu$ are called *pseudo-variables.* The pseudo-variables are the *binary* variables valid for assignments with at most one 1.

Example 9.23 *Assignments of pseudo-variables of a three-valued function of two, three, and four variables are given below:*

(a) $x_1^\circ x_2^\circ = \{00, 01, 10\}$*; given* $i = 1$, $\mu = \left\lfloor \frac{1}{2+1} \right\rfloor = 0$*;*

(b) $x_1^\circ x_2^\circ x_3^\circ = \{000, 001, 010, 100\}$*; given* $i = 20$, $\mu = \left\lfloor \frac{20}{3+1} \right\rfloor = 5$*;*

(c) $x_1^\circ x_2^\circ x_3^\circ x_4^\circ = \{0000, 0001, 0010, 0100, 1000\}$*; given* $i = 10$, $\mu = \left\lfloor \frac{10}{4+1} \right\rfloor = 2.$

Phase 3: Representation of a function by linear word-level arithmetic expression.

This phase consists of two steps:

Step 1: Forming a word-level vector $\mathbf{F}^\circ$ from subvectors $\mathbf{F}_1^\circ$, $\mathbf{F}_2^\circ$, ... $\mathbf{F}_\tau^\circ$, and

Step 2: Truncated arithmetic transform of vector $\mathbf{F}^\circ$.

Let $\mathbf{W} = \left[m^{\tau-1}\ m^{\tau-2} \cdots m^1\ m^0\right]^T$ be the weight vector. A truth vector $\mathbf{F}^\circ$ of a function f of n pseudo-variables $x_1^\circ, \ldots, x_n^\circ$ includes $n+1$ elements and is calculated by

$$\mathbf{F}^\circ = [\mathbf{F}_{\tau-1}| \ldots |\mathbf{F}_1|\mathbf{F}_0]\mathbf{W}, \tag{9.22}$$

The truncated transform of $\mathbf{F}^\circ$ yields the vector of arithmetic coefficients $\mathbf{D}$. The relationship between the $\mathbf{F}^\circ$ and vector of coefficients $\mathbf{D} = [d_0 d_1 \ldots d_n]$ is defined by the pair of transforms

$$\mathbf{D} = \mathbf{T}_{n+1} \cdot \mathbf{F}^\circ, \tag{9.23}$$

$$\mathbf{F}^\circ = \mathbf{T}_{n+1}^{-1} \cdot \mathbf{D}, \tag{9.24}$$

where $(n+1) \times (n+1)$ direct $\mathbf{T}_{n+1}$ and inverse $\mathbf{T}_{n+1}^{-1}$ truncated arithmetic transform matrices are formed by truncation of $2^n \times 2^n$ arithmetic transform matrices P_{2^n} and $P_{2^n}^{-1}$ respectively. The truncated rule is as follows:

(a) Remove all rows that contain more than one 1;
(b) Remove the remaining columns that consist of all 0s.

The vector of coefficients $\mathbf{D}$ yields the linear word-level arithmetic expression

$$D = d_0 + d_1 x_1^\circ + d_2 x_2^\circ + \cdots + d_n x_n^\circ$$

Example 9.24 *Given a two-variable three-valued function, the* 3×3 *direct and inverse arithmetic truncated matrices are equal to*

$$\mathbf{T}_3 = \begin{bmatrix} 1 & 0 & 0 \\ -1 & 1 & 0 \\ -1 & 0 & 1 \end{bmatrix}, \qquad \mathbf{T}_3^{-1} = \begin{bmatrix} 1 & 0 & 0 \\ 1 & 1 & 0 \\ 1 & 0 & 1 \end{bmatrix}.$$

Example 9.25 *(Continuation of Example 9.24) Arithmetic expressions for subvectors* $\mathbf{D_0}$, $\mathbf{D_1}$ *and* $\mathbf{D_2}$ *in Table 9.9 are calculated by the direct truncated transform (Equation 9.23). For example,* $\mathbf{D_1}$ *is calculated as follows:*

$$\mathbf{D_1} = \mathbf{T_3}\mathbf{F}_1 = \begin{bmatrix} 1 & 0 & 0 \\ -1 & 1 & 0 \\ -1 & 1 & 1 \end{bmatrix} \begin{bmatrix} 1 \\ 1 \\ 2 \end{bmatrix} = \begin{bmatrix} 1 \\ 0 \\ 1 \end{bmatrix},$$

that yields the algebraic form $d_1 = 1 + x_1^\circ$.

Example 9.26 *(Continuation of Example 9.25) The truth vector* $\mathbf{F}^\circ$ *of the two-input* $MAX(x_1, x_2)$ *function is calculated as shown in Figure 9.19. The* *direct truncated transform (Equation 9.23) is used for transformation. The final result is the linear expression*

$$
\begin{aligned}
D &= 3^2 D_2 + 3^1 D_1 + 3^1 D_0 \\
&= 21 + 5x_1^\circ + x_2^\circ
\end{aligned}
$$

TABLE 9.9
Partitioning of the truth vector **F** and deriving the linear word-level arithmetic expression for a ternary $MAX(x_1, x_2)$ function.

Function MAX			Linear model		
x_1x_2	$\mathbf{F}$	$\mathbf{F}_\mu$	$x_1^\circ x_2^\circ$	$\mathbf{D}_\mu$	D_μ
00 01 02	0 1 2	$\mathbf{F}_0 = \begin{bmatrix} 0 \\ 1 \\ 2 \end{bmatrix}$	00 01 10	$\mathbf{D}_0 = \begin{bmatrix} 0 \\ 1 \\ 2 \end{bmatrix}$;	$D_0 = x_2^\circ + 2x_1^\circ$
10 11 12	1 1 2	$\mathbf{F}_1 = \begin{bmatrix} 1 \\ 1 \\ 2 \end{bmatrix}$	00 01 10	$\mathbf{D}_1 = \begin{bmatrix} 1 \\ 0 \\ 1 \end{bmatrix}$;	$D_1 = 1 + x_1^\circ$
20 21 22	2 2 2	$\mathbf{F}_2 = \begin{bmatrix} 2 \\ 2 \\ 2 \end{bmatrix}$	00 01 10	$\mathbf{D}_2 = \begin{bmatrix} 2 \\ 0 \\ 0 \end{bmatrix}$;	$D_2 = 2$

9.6.3 Manipulation of the linear model

A linear word-level expression of elementary multivalued functions is a form of representation and computation due to the following properties:

- ▶ It is convertible to the initial function by way of an operator (a control parameter of the linear model);
- ▶ It is an intrinsically parallel model because it is at word-level; and
- ▶ It is extendable to arbitrary logic functions.

The example below illustrates some of these properties by calculation of the function using the linear model given the input assignments. Let a three-valued ($m = 3$) two-input ($n = 2$) elementary logic function be given by linear expression D. The masking operation

$$f = \Xi^\mu\{D\} \tag{9.25}$$

is used to recover the value of the logic function.

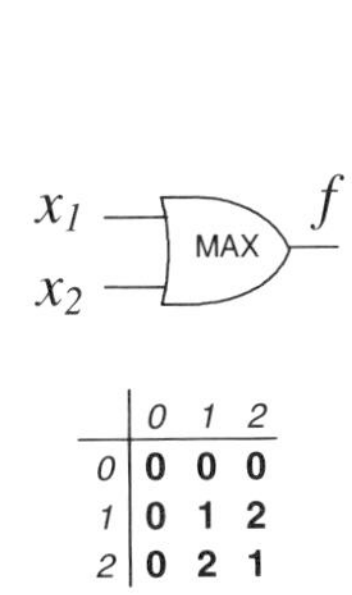

	0	1	2
0	**0**	**0**	**0**
1	**0**	**1**	**2**
2	**0**	**2**	**1**

$f = MAX(x_1, x_2)$

The truth table is partitioned to

$\tau = \lceil 3^2/(2+1) \rceil = 3$ vectors

Truth vector $\mathbf{F}^{\circ}$:

$$\mathbf{F}^{\circ} = [\mathbf{F}_2|\mathbf{F}_1|\mathbf{F}_0]\mathbf{W} = \begin{bmatrix} 2\ 1\ 0 \\ 2\ 1\ 1 \\ 2\ 2\ 2 \end{bmatrix} \begin{bmatrix} 3^2 \\ 3^1 \\ 3^0 \end{bmatrix} = \begin{bmatrix} 21 \\ 22 \\ 26 \end{bmatrix}$$

Vector of coefficients:

$$\mathbf{D} = \mathbf{T}_3 \cdot \mathbf{F}^{\circ} = \begin{bmatrix} 1\ 0\ 0 \\ -1\ 1\ 0 \\ -1\ 0\ 1 \end{bmatrix} \begin{bmatrix} 21 \\ 22 \\ 26 \end{bmatrix} = \begin{bmatrix} 21 \\ 1 \\ 5 \end{bmatrix}$$

Linear expression:

$$D = 21 + 5x_1^{\circ} + x_2^{\circ}, \quad x_1^{\circ}, x_2^{\circ} \in \{0, 1\}$$

Calculation of $f = MAX(2, 1)$, $\mu = 2$:

$$x_1 = 2 \rightarrow x_1^{\circ} = 0$$
$$x_2 = 1 \rightarrow x_2^{\circ} = 1$$

$$MAX(2,1) = \Xi^2\{21 + 5x_1^{\circ} + x_2^{\circ}\}$$
$$= \Xi^2\{22\}$$
$$= \Xi^2\{211_3\} = 2$$

FIGURE 9.19
Representation of the 3-valued 2-variable logic function $f = MAX(x_1, x_2)$ by a linear word-level arithmetic expression (Examples 9.26 and 9.27).

Example 9.27 *(Continuation of Example 9.26.) Calculation of values of* $MAX(x_1, x_2)$ *given the linear model and* $x_1 = 2$, $x_2 = 1$ *involves several steps (Figure 9.19):*

(a) *Find the index* μ *of subvector* $\mathbf{F}_{\mu}$ *in a word-level representation. Here, the parameter* μ *is determined as follows: assignment* $x_1x_2 = 21$ *corresponds to the 7-th element of the truth-vector* $\mathbf{F}$*; hence* $\mu = \lfloor 7/3 \rfloor = 2$.

(b) *Use the encoding rule given in Table 9.9:* $x_1x_2 = 21 \rightarrow x_1^{\circ}, x_2^{\circ} = 01$.

(c) *Calculate the value of* $MAX(2,1)$ *for the assignment of pseudo-variables* $x_1^{\circ} = 0, x_2^{\circ} = 1$*:* $MAX(2,1) = D_2(0,1) = 2$.

9.6.4 Library of linear models of multivalued gates

Table 9.10 contains the linear arithmetic expressions of various ternary gates from a library of gates. The linear models from Table 9.10 can be extended to an arbitrary logic function.

Example 9.28 *The ternary function* $\overline{x_1 + x_2}$ *can be represented by a linear*

TABLE 9.10
Library of linear word-level arithmetic models of three-valued gates.

Function			Vector of coefficients
$\overline{x}$	$=$	$2-x$	$\mathbf{D}=[2\ -1]^{\mathrm{T}}$
$x_1 \cdot x_2 \pmod 3$	$=$	$15x_1^{\circ}+21x_2^{\circ}$	$\mathbf{D}=[0\ 21\ 15]^{\mathrm{T}}$
$MIN(x_1,x_2)$	$=$	$21x_1^{\circ}+12x_2^{\circ}$	$\mathbf{D}=[0\ 12\ 21]^{\mathrm{T}}$
$TSUM(x_1,x_2)$	$=$	$21+5x_1^{\circ}+4x_2^{\circ}$	$\mathbf{D}=[21\ 4\ 5]^{\mathrm{T}}$
$MAX(x_1,x_2)$	$=$	$21+5x_1^{\circ}+x_2^{\circ}$	$\mathbf{D}=[21\ 1\ 5]^{\mathrm{T}}$
$TPROD(x_1,x_2)$	$=$	$21x_1^{\circ}+9x_2^{\circ}$	$\mathbf{D}=[0\ 9\ 21]^{\mathrm{T}}$
$(x_1+x_2) \pmod 3$	$=$	$21-10x_1^{\circ}-14x_2^{\circ}$	$\mathbf{D}=[21\ -14\ -10]^{\mathrm{T}}$
$x_1 \vert x_2$	$=$	$1-x_1^{\circ}+5x_2^{\circ}$	$\mathbf{D}=[1\ 5\ -1]^{\mathrm{T}}$

expression as follows

$$\begin{aligned}\overline{x_1+x_2} \pmod 3 &= \Xi^{\mu}\{3^0(2-x_2^{\circ}-2x_1^{\circ}) \\ &+ 3^1(1-x_2^{\circ}+x_1^{\circ}) \\ &+ 3^2(2x_2^{\circ}+x_1^{\circ})\} \\ &= \Xi^{\mu}\{5+14x_2^{\circ}+10x_1^{\circ}\}.\end{aligned}$$

9.6.5 Representation of a multilevel, multivalued circuit

Let D be a level of multivalued, multilevel circuit and consist of r two-input multivalued gates. The level implements n-input r-output logic function, or subcircuit over the library of gates. Since each gate is described by a linear arithmetic expression, this subcircuit can be described by a linear expression too. The strategy for representation of a multivalued logic circuit by a set of linear expressions is as follows:

$$\texttt{Gate model } D_j \Longleftrightarrow f = \Xi^{\mu}\{D\}$$

$$\texttt{Circuit level model D} \Longleftrightarrow f_j = \Xi^{3(j-1)+\mu}\{L\}$$

$$\texttt{Circuit model (set of D)} \Longleftrightarrow \texttt{ Set of level outputs}$$

To simplify the formal notation, let us consider the library of ternary gates given in Table 9.10.

Let D_j, $j=1,2,\ldots,r$, be a linear arithmetic representation of the j-th gate and its output corresponds to the j-output of a subcircuit. Assume that the order of gates in the subcircuit is fixed. A linear word-level arithmetic of an n-input r-output of a ternary subcircuit (level) is defined as

$$D = \sum_{j=1}^{r} 3^{3(j-1)} D_j. \tag{9.26}$$

Example 9.29 *Let a level of a ternary circuit be given as shown in Figure 9.20. This figure explains the calculation of the linear expression using Equation 9.26.*

To calculate the value of the j-th output f_j, $j \in \{1, \ldots, r\}$, a masking operator is utilized:

$$f_j = \Xi^{\xi}\{D\}, \tag{9.27}$$

where $\xi = 3(j-1) + \mu$. This recovers the ξ-th digit in a word-level value D.

Example 9.30 *(Continuation of Example 9.29.) Given assignment*

$$x_1x_2x_3x_4x_5x_6 = 201112 \Longrightarrow x_1^{\circ}x_2^{\circ}x_3^{\circ}x_4^{\circ}x_5^{\circ}x_6^{\circ} = 000110,$$

the outputs f_j, $j \in \{1,2,3\}$, are calculated as follows: $f_1 = 2$, $f_2 = 1$, and $f_3 = 2$.

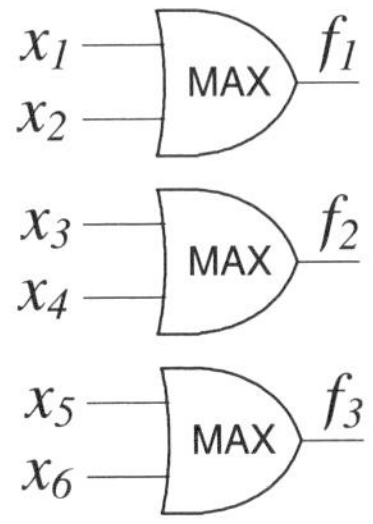

x_1, x_2 → MAX → f_1

x_3, x_4 → MAX → f_2

x_5, x_6 → MAX → f_3

$f_1 \rightarrow D_1$
$f_2 \rightarrow D_2$
$f_3 \rightarrow D_3$
$D_1 = 21 + 5x_1^{\circ} + x_2^{\circ}$
$D_2 = 21 + 5x_3^{\circ} + x_4^{\circ}$
$D_3 = 21 + 5x_5^{\circ} + x_6^{\circ}$

$$\begin{aligned} D &= \sum_{j=1}^{3} 3^{3(j-1)} D_j = 3^0 D_1 + 3^1 D_2 + 3^2 D_3 \\ &= 3^0(21 + 5x_1^{\circ} + x_2^{\circ}) \\ &+ 3^3(21 + 5x_3^{\circ} + x_4^{\circ}) \\ &+ 3^6(21 + 5x_5^{\circ} + x_6^{\circ}) \\ &= 15897 + 5x_1^{\circ} + x_2^{\circ} + 135x_3^{\circ} + 27x_4^{\circ} + 3645x_5^{\circ} + 729x_6^{\circ}. \end{aligned}$$

The relationship of the assignments of variables and pseudo-variables:

$$x_1x_2x_3x_4x_5x_6 = 201112 \Longrightarrow x_1^{\circ}x_2^{\circ}x_3^{\circ}x_4^{\circ}x_5^{\circ}x_6^{\circ} = 000110$$

Given the assignments $\mu_1 = 2$, $\mu_2 = 1$, $\mu_3 = 1$,
$D = 15897 + 5 \cdot 0 + 1 \cdot 0 + 135 \cdot 0 + 27 \cdot 1 + 3645 \cdot 1 + 729 \cdot 0 = 19569$

The outputs f_j, $j \in \{1,2,3\}$, *are recovered by*

$$f_1 = \Xi^{3\cdot 0+2}\{19569\} = \left\lfloor \frac{19569}{3^2} \right\rfloor \ (mod3) = 2$$

$$f_2 = \Xi^{3\cdot 1+1}\{19569\} = \left\lfloor \frac{19569}{3^4} \right\rfloor \ (mod3) = 1$$

$$f_3 = \Xi^{3\cdot 2+1}\{19569\} = \left\lfloor \frac{19569}{3^7} \right\rfloor \ (mod3) = 2$$

FIGURE 9.20
Recover of the MAX function from a word-level linear arithmetic expression (Examples 9.29 and 9.30).

9.6.6 Linear decision diagrams

There are three hierarchical levels in the representation of multivalued functions. The first level corresponds to description of a gate:

```
Gate ⟺
     Linear expression ⟺
                         Linear decision diagram
```

The second level corresponds to description of a level in a multilevel circuit:

```
Circuit level ⟺
              Linear expression ⟺
                                  Linear decision diagram
```

The third level corresponds to description of the circuit:

```
Circuit ⟺
      Set of linear expressions ⟺
                                 Set of linear decision diagrams
```

Based on the above statements, an arbitrary multivalued network can be modeled by a set of linear word-level decision diagrams.

Example 9.31 *The linear decision diagram and its embedding in a $\mathcal{N}$-hypercube for the ternary MAX gate is represented in Figure 9.21.*

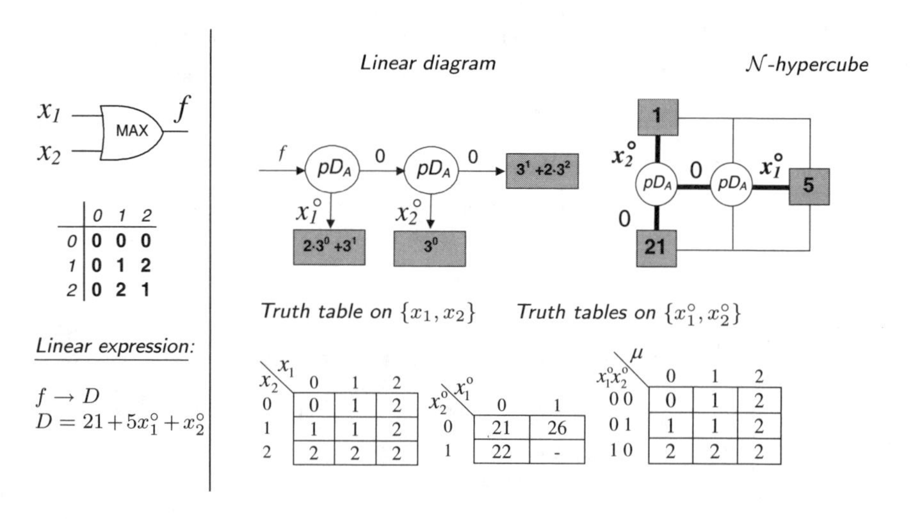

	0	1	2
0	0	0	0
1	0	1	2
2	0	2	1

$x_2 \backslash x_1$	0	1	2
0	0	1	2
1	1	1	2
2	2	2	2

$x_2^\circ \backslash x_1^\circ$	0	1
0	21	26
1	22	-

$x_1^\circ x_2^\circ \backslash \mu$	0	1	2
0 0	0	1	2
0 1	1	1	2
1 0	2	2	2

FIGURE 9.21
Representation of the ternary function MAX by linear decision diagram and $\mathcal{N}$-hypercube (Example 9.31).

9.6.7 Remarks on computing details

One of the problems of word-level representation, including linear forms, is the exponential values of terminal nodes. To calculate these, we utilize so-called Zero-suppressed BDD-like trees. A special encoding scheme must be applied in order to achieve reasonable memory usage (see additional information in the "Further Reading" Section).

9.7 Linear nonarithmetic word-level representation of multivalued functions

It has been shown in Chapter 7 that an arbitrary switching function can be represented by a linear nonarithmetic word-level expression. In this section, the extension of this technique to multivalued functions is presented.

9.7.1 Linear word-level for MAX expressions

Let us denote:

Variable x_i by $x_{i,0}$ $(q = 0)$,
Complement of variable $\overline{x}_i = (m-1) - x_i$ by $x_{i,1}(q = 1)$,
Cyclic complement of variable $\overset{\frown}{x}_i = x_i + 1$ by $x_{i,2}$ $(q = 2)$,
Integer positive values that corresponds to the i-th variable x_i by $w_{i,q}$, and
MAX function by $\vee$.

A linear word-level expression for the MAX operation of a n-variable multivalued function is defined by

$$f = \overset{\frown}{\bigvee_{i=1}^{n}} w_{i,q} x_{i,q}, \tag{9.28}$$

Example 9.32 *Examples of word-level representation are given below. For a single-output ternary MAX function of two variables* $f = x_1 \overset{\frown}{\vee} x_2 = x_1 \vee x_2$. *It is linear expression because it does not contain any product of variables. A two-output ternary function* $f_1 = \overline{x}_1 \vee \overline{x}_2$, $f_2 = x_1 \vee x_2$ *of two variables can be represented by linear expression* $f = 3\overline{x}_1 \overset{\frown}{\vee} 3\overline{x}_2 \overset{\frown}{\vee} x_1 \overset{\frown}{\vee} x_2$. *Details are given in Figure 9.22.*

To recover the initial data from the linear expression, we apply a masking operator. A value f_j of a j-th multivalued function, $j \in \{1, \ldots, r\}$, can be recovered from a linear expression (Equation 9.28) by masking operator $f_j = \Xi^{j-1}\{f\}$.

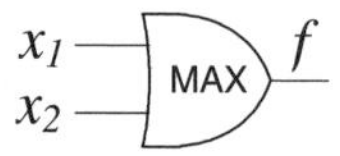

x_1 MAX f
x_2

$f = x_1 \vee x_2$

Word-level representation

$$f = w_{1,0}x_{1,0} \,\widehat{\vee}\, w_{2,0}x_{2,0} = x_1 \,\widehat{\vee}\, x_2 = x_1 \vee x_2,$$

where $q = 0,\ i = 1, 2,\ w_{1,0} = w_{2,0} = 1$

(a)

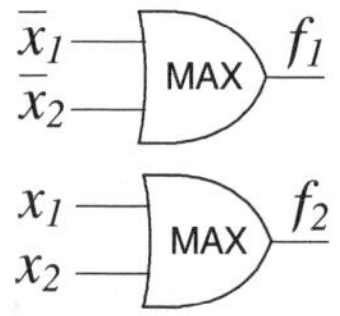

$\overline{x}_1$ MAX f_1
$\overline{x}_2$

x_1 MAX f_2
x_2

$f_1 = \overline{x}_1 \vee \overline{x}_2$
$f_2 = x_1 \vee x_2$

Word-level representation

$$f = 3(\overline{x}_1 \vee \overline{x}_2) \,\widehat{\vee}\, (x_1 \vee x_2) = w_{1,0}x_{1,0} \,\widehat{\vee}\, w_{2,0}x_{2,0}$$
$$= 3\overline{x}_1 \,\widehat{\vee}\, 3\overline{x}_2 \,\widehat{\vee}\, x_1 \,\widehat{\vee}\, x_2,$$

where $q = \{0, 1\},\ i = 1, 2,\ w_{1,0} = w_{2,0} = 1,\ w_{1,1} = w_{2,1} = 3$

(b)

FIGURE 9.22
Linear word-level nonarithmetic representation of a ternary MAX function (a) and two ternary MAX functions (Example 9.32).

Example 9.33 *(Continuation of Example 9.32).*
(a) Single-output ternary MAX function of two variables is recovered:
$f = \Xi^0\{x_1 \,\widehat{\vee}\, x_2\} = x_1 \vee x_2.$
(b) Two-output ternary function of two variables is recovered:
$f_1 = \Xi^0\{3\overline{x}_1 \,\widehat{\vee}\, 3\overline{x}_2 \,\widehat{\vee}\, x_1 \,\widehat{\vee}\, x_2\} = \overline{x}_1 \vee \overline{x}_2.$
$f_2 = \Xi^1\{3\overline{x}_1 \,\widehat{\vee}\, 3\overline{x}_2 \,\widehat{\vee}\, x_1 \,\widehat{\vee}\, x_2\} = x_1 \vee x_2.$

9.7.2 Network representation by linear models

A multilevel multivalued logic network can be described by a linear word-level logic, once each level consists of gates of the same type.

Example 9.34 *The two-input, three-output level of a ternary circuit given in Figure 9.23 is described by the linear expression* $f = 10x_1 \,\widehat{\vee}\, 4x_2 \,\widehat{\vee}\, 3\overline{x}_1 \,\widehat{\vee}\, 9\overline{x}_2$. *The linear decision diagram that corresponds to this expression consists of four nodes implementing the multiple ternary Shannon expansion. The values of the outputs given truth vectors* $\mathbf{f_1}$, $\mathbf{f_2}$ *and* $\mathbf{f_3}$ *are calculated as below*

$$[\mathbf{f_3}\ \mathbf{f_2}\ \mathbf{f_1}] = \begin{bmatrix} 10\cdot 0 \,\widehat{\odot}\, 4\cdot 0 \,\widehat{\odot}\, 3\cdot 2 \,\widehat{\odot}\, 9\cdot 2 \\ 10\cdot 0 \,\widehat{\odot}\, 4\cdot 1 \,\widehat{\odot}\, 3\cdot 2 \,\widehat{\odot}\, 9\cdot 1 \\ 10\cdot 0 \,\widehat{\odot}\, 4\cdot 2 \,\widehat{\odot}\, 3\cdot 2 \,\widehat{\odot}\, 9\cdot 0 \\ 10\cdot 1 \,\widehat{\odot}\, 4\cdot 0 \,\widehat{\odot}\, 3\cdot 1 \,\widehat{\odot}\, 9\cdot 2 \\ 10\cdot 1 \,\widehat{\odot}\, 4\cdot 1 \,\widehat{\odot}\, 3\cdot 1 \,\widehat{\odot}\, 9\cdot 1 \\ 10\cdot 1 \,\widehat{\odot}\, 4\cdot 2 \,\widehat{\odot}\, 3\cdot 1 \,\widehat{\odot}\, 9\cdot 0 \\ 10\cdot 2 \,\widehat{\odot}\, 4\cdot 0 \,\widehat{\odot}\, 3\cdot 0 \,\widehat{\odot}\, 9\cdot 2 \\ 10\cdot 2 \,\widehat{\odot}\, 4\cdot 1 \,\widehat{\odot}\, 3\cdot 0 \,\widehat{\odot}\, 9\cdot 1 \\ 10\cdot 2 \,\widehat{\odot}\, 4\cdot 2 \,\widehat{\odot}\, 3\cdot 0 \,\widehat{\odot}\, 9\cdot 0 \end{bmatrix} = \begin{bmatrix} 000 \,\widehat{\odot}\, 000 \,\widehat{\odot}\, 020 \,\widehat{\odot}\, 200 \\ 000 \,\widehat{\odot}\, 011 \,\widehat{\odot}\, 020 \,\widehat{\odot}\, 100 \\ 000 \,\widehat{\odot}\, 022 \,\widehat{\odot}\, 020 \,\widehat{\odot}\, 000 \\ 101 \,\widehat{\odot}\, 000 \,\widehat{\odot}\, 010 \,\widehat{\odot}\, 200 \\ 101 \,\widehat{\odot}\, 011 \,\widehat{\odot}\, 010 \,\widehat{\odot}\, 100 \\ 101 \,\widehat{\odot}\, 022 \,\widehat{\odot}\, 010 \,\widehat{\odot}\, 000 \\ 202 \,\widehat{\odot}\, 000 \,\widehat{\odot}\, 000 \,\widehat{\odot}\, 200 \\ 202 \,\widehat{\odot}\, 011 \,\widehat{\odot}\, 000 \,\widehat{\odot}\, 100 \\ 202 \,\widehat{\odot}\, 022 \,\widehat{\odot}\, 000 \,\widehat{\odot}\, 000 \end{bmatrix} = \begin{bmatrix} 2\ 2\ 0 \\ 1\ 2\ 1 \\ 0\ 2\ 2 \\ 2\ 1\ 1 \\ 1\ 1\ 1 \\ 1\ 2\ 2 \\ 2\ 0\ 2 \\ 2\ 1\ 2 \\ 2\ 2\ 2 \end{bmatrix}$$

In particular, given the assignment $x_1x_2\overline{x}_1\overline{x}_2 = \{0022\}$, *the outputs are equal to*

$$f_1(0022) = 0 \vee 0 \vee 0 \vee 0 = 0(x_1 \vee x_2 = 0 \vee 0 = 0),$$
$$f_2(0022) = 0 \vee 0 \vee 2 \vee 0 = 2(\overline{x}_1 \vee x_2 = 2 \vee 0 = 2),$$
$$f_3(0022) = 0 \vee 0 \vee 0 \vee 2 = 2(x_1 \vee \overline{x}_2 = 0 \vee 2 = 2).$$

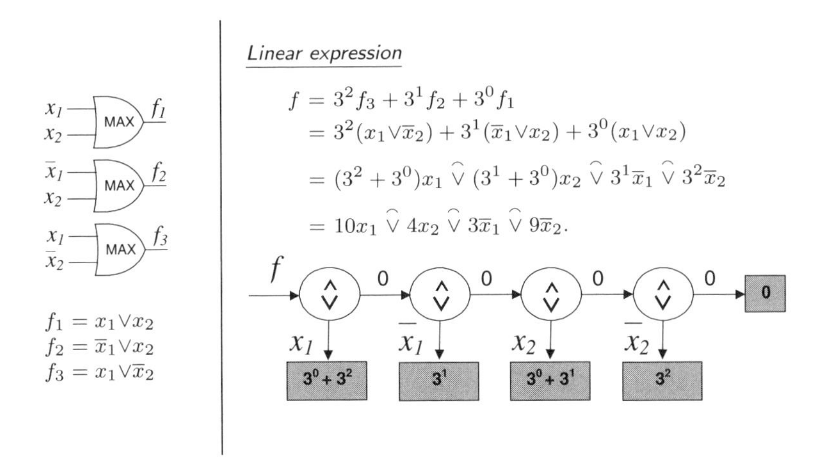

FIGURE 9.23
A two-level ternary circuit and its linear diagram (Example 9.34).

9.8 Summary

1. A multivalued signal has more information capacity per digit than has a switching function (see "Further Reading" Section). However, contemporary electronics hardly uses this great asset of multivalued logic. Quantum levelization in quantum-effect nanoelectronic opens new possibilities for design of multivalued systems.

2. Formal study of multivalued functions is mostly based on the principle of generalization of switching functions' relevant data structures. However, there are a lot of specific features of multivalued functions that require a special or innovative approach.

3. Logic difference, the formal description of a change in a multivalued system, plays the key role in the analysis of the system and relevant data structures. In a multivalued system,

- An arbitrary change in the logic value of one of the m-values can be recognized and described by logic differences;
- Circuit behavior (switching activity, power dissipation, testing, etc.) can be described in terms of change (see "Further Reading" Section);
- Important content, entropy and symmetry properties can be described by logic differences. In Chapter 12, this problem is discussed. It will be prove that the computational complexity of symmetry detection can be drastically reduced;
- Generation of logic Taylor expansions whose coefficients are logic differences; this expansion generates the family of Reed-Muller expressions with various polarities of variables. The coefficients of arithmetic Taylor expansion are arithmetic analogs of logic difference; this expansion generates the family of arithmetic expressions of a multivalued function.

4. Linearization technique aims to simplify embedded graph-based representation in spatial dimensions, and is developed toward multivalued functions. In this chapter, linear data structures, algebraic equations, decision trees, decision diagrams and $\mathcal{N}$-hypercubes are developed for elementary functions. Based on this library of linear primitives, an arbitrary multivalued circuit can be described as a set of linear word-level diagrams and mapped in spatial dimensions. Linearization technique includes:

- Linear word-level arithmetic models, and
- Linear word-level nonarithmetic models.

9.9 Problems

Problem 9.1 Use spectral transforms and calculate spectrum and derive decision diagrams for logic functions of two variables given below

(a) Reed-Muller spectrum of ternary and quaternary functions MIN, MAX, MODPROD. Follow Example 9.10 and use Tables 9.4, 9.11, and 9.12.
(b) Arithmetic spectrum ternary and quaternary functions MIN, MAX, MODPROD. Follow Example 9.12 and use Tables 9.4, 9.11, and 9.12.
(c) Arithmetic spectrum of word-level representation of ternary functions is given below. Follow Example 9.13.

(i) $f_1 =$MIN(x_1, x_2), $f_2 =$MIN$(\overline{x}_1, x_2)$
(ii) $f_1 =$MIN(x_1, x_2), $f_2 =$MAX(x_1, x_2)
(iii) $f_1 =$MIN(x_1, x_2), $f_2 =$MIN(x_1, x_2), $f_3 =$MAX(x_1, x_2)

Problem 9.2 Calculate logic differences for the quaternary logic function MIN(x_1, x_2). Follow Examples 9.15 and 9.16. Use Table 9.12.

(a) The first difference
(b) The second difference
(c) The third difference

Problem 9.3 Calculate the logic differences with respect to variable x_1 for the ternary function of two variables. Follow Example 9.19.

(a) MIN(x_1, x_2)
(b) MODSUM(x_1, x_2)
(c) MODPROD(x_1, x_2)
(d) TSUM(x_1, x_2)
(e) TPROD(x_1, x_2)

Problem 9.4 Using the logic Taylor expression, represent the ternary function of two variables by Reed-Muller expression of polarity c. Follow Examples 9.20 and 9.21.

(a) MIN(x_1, x_2) , $c = 0$
(b) MODSUM(x_1, x_2), $c = 2$
(c) MODPROD(x_1, x_2), $c = 1$
(d) TSUM(x_1, x_2) $c = 3$

Problem 9.5 Using the linearization technique, represent the ternary function of two variables by linear word-level arithmetic expression and decision diagram. Follow Example 9.27.

(a) MIN(x_1, x_2)
(b) MODSUM(x_1, x_2)

(*c*) MODPROD(x_1, x_2)
(*d*) TSUM(x_1, x_2)
(*e*) TPROD(x_1, x_2)

Problem 9.6 In Figure 9.24 a linear decision diagram for ternary MIN function is given. Justify by truth tables that this representation is correct. Construct a hypercube.

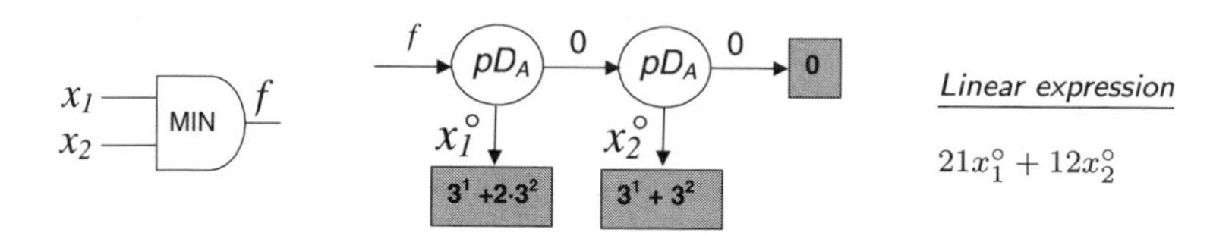

FIGURE 9.24
The linear diagrams of the ternary function MIN(x_1, x_2) (Problem 9.6).

Problem 9.7 The ternary circuit and corresponding set of two linear word-level decision diagrams are given in Figure 9.25. Justify that calculation on these decision diagrams has been done properly. For this

(*a*) Write the truth tables for the first and second levels of a circuit
(*b*) Describe the first and second levels by linear word-level expressions. Use linear models of MIN and MAX gates from Table 9.10
(*c*) Write the rule to encode ternary variables $x_1x_2x_3x_4$ into pseudo-variables $x_1^\circ x_2^\circ x_3^\circ x_4^\circ$. Follow the encoding rule for two variables given in Table 9.9
(*d*) Calculate the outputs f_1 and f_2 given the assignments $x_1^\circ x_2^\circ x_3^\circ x_4^\circ = \{0000, 0001, 1111\}$

Problem 9.8 Figure 9.26 illustrates topological representations of a three-input ternary logic function. It has been generated from the decision tree for this function. Next, a pyramid is embedded in a $\mathcal{N}$-hypercube. Evaluate (derive the formula) a diameter and the number of terminal nodes in the n-dimensional ternary hypercube.

Problem 9.9 Use material in Section 9.7 to solve the problems below:

(*a*) Given a two-output ternary function f with outputs f_1 =MIN(x_1x_2) and f_2 =MIN(x_2x_3), justify that

f_1 =MIN$(x_1x_2) = \Xi^1\{3^0\overline{x}_1 \mathbin{\widehat{\vee}} (3^0+3^1)\overline{x}_2 \mathbin{\widehat{\vee}} 3^1\overline{x}_3\}$ and

f_2 =MIN$(x_2x_3) = \Xi^2\{3^0\overline{x}_1 \mathbin{\widehat{\vee}} (3^0+3^1)\overline{x}_2 \mathbin{\widehat{\vee}} 3^1\overline{x}_3\}$.

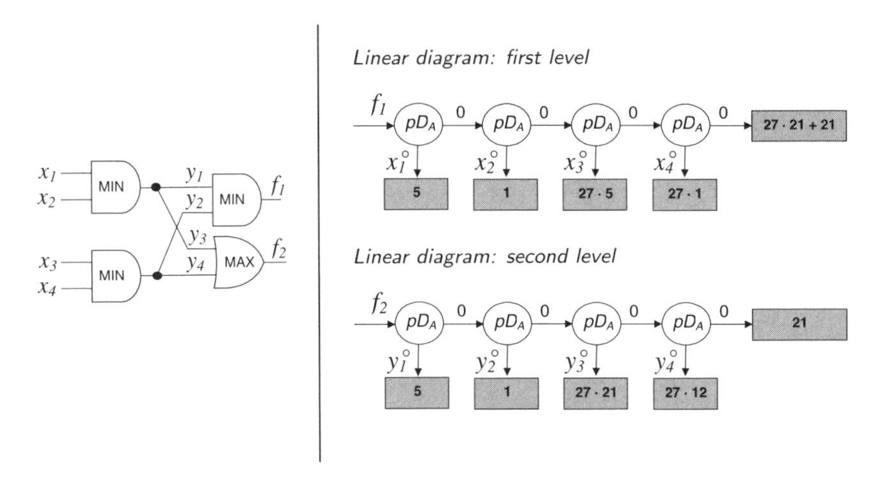

FIGURE 9.25
Two-level ternary circuit (left) and linear word-level decision diagrams (Problem 9.7).

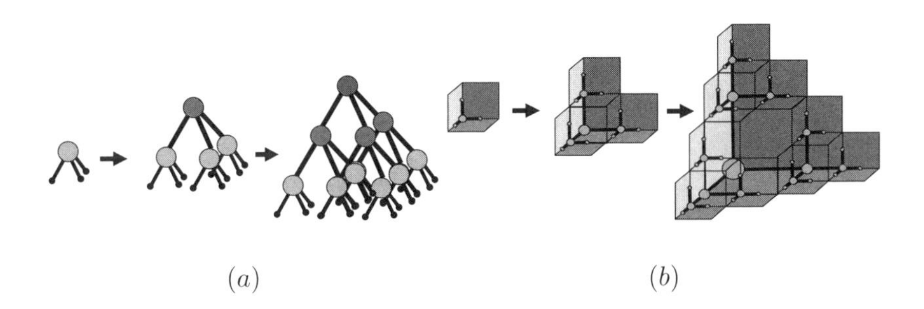

FIGURE 9.26
Representation of a logic function by decision trees (a) and hypercube-like structures (b) (Problem 9.8).

(*b*) Find linear word-level logic expression of the two-output function $f_1 = x_1 x_2,\ f_2 = x_2 x_3$.
Hint: apply DeMorgan's law and operate on $\overline{f}_1$ and $\overline{f}_2$.

(*c*) Given the two-output ternary function f: $f_1 = \text{TPROD}(x_1, x_2) = x_1 \sqcap x_2$ and $f_2 = \text{TPROD}(x_2, x_3) = x_2 \sqcap x_3$, justify that

$f_1 = \Xi^1\{\overline{x}_1 \,\widehat{\sqcup}\, (3^0 + 3^1)\overline{x}_2 \,\widehat{\sqcup}\, 3^1\overline{x}_3\}$, *and*

$f_2 = \Xi^2\{\overline{x}_1 \,\widehat{\sqcup}\, (3^0 + 3^1)\overline{x}_2 \,\widehat{\sqcup}\, 3^1\overline{x}_3\}$

Hint: apply DeMorgan's law and represent f_1 and f_2 by $\overline{f}_1 = \overline{x}_1 \sqcup \overline{x}_2$ and $\overline{f}_2 = \overline{x}_2 \sqcup \overline{x}_3$.

(*d*) Find linear word-level logic expression of the two-output ternary function

$$f_1 = x_1 \oplus x_3, \; f_2 = x_2 \oplus x_3.$$

Problem 9.10 Consider a level of a ternary circuit consisting of gates with outputs $f_1 = x_1 \odot x_2$, $f_2 = \overline{x}_1 \odot x_2$, and $f_3 = x_1 \odot \overline{x}_2$, where $\odot$ denotes an arbitrary elementary function from the set

$$\text{MAX, MIN, MODSUM, TSUM, TPROD.}$$

The linear word-level model is given below

$$[\mathbf{f_3\ f_2\ f_1}] = \begin{bmatrix} 10\cdot 0 \widehat{\odot} 4\cdot 0 \widehat{\odot} 3\cdot 2 \widehat{\odot} 9\cdot 2 \\ 10\cdot 0 \widehat{\odot} 4\cdot 1 \widehat{\odot} 3\cdot 2 \widehat{\odot} 9\cdot 1 \\ 10\cdot 0 \widehat{\odot} 4\cdot 2 \widehat{\odot} 3\cdot 2 \widehat{\odot} 9\cdot 0 \\ 10\cdot 1 \widehat{\odot} 4\cdot 0 \widehat{\odot} 3\cdot 1 \widehat{\odot} 9\cdot 2 \\ 10\cdot 1 \widehat{\odot} 4\cdot 1 \widehat{\odot} 3\cdot 1 \widehat{\odot} 9\cdot 1 \\ 10\cdot 1 \widehat{\odot} 4\cdot 2 \widehat{\odot} 3\cdot 1 \widehat{\odot} 9\cdot 0 \\ 10\cdot 2 \widehat{\odot} 4\cdot 0 \widehat{\odot} 3\cdot 0 \widehat{\odot} 9\cdot 2 \\ 10\cdot 2 \widehat{\odot} 4\cdot 1 \widehat{\odot} 3\cdot 0 \widehat{\odot} 9\cdot 1 \\ 10\cdot 2 \widehat{\odot} 4\cdot 2 \widehat{\odot} 3\cdot 0 \widehat{\odot} 9\cdot 0 \end{bmatrix} = \begin{bmatrix} 000 \widehat{\odot} 000 \widehat{\odot} 020 \widehat{\odot} 200 \\ 000 \widehat{\odot} 011 \widehat{\odot} 020 \widehat{\odot} 100 \\ 000 \widehat{\odot} 022 \widehat{\odot} 020 \widehat{\odot} 000 \\ 101 \widehat{\odot} 000 \widehat{\odot} 010 \widehat{\odot} 200 \\ 101 \widehat{\odot} 011 \widehat{\odot} 010 \widehat{\odot} 100 \\ 101 \widehat{\odot} 022 \widehat{\odot} 010 \widehat{\odot} 000 \\ 202 \widehat{\odot} 000 \widehat{\odot} 000 \widehat{\odot} 200 \\ 202 \widehat{\odot} 011 \widehat{\odot} 000 \widehat{\odot} 100 \\ 202 \widehat{\odot} 022 \widehat{\odot} 000 \widehat{\odot} 000 \end{bmatrix} = \begin{bmatrix} c_0 & b_0 & a_0 \\ c_1 & b_1 & a_1 \\ c_2 & b_2 & a_2 \\ c_3 & b_3 & a_3 \\ c_4 & b_4 & a_4 \\ c_5 & b_5 & a_5 \\ c_6 & b_6 & a_6 \\ c_7 & b_7 & a_7 \\ c_8 & b_8 & a_8 \end{bmatrix}$$

Follow Example 9.34 and calculate the outputs given the assignments $x_1 x_2 \overline{x}_3 \overline{x}_4 = \{0000,\ 1111,\ 2222\}$ if

(*a*) $f_1,\ f_2,\ f_3$ are MIN gates
(*b*) $f_1,\ f_2,\ f_3$ are TSUM gates
(*c*) $f_1,\ f_2,\ f_3$ are TPROD gates
(*d*) $f_1,\ f_2,\ f_3$ are MODSUM gates

Support data for ternary logic

The basic operations for ternary and quaternary logic are given in Tables 9.11 and 9.12.

9.10 Further reading

International Community on Multivalued Logic. Plenty of useful information on the theory and application of multivalued logic can be found in the *Proceedings of the Annual IEEE International Symposium on Multiple-Valued Logic* that has been held since 1970. The following topics are traditionally the focus of discussions at this forum: multivalued circuit design and implementation, minimization and decomposition techniques, fault modeling,

TABLE 9.11
Basic operation over a three-valued argument.

COMPLEMENT

	x=0	1	2
$\overline{x}$	2	1	0

r-CYCLIC COMPLEMENT

r \ x	0	1	2
0	0	1	2
1	1	2	0
2	2	0	1

LITERAL

	x=0	1	2
x^0	2	0	0
x^1	0	2	0
x^2	0	0	2

WINDOW LITERAL

	x=0	1	2
${}^0x^0$	2	0	0
${}^0x^1$	2	2	0
${}^0x^2$	2	2	2
${}^1x^1$	0	2	0
${}^1x^2$	0	2	2
${}^2x^2$	0	0	2

MAX

	0	1	2
0	0	1	2
1	1	1	2
2	2	2	2

$\overline{MAX}$

	0	1	2
0	2	1	0
1	1	1	0
2	0	0	0

TSUM

	0	1	2
0	0	1	2
1	1	2	2
2	2	2	2

$\overline{TSUM}$

	0	1	2
0	2	1	0
1	1	0	0
2	0	0	0

TPROD

	0	1	2
0	0	0	0
1	0	0	1
2	0	1	2

$\overline{TPROD}$

	0	1	2
0	2	2	2
1	2	2	1
2	2	1	0

MIN

	0	1	2
0	0	0	0
1	0	1	1
2	0	1	2

$\overline{MIN}$

	0	1	2
0	2	2	2
1	2	1	1
2	2	1	0

MODSUM

	0	1	2
0	0	1	2
1	1	2	0
2	2	0	1

$\overline{MODSUM}$

	0	1	2
0	2	1	0
1	1	0	2
2	0	2	1

MODPROD

	0	1	2
0	0	0	0
1	0	1	2
2	0	2	1

$\overline{MODPROD}$

	0	1	2
0	2	2	2
1	2	1	0
2	2	0	1

fault diagnostics and testing, decision diagram technique, multivalued algebras, fuzzy logic and their application, automated reasoning and complexity, theorem proving, computing paradigms, nanoICs design, neural networks and evolutionary computing, and information theory.

In addition, a good source of information on the above mentioned topics is the *International Journal on Multiple-Valued Logic and Soft Computing.* Also, there are a number of books on the applied problems of multivalued logic [4, 22, 24, 34, 47].

Logic differential calculus. Excellent contributions to the fundamentals of logic differential and integral calculus of multivalued logic have been made in [7, 8, 12, 38, 39]. Computational aspects of this theory have been studied in [27, 28, 30, 35, 45, 47].

Multivalued decision diagrams can be considered as an extension of BDD technique for multivalued functions. This technique is the focus of papers [18, 31, 32, 33]. An excellent survey of ternary decision diagram technique can be found in [26]. The paper by Kam et al. [10] summarizes the results

TABLE 9.12
Basic operations over a four-valued argument.

COMPLEMENT

	x = 0	1	2	3
$\overline{x}$	3	2	1	0

r-CYCLIC COMPLEMENT

r \ x	0	1	2	3
0	0	1	2	3
1	1	2	3	0
2	2	3	0	1
3	3	0	1	2

LITERAL

	x = 0	1	2	3
x^0	3	0	0	0
x^1	0	3	0	0
x^2	0	0	3	0
x^3	0	0	0	3

WINDOW LITERAL

	x = 0	1	2	3
${}^0x^0$	3	0	0	0
${}^0x^1$	3	3	0	0
${}^0x^2$	3	3	3	0
${}^0x^3$	3	3	3	3
${}^1x^1$	0	3	0	0
${}^1x^2$	0	3	3	0
${}^1x^3$	0	3	3	3
${}^2x^2$	0	0	3	0
${}^2x^3$	0	0	3	3
${}^2x^3$	0	0	3	3
${}^3x^3$	0	0	0	3

MAX

	0	1	2	3
0	0	1	2	3
1	1	1	2	3
2	2	2	2	3
3	3	3	3	3

$\overline{MAX}$

	0	1	2	3
0	3	2	1	0
1	2	2	1	0
2	1	1	1	0
3	0	0	0	0

TSUM

	0	1	2	3
0	0	1	2	3
1	1	2	3	3
2	2	3	3	3
3	3	3	3	3

$\overline{TSUM}$

	0	1	2	3
0	3	2	1	0
1	2	1	0	0
2	1	0	0	0
3	0	0	0	0

TPROD

	0	1	2	3
0	0	0	0	0
1	0	0	0	1
2	0	0	1	2
3	0	1	2	3

$\overline{TPROD}$

	0	1	2	3
0	3	3	3	3
1	3	3	3	2
2	3	3	2	1
3	3	2	1	0

MIN

	0	1	2	3
0	0	0	0	0
1	0	1	1	1
2	0	1	2	2
3	0	1	2	3

$\overline{MIN}$

	0	1	2	3
0	3	3	3	3
1	3	2	2	2
2	3	2	1	1
3	3	2	1	0

MODSUM

	0	1	2	3
0	0	1	2	3
1	1	2	3	0
2	2	3	0	1
3	3	0	1	2

$\overline{MODSUM}$

	0	1	2	3
0	3	2	1	0
1	2	1	0	3
2	1	0	3	2
3	0	3	2	1

MODPROD

	0	1	2	3
0	0	0	0	0
1	0	1	2	3
2	0	2	0	2
3	0	3	2	1

$\overline{MODPROD}$

	0	1	2	3
0	3	3	3	3
1	3	2	1	0
2	3	1	3	1
3	3	0	1	2

for multivalued logic decision diagram technique.

Linear word-level arithmetic representations of multivalued functions have been introduced by Dziurzanski et al. [3], Yanushkevich et al. [48] and logic ones have been studied by Tomaszewska et al. [40].

Arithmetic representations. In [36], the Pascal triangle has been used to obtain arithmetic representations of elementary multivalued functions. Techniques based on spectral transforms have been developed by many researchers. Yanushkevich has used this technique and arithmetic Taylor expansion to generate arithmetic forms of multivalued functions [44].

Spectral technique in multivalued logic is based on spectral transformations of discrete signals. Historically, the first recognizable activity on application of the spectral theory of digital signal processing to switching functions come back to the development of the fast Fourier transform (FFT). The first results did not attract researchers because a Fourier spectrum of a switching function is very difficult to interpret in terms of switching theory. Reed-Muller, arithmetic and Walsh spectra have a simple interpretation both in switching and multivalued theory. These problems are the focus of papers by Karpovsky [11]. We also recommend papers by Green [5, 6] where a very detailed study of Reed-Muller expressions of logic functions has been introduced.

The book by Hurst et al. [9] is a good contribution to the fundamentals of spectral technique in multivalued logic. In a paper by Moraga [20], formal aspects of complex representation of multivalued logic functions are developed. The spectral technique has also been introduced in [17, 25, 33].

Testing. The class of failures in multivalued circuits is much larger compared with binary circuits. This is the focus of the study of [2, 16, 41]. Generalization of the D-algorithm for multivalued combinational circuits has been proposed by Spillman and Su [29] and its extensions can be found in papers [28, 37, 46].

Logic design. While multivalued logic primarily aims at reducing the total number of transistors compared to CMOS full adder circuits, the functional integration furthermore enables it to extend digital logic. Compared to purely digital logic, multivalued logic has some advantages when designing high-speed arithmetic components in avoiding time-consuming carry propagation, inherent in switching gates. However, there are several disadvantages, since multivalued logic circuits often have to be embedded in a conventional digital system and therefore additional circuitry is needed to transform multivalued logic into digital signals and vice versa. Thus, global system performance plays an important role, too. Furthermore, when reducing the supply voltages, the noise margin for the logic levels in multivalued resonant tunneling device (RTD) circuits decreases and affects reliability. On the other hand, multivalued logic for storing synaptic weights in neural circuitry preserves robust information processing and reduces the number of circuit components per artificial synaptic circuit. Especially for monolithically integrated neural systems, multistate RTD-memory cells are a promising way to implement area efficient multivalued logic circuits. The hope is that the fault tolerance of neural circuits will compensate the errors caused by smaller noise margins.

Different aspects of logic design of multivalued circuits and systems have been developed, in particular, decomposition [15], linearly independent logic [23], and minimization of incompletely specified multivalued logic functions [47, 49]. Moraga has shown that multivalued functions can be processed by systolic arrays [21]. Linear systolic arrays have been developed in [42, 43]. Computer aided design (CAD) tools of multivalued systems have been discussed by Miller [17].

Reversible logic and molecular devices. Since reversible computation has created a solid background for quantum computing paradigm, study of the implementation of multivalued functions on quantum-effect devices is interdisciplinary and a beneficial direction indeed. An m-input, m-output totally-specified multivalued logic function is reversible if it maps each input assignment to a unique output assignment. So-called ternary field sum-of-products expressions and decision diagrams are introduced for the design of ternary reversible devices by Khan et al. [14]. The heuristic algorithm for synthesis of multivalued reversible circuits over the library of reversible gates is developed by Miller et al. [19]. Some properties of multivalued reversible gates related to completeness are studied by Kerntopf et al. [13] (a set of m-valued logic functions (or elementary functions) is called universal (or completeness) if an arbitrary m-valued logic function can be represented by this finite set of elementary functions). In addition, see references in Chapter 1.

The usefulness of multivalued logic models in design of molecular devices has been shown, in particular, by Aoki et al. [1].

9.11 References

[1] Aoki T, Kameyama M, and Higuchi T. Design of interconnection-free biomolecular computing systems. In *Proceedings 21st IEEE International Symposium on Multiple-Valued Logic*, pp. 173–180, 1991.

[2] Coy W, and Moraga C. Description and detection of faults in multiple-valued logic circuits. In *Proceedings 19th IEEE International Symposium on Muliple-Valued Logic*, pp. 74–81, 1979.

[3] Dziurzanski P, Malyugin VD, Shmerko VP, and Yanushkevich SN. Linear models of circuits based on the multivalued components. *Automation and Remote Control*, Kluwer/Plenum Publishers, 63(6):960–980, 2002.

[4] Epstein G. *Multi-Valued Logic Design.* Institute of Physics Publishing, London, UK, 1993.

[5] Green DH. Ternary Reed-Muller switching functions with fixed and mixed polarity. *International Journal of Electronics*, 67:761–775, 1989.

[6] Green DH. Families of Reed-Muller canonical forms. *International Journal of Electronics*, 70(2):259–280, 1991.

[7] Guima TA, and Katbab A. Multivalued logic integral calculus. *International Journal of Electronics*, 65:1051–1066, 1988.

[8] Guima TA, and Tapia MA. Differential calculus for fault detection in multivalued logic networks. In *Proceedings 17th IEEE International Symposium on Multiple-Valued Logic*, pp. 99–108, 1987.

[9] Hurst S, Miller D, and Muzio J. *Spectral Technique in Digital Logic.* Academic Press, New York, 1985.

[10] Kam T, Villa T, Brayton RK, and Sagiovanni-Vincentelli AL. Multivalued decision diagrams: theory and applications. *International Journal on Multiple-Valued Logic*, 4(1-2):9–62, 1998.

[11] Karpovsky MG., Ed., *Spectral Techniques and Fault Detection.* Academic Press, New York, 1985.

[12] Katbab A, and Guima T. On multi-valued logic design using exact integral calculus. *International Journal of Electronics*, 66(1):1–18, 1989.

[13] Kerntopf P, Perkowski M, and Khan MHA, On universality of general reversible multi-valued logic gates, In *Proceedings 34th IEEE International Symposium on Multiple-Valued Logic*, pp. 68–73, 2004

[14] Khan MHA, Perkowski M, and Khan MR. Ternary Galois field expansions for reversible logic and Kronecker decision diagrams for ternary GFSOP minimization. In *Proceedings 34th IEEE International Symposium on Multiple-Valued Logic*, pp. 58–67, 2004

[15] Luba T. Decomposition of multiple-valued functions. In *Proceedings 25th IEEE International Symposium on Multiple-Valued Logic*, pp. 256–263, 1995.

[16] Miller DM. Spectral signature testing for multiple-valued combinational networks. In *Proceedings 12th IEEE International Symposium on Multiple-Valued Logic*, pp. 152–158, 1982.

[17] Miller DM. Multiple-valued logic design tools. In *Proceedings 23rd IEEE International Symposium on Multiple-Valued Logic*, pp. 2–11, 1993.

[18] Miller DM. Spectral transformation of multiple-valued decision diagrams. In *Proceedings 24th IEEE International Symposium on Multiple-Valued Logic*, pp. 89–96, 1994.

[19] Miller DM, Dueck GW, and Maslov D. A synthesis method for MVL reversible logic. In *Proceedings 34th IEEE International Symposium on Multiple-Valued Logic*, pp. 74–80, 2004.

[20] Moraga C. Complex spectral logic. In *Proceedings 8th IEEE International Symposium on Multiple-Valued Logic*, pp. 149–156, 1978.

[21] Moraga C. Systolic systems and multiple-valued logic. In *Proceedings 14th IEEE International Symposium on Multiple-Valued Logic*, pp. 98–108, 1984.

[22] Muzio JC, and Wesselkamper TS. *Multiple-Valued Switching Theory.* Adam Higler Ltd., Bristol and Boston, 1986.

[23] Perkowski M, Sarabi A, and Beyl F. Fundamental theorems and families of forms for binary and multiple-valued linearly independent logic. *Proceedings IFIP WG 10.5 International Workshop on Applications of the Reed-Muller Expansions in Circuit Design*, pp. 288–299, Japan, 1995.

[24] Rine DC., Ed., *Computer Science and Multiple-Valued Logic. Theory and Applications.* 2nd ed., North-Holland, Amsterdam, 1984.

[25] Sadykhov R, Chegolin P, and Shmerko V. *Signal Processing in Discrete Bases.* Publishing House "Science and Technics", Minsk, Belarus, 1986.

[26] Sasao T. Ternary decision diagram – survey. *Proceedings 27th IEEE International Symposium on Multiple-Valued Logic*, pp. 241–250, 1997.

[27] Shmerko VP, Yanushkevich SN, and Levashenko VG. Techniques of computing logical derivatives for MVL functions. In *Proceedings 26th IEEE Intenational Symposium on Multiple-Valued Logic*, pp. 267–272, 1996.

[28] Shmerko VP, Yanushkevich SN, and Levashenko VG. Test pattern generation for combinational MVL networks based on generalized *D*-algorithm. In *Proceedings 22nd IEEE International Symposium on Multiple-Valued Logic*, pp. 139–144, 1997.

[29] Spillman RJ, and Su SYH. Detection of single, stuck-type failures in multivalued combinational networks. *IEEE Transactions on Computers*, 26(12):1242–1251, 1977.

[30] Stanković RS. Some remarks on fourier transforms and differential operators for digital functions. In *Proceedings 22nd IEEE International Symposium on Multiple-Valued Logic*, pp. 365–370, 1992.

[31] Stanković R, Stankovic M, Moraga C, and Sasao T. Calculation of Reed-Muller-Fourier coefficients of multiple-valued functions through multiple-place decision diagrams. *Proceedings 24th IEEE International Symposium on Multiple-Valued Logic*, pp. 82–87, 1994.

[32] Stanković RS. Functional decision diagrams for multiple-valued functions. In *Proceedings 25th IEEE International Symposium on Multiple-Valued Logic*, pp. 284–289, 1995.

[33] Stanković RS, Sasao T, and Moraga C. Spectral transform decision diagrams. In Sasao T., Ed., *Representations of Discrete Functions*, pp. 55–92. Kluwer, Dordrecht, 1996.

[34] Stanković RS., Ed., *Recent Developments in Abstract Harmonic Analysis with Applications in Signal Processing.* Nauka, Belgrade, Yugoslavia, 1996.

[35] Stanković RS, Moraga C, and Astola JT. Derivatives for multiple-valued functions induced by Galois field and Reed-Muller-Fourier expressions. In *Proceedings 34th IEEE International Symposium on Multiple-Valued Logic*, pp. 184–189, 2004.

[36] Strazdins I. The polynomial algebra of multivalued logic. *Algebra, Combinatorics and Logic in Computer Science*, 42:777–785, 1983.

[37] Tabakow IG. Using D-algebra to generate tests for m-logic combinational circuits. *International Journal of Electronics*, 75(5):897–906, 1993.

[38] Tapia MA, Guima TA, and Katbab A. Calculus for a multivalued logic algebraic system. *Applied Mathematics and Computation*, pp. 225–285, 1991.

[39] Thayse A. Differential calculus for functions from $GF(p)$. *Philips Research Reports*, 29:560–586, 1974.

[40] Tomaszewska A, Yanushkevich SN, and Shmerko VP. Word-level models for efficient computation of multiple-valued functions. Part 2: LWL based models. In *Proceedings IEEE 32nd International Symposium on Multiple-Valued Logic*, pp. 209–214, Boston, 2002.

[41] Wang HM, Lee CL, and Chen JE. Complete test set for multiple-valued logic networks. In *Proceedings 24th IEEE International Symposium on Multiple-Valued Logic*, pp. 289–296, 1994.

[42] Yanushkevich SN. Systolic arrays for multivalued logic data processing. In Kukharev GA, Shmerko VP, and Zaitseva EN, *Algorithms and Systolic Processors for Multi-Valued Data Processing*, pp. 157–252. Publishing House Science and Technics, Minsk, Belarus, 1990.

[43] Yanushkevich SN. Systolic algorithms to synthesize arithmetical polynomial forms for k-valued logic functions. *Automation and Remote Control.* Kluwer/Plenum Publishers, 55(12):812–1823, 1994.

[44] Yanushkevich SN. Arithmetical canonical expansions of Boolean and MVL functions as generalized Reed-Muller series. *Proceedings IFIP WG 10.5 Workshop on Applications of the Reed-Muller Expansions in Circuit Design*, pp. 300–307, Japan, 1995.

[45] Yanushkevich SN. Matrix method to solve logic differential equations. *International Journal, IEE Proceedings, Pt.E, Computers and Digital Technique*, UK, 144(5):267–272, 1997.

[46] Yanushkevich SN, Levashenko VG, and Moraga C. Fault models for multiple-valued combinational circuits. In *Proceedings International Conference on Applications of Computer System*, pp. 309–314. Szczecin, Poland, 1997.

[47] Yanushkevich SN. *Logic Differential Calculus in Multi-Valued Logic Design.* Technical University of Szczecin Academic Publishers, Szczecin, Poland, 1998.

[48] Yanushkevich SN, Dziurzanski P, and Shmerko VP. Word-level models for efficient computation of multiple-valued functions. Part 1: LAR based models. In *Proceedings IEEE 32nd International Symposium on Multiple-Valued Logic*, pp. 202–208, Boston, 2002.

[49] Zilic Z, and Vranesic Z. Multiple valued Reed-Muller transform for incomplete functions. *IEEE Transactions on Computers*, 44(8):1012–1020, 1995.

10

Parallel Computation in Nanospace

In this chapter the technique for designing arrays for switching function computing is introduced. Four types of arrays are considered here:

- 1-D, or linear arrays,
- 2-D arrays; in particular, tree-structured networks, and
- $\mathcal{N}$-hypercube arrays.

These architectures are implemented in a pipelined logic style that avoids long range interconnections and accelerates computations. Independent of the technological realization and the operating principles of the different families of ultra-small devices with extreme circuit-level scalability, important design principles are:

- *A regular layout* with a small number of different circuit modules (cells);
- *Local interconnections* on the circuit and the system level, and
- *Concurrent computation and pipelining* at the bit-level to achieve a low latency.

In addition, nanoscale devices, such as single-electron ones, possess

- *A stochastic character* of computing due to quantum effect phenomena, and
- *Fault tolerant logic schemes* to compensate for fabrication tolerances and background charges.

The material is introduced as follows. In Sections 10.1 and 10.2, the characteristics of array architectures are introduced. In Section 10.3, the design style of linear systolic arrays is given. Sections 10.4, 10.6, and 10.7 focus on design of linear systolic arrays for computing Reed-Muller, arithmetic and Walsh expressions. Section 10.8 introduces tree-network design. Section 10.9 describes $\mathcal{N}$-hypercube array design. In Section 10.11, we propose to solve several problems of array computer architecture design. Finally, in Section 10.12, recommendations for "Further Reading" are given.

10.1 Data structures and massive parallel computing

Data structure is an important attribute, highly relevant to implementation of the algorithm and organizing information flow in parallel structures, i.e., mapping algorithms into computing tools.

With respect to capacity of data transmission/storage and speed of calculations, data structures are classified as:

- *Bit-level representations.*
- *Word-level representations.* Manipulation of logic functions and spectral transforms can be implemented on words. However, among word-level representations, only the arithmetic one satisfies the principles of linearity and superposition that are inherent properties of spectral transformation.
- *Linear word-level representations* are a boundary case of word-level representation of logic functions. Note that linearization of word-level expressions is not relevant to the above mentioned properties of linearity and superposition of spectral transform. Linearization of word-level expression is a method of hiding information: a special hiding technique is used on spectral components that corresponds to product terms with more than one literal in an algebraic expression. It is possible to approximate the spectrum using linear pieces without errors.

Bit-level representation, as described further in this chapter, is utilized for edge bit-serial input/output organization while processing is implemented in parallel. This paradigm is called *parallel-pipelined* or *systolic* computation. Word-level enhances the parallelism at input/output and processing levels, while preserving the parallel-pipelined logic.

Data structures compatible with the parallel-pipelined method are:

- *Algebraic descriptions* at the bit-level or word-level of computation (e.g., algebraic calculation of a switching function).
- *Arrays*, i.e., 1-D (vectors, e.g., the truth table of a switching function), 2-D (matrices), or multidimension arrays of logic values).
- *Decision trees and diagrams.* These data structures represent many optimization algorithms, such as branch-and-bound search, recursive procedures, etc. Regarding logic functions, decision diagrams are tools for spectral representation and calculation (Chapter 4).
- *Spatial representations* are defined as multidimensional topological structures. In this book, we focus on hypercube structures that are the result of embedding of binary or multivalued decision trees, including spectral ones. It should be noted that the $\mathcal{N}$-hypercube we introduced for logic functions does not perform the task of optimization of distributed processing on a macrolevel as does communication theory, but resembles it in a certain sense.

Finally, at the algorithmic level, the most appropriate representation forms of the parallel computing include:

- *Matrix transforms,* as an algebraic model, and
- *Signal graph* derived directly from the matrix transform, and the information flow model of the algorithm, which describes transmission of information from input to output, involving algebraic operations, etc.

10.2 Arrays

In microelectronics, computing on arrays is a solution to the problem of wiring, by the introduction of local interconnections. In this section we focus on nanoscale arrays that are supposed to fully exploit extreme scaling, e.g. with single-electron devices. Such circuits are suited to implementing bit-level processing, since they are locally interconnected, i.e., the processing elements are integrated into the nanowires so that local connection of neighbor elements is the optimal solution.

10.2.1 Cellular arrays

Cellular array refers to a network composed of some regular interconnection of logic cells. Cellular arrays have a number of properties useful for computing in nanodimensions:

- They are designed using a small set of simple processing elements (PEs), or cells, locally connected.
- They are locally controlled (e.g. Coloumb blockade devices by biasing in the single-electron circuits).
- Input-output is organized on edge-fed basis.
- Their topology can be embedded into many dimensions (1-D, 2-D, 3-D as shown in Figure 10.1).

The theory of cellar automata deals with large collections of interconnected finite automata, each finite automaton being thought of a cell. The location of each cell is specifiable by its Cartesian coordinates with respect to some arbitrary chosen origin and set of axes. Each cell contains an identical copy of the finite automaton, and state of a cell at time t is precisely the state of the its associated automaton at time t. Each cell is connected to the neighboring cells. An allowable assignment of states to all cells in the space is called a *configuration.*

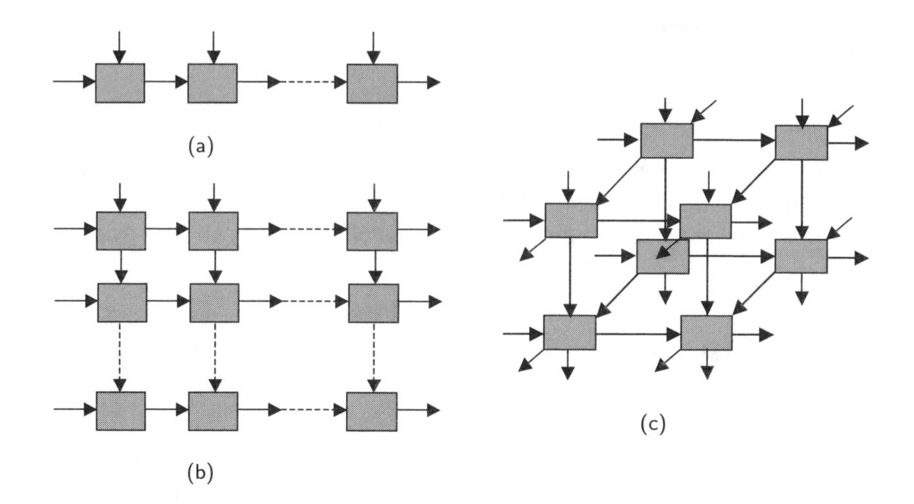

FIGURE 10.1
The structures of 1-D (a), 2-D (b) and 3-D (c) cellular arrays.

10.2.2 Systolic arrays

1-D and 2-D arrays implementing parallel-pipelined computation are known as *linear systolic arrays*. It is a subclass of cellular arrays.

Linear systolic arrays are defined as arrays of synchronized processors that process data in parallel by transmitting them, in rhythmic fashion, from one processor to the ones to which it is connected. For example, the implementation of a signal processing algorithm such as a fast Fourier transform (FFT), means one iteration ("butterfly" operation) per processor, so that an n-iteration fast transform is implemented on an array of n processors.

Linear systolic arrays can be divided into two classes:

- *Linear arrays* with bit-serial input/output, and processing elements (PEs) performing simple Boolean computations, and
- *Linear arrays* with word-level input/output and more sophisticated PE structure.

Linear arrays have a number of useful properties, in addition to those typical of cellular design in general:

- Simple input/output organization; and
- Easy embedding into hypercube-like topology.

In addition, the linear arrays can be optimized for a specific task; for example, for spectral transformations, simulation of cellular automata, cellular neural networks etc.

10.2.3 Tree-structured networks

Many search and recursive algorithms are tree-structured, and so too are the binary decision trees and diagrams for representation and manipulation of logic functions considered in this book. At the hardware-level, they can be mapped into tree-like arrays. The latter are remarkable for the following properties:

- ▶ They can be designed directly from decision trees (Shannon, Davio, Walsh, etc.);
- ▶ They can be embedded in spatial topologies, in particular, $\mathcal{N}$-hypercubes;
- ▶ Computing in tree-like arrays can be organized in parallel-pipelined fashion;
- ▶ These arrays have sequential and parallel inputs and outputs, due to their levelized structure.

Example 10.1 *A switch-type tree is designed from a reduced ordered binary decision diagram (ROBDD) by replacing Shannon nodes with demultiplexors. This is a particular ability of tree-like topology.*

10.3 Linear systolic arrays for computing logic functions

In this section, our approach to designing linear systolic arrays for manipulation of logic functions is introduced. This approach is appropriate for both switching and multivalued logic functions, so we will consider primarily switching functions, without losing generality.

10.3.1 Design technique

Designing the linear systolic arrays is accomplished by:

- ▶ Representation of the algorithm in recursive form, referred to as factorized form of matrix transforms, e.g., FFT, convolution, etc.;
- ▶ Mapping matrix equations into linear arrays; and
- ▶ Designing the PE-based on their formal model (size of memory for input and intermediate data, their operation and synchronization, etc.).

Corresponding signal flowgraphs (bitwise and word-wise) represent the iterative processing of the components of 1-D arrays (vectors) of input data in a parallel way.

In order to accomplish n-iteration, or recursive, computing, the linear systolic array consists of n PEs with memory organized as a first-in-first-out (FIFO) register (Figure 10.2a). Each PE includes a computing cell (CC) and a storage cell (SC). An example of a PE is given in Figure 10.2b.

The control of this element is very simple, though it must be carefully designed in the case of ultra-small devices (see "Further Reading" Section).

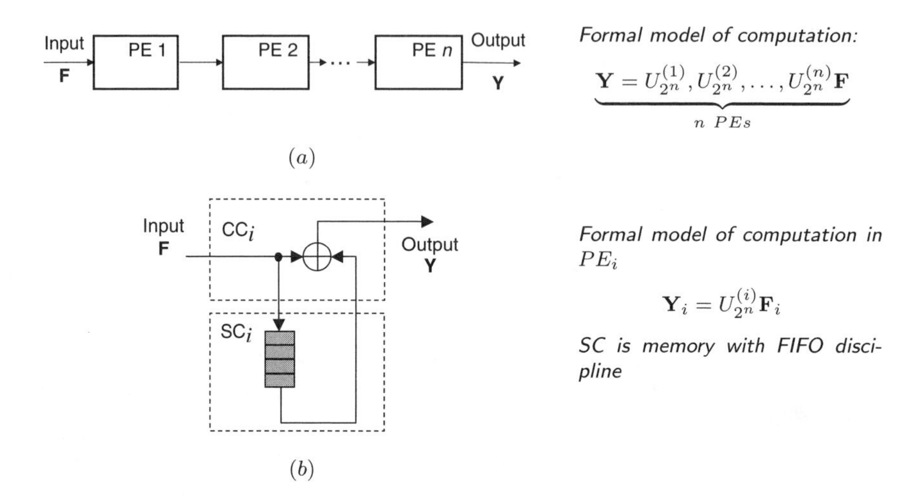

FIGURE 10.2
Linear systolic array (a) and PE (b).

10.3.2 Formal model of computation in a linear array

The formal model of computation in the i-th PE of the linear array is the matrix eqaution

$$\mathbf{Y}_i = U_{2^n}^{(i)}\mathbf{F}_i \tag{10.1}$$

where $2^n \times 2^n$ matrix $U_{2^n}^{(i)}$ is a Kronecker product

$$U_{2^n}^{(j)} = \underbrace{I_{2^{n-j}} \otimes U_2 \otimes I_{2^{j-1}}}_{i-th\ PEi},$$

U_2 is a 2×2 transform matrix, and $I_{2^{n-j}}$ and $I_{2^{n-j}}$ are identity matrices.

The formal model of computation in the array of n PEs corresponds to the product of n matrices

$$\underbrace{\mathbf{Y} = U_{2^n}^{(1)} \times U_{2^n}^{(2)} \times \cdots \times U_{2^n}^{(n)} \times \mathbf{F}}_{n\ processing\ elements}$$

The input data for the device is the $2^n \times 1$ coefficient vector $\mathbf{F}$, its l-th components $f(l)$ are loaded at l-th time $(l = 1, ..., 2^n - 1)$. Then, CCs provide the transform of the vector **F**, and the resulting coefficient vector **Y** appears as an output of n-th CC.

Example 10.2 *The design of the first PE_1 $(i = 1)$ of a linear systolic array given $n = 3$ and*

$$U = \begin{bmatrix} 1 & 1 \\ 1 & 1 \end{bmatrix}$$

is shown in Figure 10.3. The structure of the PE is identical to one depicted in Figure 10.2, and the SC_i includes four-bit FIFO register.

In the above example, the structure of matrix and the corresponding signal graph are the basis for defining FIFO discipline and synchronization.

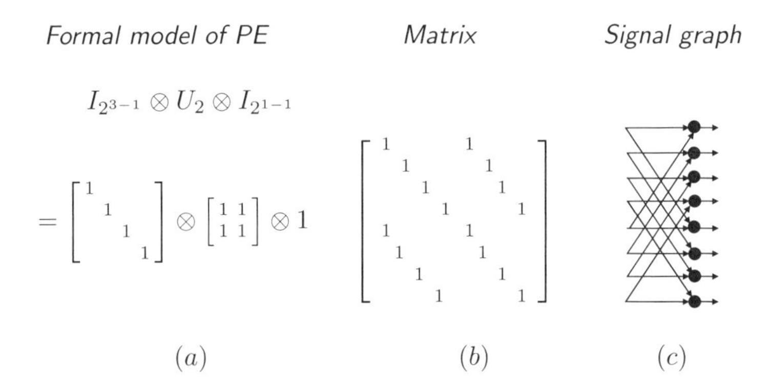

FIGURE 10.3
A PE_i model given $i = 1$: matrix eqaution (a), matrix (b), and "butterfly" flowgraph (c) (Example 10.2).

10.3.3 Parallel-pipelined computing

A linear systolic array processes data similarly to an assembly line:

- *The first phase* is called *speeding-up* (acceleration) a process whereby input data are processed by the first PE and the result is passed to the second PE, etc. In this phase, the computing resources of systolic processor are partially used;
- *The second phase* is called *stationary* processing of data. At this phase all PEs are used, i.e., the computing resources of a systolic processor are used totally;
- *The third phase* is called a *slowdown* (deceleration) process. At this phase, there are no input data but the rest of PEs continue the processing.

10.4 Computing Reed-Muller expressions

We introduce two approaches for designing linear systolic arrays for computing Reed-Muller expressions:

- Design based on factorization of the transform matrix. This technique utilizes the multiplicative property of matrices of direct and inverse transforms of logic functions. The transform matrix is represented by multiplication of factorized matrices. The problem is formulated as mapping a factorized matrix into the structure of a processing element of a linear array;
- Design based on Taylor expansion utilizes the properties of logic Taylor expansion to represent an arbitrary switching function in terms of Boolean differences. From the point view of computation, the complexity of this approach is the same as in factorization-based technique. However, it has several useful properties, in particular, information about the coefficients in terms of Boolean differences is given.

This technique can be applied to design of linear systolic arrays for computing arithmetic and Walsh expressions and inverse transforms. This is also suitable for processing multiple-valued functions unless the model of computation is matrix based.

10.4.1 Factorization of transform matrix

To design the flowgraph of an algorithm, the matrix $\mathbf{R}_{2^n}$ must be represented in the factorized form

$$\mathbf{R}_{2^n} = \underbrace{\mathbf{R}_{2^n}^{(1)}\mathbf{R}_{2^n}^{(2)}\cdots\mathbf{R}_{2^n}^{(n)}}_{n\ PEs}, \tag{10.2}$$

where $\mathbf{R}_{2^n}^{(i)}$, $i = 1, 2, \ldots, n$, is formed by the Kronecker product

$$\mathbf{R}_{2^n}^{(i)} = \underbrace{\mathbf{I}_{2^{n-i}} \otimes \mathbf{R}_2 \otimes \mathbf{I}_{2^{i-1}}}_{PEi}. \tag{10.3}$$

Hence, Reed-Muller coefficients are computed in n iterations.

Example 10.3 *Given $n = 3$, the flowgraph includes three iterations according to the factorization relation:*

$$
\begin{aligned}
\mathbf{R}_{2^3} &= \mathbf{R}_{2^3}^{(1)}\mathbf{R}_{2^3}^{(2)}\mathbf{R}_{2^3}^{(3)} \\
&= \underbrace{(\mathbf{I}_{2^{3-1}} \otimes \mathbf{R}_2 \otimes \mathbf{I}_{2^{1-1}})}_{PE1}\underbrace{((\mathbf{I}_{2^{3-2}} \otimes \mathbf{R}_2 \otimes \mathbf{I}_{2^{2-1}})}_{PE2}\underbrace{((\mathbf{I}_{2^{3-3}} \otimes \mathbf{R}_2 \otimes \mathbf{I}_{2^{3-1}})}_{PE3} \\
&= \underbrace{((\mathbf{I}_{2^2} \otimes \mathbf{R}_2 \otimes 1)}_{PE1}\underbrace{((\mathbf{I}_2 \otimes \mathbf{R}_2 \otimes \mathbf{I}_2)}_{PE2}\underbrace{((1 \otimes \mathbf{R}_2 \otimes \mathbf{I}_{2^2})}_{PE3} \\
&= \left[\begin{array}{cc|cc} \mathbf{R}_2 & & & \\ & \mathbf{R}_2 & & \\ \hline & & \mathbf{R}_2 & \\ & & & \mathbf{R}_2 \end{array}\right]
\left[\begin{array}{cc|cc} \mathbf{I}_2 & & & \\ \mathbf{I}_2 & \mathbf{I}_2 & & \\ \hline & & \mathbf{I}_2 & \\ & & \mathbf{I}_2 & \mathbf{I}_2 \end{array}\right]
\left[\begin{array}{c|c} \mathbf{I}_4 & \\ \hline \mathbf{I}_4 & \mathbf{I}_4 \end{array}\right] \\
&= \underbrace{\begin{bmatrix} 1 &&&&&&& \\ 1 & 1 &&&&&& \\ && 1 &&&&& \\ && 1 & 1 &&&& \\ &&&& 1 &&& \\ &&&& 1 & 1 && \\ &&&&&& 1 & \\ &&&&&& 1 & 1 \end{bmatrix}}_{FIFO_{SC_1}}
\underbrace{\begin{bmatrix} 1 &&&&&&& \\ & 1 &&&&&& \\ 1 && 1 &&&&& \\ & 1 && 1 &&&& \\ &&&& 1 &&& \\ &&&&& 1 && \\ &&&& 1 && 1 & \\ &&&&& 1 && 1 \end{bmatrix}}_{FIFO_{SC_2}}
\underbrace{\begin{bmatrix} 1 &&&&&&& \\ & 1 &&&&&& \\ && 1 &&&&& \\ &&& 1 &&&& \\ 1 &&&& 1 &&& \\ & 1 &&&& 1 && \\ && 1 &&&& 1 & \\ &&& 1 &&&& 1 \end{bmatrix}}_{FIFO_{SC_3}}
\end{aligned}
$$

Figure 10.4 illustrates designing PEs of a linear systolic array for computing Reed-Muller expressions of three variables. The following matrix model corresponds to this iterative process:

$$\mathbf{R} = R_{2^3}^{(1)} R_{2^3}^{(2)} R_{2^3}^{(3)} \mathbf{F} \quad (mod\ 2)$$

A systolic array for computing a 0-polarity Reed-Muller expression of n variables is shown in Figure 10.5a. Figure 10.5b shows the SC's content at the first three and the last two cycles of calculation in the arrays.

10.4.2 Design based on logic Taylor expansion

The *l*ogic Taylor series for a Boolean function f of n variables at point $c \in 0, 1, \ldots, 2^n - 1$ is defined as

$$f = \bigoplus_{i=0}^{2^n-1} f_i^{(c)} \underbrace{(x_1 \oplus c_1)^{i_1} \ldots (x_n \oplus c_n)^{i_n}}_{i-th\ product},$$

where the i-th coefficient is the Boolean difference

$$f_i^{(c)}(d) = \left. \frac{\partial^n f(c)}{\partial x_1^{i_1} \partial x_2^{i_2} \ldots \partial x_n^{i_n}} \right|_{d=c} \quad \text{and} \quad \partial x_i^{i_j} = \begin{cases} 1, & i_j = 0 \\ \partial x_j, & i_j = 1 \end{cases},$$

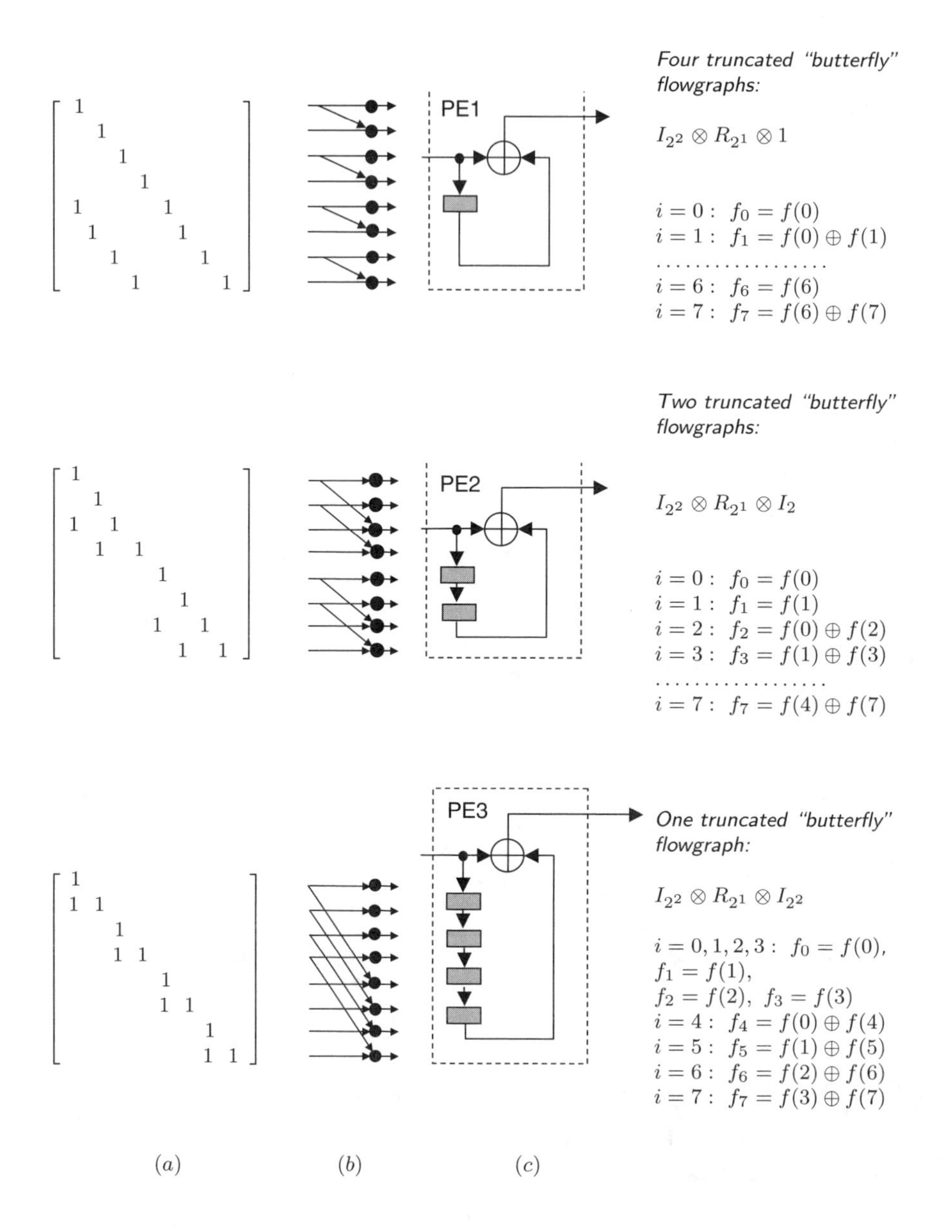

FIGURE 10.4
Design of PEs for computing Reed-Muller expressions: transform matrix (a), 4-, 2-, and 1-truncated "butterfly" flowgraphs (b), and PE's structure and model (c) (Example 10.3).

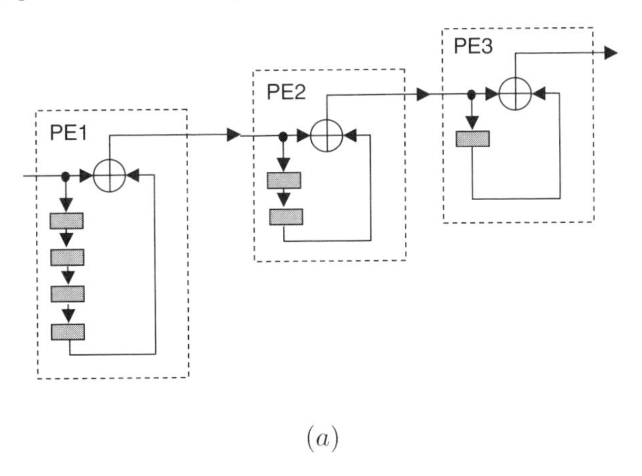

(a)

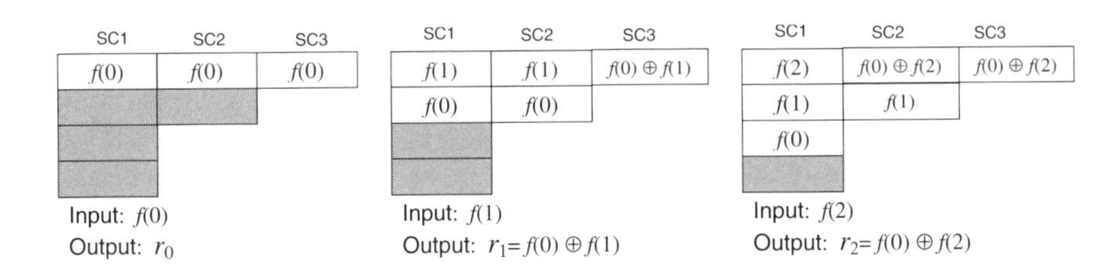

(b)

FIGURE 10.5
Design of a linear systolic array (Example 10.3).

which is a value of the n-ordered Boolean difference of f where $x_1 = c_1$, $x_2 = c_2, \ldots, x_n = c_n$. It follows from this that deriving Reed-Muller coefficients can be accomplished by using processors for calculation of Boolean differences (see Section 10.5).

10.5 Computing Boolean differences

In Chapter 8, the matrix form of a Boolean difference with respect to the i-th variable x_i was defined by the equation

$$\frac{\partial \mathbf{F}}{\partial x_i} = D_{2^n}^{(i)} \mathbf{F} \quad (mod\ 2).$$

Matrix $D_2^{(i)}$ is derived using the equation

$$D_{2^n}^{(i)} = I_{2^{i-1}} \otimes \begin{bmatrix} 1 & 1 \\ 1 & 1 \end{bmatrix} \otimes I_{2^{n-i}}.$$

This equation is one iteration in a calculation of m-order Boolean difference in a linear systolic array consisting of m PEs. The m-order Boolean difference is already factorized representation. A PE must implement the matrix-vector multiplication $D_2^{(i)}\mathbf{F}$, and the structure of the PE is similar to one considered for calculation of Reed-Muller transform, except that two switches are required instead of one.

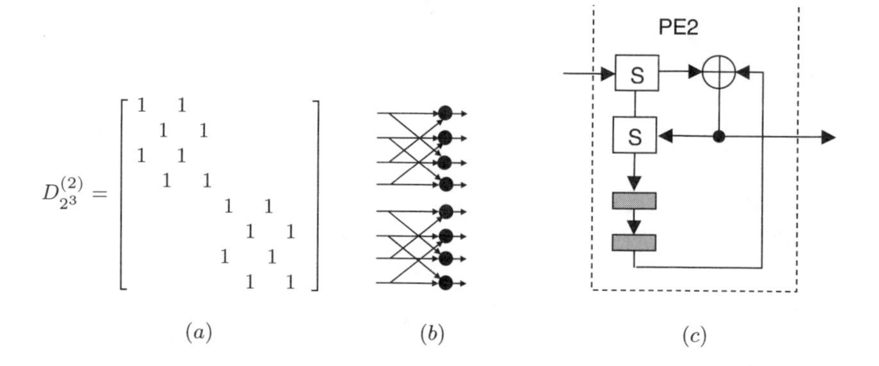

FIGURE 10.6
Computing of Boolean differences of a switching function of three variables with respect to variable x_2: transform matrix (a), "butterfly" flowgraph (b), and PE (c) (Example 10.4).

Example 10.4 *Figure 10.6 illustrates designing a PE given $i = 2$ of a linear systolic array for computing a multiple Boolean difference of a switching function of three variables.*

10.6 Computing arithmetic expressions

The computing of arithmetic expressions discussed below is applicable to

- Single-output switching functions; and
- Multioutput switching functions, or word-level representations.

To map the arithmetic transform algorithm to a linear systolic array, the matrix $\mathbf{P}_{2^n}$ has to be factorized

$$\mathbf{P}_{2^n} = \mathbf{P}_{2^n}^{(1)}\mathbf{P}_{2^n}^{(2)} \cdots \mathbf{P}_{2^n}^{(n)}, \tag{10.4}$$

where $\mathbf{P}_{2^n}^{(i)}$, $i = 1, 2, \ldots, n$, is formed by the Kronecker product

$$\mathbf{P}_{2^n}^{(i)} = \underbrace{\mathbf{I}_{2^{n-i}} \otimes \mathbf{P}_2 \otimes \mathbf{I}_{2^{i-1}}}_{PEi}.$$

Example 10.5 *The flowgraph includes two iterations accordingly to factorization relations*

$$\begin{aligned}
\mathbf{P}_{2^3} &= \mathbf{P}_{2^3}^{(1)}\mathbf{P}_{2^3}^{(2)}\mathbf{P}_{2^3}^{(3)} \\
&= \underbrace{(\mathbf{I}_{2^{3-1}} \otimes \mathbf{P}_2 \otimes \mathbf{I}_{2^{1-1}})}_{PE1}\underbrace{(\mathbf{I}_{2^{3-2}} \otimes \mathbf{P}_2 \otimes \mathbf{I}_{2^{2-1}})}_{PE2}\underbrace{(\mathbf{I}_{2^{3-3}} \otimes \mathbf{P}_2 \otimes \mathbf{I}_{2^{3-1}})}_{PE3} \\
&= \underbrace{(\mathbf{I}_{2^2} \otimes \mathbf{P}_2 \otimes 1)}_{PE1}\underbrace{(\mathbf{I}_2 \otimes \mathbf{P}_2 \otimes \mathbf{I}_2)}_{PE2}\underbrace{(1 \otimes \mathbf{P}_2 \otimes \mathbf{I}_{2^2})}_{PE3} \\
&= \left[\begin{array}{cc|cc} \mathbf{P}_2 & & & \\ & \mathbf{P}_2 & & \\ \hline & & \mathbf{P}_2 & \\ & & & \mathbf{P}_2 \end{array}\right]
\left[\begin{array}{cc|cc} \mathbf{I}_2 & & & \\ \mathbf{I}_2 & \mathbf{I}_2 & & \\ \hline & & \mathbf{I}_2 & \\ & & \mathbf{I}_2 & \mathbf{I}_2 \end{array}\right]
\left[\begin{array}{c|c} \mathbf{I}_4 & \\ \hline \mathbf{I}_4 & \mathbf{I}_4 \end{array}\right]
\end{aligned}$$

$$
= \underbrace{\begin{bmatrix} 1 &&&&&&& \\ -1 & 1 &&&&&& \\ && 1 &&&&& \\ && -1 & 1 &&&& \\ &&&& 1 &&& \\ &&&& -1 & 1 && \\ &&&&&& 1 & \\ &&&&&& -1 & 1 \end{bmatrix}}_{FIFO_{SC_1}}
\underbrace{\begin{bmatrix} 1 &&&&&&& \\ & 1 &&&&&& \\ -1 && 1 &&&&& \\ & -1 && 1 &&&& \\ &&&& 1 &&& \\ &&&&& 1 && \\ &&&& -1 && 1 & \\ &&&&& -1 && 1 \end{bmatrix}}_{FIFO_{SC_2}}
\underbrace{\begin{bmatrix} 1 &&&&&&& \\ & 1 &&&&&& \\ && 1 &&&&& \\ &&& 1 &&&& \\ -1 &&&& 1 &&& \\ & -1 &&&& 1 && \\ && -1 &&&& 1 & \\ &&& -1 &&&& 1 \end{bmatrix}}_{FIFO_{SC_3}}
$$

A linear systolic array for computing arithmetic expressions of n-variable switching function includes n PEs. Computing of identical transform for many functions is organized in a parallel-pipelined linear array by:

- ▶ Sequential feeding of the input with elements (truth-vectors) of the function in a particular order, or
- ▶ Parallel input/output of words of elements.

The first does not need any modification of bit-level linear systolic array design. The second requires significant modification to afford parallelization on the PE level.

10.7 Computing Walsh expressions

One approach to mapping the Walsh transform in linear systolic array design is similar to the one for Reed-Muller and arithmetic transforms. First, the matrix $\mathbf{W}_{2^n}$ must be represented in the factorized form

$$
\mathbf{W}_{2^n} = \underbrace{\mathbf{W}_{2^n}^{(1)} \mathbf{W}_{2^n}^{(2)} \cdots \mathbf{W}_{2^n}^{(n)}}_{n\ PEs}, \tag{10.5}
$$

where $\mathbf{W}_{2^n}^{(i)}$, $i = 1, 2, \ldots, n$, is formed by the Kronecker product

$$
\mathbf{W}_{2^n}^{(i)} = \underbrace{\mathbf{I}_{2^{n-i}} \otimes \mathbf{W}_2 \otimes \mathbf{I}_{2^{i-1}}}_{PEi}.
$$

The flowgraph of the algorithm corresponds to the n-iteration processing of the input data (components of the truth table). The corresponding systolic array includes n PEs; each PE implements one iteration. The structure of the SC and combinational part in the PE is similar to one for calculation of Boolean difference, except that control is organized in a different way.

Example 10.6 *Given a three-variable switching function, calculation of Walsh coefficients includes three iterations according to the factorization relations. The first, second and third iteration are implemented on the PE with four, two and one register SC respectively:*

$$
\begin{aligned}
\mathbf{W}_{2^3} &= \mathbf{W}_{2^3}^{(1)}\mathbf{W}_{2^3}^{(2)}\mathbf{W}_{2^3}^{(3)} \\
&= \underbrace{(\mathbf{I}_{2^{3-1}} \otimes \mathbf{W}_2 \otimes \mathbf{I}_{2^{1-1}})}_{PE1}\underbrace{(\mathbf{I}_{2^{3-2}} \otimes \mathbf{W}_2 \otimes \mathbf{I}_{2^{2-1}})}_{PE2}\underbrace{(\mathbf{I}_{2^{3-3}} \otimes \mathbf{W}_2 \otimes \mathbf{I}_{2^{3-1}})}_{PE3} \\
&= \underbrace{(\mathbf{I}_{2^2} \otimes \mathbf{W}_2 \otimes 1)}_{PE1}\underbrace{(\mathbf{I}_2 \otimes \mathbf{W}_2 \otimes \mathbf{I}_2)}_{PE2}\underbrace{(1 \otimes \mathbf{W}_2 \otimes \mathbf{I}_{2^2})}_{PE3} \\
&= \left[\begin{array}{cc|cc}
\mathbf{W}_2 & & & \\
 & \mathbf{W}_2 & & \\ \hline
 & & \mathbf{W}_2 & \\
 & & & \mathbf{W}_2
\end{array}\right]
\left[\begin{array}{cc|cc}
\mathbf{I}_2 & & & \\
\mathbf{I}_2 & \mathbf{I}_2 & & \\ \hline
 & & \mathbf{I}_2 & \\
 & & \mathbf{I}_2 & \mathbf{I}_2
\end{array}\right]
\left[\begin{array}{c|c}
\mathbf{I}_4 & \\ \hline
\mathbf{I}_4 & \mathbf{I}_4
\end{array}\right] \\
&= \underbrace{\begin{bmatrix}
1 & 1 & & & & & & \\
-1 & 1 & & & & & & \\
 & & 1 & 1 & & & & \\
 & & -1 & 1 & & & & \\
 & & & & 1 & 1 & & \\
 & & & & -1 & 1 & & \\
 & & & & & & 1 & 1 \\
 & & & & & & -1 & 1
\end{bmatrix}}_{FIFO_{SC_1}}
\underbrace{\begin{bmatrix}
1 & & 1 & & & & & \\
 & 1 & & 1 & & & & \\
-1 & & 1 & & & & & \\
 & -1 & & 1 & & & & \\
 & & & & 1 & & 1 & \\
 & & & & & 1 & & 1 \\
 & & & & -1 & & 1 & \\
 & & & & & -1 & & 1
\end{bmatrix}}_{FIFO_{SC_2}}
\underbrace{\begin{bmatrix}
1 & & & & 1 & & & \\
 & 1 & & & & 1 & & \\
 & & 1 & & & & 1 & \\
 & & & 1 & & & & 1 \\
-1 & & & & 1 & & & \\
 & -1 & & & & 1 & & \\
 & & -1 & & & & 1 & \\
 & & & -1 & & & & 1
\end{bmatrix}}_{FIFO_{SC_3}}
\end{aligned}
$$

10.8 Tree-based network for manipulating a switching function

The tree-based structure is remarkable due to the fact that:

- The tree, embedded into a hypercube, will form a multidimensional architecture for parallel-pipelined computing.
- The tree-based structure can be used for spectral transforms as well. This is due to the Taylor logic series interpretation: the values of the Boolean differences are the coefficients of the Reed-Muller expansion. This means that calculation of other transforms (arithmetic, Walsh) is also possible on that architecture.

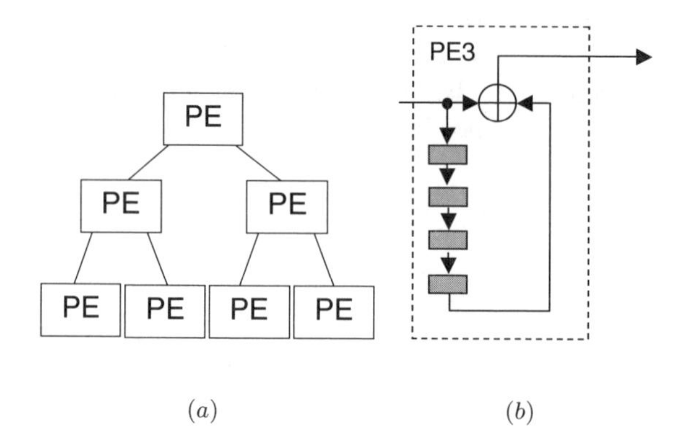

FIGURE 10.7
Tree-network systolic array for implementation of Davio decision tree of a three-variable functions (a), and PE for Davio expansion with respect to the first variable (b).

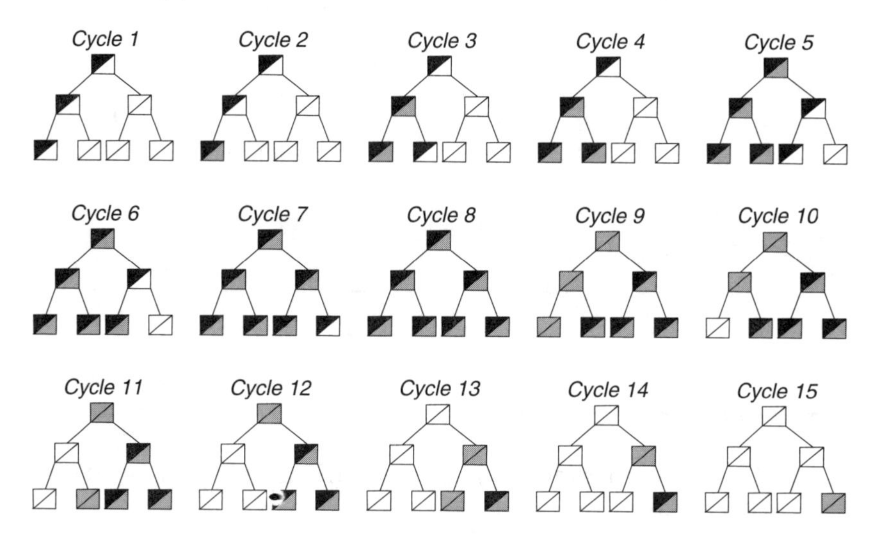

FIGURE 10.8
Fifteen cycles of a tree-network systolic array (black, gray and white state correspond to loading, computing and noncomputing processes respectively).

10.9 Hypercube arrays

In previous chapters, different aspects of computations using hypercubes and $\mathcal{N}$-hypercubes were introduced. These models can be mapped to the so-called *hypercube arrays* since they have

- A homogeneous structure (a network of locally connected identical nodes);
- The functions of nodes are the same;
- The information flow is a parallel-pipelined; and
- These arrays have sequential and parallel inputs and outputs, due to a certain lack of constraints on locality in 3-D space.

We consider two approaches to designing $\mathcal{N}$-hypercube cellular arrays:

- Mapping the $\mathcal{N}$-hypercubes that are derived from embedding decision trees and diagrams that represent switching or multivalued logic functions; and
- Mapping the binary tree-based structures, considered in this chapter above, in particular, tree-based arrays calculation of Reed-Muller spectrum, arithmetic and Walsh transforms and Boolean differences.

The first class of the hypercube arrays are essentially a direct mapping of the $\mathcal{N}$-hypercube topology into an array, with adequate data flow organization. They are based on switch-type PEs. However, the principal difference is that in parallel-pipelined computation, PEs are usually interspersed with memory, or delay elements, so $\mathcal{N}$-hypercube topology must be modified taking into account this requirement.

Example 10.7 *Let us implement a two-variable switching function, represented by a binary decision tree (Figure 10.9) on a cellular array. First, we embed the tree in the $\mathcal{N}$-hypercube topology. Next, we map the $\mathcal{N}$-hypercube into 2-D array topology (since $n = 2$). The PE is a switch (the simplest demultiplexer), and PEs are also divided by delay element, which maintains pipeline processing paradigm in the cellular arrays.*

The hypercube arrays that belong to the second class are designed by embedding (reconfigurating) the tree-based arrays introduced in Section 10.8. The PEs and data flow are identical to the ones used in the tree-based structure.

Example 10.8 *To implement a 0-polarity Reed-Muller transform for a two-variable switching function, we:*

(a) Design a Davio decision tree,
(b) Embed the tree in a 2-D hypercube,
(c) Create a matrix model of the transform (two-iteration matrix-vector multiplication), and
(d) Design two PEs so that they implement the first and the second iteration of the matrix transform; these PEs are the central and the intermediate nodes of the 2-D hypercube correspondingly (Figure 10.10).

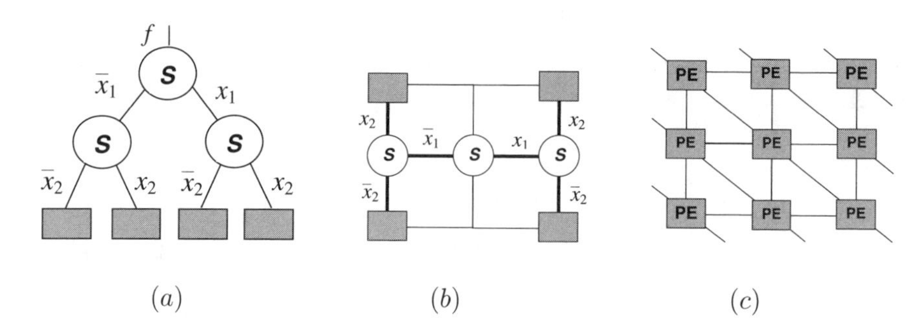

FIGURE 10.9
Graphical representation of two-input function in sum-of-products form: decision tree (a), decision tree embedded in a hypercube (b), and cellular topology (c).

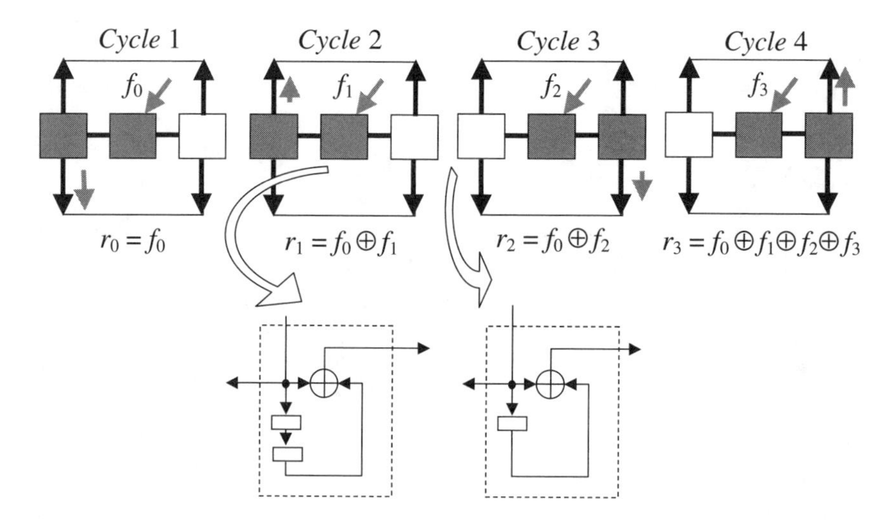

FIGURE 10.10
Design of an $\mathcal{N}$-hypercube array (Example 10.8).

10.10 Summary

1. Arrays of regularly and locally interconnected logic cells, called *cellular systolic arrays*, are a prospective solution for nanocircuit design because:

   - The *pipelined logic style* implemented in the cellular systolic arrays is well-suited for ultimate scalability of nanostructures;

- They avoid long range interconnections, which is consistent with the restrictions of nanotechnolgies, i.e. nanowire paradigm,
- Signal flow in the cellular arrays is suited to manipulation with single bits, or bitwise words. This is consistent with elementary carries of information in nanoscale devices, e.g. single electrons; and
- 1-D and 2-D cellular arrays can be easily embedded in hypercube-like structure using the technique proposed in this book and called *embedding in the $\mathcal{N}$-hypercube.*

2. Systolic arrays utilize parallelism on two levels:

- *Global* (cells perform parallel to each other), and
- *Local* (each cell explores inherent properties of parallel processing).

This parallelism is not in contradiction with nanoelectronic design principles, unless physical requirements are satisfied (such as screening length that must be short compared to the spacing between electrons in a cellular single-electronic circuit where more than one electron is used in processing at the same time, in contrast with noncellular single-electron devices).

3. Linear nanoscale systolic arrays can be used for computing

- Switching functions represented in sum-of-products form;
- Reed-Muller, arithmetic, and Walsh forms of logic functions;
- Multivalued logic functions;
- Boolean and logic differences;
- Switching and multivalued arithmetic word-level forms, utilizing extra resources for parallelization; and
- Matrix computing.

10.11 Problems

Problem 10.1 The 0-polarity Reed-Muller transform for a two-variable switching function has been mapped in a hypercube-like structure: (Figure 10.11). The function of PEs is given in an order different from that considered in the sections above. Prove that the results must be the same for this order of iterations, compared to the opposite order.

Problem 10.2 Given a Shannon tree for a two-input function NAND (Figure 10.12), find the design of the cellular topology and show the data flow during cycles of acceleration, computing, and deceleration.

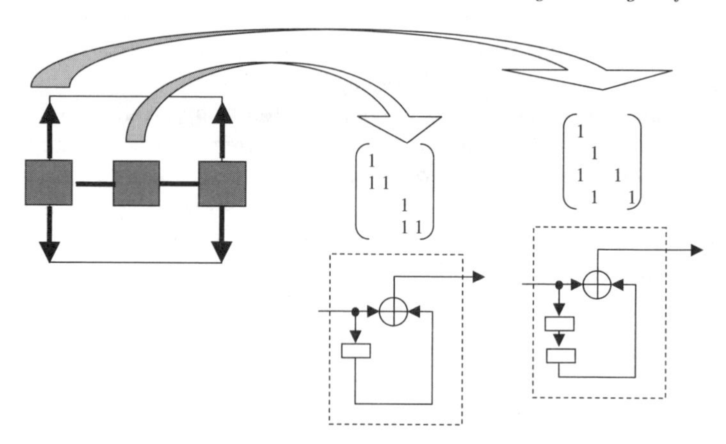

FIGURE 10.11
Design of an $\mathcal{N}$-hypercube array (Problem 10.1).

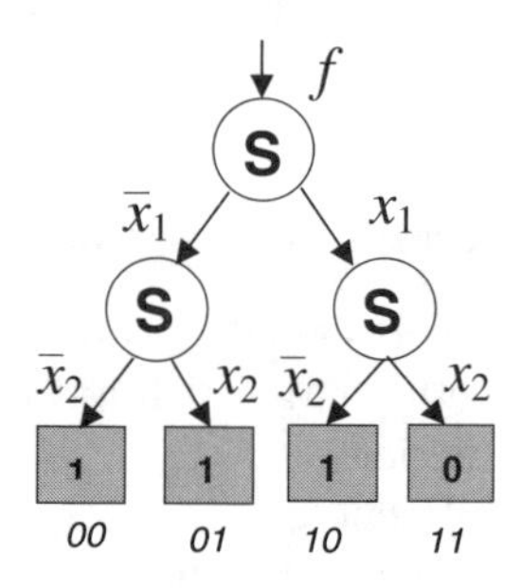

FIGURE 10.12
Shannon decision tree of two variables (Problem 10.2).

Problem 10.3 Propose the structure of the first PE (implementing first iteration) in a linear systolic array for computing the Walsh spectrum of a three variable switching function.

Problem 10.4 Propose the structure of the third PE (implementing third iteration) in a linear systolic array for computing the arithmetic spectrum of a three variables switching function.

Problem 10.5 Propose a single-electron solution of PE design with one FIFO register. Use a pump-based switch, memory element, and EXOR gate implementation considered in Chapter 2.

Problem 10.6 Let a switching function of two variables be represented by a Davio decision tree. Interpret its right branch in terms of Boolean differences (Figure 10.13a). The tree embedded into the $\mathcal{N}$-hypercube is shown in Figure

10.13b. Show the data flow during cycles of acceleration, computing, and deceleration.

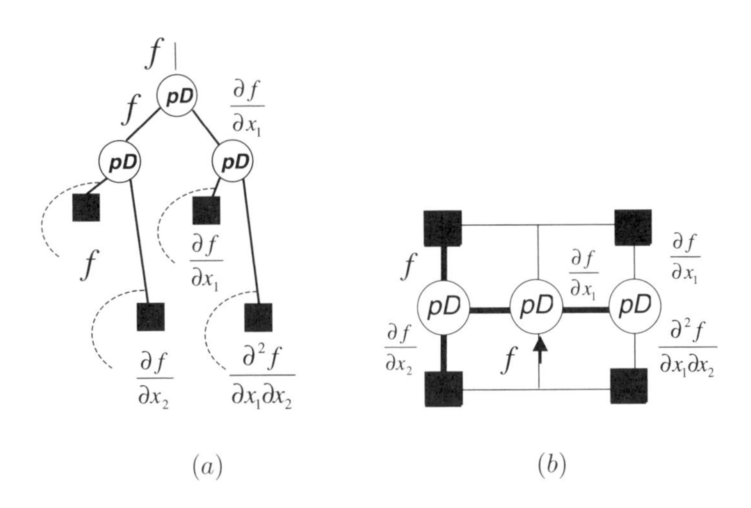

FIGURE 10.13
Embedding a Davio decision tree of two variables interpreted in terms of Boolean differences in the $\mathcal{N}$-hypercube (Problem 10.6).

10.12 Further reading

Parallel architectures. Leighton's book [12] is a comprehensive general reference to parallel algorithms and architectures.

Cellular arrays. A class of cellular parallel-pipelined arrays, cellular automata were introduced by John von Neumann in his paper "The Theory of Automata: Construction, Reproduction, Homogeneity" in trying to develop an abstract model of self-reproduction in biology – a topic which had emerged from investigations in cybernetics. Self-reproduction in cellular space is defined as a special case of configuration. The configuration C is self-reproducing if there exists a transition δ such that C construct C_δ.

By the end of the 1950s, cellular automata could be viewed as parallel computers. Around 1961, Edward Fredkin simulated the 2-D analog of rule 90 on a PDP-1 computer, and noted its self-reproduction properties.

Quite disconnected from all this, even in the 1950s, specific types of 1-D and 2-D cellular automata were already being used in various electronic devices and special-purpose computers. In fact, when digital image processing began to be done in the mid-1950s (for such applications as optical character recognition and microscopic particle counting) 2-D cellular automaton rules were usually what was used to remove noise. And for several decades, starting in 1960, a long line of so-called cellular logic systems were built to implement 2-D cellular automata (see, for example, [2, 4, 14]), mainly for image processing. Ever since the 1960s the idea of making array or parallel computers has nevertheless resurfaced repeatedly, notably in systems like the ILLIAC IV from the 1960s and 1970s, and systolic arrays and various massively parallel computers from the 1980s. From the 1960s onward simulations of idealized neural networks sometimes had neurons connected to neighbors on a grid, yielding a 2-D cellular automaton.

Systolic arrays. Systolic 1-D and 2-D arrays have been proposed by Kung [10, 11]. In a certain sense, they resemble cellular automata. This idea, combined with RISC architectures based on pipeline paradigm, gave birth to parallel-pipelined computations. The processors that implement such computing have been called *systolic* by Kung in 1980. Later, however, the modest term *parallel-pipelined* has been rehabilitated. A 2-D architecture has been called *matrix systolic arrays*, or simply systolic arrays. They can be two-dimensional (rectangular, triangular, tree-like) and data may flow between the cells (which can be programmed) in different directions and at different speeds. The algorithms that specify operations of each cell in systolic arrays are called *systolic algorithms.* These algorithms have been suggested for solving many problems such as binary and polynomial arithmetic, solution of linear systems, geometric problems, and matrix operations. The latter, especially matrix multiplication, filtering and convolution, have become the basis for systolic signal processors. These processors are based on the mapping of fast digital signal processing algorithms (FFT) to parallel-pipelined structures of the single-input and single-output systolic arrays.

Linear systolic arrays. A pipeline is an example of a *linear systolic array* in which data flows only in one direction. A design of linear systolic arrays have been considered in [9]. Also a multidimensional analysis of signal based on multiple application of one-dimensional analysis has been introduced in by Kukharev et al. [7, 8]. Linear systolic arrays have also been developed for transforms of Boolean functions [7, 16], including solutions of Boolean equations [13] and arithmetic and other transforms [19].

Array arithmetic circuits. A multiplier can be considered as matrix array (Figure 10.14). The formal description is based on a matrix of partial products that is implemented by two-dimensional arrays of full adders (to sum the rows of partial products). In Figure 10.14, the 8×6 structure of a multiplier is shown that is known as an *array multiplier*. The multiplier consists of $m - 1 = 5$ $n = 8$-bit carry-ripple adders and $m = 6$ arrays of n AND gates. The delay of the multiplier is defined as the critical path equal to sum of delay of the buffer circuit connecting the input signal and the AND gates, the delay of AND gate, and the delay of the adders. Systolic architecture of a multiplier utilizes the natural properties of a multiplication, namely, a regularity of structure and a local interconnection (each cell is connected only to its neighbor's).

A design of a multiplier based on systolic paradigm has been discussed, in particular, by Sinha and Srimani [17] and Danielsson [5]. The homogeneous highly parallel arithmetic circuits developed in last decade, are the focus in nanotechnology. For example, the logarithmic multiplier can be considered as a good candidate for nanotechnology. This multiplier computes the product of two terms. The property used is $\log(A \times B) = \log(A) + \log(B)$. To obtain the logarithm of a number, the look-up tables, recursive algorithms or the segmentation of the logarithmic curve can be used [3].

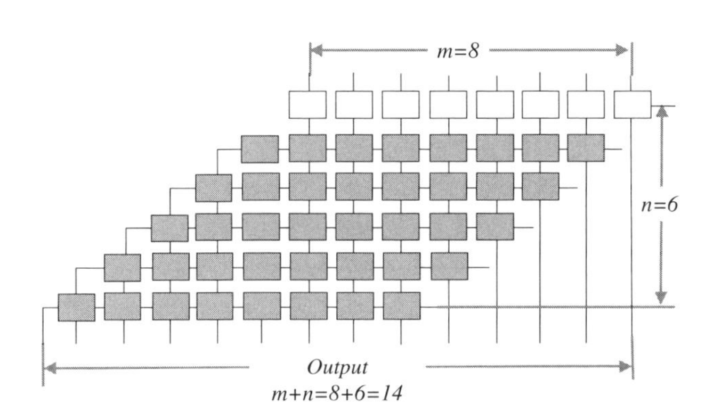

a_7b_0 a_6b_0 a_5b_0 a_4b_0 a_3b_0 a_2b_0 a_1b_0 a_0b_0
a_7b_1 a_6b_1 a_5b_1 a_4b_1 a_3b_1 a_2b_1 a_1b_1 a_0b_1
a_7b_2 a_6b_2 a_5b_2 a_4b_2 a_3b_2 a_2b_2 a_1b_2 a_0b_2
a_7b_3 a_6b_3 a_5b_3 a_4b_3 a_3b_3 a_2b_3 a_1b_3 a_0b_3
a_7b_4 a_6b_4 a_5b_4 a_4b_4 a_3b_4 a_2b_4 a_1b_4 a_0b_4
a_7b_5 a_6b_5 a_5b_5 a_4b_5 a_3b_5 a_2b_5 a_1b_5 a_0b_5

FIGURE 10.14
A 8×6 array multiplier.

Addition and multiplication in Galois fields, $GF(2^n)$ plays an important role in coding theory and is widely used in digital computers and data transmission or storage systems. The group theory is used to introduce the algebraic system, called a *field*. A field is a set of elements in which we can do addition, subtraction, multiplication and division without leaving the set. Systolic arrays for addition and multiplication in Galois fields have been proposed in many papers on digital signal processing, in particular, in [15, 18].

A design of systolic array using SET has been introduced by Ancona [1].

Additional references on systolic arrays and fault-tolerance cellular arrays are provided in Chapter 11.

10.13 References

[1] Ancona MG. Systolic processor design using single-electron digital circuits. *Superlattices and Microstructures*, 20(4):461–472, 1996.

[2] Chaudhuri PP, Chowdhury DP, Nandi S, and Chattopadhyay S. *Additive Cellular Automata: Theory and Applications.* IEEE Computer Society – Wiley, USA, 1997.

[3] Coleman JN, Chester CI, Softley CI, and Kaldec J. Arithmetic on the European logarithmic microprocessor. *IEEE Transactions on Computers*, 49(7):702–715, 2000.

[4] Delorme M, and Mazoyer J. *Cellular Automata: A Parallel Model.* Kluwer Academic Publishers, Dordrecht, 1998.

[5] Danielsson PE. Serial parallel convolvers. In Swartzlander EE Jr., Ed., *Systolic Signal Processing Systems.* Dekker, NY, 1987.

[6] Jones SR, Sammut KM, and Hunter J. Learning in linear systolic neural network engines: analysis and implementation. *IEEE Transactions on Neural Networks*, 5(4):584-593, 1994.

[7] Kukharev GA, Tropchenko AY, and Shmerko VP. *Systolic Signal Processors.* Publishing House "Belarus", Minsk, Belarus, 1988.

[8] Kukharev GA, Shmerko VP, and Zaiseva EN. *Processing Multi-Valued Data.* Publishing House "Science and Technics", Minsk, Belarus, 1990.

[9] Kumar VKP, and Tsai Y-C. Designing linear systolic arrays. *Journal of Parallel and Distributed Computing*, 7:441–463, 1989.

[10] Kung SY. *VLSI Array Processors.* Prentice Hall, New York, 1988.

[11] Kung HT, and Leiserson CE. Systolic arrays (for VLSI). In *Sparse Matrix Proceedings*, pp. 256–282. SIAM, Philadelphia, 1978.

[12] Leighton FT. *Introduction to Parallel Algorithms and Architectures: Arrays, Trees and Hypercubes.* Kaufmann, San Mateo, CA, 1992.

[13] Levashenko VG, Shmerko VP, and Yanushkevich SN. Solution of Boolean Differential Equations on Systolic Arrays. *Cybernetics and Systems Analysis*, Kluwer/Plenum Publishers, 32(1):26–40, 1996.

[14] Lohn JD, and Reggia JA. Automatic discovery of self replicating structures in cellular automata. *IEEE Transactions on Evolutionary Computation*, 1(3):165–178, 1997.

[15] Popovici E, and Fitzpatrick P. Algorithm and architecture for a multiplicative Galois field processor, *IEEE Transactions on Information Theory*, 49:3303–3307, 2003.

[16] Shmerko VP, and Yanushkevich SN. Algorithms of Boolean differential calculus for systolic arrays. *Cybernetics and Systems Analysis*, Kluwer/Plenum Publishers, 3:38–47, 1990.

[17] Sinha BP, and Srimani PK. Fast parallel algorithms for binary multiplication and their implementation on systolic architectures. *IEEE Transactions on Computers*, 38(3):424–431, 1989.

[18] Wang CL, and Lin JL. Systolic array implementation of multipliers for finite fields GF(2^m). *IEEE Transactions on Circuits and Systems*, 38(7):796–800, 1991.

[19] Yanushkevich SN. Systolic algorithms to synthesize arithmetical polynomial forms for k-valued logic functions. *Automation and Remote Control*, Kluwer/Plenum Publishers, 55(12):1812–1823, 1994.

11

Fault-Tolerant Computation

This chapter contributes to fault-tolerant nanoICs design. In the deterministic models of gates and circuits that were considered in previous chapters, the basic statements are

- The input and output signals are deterministic, and
- The implemented logic function is performed correctly.

> Physical perfection in nanocircuits is hard to achieve: defects and faults arise from instability and noise-proneness on nanometer scales. This leads to unreliable and undesirable results of computation. In order to ensure more reliable computation, techniques are necessary to cope with such errors.

This can be achieved in nanotechnology using probabilistic models. In these models, it is assumed that

- The input and output signals are performed within some probability because of noise signals, and
- The implemented logic function is performed within some probability because of the nature of nanodevices.

The nature of noise signals in nanocircuits varies from thermal fluctuation to wave interface. Hence, different models are needed for investigation of the effects of noise, and development of methods for protection. For example, a model can be developed based on the assumption that desired signals in circuits are very noisy. In this model, the signals are modeled by the average of stochastic pulses generated by special devices.

This chapter is organized as follows. After basic definitions of fault-tolerance computing (Section 11.1), the probability behavior of nanodevices is considered in Section 11.2. Stochastic neural networks are discussed in Section 11.3. Stochastic computing is the focus of Section 11.4. In Section 11.5, Von Neumann's model of fault-tolerance computing is introduced. Computing with faulty-hypercubes is discussed in Section 11.6.

11.1 Definitions

The basic terminology of fault tolerance computing includes:

- *Robustness to errors* is the ability of a computer system to operate correctly in the presence of errors.
- *Fault tolerance* is the ability of a computer system to recover from transient errors during computing.
- *Defect tolerance* is the ability of a computer system to operate correctly in the presence of permanent hardware errors that emerged in the manufacturing process.

The above characteristics can be modified with respect to the particular tasks, usage technique, and architecture of any computer system.

11.2 Probabilistic behavior of nanodevices

This section is a brief introduction to the problem of the probabilistic behavior of nanodevices.

11.2.1 Noise

Noise in digital circuits is defined as any deviation of a signal from its stable value that can stem from sources as varied as physical and chemical processes in devices, measurement limitations, stochastic simulation procedures, etc. The noise can

- Affect timing, causing a delay failure,
- Increase power consumption, and
- Cause function failure because of signal deviation.

It is important to understand and predict the effects of noise. As noise can have a variety of sources, different noise models that are effective in different situations are desirable.

Sources of noise in nanodevices include:

- Thermodynamic fluctuations,
- Electromagnetic interference,
- Radiation,
- Quantum tunneling, and
- Parameter fluctuations.

Example 11.1 *In the above cases different noise models are needed.*

(*a*) *Mutual inductance noise: when signal switching causes transient current to flow through the loop formed by the signal wire and current return path, a changing magnetic field is created and mutual inductance noise occurs.*

(*b*) *Thermal noise: electrical power distribution and signal transmission through interconnections are always accompanied by thermal noise due to self-heating caused by the current flow. Thermal noise affects both interconnect design and reliability.*

It is reasonable to distinguish effects of noise in

- ▶ Transmission and storage of information, and
- ▶ Processing of information.

11.2.2 Nanogates

Figure 11.1 illustrates the random factors that influence the performance of a nanodevice:

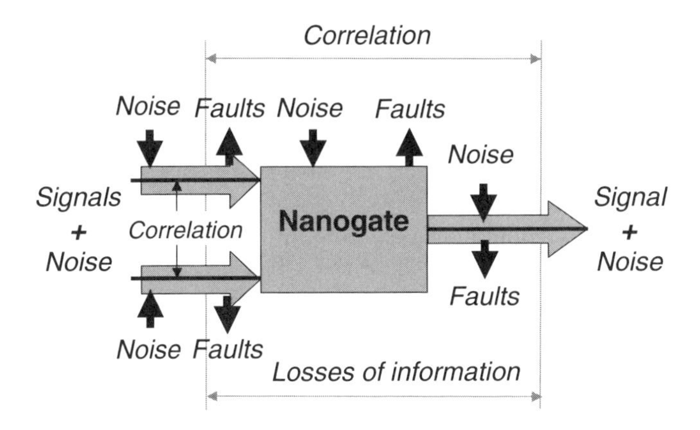

FIGURE 11.1
Random factors that influence the performance of a nanodevice.

- ▶ The input and output signals are additive or multiplicative of noise with respect to the desired signal. There are critical parameters which can be referred to, to distinguish when it is impossible and to extract the desired signal from the noisy signal.
- ▶ There are different faults that input and output wires can generate (open and short wires, multiple faults) in nanogates.

- There are different methods for detecting, recognizing, and correcting faults.

11.2.3 Noise models

Deterministic models operate with

- Noise-free signals,
- Noise-free gates,
- Noise-free networks, and
- Fault-free hypercube structures.

Probabilistic models assume that

- Input signals are applied to gates with some level of probability.
- Correct output signals are calculated with some level of probability.

When noise is allowed, the switching function is replaced with a random function and the configuration is a set of random variables.

There are different approaches to the development of probabilistic models, in particular:

- *Stochastic models* for noise-making signals, in particular, Markov chain models and stochastic pulse stream models.
- *Neural networks* that use resources for optimization, and feedforward networks for computing logic functions over threshold elements.
- *Computational techniques that are inspired by biology.* Some common examples include networks, evolutionary algorithms and artificial immune systems (immunological computation). The similarity between all known applications of algorithms based on biological paradigms is that they utilize the pattern-matching and learning mechanisms of the immune system to perform desired system functions. Biological immune system models are parallel and distributed structures that can be viewed as a multiagent system (separate functions are carried out by individual agents). The immune system model is a model of adaptive processes at the *local level*, resulting in useful behavior at the global level.
- *Fuzzy system technology* can be used in stochastic models, neural networks and evolutionary algorithms. The formal basis of fuzzy logic technique is a fuzzy that is an extension of the classical set. In classical set theory, there are binary logic operators AND, OR, and NOT. The corresponding fuzzy logic operators exist in fuzzy theory. Unlike the binary AND and OR operators whose operations are uniquely defined, their fuzzy counterparts are nonunique.

Example 11.2 *Based on fuzzys, fuzzy switching functions, fuzzy trees, fuzzy decision diagrams, fuzzy hypercube, and fuzzy spaces can be defined.*

Below, some of state-of-the-art models based on the assumption of random factors are listed:

(*a*) *Models for faults detection in wires.*

Example 11.3 *Stuck-at-0 or stuck-at-1 is a fault type that causes a wire to be stuck at zero or one respectively. The conditions for observing the fault at input x_i and its transportation to output are described by the Boolean equations. Solutions to the equations specify the tests for detecting both stuck-at-0 and stuck-at-1 faults.*

(*b*) *Stochastic* (probabilistic) models of behavior of gates and circuits. In these models, to estimate signal probabilities, one has to calculate the switching activity (see Chapter 8) of the internal nodes of the circuit. Computing is performed under the assumption that the values applied to each circuit input are temporally independent. Other algorithms are based on computing the lower and upper bounds of probabilities. In the above approaches, different data structures can be used: algebraic equations, circuit schemes, trees, and decision diagrams.

(*c*) *Error correction codes* correct errors in order to ensure data fidelity. The random error correction codes refers to its ability to correct random bit errors within a code word. While error position can be random, the number of error bits within one code word that can be corrected, referred to as the random error correction capability of the code, is critical. The frequently used block codes are often denoted by a pair of two integers (n, k), and one block code is completely defined by 2^k binary sequences, each an n-tuple of bits, called a *code word. Bursts* (clusters) of errors are defined as a group of consecutive error bits.

Example 11.4 *Below, the main characteristic of a code, the correction capability, is explained in more detail.*

(*i*) *The notation of Bose-Chaudhuri-Hochquenghem code BCH* $(31, 6)$ *indicates that there are at most* 2^6 *distinct messages, each represented by 6 bits and encoded by a code word consisting of 31 bits.*

(*ii*) *The BCH* $(31, 6)$ *code can correct 63 errors in a 255-bit code word. However, the correction capability at the cost of high redundancy has been wasted because each code word consists of 255 bits.*

(*d*) *Model of switching activity,* or transition density model is based on the concept of change. The model is represented in terms of Boolean (logic) differences and Boolean (logic) equations. Note that switching activity at the output of a gate depends not only on the switching activities at the inputs and the logic function of gate, but also on the spatial and temporal dependencies among the gate inputs. This model is often used in the estimation of power dissipation and delays in a network.

It is essential that input and output signals of nanogates are described by additive or multiplicative expression of both noise and the desired signal. Often the desired signal is very noisy. This means that a special technique must be utilized to extract the desired signal from the noisy signal. These methods are well known and widely used in communication. However, these are costly methods, and their technical implementation is complicated. It is rather impractical to apply these methods in nanocircuit design. Therefore, other models are needed to solve the problem of fault tolerant computation in nanocircuits, even at the level of a single nanogate.

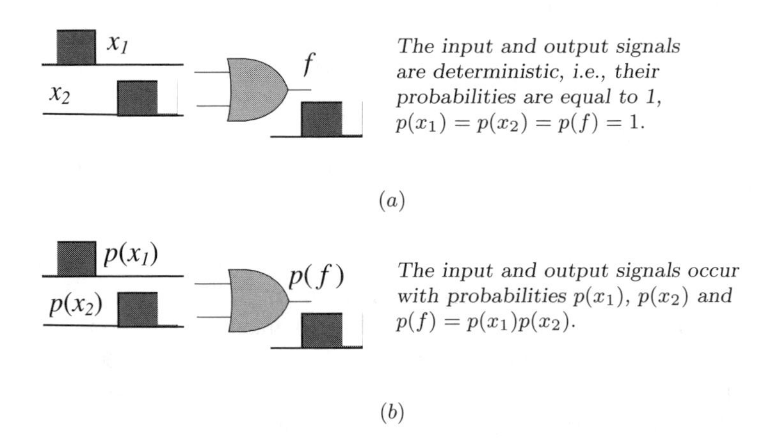

FIGURE 11.2
Deterministic (a) and probabilistic (b) models of computing.

Example 11.5 *Let the inputs of OR gate $x_1 \in \{0,1\}$ and $x_2 \in \{0,1\}$ be mutually independent with probabilities $p_1 = p(x_1)$ and $p_2 = p(x_2)$ correspondingly (Figure 11.2). The output probability can be evaluated as the probability of at least one event x_1 and x_2, i.e.,*

$$p = 1 - (1 - p_1)(1 - p_2) = p_1 + p_2 - p_1 p_2.$$

Supposing $p_1 = 0.8$, $p_2 = 0.9$, correct output is produced with a probability of $p = 0.8 + 0.9 - 0.8 \cdot 0.9 = 0.98$. If $p_1 = p_2 = 1$, the inputs become deterministic and $f = x_1 + x_2 - x_1 x_2$, i.e., $f = x_1 \vee x_2$.

Note that in the above example, in general, the mean $E(f)$ and variance $D(y)$ of the output y are equal to

$$E(f) = p_1 + p_2 - p_1 p_2 = p$$
$$D(f) = p \cdot (1 - p).$$

11.2.4 Fault-tolerant computing

There are several aspects to the design of fault-tolerant nanodevices, and circuits and systems over the library of these nanodevices.

Hierarchical levels of fault-tolerant computing consist of:

- The basic primitives of a system. In the simplest case, the primitives are similar to a library of cells in conventional design. However, the complexity of primitives depends on the technology.
- A finite set of primitives makes up macroprimitives, which are the smallest processors possible within the associated memory. This is similar to the microprocessor in conventional systems.
- A finite set of macroprimitives makes up a system similar to the organization of multiprocessor systems.
- The system makes up a distributed set of systems. This is the highest level of organization of conventional computer systems.

Example 11.6 *A 4-D hypercube can be recognized as a distributed (two connected 3-D hypercubes) and multiprocessor system (each node corresponds to processor).*

The next level of this hierarchy is the level of *self-replicated*, *self-repairing* and *self-assembling* systems. At this level, a system, for example, can replicate itself, giving rise to a population of identical systems.

From the above follow different fault-tolerant computing models:

- *Probabilistic fault-tolerant computing models of nanogates* which rely on the observation of the mechanism of nanodevices, based on the probabilistic behavior of nanostructures (electrons, molecules).
- *Probabilistic behavior of circuits and systems.* At these levels the probability of getting failed components becomes higher. This approach is based on the idea of incorporating into the circuit and system a "guard" against failures.

Example 11.7 *In the presence of faults, a fault-tolerant circuit or system reconfigures itself to exclude the faulty elements. Normally, it is expected for a circuit and system, upon reconfiguration, to encompass all the healthy elements whenever possible. A system so reconfigured may or may not change its topology. Ideally, a fault-tolerant design retains the same system topology after faults arise.*

Noise is but one aspect of the effect of errors on the practical implementation of computing circuits and systems. *Permanent defects* affecting computing resources during the manufacture of the system and within their subsequent lifetime are an engineering problem. Reconfigurable and self-repairing architectures are used to solve this problem.

Techniques. Fault-tolerant techniques are basically built on two approaches:

▶ Redundancy (R-fold modular and reconfiguration) achieves tolerance to faults by employing R copies of a unit.

Example 11.8 *Let $R = 3$. This type of redundancy is called "triple modular." Combining the outputs of three units by a majority gate in each level, cascaded triple modular is obtained. If the output voted at least two times out of three for each sub-unit, this output is considered correct.*

▶ Stochastic computing achieves tolerance to faults by employing statistical models in which deterministic logic signals are replaced by random variables.

Example 11.9 *Let Boolean variables x_1 and x_2 correspond to stochastic pulse signals with averages $E(x_1)$ and $E(x_2)$. Suppose these pulse streams are independent. It is possible to find logic operations that correspond to the sum*

$$E(x_1) + E(x_2)$$

and product

$$E(x_1) \times E(x_2)$$

of these averages.

Note that reliable computations are always considered with respect to data structure. Data structure influences hardware optimization, testing, verification, power dissipation, interconnection, and finally, implementation.

Example 11.10 *Interconnection plays an important role in fault-tolerant computing. The interconnection determines various parameters of nanocircuit, total area and capacity, delay, and power dissipation. Hence, it must be accounted for as early as possible during the design process. Approximate estimates can be obtained by using so-called stochastic interconnect models.*

11.3 Neural networks

Neural networks can be defined as a computational paradigm alternative to the conventional von Neumann model. The computational potential and limits of conventional computing models are well understood in terms of classical models such as the Turing machine. Many important results have been achieved in investigation of the computational power of neural networks by comparison with conventional computational tools such as finite automata, Turing machines, and logic circuits.

11.3.1 Threshold networks

Logic circuit design based on threshold gates can be considered as an alternative to traditional logic gate design procedure. The implementation of a massively interconnected network of threshold gates is possible. Formally, a threshold gate is described by a threshold decision (linearly separable) function. This principle is a general one in and of itself, so simple logic gates, such as AND and OR gates, are merely special cases of the threshold gate. The power of the threshold gate design style lies in the intrinsic complex functions implemented by such gates, which allow implementations with less threshold gates or gate levels than does design with standard logic gates.

Example 11.11 *Multiple-addition, multiplication, division and sorting can be implemented by polynomial-size threshold circuits of small depth.*

Feedforward neural networks are modeled by a direct acyclic graph (similar to the switching network). Units in this network can be grouped in a unique minimal way into disjointed layers so that neurons in any layer are connected only to neurons in other layers. Computation proceeds from the input layer to the output layer (this architecture is similar to the multilevel switching network). This feedforward neural network coincides with a switching circuit designed over the threshold elements.

11.3.2 Stochastic feedforward neural networks

An example of these is a stochastic feedforward neural network built on *noisy spiking neurons*, which have been developed to model biological neurons. The important properties of these networks are as follows:

- An arbitrary switching function can be implemented by a sufficient large network of noisy spiking neurons with an arbitrarily high probability of correctness.
- An arbitrary deterministic finite automation can be simulated by a network of noisy spiking neurons with an arbitrarily high probability of correctness.

11.3.3 Multivalued feedforward networks

Multivalued feedforward networks are defined as models of multilevel combinational multivalued logic circuit with no feedback and no learning. This model includes neuron-like gates, each representing a level of a multivalued circuit, so that the number of gates in the network is equal to the number of levels in the circuit. The formal description of a gate is a linear arithmetic expression that is directly mapped to the linear word-level decision diagram introduced in Chapter 9. Thus, an l-level circuit is described by a set of l linear word-level diagrams.

11.4 Stochastic computing

In this section, models based on reliable gates with stochastic input streams are considered. Figure 11.3 illustrates this model. If the input stochastic streams are independent (technically this means that independent generators of random pulses are used with some additional tools for decorrelation of signals) with $E(x_1)$ and $E(x_2)$, the output is described by the equation $E(f) = E(x_1) \times E(x_2)$. Then follows the transformation of the values to the range $[0, 1]$.

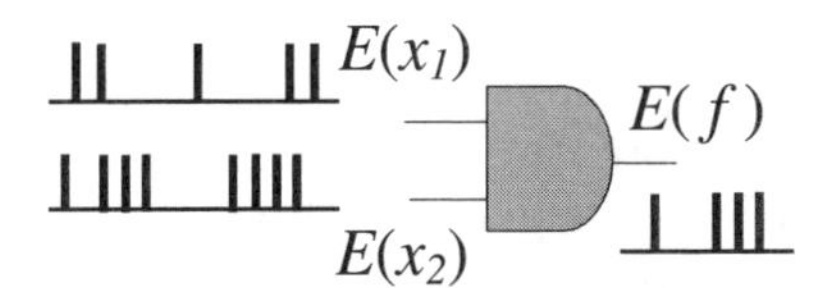

The averages of the stochastic pulse stream of input and output signals are $E(x_1)$, $E(x_2)$ and $E(f) = E(x_1) \times E(x_2)$.

FIGURE 11.3
Stochastic pulse model of computing.

Example 11.12 *Given the deterministic signal $x \in \{0, 1\}$, generate this signal with probability $p(x)$. The simplest model is $p(x) = x \cdot r$ where $r \in \{0, 1\}$ is a random variable with probability $p(r)$.*

Example 11.13 *Implementation of the model for generating a signal with a given probability is shown in Figure 11.4 (synchronization is not shown).*

Example 11.13 illustrates the possibility of generating a signal with a given probability. For example, if $p(r) = 1$, the output is a signal x with probability $p(x) = 1$. Based on this model it is possible to study the simplest features of probabilistic computation.

11.4.1 The model of a gate for input random pulse streams

Let us analyze the output of a gate that implements an elementary switching function for input independent random pulse streams as 0s and 1s. A binary stochastic pulse stream is defined as *a sequence of binary digits, or bits.* The information in a pulse stream is contained in the *primary statistics of the bit stream,* or the probability of any given bit in the stream being a logic 1. Hence, the output of a gate will generally be in the form of a nonstationary Bernoulli sequence. Such a sequence can be considered in probabilistic terms as a *deterministic signal with superimposed noise.* Suppose that statistical

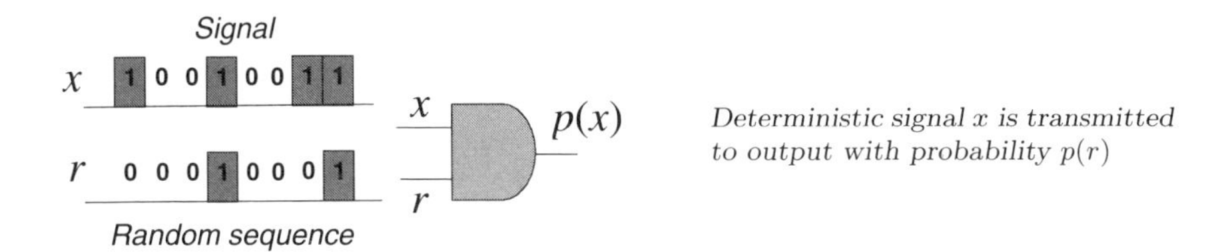

FIGURE 11.4
Model for generating the signal binary x with probability $p(r)$.

characteristics of these streams are known, i.e., can be measured. In other words, these streams carry a signal by statistical characteristics (a single event carries very little information, it is not enough for decision making).

The stochastic pulse stream model states that

- Input signals are modeled by stochastic pulse streams with known characteristics, and
- The output signals are calculated as an average of statistical characteristics.

Example 11.14 *Consider the AND gate. Let r_1 and r_2 are independent random input signals with probabilities $p(r_1)$ and $p(r_2)$ are transmitted to the output with probability $p(f) = p(r_1)p(r_2)$. In Figure 11.5 details of the mean $E(f)$ and variance $D(f)$ calculation are given. Note that the property*

$$\int_{-\infty}^{\infty} \delta(r-a)f(r)dr = f(a),$$

and the property of joint probability distribution for independent random variables r_1 and r_2, $f(r_1, r_2) = f(r_1)f(r_2)$, are used.

The generated probability of a sequence of logic levels corresponds to the *relative frequency* of 1 logic levels in a sufficiently long sequence. A probability cannot be measured exactly but only estimated as the relative frequency of 1 logic level in a sufficiently long sample.

The stochastic computer introduces its own errors in the form of *random variance.* If we observe a sequence of N logic levels and k of them are 1, then the estimated generating probability is $\hat{p} = k/N$. The sampling distribution of the value of k is binomial, and hence the standard deviation of the estimated probability $\hat{p}$ from the true probability p is

$$\sigma(\hat{p}) = [p(1-p)/N]^{1/2}.$$

Hence the accuracy in estimation of a generated probability increases as the square root of the length of the sequence examined, i.e., the square root of the length, or time, of computation.

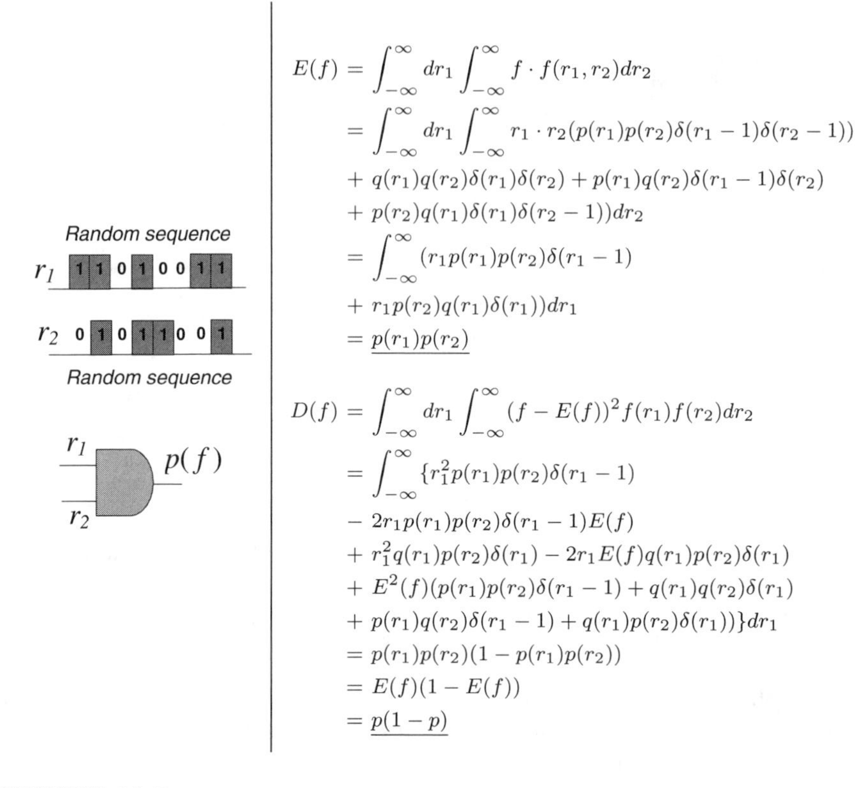

$$
\begin{aligned}
E(f) &= \int_{-\infty}^{\infty} dr_1 \int_{-\infty}^{\infty} f \cdot f(r_1, r_2) dr_2 \\
&= \int_{-\infty}^{\infty} dr_1 \int_{-\infty}^{\infty} r_1 \cdot r_2 (p(r_1)p(r_2)\delta(r_1 - 1)\delta(r_2 - 1)) \\
&+ q(r_1)q(r_2)\delta(r_1)\delta(r_2) + p(r_1)q(r_2)\delta(r_1 - 1)\delta(r_2) \\
&+ p(r_2)q(r_1)\delta(r_1)\delta(r_2 - 1))dr_2 \\
&= \int_{-\infty}^{\infty} (r_1 p(r_1)p(r_2)\delta(r_1 - 1) \\
&+ r_1 p(r_2)q(r_1)\delta(r_1))dr_1 \\
&= \underline{p(r_1)p(r_2)}
\end{aligned}
$$

$$
\begin{aligned}
D(f) &= \int_{-\infty}^{\infty} dr_1 \int_{-\infty}^{\infty} (f - E(f))^2 f(r_1) f(r_2) dr_2 \\
&= \int_{-\infty}^{\infty} \{r_1^2 p(r_1)p(r_2)\delta(r_1 - 1) \\
&- 2r_1 p(r_1)p(r_2)\delta(r_1 - 1)E(f) \\
&+ r_1^2 q(r_1)p(r_2)\delta(r_1) - 2r_1 E(f)q(r_1)p(r_2)\delta(r_1) \\
&+ E^2(f)(p(r_1)p(r_2)\delta(r_1 - 1) + q(r_1)q(r_2)\delta(r_1) \\
&+ p(r_1)q(r_2)\delta(r_1 - 1) + q(r_1)p(r_2)\delta(r_1))\}dr_1 \\
&= p(r_1)p(r_2)(1 - p(r_1)p(r_2)) \\
&= E(f)(1 - E(f)) \\
&= \underline{p(1 - p)}
\end{aligned}
$$

FIGURE 11.5
Model of AND gate for the input random pulse streams.

11.4.2 Data structure

There are several features that distinguish classical computation and stochastic computation:

- A signal is represented by the probability that a logic level be 1 or 0 at a clock pulse.
- Random noise is being deliberately introduced into the data; usually, noise distribution is normal.
- A quantity is represented by a clocked sequence of logic levels generated by a random process: the successive levels are statistically independent, and the probability of the logic level being ON is a measure of that quantity.
- Arithmetic operations are performed via the completely random data, and probability that a logic level will be ON or OFF is determined. Its actual value is

  (*a*) A chance event which cannot be predicted, and

(*b*) Repetition of a computation will give rise to a different sequence of logic levels.

In a conventional computer, logic levels represent data change deterministically from value to value as the computation proceeds. If the computation is repeated, the same sequence of logic levels will occur.

Note that unlike binary radix arithmetic, stochastic arithmetic is robust in the presence of noise/single bit fault, and accuracy may be controlled using the dimension of time.

11.4.3 Primary statistics

If the input distribution is unconstrained, Bernoulli sequences can be used for formal modeling. This means that

- The probability of a given bit being a 1 is independent of the values of any previous bits.
- Elements' processing functions are evaluated only with respect to their outputs' primary statistics. The outputs are not, in general, Bernoulli sequences.
- In the case of processing elements with multiple inputs, the inputs are uncorrelated with each other.

Example 11.15 *In Table 11.1 probabilistic parameters of elementary switching functions for stochastic computing are given, where autocorrelation function is defined as* $K_f(\tau) = E[(f(t) - E(f))(f(t-\tau) - E(f))]$.

It is follow from Table 11.1 that, for example, for stochastic computing of a NOT function, we assume that input signal x is a stochastic pulse stream characterized by probability $p(x)$ and the autocorrelation function $K_x(\tau_x)$. The mean of the output signal is $E(f) = p(f)$, and hence $p(\overline{x}) = 1 - p(x) = 1 - E(x)$. By analogy, if the input pulse streams are independent

- AND gate $f = x_1 x_2$ is modeled by $E(f) = p_1 p_2$
- OR gate $f = x_1 \vee x_2$ is modeled by $E(f) = p_1 + p_2 - p_1 p_2$
- NOR gate $f = \overline{x_1 \vee x_2}$ is modeled by $E(f) = 1 - p_1 - p_2 + p_1 p_2$
- EXOR gate $f = x_1 \oplus x_2$ is modeled by $E(f) = p_1 + p_2 - 2p_1 p_2$
- NOT-EXOR gate $f = \overline{x_1 \oplus x_2}$ is modeled by $E(f) = 1 - p_1 - p_2 + 2p_1 p_2$

where $p_1 = E(x_1)$ and $p_2 = E(x_2)$.

Stochastic computing can be interpreted by decision trees and $\mathcal{N}$-hypercubes.

Example 11.16 *In Figure 11.6, stochastic computing of AND function is interpreted by decision trees and* $\mathcal{N}$*-hypercubes.*

TABLE 11.1
Stochastic computing of elementary switching functions.

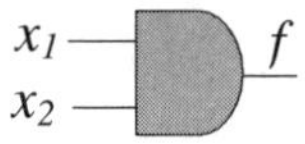

$$E(f) = \begin{cases} E(x_1)E(x_2) + K_{x_1x_2} & x_1 \text{ and } x_2 \text{ are dependent} \\ E(x_1)E(x_2) & \text{otherwise.} \end{cases}$$

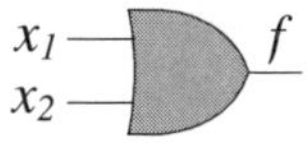

$$E(f) = \begin{cases} E(x_1) + E(x_2) - E(x_1)E(x_2) - K_{x_1x_2} & x_1 \text{ and } x_2 \text{ are dependent} \\ E(x_1) + E(x_2) - E(x_1)E(x_2) & \text{otherwise.} \end{cases}$$

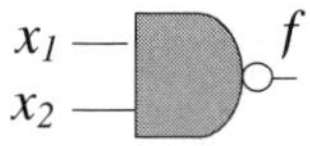

$$E(f) = \begin{cases} 1 - E(x_1)E(x_2) - K_{x_1x_2} & x_1 \text{ and } x_2 \text{ are dependent} \\ 1 - E(x_1)E(x_2) & \text{otherwise.} \end{cases}$$

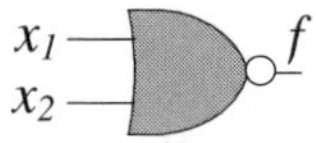

$$E(f) = \begin{cases} 1 - E(x_1) - E(x_2) + E(x_1)E(x_2) + K_{x_1x_2} & x_1 \text{ and } x_2 \text{ are dependent} \\ 1 - E(x_1) - E(x_2) + E(x_1)E(x_2) & \text{otherwise.} \end{cases}$$

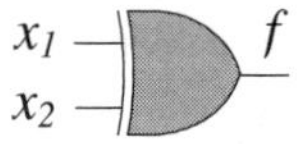

$$E[f] = \begin{cases} E(x_1) + E(x_2) - 2E(x_1)E(x_2) - 2K_{x_1x_2}, & x_1 \text{ and } x_2 \text{ are dependent} \\ E(x_1) + E(x_2) - 2E(x_1)E(x_2) & \text{otherwise.} \end{cases}$$

11.4.4 Stochastic encoding

A binary number X is compared with a uniform random number generated by a generator of random numbers (Figure 11.7). The upper limit of numbers is X_{max}. The firing probability P_f of the comparator output is equal to X/X_{max}, so the output value $\widehat{X}$ which is obtained by accumulating the pulse times follows the binomial distribution.

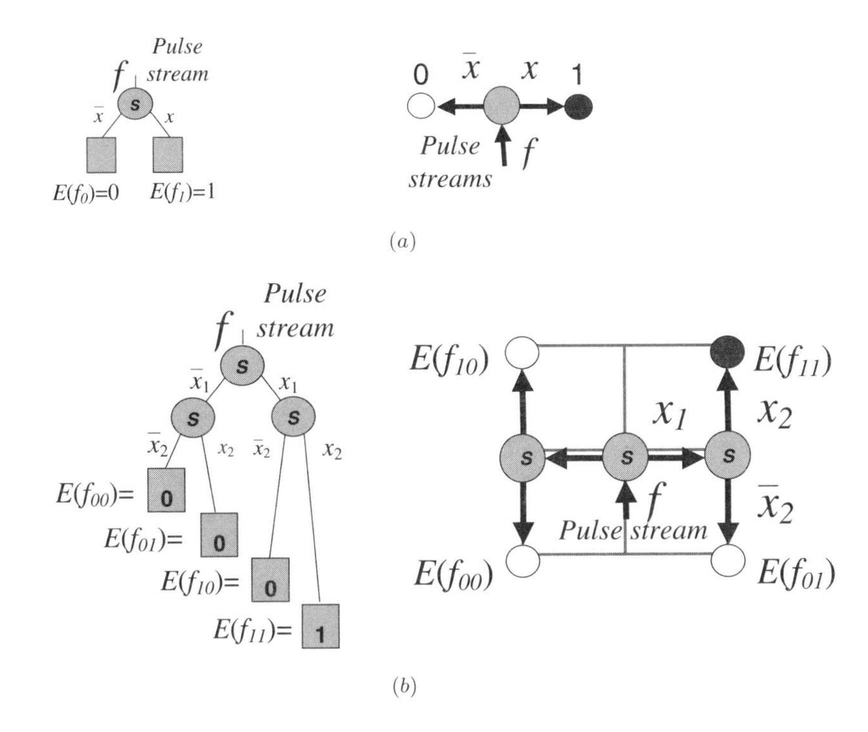

FIGURE 11.6
Interpretation of stochastic computing by decision trees and $\mathcal{N}$-hypercube of two-input (a) and three-input (b) gate AND (Example 11.16).

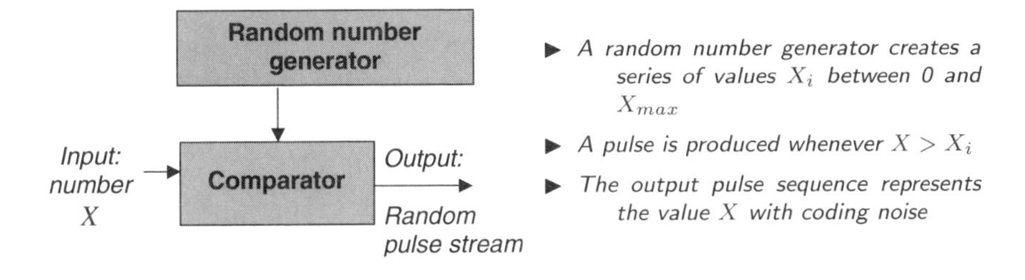

FIGURE 11.7
A coding circuit for generating random pulse sequences

11.5 Von Neumann's model on reliable computation with unreliable components

The study of reliable computation by unreliabledevices originates with von Neumann. He developed the multiplexing technique known as *von Neumann's model of computing.* In this model, each wire in a circuit is replaced by a bundle of wires on which a majority vote is conducted to establish its value. In this section, von Neumann's multiplexing technique is briefly reviewed. For this, an unreliable NAND gate is chosen.

11.5.1 Architecture

Let NAND be an unreliable gate. In order to increase its functional probability, let us use von Neumann's model of computing:

- ▶ Replace each input of the NAND gate as well as its output by a bundle of N lines,
- ▶ Duplicate the NAND N times.
- ▶ Perform a random permutation of the input signals: each signal from the first input bundle is randomly paired with a signal from the second input bundle to form the input pair of one of the duplicated NANDs.

Figure 11.8 illustrates the architecture where U is a random permutation. The key to this approach is modifying the NAND gate (an arbitrary network, in general) by replacing each interconnect with a parallel bundle of interconnects and a strategy of random interconnections that prevents the propagation of errors. In other words, parallelization by bundles and random interconnections can be viewed as a method for increasing the reliability of the NAND element. This phenomenon can be explained in terms of information theory and error correcting codes (see "Further Reading" Section).

11.5.2 Formalization

Formally, von Neumann model is based on error correction code known as *repetition* code. To form the redundancy in this code, each bit (message) is repeated many times. Let

- ▶ X be the set of lines in the first input bundle being stimulated (a logic 1),
- ▶ Y be the corresponding set for the second input bundle,

Z be the corresponding set for the output bundle,

- ▶ ε be the faulty probability of a NAND gate,
- ▶ (X, Y, Z) have $(\overline{x} \cdot N,\ \ \overline{y} \cdot N,\ \ \overline{z} \cdot N)$ elements, respectively,
- ▶ $(\overline{x}, \overline{y}, \overline{z})$ are relative levels of excitation of the two input bundles and of the output bundle, respectively,

Then, with an extremely large N:

- ▶ $\overline{z}$ is a stochastic variable, approximately normally distributed,
- ▶ The upper bound for the failure probability per gate that can be tolerated: $\varepsilon_0 = 0.0107$.

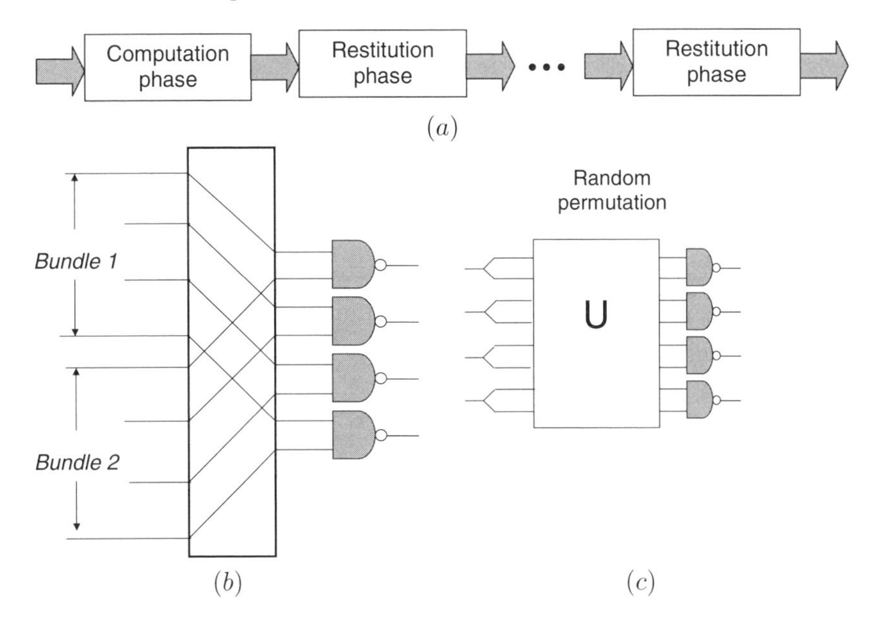

FIGURE 11.8
Von Neumann model of fault-tolerance computing (a), computation phase (b), and restitution phase (c).

The mean of the a stochastic variable $\overline{z}$, approximately normally.

Note that in fault-tolerance computing based on error correcting codes can be used different architectures, for example, 1-D, 2-D, and 3-D cellular systolic arrays and systolic arrays.

11.6 Faulty hypercube-like computing structures

In this section, the focus is fault-tolerance hypercube-like computing structures. Fault-tolerance properties of a hypercube-like structures are well studied and used in computing systems. Here, we briefly review the basic principles of fault-tolerance computing via hypercube-like structures.

11.6.1 Definitions

Below, the basic definitions of fault tolerance computing in hypercube and hypercube-like structures are given:

▶ A hypercube computing structure is called a *faulty* hypercube if it contains any faulty node (computing device) or communication link. For hypercube-like structures of large dimensions, the number of processing elements is very large and hence the probability of occurrence of faults increases.

- ▶ A network is *robust* if its performance does not decrease significantly when its topology changes.
- ▶ Since efficient cooperation between nonfaulty computing devices is desirable, one measure for robustness is the network *connectivity*, which is defined as the number of node or link failures that can be allowed without disrupting the system.
- ▶ *Fault models* of a hypercube computing system are defined from subcube and node reliability.
- ▶ *Multiple fault models* of a hypercube computing system are calculated based on the probability that many faults in the node or subcube exist.
- ▶ The *reliability* of a hypercube-based computing system is defined as the probability that the system has survived the interval $[0, t]$ given that it was operational at time $t = 0$, where t is the time. Usually, reliability is used in models of computing systems in which repair cannot take place.
- ▶ The *terminal reliability* of a computer system is defined as the reliability of computer devices in nodes of a hypercube computer system. Terminal reliability can be also defined as *task-based reliability*, which is the probability that some minimal set of connected nodes are available in the hypercube structure.
- ▶ The *fault-tolerance computing* of a hypercube-based computing system is provided by *reconfiguration* and application of *error correcting codes.*

There are several assumptions and additional data that must be known to apply the above characteristics: in particular, reliability function of node computing devices (usually assumed homogeneous), and the characteristic of statistical dependence of failures of computing devices in nodes (usually assumed independent).

The hypercube-like network has been proved to be very robust and to divide it into two components requires at least n faults.

In the probability fault model, the reliability of each node at time t is a random variable. The probability that a hypercube-like network is operational is represented by the reliability of the computing devices in the hypercube-like network. The reliability of computing hypercube-like structure can be formulated as the union of probabilistic events that all the possible hypercubes of lower dimensions are operational.

Example 11.17 *In Figure 11.9a, the node 001 is faulty, and one link marked is faulty too. In Figure 11.9b, the node 0010 is faulty.*

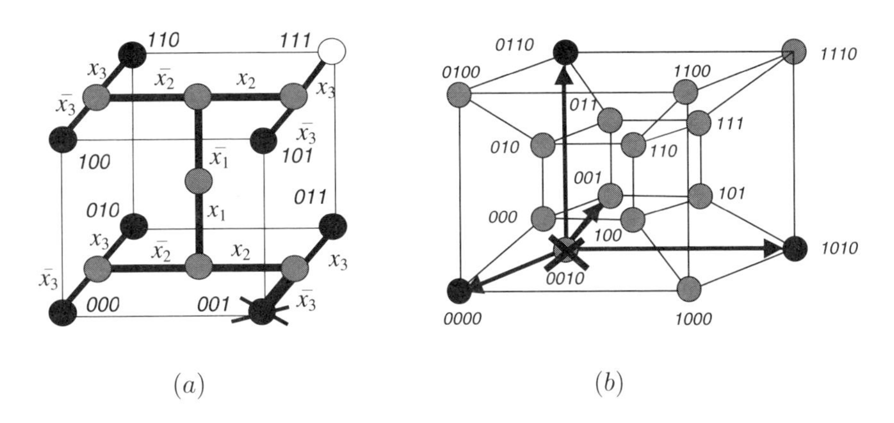

(a) (b)

FIGURE 11.9
Faulty 3-D $\mathcal{N}$-hypercube (a) and faulty 4-D hypercube(b) (Example 11.17).

11.6.2 Fault-tolerance technique

Fault-tolerance technique for hypercube and hypercube-like structures is based on the principle of

- Reconfiguration, and
- Error correcting.

These techniques are well studied and widely used in computer system design. Several algorithms have been developed for reconfiguring a hypercube with faults. These algorithms aim to achieve different characteristics after reconfiguration, in particular, acceptable performance and connectivity. The crucial idea is to identify maximum subcubes in a faulty hypercube, retaining as many healthy nodes as possible in order to keep performance degradation to a minimum.

Example 11.18 *Let 3-D hypercube-like structure be a 3-D cellular array. The basic component of this structure is a cell that includes computing device and a memory. The discipline of information flows is defined by topology of computing structure. Reconfiguration of cellular array is the change of this discipline. The global and local*

(a) *Reconfiguration,*

(b) *Error correction, and*

(c) *Control*

of cells are used in 3-D cellular arrays.

11.7 Summary

1. Fault-tolerance computing is the central problem of nanosystem design because of the probabilistic nature of nanodevices.
2. There are many methods in state-of-the-art of logic design that deal with computation with unreliable elements and noise signals, in particular, methods based on:
   - Probabilistic description of signals,
   - Stochastic pulse streams organization,
   - Fuzzy logic,
   - Probabilistic logic,
   - Residue number system,
   - Paradigms inspired by biological systems,
   - Stochastic logic neural networks,
   - Von Neumann multiplexing,
   - R-fold modular, and
   - Error-correcting codes (for dynamic computation in the presence of noise based on information theoretical measures).
3. Methods listed in item 2 can be applied to design of nanosystems in spatial dimensions. For this, the topological properties of hypercubes and hypercube-like structures must be used.
4. Fault-tolerance computing based on stochastic pulse streams is related to paradigms inspired by biological systems with respect to data structure, i.e., stochastic pulse streams.

11.8 Further reading

Faulty hypercubes and fault-tolerance computing via hypercube computing structures are studied in [1, 4, 5].

Stochastic computing based on stochastic pulse streams. In [8] the basics of stochastic computing have been developed. Yakovlev and Fedorov introduced the fundamentals of reliable computation based on stochastic principles [39]. This technique is often used in design of stochastic neural networks. In modeling, a pseudo-random number generator is used. This is generator is a deterministic method, usually described with a mapping, to produce from a small set of numbers (seed) a larger set of random-looking numbers.

Probabilistic (stochastic) networks can be defined by augmenting the respective deterministic model with additional random binary input units whose states in time establish independent identically distributed binary sequences. This model of probabilistic networks relates to neural networks with other stochastic behavior including those which are unreliable in computing states and connecting units. In stochastic networks, probabilistic feedforward networks based on deterministic threshold circuits are well studied.

Birge et al. studied the effects of statistical fluctuations in computation [2]. The authors found that redundancies in the thousands were needed for reliable molecular computation. There are many related problems which can be solved by probabilistic methods, for example, verification [13, 16, 19, 24, 25, 32, 38].

Stochastic neural networks. The computational properties of feedforward neural networks have been studied, in particular, by Wegener [36]. Maass has shown that arbitrary switching functions and deterministic finite automata can be implemented by a sufficiently large network of noisy spiking neurons with an arbitrarily high probability of correctness [17]. A model of a multilevel combinational multivalued logic circuit with no feedback and no learning is introduced in [40]. In addition, see [30, 35]. The classic book by Muroga [23] can be recommended for study logic circuit design based on threshold gates. Recently, interest in study of threshold gates computing of logic functions has resurged, since the advances in technologies [35].

Information-theoretical approach. Information theory sets bounds on the tolerable noise during the transmission and storage of information. Numerous error-correcting codes have been developed based on the information theoretical approach. Information theoretical measures are useful in noise tolerance computing, including dynamical computation in the presence of noise.

Evolutionary computation. The principle of evolution is the primary concept of biology, linking every organism together in a historical chain of events. For example, every creature (circuit) in the chain is the product of a series of events (subcircuits) that have been sorted out thoroughly under selective pressure (correct or noncorrect subcircuits) from the environment (design technique). Over many generations, random variation and natural selection shape the behaviors of individuals (circuits) to fit the demands of their surroundings (principles of circuit design) [7, 10, 21].

The reader can find much useful information on computational techniques that look to biology for inspiration in journals *IEEE Trans. on Evolutionary Computing.* In particular, the collection of papers by Dasgupta covers various computational aspects of the immune system [6]. The natural immune system is a complex adaptive pattern-recognition system that defends the body from foreign pathogens. Rather than rely on any central control, it has a distributed task force that has the intelligence to take action from a local and also global perspective using its network of messengers for communication. The immune system has evolved innate (nonspecific) immunity and adaptive (specific) immunity. From the computation point of view, the natural immune system is a parallel and distributed adaptive system. The immune system uses learning, memory, and associative retrieval to solve recognition and classification tasks. Specifically, it learns to recognize relevant patterns, remember patterns that have been seen previously, and use combinatorics to construct pattern detectors efficiently.

Details of applications of an evolutionary technique can be found in [12], and references in Chapter 12.

Von Neumann model. Boolean circuits were the first model of computation in which even the most elementary components of the computer were assumed noisy. This topic was first investigated by von Neumann. He showed that it is possible to perform fault-tolerant computation with switching circuits. Recently it was shown that if each NAND gate fails independently, the tolerable threshold probability of each gate will be $\varepsilon_0 = 0.08856$. However, this result was obtained by formulas constructed from noisy NAND gates rather than circuits. In other words, according to von Neumann, if $\varepsilon \geq \varepsilon_0$, the failure probability of the NAND multiplexing network will be larger than a fixed, positive lower bound, no matter how large a bundle size N is used. Note that von Neumann model is based on so-called *repetition* error correction code [18, 37].

In addition, the works by von Neumann on realization of a self-replicated automation endowed with the properties of universal computation and construction are useful.

In [9, 28], von Neumann's technique has been improved. Authors have proposed generalization of the restitution phase. Parallel restitution is defined as a specific method of parallelization. Peper et al. studied the asynchronous cellular systolic arrays that are tolerant to transient errors [26]. Asynchronous cellular arrays have some advantages for design, prototyping and manufacturing methods.

Some results on defect-tolerance have been obtained by building the Teramac [11], a parallel computer based on FPGAs that is able to achieve high-performance computing, even if a significant number of its components are defective.

Error correction codes. Homogenous highly parallel arithmetic circuits, in particular, systolic structures [29] and error correction coding [27] developed in the last decade, are the focus in nanotechnology. Scenarios where errors may occur include nanodevices, data storage, interconnections, etc. [37]. The design technique of 1-D interleaving and 2-D error burst correction codes have been well studied and documented. The multidimensional interleaving technique followed by a random error correction code has become the most common approach to correcting multidimensional error bursts [3]. For example, size 2 means that the error burst is a 3-D hypercube volume $2 \times 2 \times 2$ blocks. Random error correction codes can be used efficiently to correct bursts of errors.

Residue number systems. The hardware implementation of an arithmetic algorithm is largely affected by a choice of a specific numbering system. The attractive properties of residue number systems (RNS) include

- Carry-free,
- Fault isolating, and
- Modular characteristics,

are widely used, in particular, in high-speed digital signal processing. The most attractive property of RNS is that there are no carry propagations inside the set. In RNS-based system, conversion procedures from conventional binary representation to residue format, and vise versa, are used.

In RNS, an integer is represented as a set of residues with respect to a set of relatively prime integers called *moduli*. An RNS is defined in terms of a set of relatively prime moduli $\{r_1, r_2, \ldots, r_s\}$ where the greatest common divisor is equal to 1 for each pair of the moduli.

- If a and b are integers and m is a natural number, the statement $a \equiv b \pmod{m}$ (a is *congruent* to b modulo m) means that the difference, $a - b$, is exactly divisible by the positive integer m.
- If a and b are two integers and $a \equiv b \pmod{m}$, then b is a *residue* of a modulo m.

Two or more congruences may be added, subtracted, multiplied, provided the same modulus is used throughout, i.e., congruences behave like ordinary equations in algebra. While in ordinary arithmetic, there are an infinite number of integers $0, 1, 2, \ldots$, in the modular arithmetic there are essentially only a finite number of integers.

Kinoshita studied RNS floating-point arithmetic (addition, subtraction, multiplication, division, and square root) for an interval number with the goal to achieve reliable computation when hardware representations of numbers have inadequate precision [15]. For example, a double-base representation ($m_1 = 2, m_2 = 3, n = 2$) is $x = \sum_{i,j} w_{i,j} 2^i 3^j$, where i and j are positive integers. For $j = 0$ and $i = 0$, this equation becomes a binary and ternary system representation respectively.

Adder based residue-to-binary number converters have been reported by Vinnakota and Rao [33]. Wang designed residue-to-binary number converters for the RNS $\{2^n - 1,\ 2^n,\ 2^n + 1\}$ using $2n$-bit or n-bit adders, that are twice as fast as generic ones, and achieve improvement in area and dynamic range as well [34].

Fuzzy logic technique and application to logic design have been discussed in [14, 20, 22, 31].

11.9 References

[1] Becker B, and Simon HU. How robust is the n-cube? *Information and Computation*, 77:162–178, 1988.

[2] Birge RR, Lawrence AF, and Tallent JB. Quantum effects, thermal statistics and reliability of nanoscale molecular and semiconductor devices. *Nanotechnology*, 2:73–87, 1991.

[3] Blaum M, Bruck J, and Vardy A. Interleaving schemes for multidimensional cluster errors. *IEEE Transactions on Information Theory*, 44(2):730–743, 1998.

[4] Chen HL, and Tzeng NF. A Boolean expression-based approach for maximum incomplete subcubes identification in faulty hypercubes, *IEEE Transactions on Parallel and Distributed Systems*, 8(11):1171–1183, 1997.

[5] Chen HL, and Tzeng NF. Subcube determination in faulty hypercubes, *IEEE Transactions on Computers*, 46(8):871–879, 1997.

[6] Dasgupta D. *Artificial Immune Systems and Their Applications* Springer-Verlag, Heidelberg, 1999.

[7] Evans WS, and Schulman LJ. Signal propagation and noisy circuits. *IEEE Transactions on Information Theory*, 45(7):2367–2373, 1999.

[8] Gaines BR. Stochastic computing systems. In: Tou JT., Ed., *Advances in Information Systems Science*, Plenum, New York, vol. 2, chap. 2, pp. 37–172, 1969.

[9] Han J, and Jonker P. A system archtecture solution for unreliable nanoelectronic devices. *IEEE Transactions on Nanotechnology*, 1(4):201–208, 2002.

[10] Hansson H. *Time and Probability in Formal Design of Distributed Systems*. Elsevier, 1994.

[11] Heath JR, Kuekes PJ, Snider GS, and Williams RS. A defect-tolerant computer architecture: opportunities for nanotechnology. *Science*, 280:1716-1721, 1996.

[12] Iba H, Iwata M, and Higuchi T. Machine learning approach to gate level evolvable hardware. In *Lecture Notes in Computer Science*, 1259:327–393, Springer-Verlag, Heidelberg, 1997.

[13] Jain J, Bitner J, Fussell DS, and Abraham JA. Probabilistic verification of Boolean functions. In *Formal Methods in System Design*, 1:61–115, Kluwer Academic Publishers, 1992.

[14] Kamiura N, Hata Y, and Yamato K. On concurrent tests of fuzzy controllers. In *Proceeding IEEE 28th International Symposium on Multiple-Valued Logic*, pp. 356–361, 1998.

[15] Kinoshita E, and Lee Ki-Ja. A residue arithmetic extension for reliable scientific computation. *IEEE Transactions on Computers*, 46(2):129-138, 1997.

[16] Kumar SK, and Breuer MA. Probabilistic aspects of Boolean switching functions via a new transform. *Journal of ACM*, 28:502–520, 1981.

[17] Maass W. Lower bounds for the computational power of networks of spiking neurons. *Neural Computation*, 8(1):1–40, 1996.

[18] MacWilliams FJ, and Sloan NJ. *The Theory of Error-Correcting Codes.* North Holland, Amsterdam, 1978.

[19] Majumdar A, and Vrudhula SBK. Analysis of signal probability in logic circuits using stochastic models. *IEEE Transactions on VLSI*, (3):365–379, 1993.

[20] Marinos PN. Fuzzy logic and its application to switching systems. *IEEE Transactions on Computers*, 18:343–348, 1969.

[21] Mitra S, Saxena NR, and McCluskey EJ. Common-mode failures in redundant VLSI systems: a survey. *IEEE Transactions on Reliability*, 49:285–295, 2000.

[22] Moraga C, Trillas E, and Guadarrama S. Multiple-valued logic and artificial intelligence fundamentals of fuzzy control revisited. In: Yanushkevich SN., Ed., *Artificial Intelligence in Logic Design.* Kluwer, Dordrecht, pp. 9–37, 2004.

[23] Muroga S. *Threshold Logic and its Applications.* Wiley-Interscience, New York, 1971.

[24] Parker KP, and McCluskey EJ. Analysis of logic circuits with faults using input signal probabilities. *IEEE Transactions on Computers*, 24(5):573–578, 1975.

[25] Parker KP, and McCluskey EJ. Probabilistic treatment of general combinational networks. *IEEE Transactions on Computers*, 24(6):668–670, 1981.

[26] Peper F, Lee J, Abo F, Isokawa T, Adachi S, Matsui N, and Mashiko S. Fault-tolerance in nanocomputers: a cellular array approach. *IEEE Transactions on Nanotechnology*, 3(1):187–201, 2004.

[27] Rao TRN. *Error Coding for Arithmetic Processors,* Academic Press, New York, 1974.

[28] Sadek AS, Nikolić K, and Forshaw M. Parallel information and compuation with restriction for noise-tolerant nanoscale logic networks. *Nanotechnology,* 15:192–210, 2004.

[29] Sinha BP, and Srimani PK. Fast parallel algorithms for binary multiplication and their implementation on systolic architectures. *IEEE Transactions on Computers,* 38(3):424–431, 1989.

[30] Siu KY, Roychowdhury VP, and Kailath T. Depth-size tradeoffs for neural computation. *IEEE Transactions on Computers,* 40(12):1402–1411, 1991.

[31] Strehl K, Moraga C, Temme KH, and Stanković RS. Fuzzy decision diagrams for the representation, analysis and optimization of rule bases. In *Proceedings IEEE 30th International Symposium on Multiple-Valued Logic,* pp. 127–132, 2000.

[32] Thornton M, and Nair V. Efficient calculation of spectral coefficients and their applications. *IEEE Transactions on Computer-Aided Design of Integrated Circuits and Systems,* 14(11):1328–1341, 1995.

[33] Vinnakota B, and Rao VVB. Fast conversion technique for binary-residue number systems. *IEEE Transactions on Circuits and Systems/I,* 41(12):927–929, 1994.

[34] Wang Y. Residue-to-binary converters based on new Chinese remainder theorems. *IEEE Transactions on Circuits and Systems/II,* pp. 197–206, Mar., 2000.

[35] Webster J., Ed., *Encyclopedia of Electrical and Electronics Engineering. Threshold Logic,* Vol. 22, John Wiley & Sons, New York, pp. 178–190, 1999.

[36] Wegener I. *The Complexity of Boolean Functions,* John Wiley & Sons, New York, 1987.

[37] Wicker SW. *Error Control Systems for Digital Communication and Storage.* Prentice-Hall, New York, 1995.

[38] Winograd S, and Cowan JD. *Reliable Computation in the Presence of Noise.* MIT Press, Cambridge, MA, 1963.

[39] Yakovlev VV, and Fedorov RF. *Stochastic Computing,* Mashinostroenie Publishers, Moscow, 1974.

[40] Yanushkevich SN, Dziurzanski P, and Shmerko VP. Word-level models for efficient computation of multiple-valued functions. Part 1: LAR based models. In *Proceedings IEEE 32nd International Symposium on Multiple-Valued Logic,* pp. 202–208, Boston, 2002.

12

Information Measures in Nanodimensions

This chapter focuses on information measures in nanosystems. Applying the notation to a physical system (hardware), information, in a certain sense, is a measurable quantity, which is independent of the physical medium by which it is conveyed. The most appropriate measure of information is mathematically similar to the measure of entropy, but there is good reason for reversing the sign and stating that information is the negative of entropy in nature as well as in mathematical formulation. The technique of information theory is applied to problems of the extraction of information from systems containing an element of randomness.

Two aspects of the information-theoretical approach are the focus of this chapter:

- The measures of logic circuits, and
- The measures of decision trees and diagrams.

In Section 12.1 we start with the information-theoretical basis of entropy calculation on logic networks and decision trees as used in general logic design. Entropy of spatial measures is introduced, and its calculation on the $\mathcal{N}$-hypercube is discussed. The estimated attributes here are information flow, information amount and entropy measures on the $\mathcal{N}$-hypercube. In Section 12.2, the information theoretical measures used in logic design are introduced. In Section 12.3, the information measures of a typical of library of gates are given. Finally, we consider application of information theoretical measurement to the problem of synthesis in spatial dimensions in Sections 12.4, 12.5, and 12.6.

12.1 Information-theoretical measures at various levels of design in nanodimensions

Nanodimensional structures are characterized by certain attributes that can be divided into:

- *Static*,
- *Dynamic*, and
- Combinations of both *static* and *dynamic* characteristics.

Static characteristics are evaluated by examining the topology or static structure of the circuit. Dynamic characteristics are relevant to analysis of circuit behavior or dynamic usage. Compared to logic design of very large scale integration (VLSI), static characteristics are those which characterize circuit structure, while dynamic attributes include observability and controllability in test generation.

Combination of both static and dynamic characteristics is an important attribute of deep-submicron integrated circuits (ICs) design, and the same can be said regarding nanoscale design.

12.1.1 Static characteristics

Static characteristics use information from circuit topology that often depends on technology. In contrast, variations of input assignments have much less impact. This is why usage of dynamic information along with the static structure must be combined in comprehensive analysis of circuits. Static characteristics provide useful information about

- Circuit structure, and
- Connectivity.

However, they lack information about how the various parts of the circuit are utilized with respect to different input combinations. The latter is important for evaluating the amount of switching that takes place in the circuit as the nodes change states. Therefore,

- The static characteristics are computed on circuit topology, and as a result its value is often influenced by the implementation itself.
- The static characteristics may become a drawback when the characteristic is trying to capture a fundamental property of the function that is expected to remain invariant throughout different implementations of the function.

12.1.2 Dynamic characteristics

Dynamic characteristics refer to:

- Different input combinations, and
- The way in which these combinations cause data to be communicated through the structure.

Testability analysis and circuit simulation inherently exploit dynamic metrics. This requires the circuit to be exercised for several input assignments, and this usually requires more (time and space) resources than static attribute computation.

12.1.3 Combination of static and dynamic characteristics

Evaluation of a combination of both static and dynamic characteristics provides information about

- The function, and
- Its implementation.

An example of usage of static characteristics is optimization via binary decision diagrams (BDDs). Such optimizations is based on variable ordering and is performed using static attributes (node levels, depth), and may not always be satisfactory, or compliant with other criteria, since decision diagrams represent only one of several possible interpretations of a given switching function.

In both static and dynamic attributes, the issue of communication plays an essential role. This is because in nanodimensions, area, time and power performance are dominated less by logic in their circuit and more by communication between logic. In VLSI design, the complexity of VLSI implementation is measured in terms of bounds on chip area and computation time, taking into account partitioning of the input set as a method to evaluate the lower bound of the complexity. For example, communication complexity can be measured as the maximum number of bits of information exchanged between partitions over all input values to correctly compute the function. The communication complexity can be also evaluated by computing the number of compatible classes of a given function.

12.1.4 measures on data structures

The above leads to the conclusion that static circuit structure along with dynamic metrics is a trade-off solution to provide accurate analysis of nanoscale structures. For example, switching energy of a circuit is a function of static structure and connectivity, as well as the dynamic behavior of the functions to be implemented on the structure. Thus, a quantified measure of the energy consumption in the circuit can be made by unification of both characteristics.

However, it is difficult to find a measure that can unify static and dynamic characteristics. This is because of the different levels of abstraction that distort the flow of information. This applies to contemporary logic design of VLSI circuits as well. In the case of nanometric electronic devices, the problem of random logic (by nature) arises. The design of such devices must be capable of dealing with random logical effects. This means that probability evaluation is essential in the evaluation of dynamic characteristics of the circuits.

All the required virtues are successfully combined in information theory measures such as

- *Entropy*, and
- *Amount of information.*

The measures can be made on different data structures that carry information about a logic function: formal representation (sum-of-products, Reed-Muller, arithmetic, and Walsh spectrum, and word-level forms), logic network, flowgraphs, and decision trees and diagrams, including spatial representations. Information-theory measures are sensitive to data structure. This serves as a motivation to distinguish two kinds of measures:

- *Spatial* measures on hypercube and hypercube-like structures derived from a decision tree. Spatial measurement relates to the function given by the decision tree, and, thus, captures the communication between minterms of the function in the implementation. The entropy of spatial measurement is a measure of the effort needed to transmit data, which is a *dynamic* communication effort in a circuit.
- *Logic* measures on a logic networks. Logic measurement is based on logic network (netlist) and is relevant to the gate-count complexity of the original network, and thus more technology-dependent. The entropy of logic measurement depends on the number of logic operations required to perform a computation which is a *static* component of a circuit.

In this chapter, we focus on the entropy of spatial measurement in $\mathcal{N}$-hypercube space. The information content of a switching function is an inherent attribute of a function and is technology-independent. Information content defines the complexity of function implementation and can be used to estimate a lower bound on some physical (topological) parameters with respect to various implementations. Therefore, it reflects the fundamental characteristic of function behavior. Entropy of spatial measurement in $\mathcal{N}$-hypercube space can be viewed as a contribution to information content, over all nodes of the embedded decision diagram.

The above is the basis for approaches to estimate the complexity involved when data are transmitted from various points in a circuit. The estimated attributes here are information flow, information amount and entropy measures on the hypercube. Finally, we describe other information-theoretical definitions and outline their application to the problem of synthesis of 3-D structures.

12.2 Information-theoretical measures in logic design

In this section, the basics of information-theoretical measures are introduced. The most basic measure is entropy. Many useful additional characteristics are derived from the entropy, namely, the conditional entropy, mutual information, joint information, and relative information. Figure 12.1 illustrates the basic principles of input and output information measures in a logic circuit, where the shared arrows mean that the value of $x_i(f)$ carries the information, the shared arrow therefore indicating the direction of the information stream. Obviously, we can compare the results of the input and output measures and calculate the loss of information.

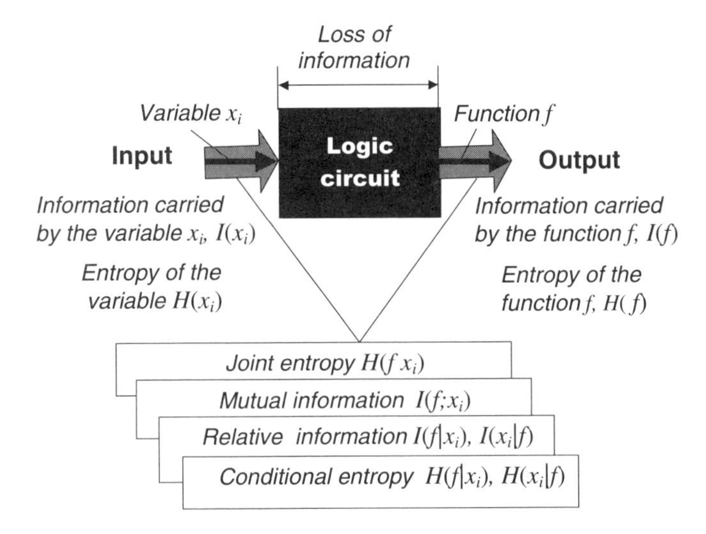

FIGURE 12.1
Information measures at the input and output of a logic circuit, and computing input/output relationships of information.

12.2.1 Information-theoretical standpoint

A computing system can be seen as a process of communication between computer components. The classical concept of information advocated by Shannon is insufficient to capture a number of features of the design and processing of a computing system.

The information-theoretical standpoint on computing is based on the following notations:

- *Source of information*, a stochastic process where an event occurs at time point i with probability p_i. In other words, the source of information is defined in terms of the probability distribution for signals from this source. Often the problem is formulated in terms of sender and receiver of information and used by analogy with communication problems.
- *Information engine*, the machine that deals with information.
- *Quantity of information*, a value of a function that occurs with the proba-

bility p carries a quantity of information equal to $(-\log_2 p)$.

- *Entropy*, $H(f)$, the measure of the information content of X. The greater the uncertainty in the source output, the higher is its information content. A source with zero uncertainty would have zero information content and, therefore, its entropy would be likewise equal to zero.

The information and entropy, in their turn, can be calculated with respect to the given sources:

Information carried by the value of a variable or function,
Conditional entropy of function f values given function g,
Relative information of the value of a function given the value of a variable,
Mutual information between the variable and function,
Joint entropy over a distribution of jointly specified functions f and g.

12.2.2 Quantity of information

Let us assume that all combinations of values of variables occur with equal probability. A value of a logic function that occurs with the probability p carries a quantity of information equal to

$$< Quantity\ of\ information > \quad = \quad -\log_2 p \;\; bit,$$

where p is the probability of that value occuring.

The information carried by the value of a of x_i is equal to

$$I(x_i)_{|x_i=a} = -\log_2 p \;\; bit,$$

where p is the quotient between the number of tuples whose i-th components equal a and the total number of tuples. Similarly, the information carried by a value b of f is

$$I(f)_{|f=b} = -\log_2 q \;\; bit,$$

where q is the quotient between the number of tuples in the domain of f and the number of tuples for which f takes the value b.

Example 12.1 *The information carried by the values of variable x_i and function f for a switching function given by a truth table is calculated in Figure 12.2.*

12.2.3 Conditional entropy and relative information

Conditional entropy is a measure of a random variable f given a random variable x. To compute the conditional entropy, the conditional probability of f must be calculated. The conditional probability of value b of logic function f, the input value a of x_i being known, is

$$p(f = b|x_i = a) = \frac{p_{|f=b \atop |x_i=a}}{p_{|x_i=a}}$$

Probabilities of the values of variable x_i:

$p(x_i = 0) = {}^3/_5,\ p(x_i = 1) = {}^2/_5$

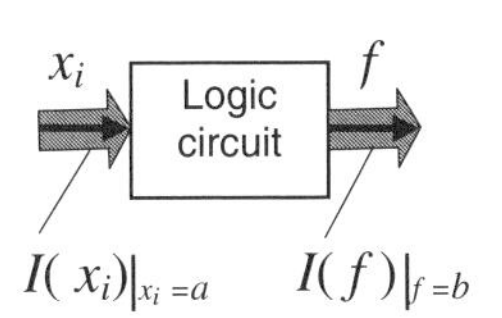

The information carried by the variable x_i:

$$\begin{cases} I(x_i)_{|x_i=0} = -\log_2 {}^3/_5 = 0.737 \ bit \\ I(x_i)_{|x_i=1} = -\log_2 {}^2/_5 = 1.322 \ bit \end{cases}$$

Probabilities of the values of function f:

$p(f = 0) = {}^4/_5,\ p(f = 1) = {}^1/_5$

while

$$\begin{cases} p(f=0)|_{x_i=0} = {}^3/_5 & p(f=0)|_{x_i=1} = {}^1/_5 \\ p(f=1)|_{x_i=0} = 0 & p(f=1)|_{x_i=1} = {}^1/_5 \end{cases}$$

Information carried by the function f:

$$\begin{cases} I(f)_{|f=0} = -\log_2 {}^4/_5 = 0.322 \ bit \\ I(f)_{|f=1} = -\log_2 {}^1/_5 = 2.322 \ bit \end{cases}$$

Input x_i	*Output* f
0	0
1	0
0	0
1	1
0	0

FIGURE 12.2
The information carried by values of variable x_i and switching function f (Example 12.1).

Similarly, the conditional probability of value a of x_i given value b of the function f is

$$p(x_i = a|f = b) = \frac{p_{|f=b \atop |x_i=a}}{p_{|f=b}}$$

Conditional entropy $H(f|g)$ of function f values given logic function g is

$$H(f|g) = H(f, g) - H(g). \tag{12.1}$$

In circuit analysis and decision tree design, the so-called *chain rule* is useful

$$H(f_1, \ldots, f_n|g) = \sum_{i=1}^{n} H(f_i|f_1, \ldots, f_{i-1}, g). \tag{12.2}$$

The *relative* information of value b of logic function f given value a_i of the input variable x_i is

$$I(f = b|x_i = a) = -\log_2 p\ (f = b|x_i = a).$$

The relative information of value a_i of the input variable x_i given value b of the logic function f is

$$I(x_i = a|f = b) = -\log_2 p\ (x_i = a|f = b).$$

Once the probability is equal to 0, we suppose that the relative information is equal to 0.

Example 12.2 *Figure 12.3 illustrates the calculation of the conditional and relative information given the truth table of a switching function.*

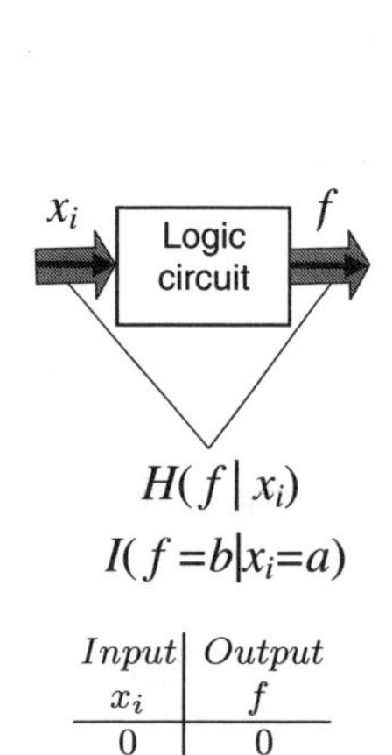

Input x_i	*Output* f
0	0
1	0
0	0
1	1
0	0

Conditional probabilities:

$$p_{\substack{|f=0\\|x_i=0}} = {}^3/_5 \quad p_{\substack{|f=0\\|x_i=1}} = {}^1/_5 \quad p_{\substack{|f=1\\|x_i=0}} = 0 \quad p_{\substack{|f=1\\|x_i=1}} = {}^1/_5$$

Then

$$p(f=0|x_i=0) = p_{\substack{|f=0\\|x_i=0}} : p_{|x_i=0} = {}^3/_5 : {}^3/_5 = 1$$
$$p(f=0|x_i=1) = p_{\substack{|f=0\\|x_i=1}} : p_{|x_i=1} = {}^1/_5 : {}^2/_5 = {}^1/_2$$
$$p(f=1|x_i=0) = p_{\substack{|f=1\\|x_i=0}} : p_{|x_i=0} = 0$$
$$p(f=1|x_i=1) = p_{\substack{|f=1\\|x_i=1}} : p_{|x_i=1} = {}^1/_5 : {}^2/_5 = {}^1/_2$$

Conditional entropy $H(f|x_i)$:

$$\begin{aligned} H(f|x_i) &= -p(f=0|x_i=0)\log p(f=0|x_i=0) \\ &\quad - p(f=0|x_i=1)\log p(f=0|x_i=1) \\ &\quad - p(f=1|x_i=0)\log p(f=1|x_i=0) \\ &\quad - p(f=1|x_i=1)\log p(f=1|x_i=1) \\ &= -1\log 1 - {}^1/_2 \log {}^1/_2 - 0\log 0 - {}^1/_2 \log {}^1/_2 \\ &= 1 \end{aligned}$$

Relative information $I(f=b|x_i=a)$:

$$I(f=0|x_i=0) = -\log_2 1 = 0$$
$$I(f=0|x_i=1) = -\log_2 {}^1/_2 = 1$$
$$I(f=1|x_i=0) = 0$$
$$I(f=1|x_i=1) = -\log_2 {}^1/_2 = 1$$

FIGURE 12.3
Computing conditional entropy and relative information (Example 12.2).

12.2.4 Entropy of a variable and a function

Let the input variable x_i be the outcome of a probabilistic experiment, and the random logic function f represent the output of some step of computation. Each experimental outcome results in different conditional probability distributions on the random f. Shannon's entropy of the variable x_i is defined as

$$H(x_i) = \sum_{l=0}^{m-1} p_{|x_i=a_l} \log_2 p_{|x_i=a_l}, \tag{12.3}$$

where m is the number of distinct values assumed by x_i. Shannon's entropy of the logic function f is

$$H(f) = \sum_{k=0}^{n-1} p_{|f=b_k} \log_2 p_{|f=b_k}, \tag{12.4}$$

where n is the number of distinct values assumed by f.

This definition of the measure of information implies that the greater the uncertainty in the source output, the smaller is its information content. In a similar fashion, a source with zero uncertainty would have zero information content and, therefore, its entropy would be likewise equal to zero.

Example 12.3 *Figure 12.4 illustrates calculation of entropy of the variable and function. The entropy of the variable x_i and switching function f is 0.971 bits and 0.722 bits.*

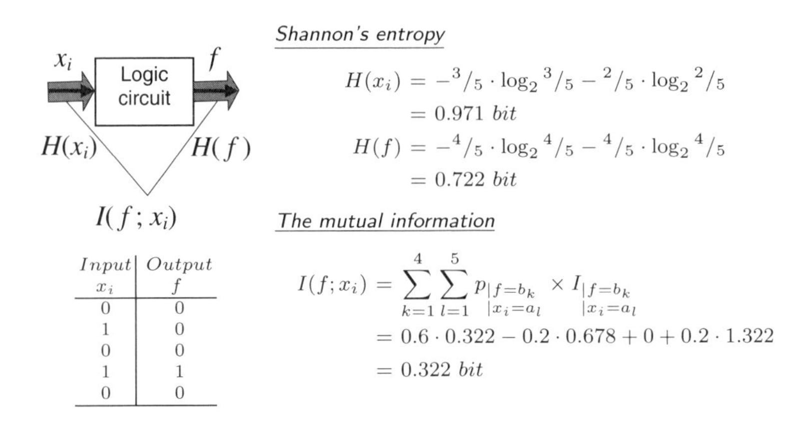

FIGURE 12.4
Shannon's entropy and mutual information (Examples 12.3 and 12.4).

Therefore,

- For any variable x_i it holds that $0 \leq H(x_i) \leq 1$; similarly, for any function f, $0 \leq H(x_i) \leq 1$.
- The entropy of any variable in a completely specified function is 1.
- The entropy of a constant is 0.

12.2.5 Mutual information

The *mutual* information is used to measure the dependence of the function f on the values of the variable x_i and vice-versa, i.e. how statically distinguishable distributions of f and x_i are. If the distributions are different, then the amount of information f carries about x_i is large. If f is independent of x_i, then f carries zero information about x_i.

The mutual information between the value b of the function and the value a of the input variable x_i is:

$$
\begin{aligned}
I(f;x_i) &= I(f;x_i)_{|f=b} - I(f=b|x_i=a) \\
&= -\log_2 p_{|f=b} + \log_2 \frac{p_{\substack{|f=b \\ |x_i=a}}}{p_{|x_i=a}}.
\end{aligned}
$$

By analogy, the mutual information between the input variable x_i and the function f is

$$
\begin{aligned}
I(f;x_i) &= \sum_k \sum_l p_{\substack{|f=b_k \\ |x_i=a_l}} \times I(f;x_i)_{\substack{|f=b_k \\ |x_i=a_l}} \\
&= \sum_k \sum_l p_{\substack{|f=b_k \\ |x_i=a_l}} \times \log_2 \frac{p_{\substack{|f=b_k \\ |x_i=a_l}}}{p_{|x_i=a_l}}.
\end{aligned}
$$

Useful relationships are

$$
\begin{aligned}
I(g;f) = I(f;g) &= H(f) - H(f|g) \\
&= H(g) - H(g|f) \\
&= H(f) + H(g) - H(f,g); \\
I(g;f_1,\ldots,f_n|z) &= \sum_{i=1}^{n} I(g;f_i|f_1,\ldots,f_{i-1},z),
\end{aligned}
$$

where $I(g;f|z)$ is the *conditional mutual* information between g and f given z. If g and f are independent, then $I(g;f) \geq 0$. The mutual information is a measure of the correlation between g and f. For example, if g and f are equal with high probability, then $I(g;f)$ is large. If f_1 and f_2 carry information about g and are independent given g then $I(z(f_1,f_2);g) \leq I(f_1;g) + I(f_2;g)$ for any switching function z.

Example 12.4 *Figure 12.4 illustrates the calculation of the mutual information. The variable x_i carries 0.322 bits of information about the switching function f.*

12.2.6 Joint entropy

Let the distribution of jointly specified functions f and g's values be known. Then, the *joint entropy* $H(f,g)$ given this distribution is defined as follows:

$$H(f,g) = -\sum_{a=0}^{m-1}\sum_{b=0}^{m-1} p_{\substack{|f=a\\|g=b}} \cdot \log p_{\substack{|f=a\\|g=b}}, \tag{12.5}$$

where $p_{\substack{|f=a\\|g=b}}$ denotes the probability that f takes value a and g takes value b, simultaneously.

12.3 Information measures of elementary switching functions

There are two approaches to information measures of elementary functions of two variables:

- ▶ The values of input variables are considered as random patterns; for a two-input elementary function there are four random patterns $x_1x_2 \in \{00.01, 10, 11\}$.
- ▶ The values of input variables are considered as noncorrelated random signals; for a two-input elementary function there are random signals $x_1 \in \{0, 1\}$ and $x_2 \in \{0, 1\}$.

Information measures based on pattern. Consider a two-input AND function with four random combinations of input signals: 00 with probability p_{00}, 01 with probability p_{01}, 10 with probability p_{10}, and 11 with probability p_{11} (Figure 12.5a).

Using Shannon's formula (Equation 12.3), we can calculate the entropy of the input signals, denoted by H_{in} as follows

$$\begin{aligned} H_{in} = &- p_{00} \times \log_2 p_{00} - p_{01} \times \log_2 p_{01} \\ &- p_{10} \times \log_2 p_{10} - p_3 \times \log_2 p_{11} \ \textit{bit/pattern}. \end{aligned}$$

Maximum entropy of the input signals can be calculated by inserting into the above equation $p_i = 0.25,\ i = 0, 1, 2, 3$ (Figure 12.6).

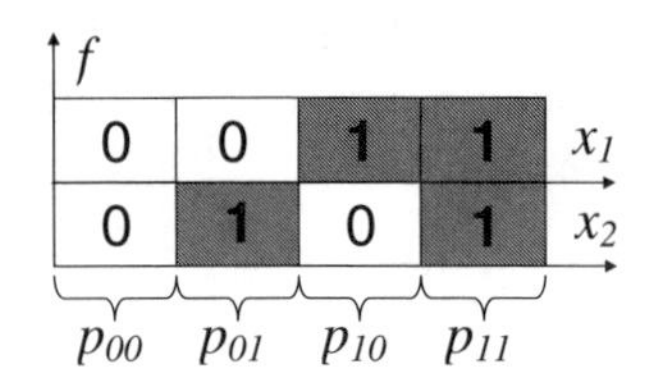

Patterns

Four random combinations of the signals and their probabilities:
$00 \rightarrow p_{00}$
$01 \rightarrow p_{01}$
$10 \rightarrow p_{10}$
$11 \rightarrow p_{11}$

(a)

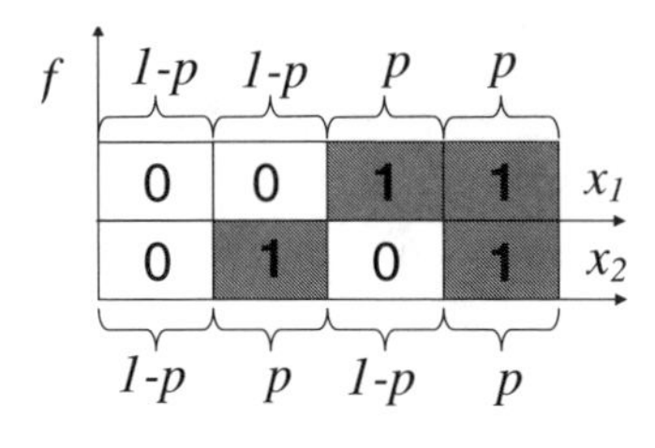

Probabilities of noncorrelated signals

$1 \rightarrow p$
$0 \rightarrow 1 - p$

(b)

FIGURE 12.5
Measurement of probabilities: random patterns (a) and noncorrelated signals (b).

The output of the AND function is equal to 0 with probability 0.25, and equal to 1 with probability 0.75. The entropy of the output signal, H_{out}, is calculated by Equation 12.4

$$H_{out} = -0.25 \times \log_2 0.25 - 0.75 \times \log_2 0.75 = 0.81 \ \textit{bit/pattern}.$$

The example below demonstrates a technique of computing information measures that input signal are not correlated.

Information measures based on noncorrelated signals. Let the input signal be equal to 1 with probability p, and 0 with probability $1 - p$ (Figure 12.5b). The entropy of the input signals is

$$\begin{aligned} H_{in} &= -(1-p)^2 \times \log_2 (1-p)^2 - 2(1-p) \times \log_2 (1-p)p - p^2 \times \log_2 p^2 \\ &= -2(1-p) \times \log_2 (1-p) - 2p \times \log_2 p \ \ \textit{bit}. \end{aligned}$$

The output of the AND function is equal to 1 with probability p^2, and equal to 0 with probability $1 - p^2$. Hence, the entropy of the output signal is

$$H_{out} = -p^2 \times \log_2 p^2 - (1-p)^2 \times \log_2 (1-p)^2 \ \ \textit{bit}.$$

The maximum value of the output entropy is equal to 1, when $p = 0.707$. Hence, the input entropy of the AND function is 0.745 *bit* (Figure 12.6). We observe that in the case of noncorrelated signals, information losses are less.

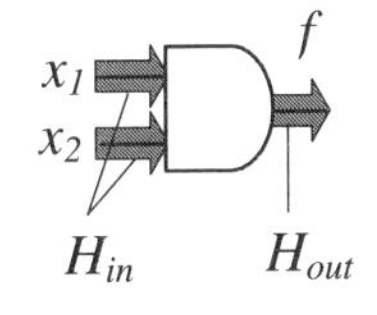

$f = x_1 x_2$

	x_1	x_2	f
$Pattern1$	0	0	0
$Pattern2$	0	1	0
$Pattern3$	1	0	0
$Pattern4$	1	1	1

Method 1:

The entropy of the input pattern (probability $p_i = 0.25$*)*

$H_{in} = -4 \times 0.25 \times \log_2 0.25 = 2\ bit/pattern$

Entropy of the output signal

$H_{out} = -0.25 \times \log_2 0.25 - 0.75 \times \log_2 0.75 =$ *0.81 bit/pattern*

Loss of information

$H_{loss} = H_{out} - H_{in} = 2.0 - 0.81 = 1.189\ bit$

Method 2:

Entropy of the input signal (probability $p = 0.707$*)*

$$
\begin{aligned}
H_{in} &= -2(1-p) \times \log_2(1-p) - 2p \times \log_2 p \\
&= -2(1-0.707) \times \log_2(1-0.707) - 2 \times 0.707 \times \log_2 0.707 \\
&= 1.745\ bit
\end{aligned}
$$

Output entropy

$$
\begin{aligned}
H_{out} &= -0.707^2 \times \log_2 0.707^2 \\
&\quad - (1-0.707)^2 \times \log_2(1-0.707)^2 = 0.804\ bit
\end{aligned}
$$

Loss of information

$$
\begin{aligned}
H_{loss} &= H_{out} - H_{in} \\
&= 1.745 - 0.804 = 0.941\ bit
\end{aligned}
$$

FIGURE 12.6
Information measures of AND functions of two variables.

Information measures in combinations of elementary switching functions. Information measures of the two-variable functions AND, OR, EXOR and NOT are given in Table 12.1 for $p(x_1) = p(x_2) = 0.5$. Note that the first approach, i.e., we suppose that information is generated by patterns. Different techniques are developed to measure the information in combinational circuits. One of the approaches is demonstrated by the following example.

Example 12.5 *Calculation of output entropy for a given circuit is shown in Figure 12.7.*

Useful properties. In Table 12.2, the entropies of elementary switching functions are grouped and calculated. We observe that

▶ A large group of the functions is characterized by the same output entropy

TABLE 12.1
Information measures of elementary switching functions of two variables.

Function	Information estimations
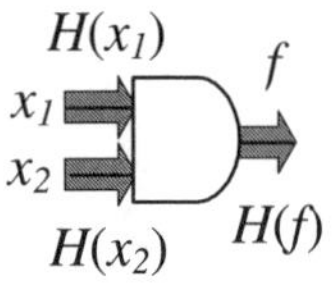 $H(x_1)$, x_1, x_2, $H(x_2)$, f, $H(f)$	$f = x_1x_2$ $H(f) = -p_{\|f=0} \cdot \log_2 p_{\|f=0} - p_{\|f=1} \cdot \log_2 p_{\|f=1}$ $p(f) = 0.5 \cdot 0.5 = 0.25$ $H(f) = -0.25 \cdot \log_2 0.25 - 0.75 \cdot \log_2 0.75 = 0.8113$ *bit*
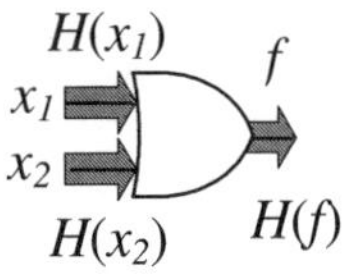 $H(x_1)$, x_1, x_2, $H(x_2)$, f, $H(f)$	$f = x_1 \vee x_2$ $H(f) = -p_{\|f=0} \cdot \log_2 p_{\|f=0} - p_{\|f=1} \cdot \log_2 p_{\|f=1}$ $p(f) = 1 - (1 - 0.5) \cdot (1 - 0.5) = 0.75$ $H(f) = -0.75 \cdot \log_2 0.75 - 0.25 \cdot \log_2 0.25 = 0.8113$ *bit*
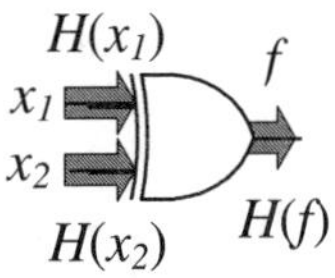 $H(x_1)$, x_1, x_2, $H(x_2)$, f, $H(f)$	$f = x_1 \oplus x_2$ $H(f) = -p_{\|f=0} \cdot \log_2 \|f = 0 - p_{\|f=1} \cdot \log_2 p_{\|f=1}$ $p(f) = 0.5 \cdot 0.5 + 0.5 \cdot 0.5 = 0.5$ $H(f) = -0.5 \cdot \log_2 0.5 - 0.5 \cdot \log_2 0.5 = 1$ *bit*
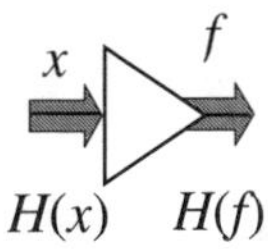 x, $H(x)$, f, $H(f)$	$f = \overline{x}$ $H(f) = p_{\|f=0} \cdot \log_2 p_{\|f=0} - p_{\|f=1} \cdot \log_2 p_{\|f=1}$ $p(f) = 1 - 0.5 = 0.5$ $H(f) = -0.5 \cdot \log_2 0.5 - 0.5 \cdot \log_2 0.25 = 1$ *bit*

$H(f) = 0.81$, conditional entropies $H(f|x_1) = 0.5$ and $H(f|x_2) = 0.5$, mutual information $I(f; x_1) = 0.31$ and $I(f; x_2) = 0.31$.

- EXOR and EQUIVALENCE functions have maximum values of entropies $H(f) = H(f|x_1) = H(f|x_2) = 1$.
- The entropy measures for a constant function, logical 1 and 0, are equal to 0.

In addition, a single-input single-output function (gate) does not lose information. Any many-input single-output logic function always results in a loss of information. An n-input n-output reversible logic function (for each input combination there is exactly one output combination, and vise versa). These fundamental properties are utilized in a method called *set of pairs of functions to be distinguished* (SPFD) and reversible nanocomputing (see "Further Reading" Section).

Information measures in elementary multivalued functions. Table 12.3 shows that transmission of information is dependent on m for the PROD

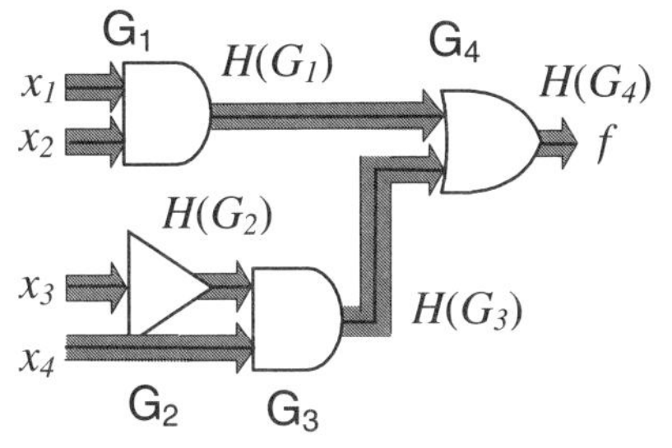

G1 H(G1) G4 H(G4)
x1 x2 f
H(G2)
x3 x4 H(G3)
G2 G3

Gate G_1:
$H(G_1) = -^1/_4 \cdot \log_2 {^1/_4} - {^3/_4} \cdot \log_2 {^3/_4}$
$= 0.5625$ *bit*
Gate G_2:
$H(G_2) = -^1/_2 \cdot \log_2 {^1/_2} - {^1/_2} \cdot \log_2 {^1/_2}$
$= 1.0$ *bit*
Gate G_3:
$H(G_3) = -^1/_4 \cdot \log_2 {^1/_4} - {^3/_4} \cdot \log_2 {^3/_4}$
$= 0.5625$ *bit*
Gate G_4:
$H(G_4) = (1 - (1 - {^1/_4})(1 - {^1/_4})$
$= 1 - {^3/_4} \cdot {^3/_4}$
$= 0.9375$ *bit*

FIGURE 12.7
Information measures in a three-level circuit (Example 12.5).

gate $(x_1x_2)_{mod\ k}$. For example, $H(f) = 0.906$ for a ternary function and $H(f) = 0.957$ for a 5-valued logic function.

Information measures in symmetric functions. The properties of logic functions can be used for the simplification of information estimation. Symmetries in circuits are classified as *structural symmetries*, arising from similarities in circuit structure and topology, and *data symmetries*, arising from similarities in the handling of data values. In the example below, we demonstrate how symmetric properties can be used for information theoretical measures.

Let f be a totally symmetric function that is 1 if m of its n variables are 1, where $m \in A \subseteq \{0, 1, \ldots, n\}$. For example, NAND of three variables is a symmetric function that is 1 iff 0, 1, and 2 of its variables are 1.

Example 12.6 *The probabilities associated with 0 and 1 of n variables of an AND function are*

$$p_{|f=0} = \frac{2^n - 1}{2^n}, \quad p_{|f=1} = \frac{1}{2^n}.$$

By analogy, for an OR function:

$$p_{|f=0} = \frac{1}{2^n}, \quad p_{|f=1} = \frac{2^n - 1}{2^n}.$$

The entropy of n-variable AND and OR functions is equal to

$$H(f) = -\frac{2^n - 1}{2^n} \cdot \log_2 \frac{2^n - 1}{2^n} - \frac{1}{2^n} \cdot \log_2 \frac{1}{2^n}$$
$$= \frac{n}{2^n} - (1 - \frac{1}{2^n}) \log(1 - \frac{1}{2^n})$$

TABLE 12.2
Information measures of elementary switching functions of two variables.

f	Entropy $H(f)$	$H(f\|x_1)$	$H(f\|x_2)$	Mutual information $I(f;x_1)$	$H\|f;x_2$
x_1x_2	0.81	0.5	0.5	0.31	0.31
$x_1\overline{x}_2$	0.81	0.5	0.5	0.31	0.31
$\overline{x}_1x_2$	0.81	0.5	0.5	0.31	0.31
$x_1 \vee x_2$	0.81	0.5	0.5	0.31	0.31
$x_1 \uparrow x_2$	0.81	0.5	0.5	0.31	0.31
$x_1 \rightarrow x_2$	0.81	0.5	0.5	0.31	0.31
$x_2 \rightarrow x_1$	0.81	0.5	0.5	0.31	0.31
$x_1 \mid x_2$	0.81	0.5	0.5	0.31	0.31
$x_1 \oplus x_2$	1	1	1	0	0
$x_1 \sim x_2$	1	1	1	0	0
x_1	1	0	1	1	0
x_2	1	0	0	0	1
$\overline{x}_1$	1	1	0	0	1
$\overline{x}_2$	1	0	1	1	0
const 1	0	0	0	0	0
const 0	0	0	0	0	0

TABLE 12.3
Information measures of elementary multivalued functions of two variables.

Function f	Entropy $H(f)$	$H(f\|x_1)$	$H(f\|x_2)$	Mutual information $I(f;x_1)$	$H\|f;x_2$
x_1x_2 $(mod\ 3)$	0.906	1.665	0.665	0.759	0.241
x_1x_2 $(mod\ 4)$	0.876	1.625	0.625	0.750	0.250
x_1x_2 $(mod\ 5)$	0.957	1.800	0.800	0.843	0.157
x_1+x_2 $(mod\ 3)$	1	2	1	1	0
x_1+x_2 $(mod\ 4)$	1	2	1	1	0

12.4 Information-theoretical measures in decision trees

In this section, we address the design of decision trees with nodes of three types: Shannon (S) positive Davio (pD) and negative Davio (nD) based on the information theoretical approach. An approach revolves around choosing the "best" variable and the "best" expansion type with respect to this variable for any node of the decision tree in terms of information measures. This means that in any step of the decision making strategy, we have an opportunity to choose both

- Variable, and
- Type of expansion,

based on the criterion of minimum entropy.

The entropy-based optimization strategy can be described as the generating of the *optimal paths* in a decision tree, with respect to the minimum entropy criterion.

12.4.1 Decision trees

Calculation of entropy and information on decision trees is best known as the induction of decision trees (ID3) algorithm for optimization.

Example 12.7 *Figure 12.8 illustrates calculation of entropy on a decision tree.*

Free binary decision trees are derived by permitting permutation of variables in a subtree independently of the order of variables in the other subtree related to the same nonterminal node. Another way of generalizing decision trees is to use different expansions at the nodes in the decision tree. This decision tree is designed by arbitrarily choosing any variable and any of the S, pD or nD expansions for each node.

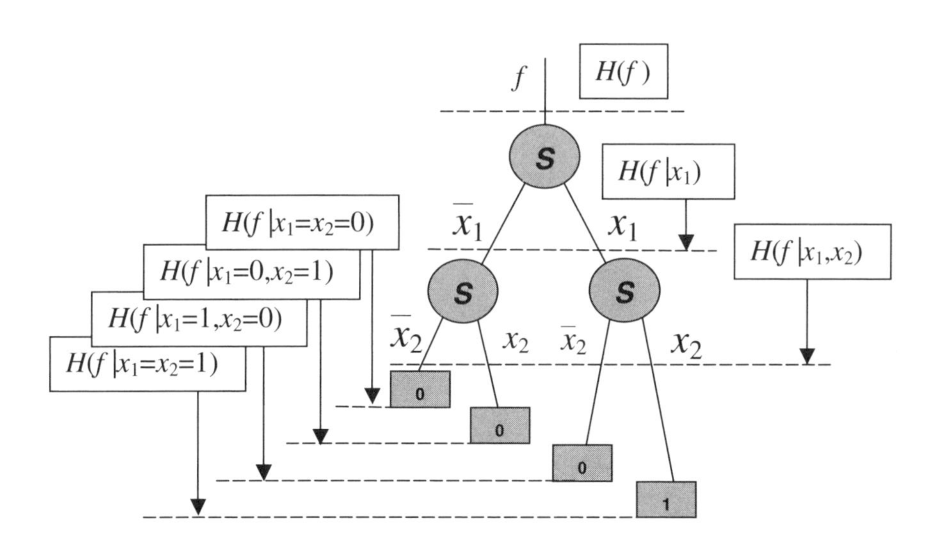

FIGURE 12.8
measure of entropy on a decision tree (Example 12.7).

12.4.2 Information-theoretical notation of switching function expansion

In the process of decision tree design two information measures are used:

$$\begin{aligned} <\textit{Conditional entropy}> &= H(f|Tree)\\ <\textit{Mutual information}> &= I(f;Tree) \end{aligned}$$

The *initial state* of this process is characterized by the maximum value for the conditional entropy

$$H(f|Tree) = H(f).$$

Nodes are recursively attached to the decision tree by using the top-down strategy. In this strategy the entropy $H(f|Tree)$ of the function is reduced, and the information $I(f;Tree)$ increases, since the variables convey the information about the function. Each *intermediate state* can be described in terms of entropy by the equation

$$I(f;Tree) = H(f) - H(f|Tree). \tag{12.6}$$

We maximize the information $I(f;Tree)$ that corresponds to the minimization of entropy $H(f|Tree)$, in each step of the decision tree design. The final state of the decision tree is characterized by $H(f|Tree) = 0$ and $I(f;Tree) = H(f)$, i.e., $Tree$ represents the switching function f (Figure 12.9).

$H(f|Tree_1)$ $H(f|Tree_3)$

IN OUT

$H(f|Tree_2)$ $H(f|Tree)=0$

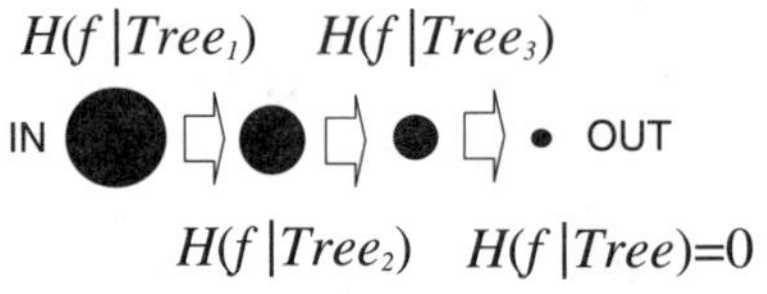

Maximizing the information

$$I(f;Tree_i)$$

corresponds to the minimization of entropy $H(f|Tree_i)$.

The final state of the decision tree design corresponds to

$$\begin{aligned} H(f|Tree3) &= 0\\ I(f;Tree) &= H(f), \end{aligned}$$

i.e., $< Tree >$ *represents the switching function f*

FIGURE 12.9
Four steps of minimization of entropy $H(f|Tree)$ in decision tree design for reduction of uncertainty.

The decision tree design process is a recursive decomposition of a switching function. A step of this recursive decomposition corresponds to the expansion of switching function f with respect to the variable x. Assume that the variable x in f conveys information that is, in some sense, the rate of influence of the input variable on the output value for f.

The initial state of the expansion $\omega \in \{S, pD, nD\}$ can be characterized by the entropy $H(f)$ of f, and the final state by the conditional entropy $H^{\omega}(f|x)$. The ω expansion of the switching function f with respect to the variable x is described in terms of entropy as follows

$$I^{\omega}(f;x) = H(f) - H^{\omega}(f|x). \tag{12.7}$$

A formal criterion for completing the sub-tree design is $H^{\omega}(f|x) = 0$, which means return from the recursion of decision tree design.

Information notation of S expansion

The designed decision tree based on the S expansion is mapped into a sum-of-products expression as follows: a leaf with the logic value 0 is mapped into $f = 0$, and with the logic value 1 into $f = 1$; a nonterminal node is mapped into $f = \overline{x} \cdot f_{|x=0} \vee x \cdot f_{|x=1}$. The information measure of S expansion for a switching function f with respect to the variable x is represented by the equation

$$H^{S}(f|x) = p_{|x=0} \cdot H(f_{|x=0}) + p_{|x=1} \cdot H(f_{|x=1}). \tag{12.8}$$

The information measure of S expansion is equal to the conditional entropy $H(f|x)$:

$$H^{S}(f|x) = H(f|x). \tag{12.9}$$

Information notation of pD and nD expansion

The information measure of pD expansion of a switching function f with respect to the variable x is represented by

$$H^{pD}(f|x) = p_{|x=0} \cdot H(f_{|x=0}) + p_{|x=1} \cdot H(f_{|x=0} \oplus f_{|x=1}). \tag{12.10}$$

The information measure of the nD expansion of a switching function f with respect to the variable x is

$$H^{nD}(f|x) = p_{|x=1} \cdot H(f_{|x=1}) + p_{|x=0} \cdot H(f_{|x=0} \oplus f_{|x=1}). \tag{12.11}$$

Theorem 12.1 *The information merit (efficiency) in choosing the pD or nD nodes for a decision tree design in comparison to the S nodes is calculated as follows:*

$$\triangle I^{pD} = p_{|x=1} \cdot (H(f_{|x=1}) - H(f_{|x=0} \oplus f_{|x=1})), \tag{12.12}$$

and

$$\triangle I^{nD} = p_{|x=0} \cdot (H(f_{|x=0}) - H(f_{|x=0} \oplus f_{|x=1})), \tag{12.13}$$

respectively.

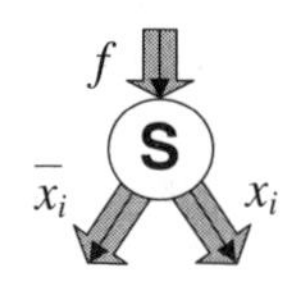

Shannon expansion

$$f = \overline{x} \cdot f_{|x=0} \oplus x \cdot f_{|x=1}$$

Information-theoretical notation

$$H^S(f|x) = \underbrace{p_{|x=0}H(f_{|x=0})}_{Left\ leaf} + \underbrace{p_{|x=1}H(f_{|x=1})}_{Right\ leaf}$$

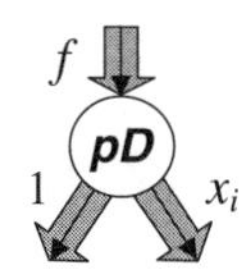

Positive Davio expansion

$$f = f_{|x=0} \oplus x \cdot (f_{|x=0} \oplus f_{|x=1})$$

Information-theoretical notation

$$H^{pD}(f|x) = \underbrace{p_{|x=0}H(f_{|x=0})}_{Left\ leaf} + \underbrace{p_{|x=1}H(f_{|x=0} \oplus f_{|x=1})}_{Right\ leaf}$$

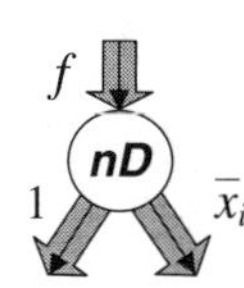

Negative Davio expansion

$$f = f_{|x=1} \oplus \overline{x} \cdot (f_{|x=0} \oplus f_{|x=1})$$

Information-theoretical notation

$$H^{nD}(f|x) = \underbrace{p_{|x=1}H(f_{|x=1})}_{Left\ leaf} + \underbrace{p_{|x=0}H(f_{|x=0} \oplus f_{|x=1})}_{Right\ leaf}$$

FIGURE 12.10
Shannon and Davio expansions and their information measures for a switching function.

PROOF Because $H^S(f|x) = H(f|x)$ and $I^S(f;x) = I(f;x)$, thus $I(f;x) = H(f) - H(f|x)$. Denote the information merit by

$$I^{\omega}(f;x) = I(f;x) + \triangle I^{\omega}. \tag{12.14}$$

Since

$$\begin{aligned}
H(f|x) &= p_{|x=0} \cdot H(f_{|x=0}) + p_{|x=1} \cdot H(f_{|x=1}) + p_{|x=1} \cdot H(f_{|x=0} \oplus f_{|x=1}) \\
&\quad - p_{|x=1} \cdot H(f_{|x=0} \oplus f_{|x=1}) \\
&= H^{pD}(f|x) + p_{|x=1} \cdot (H(f_{|x=1}) - H(f_{|x=0} \oplus f_{|x=1})) \\
I(f;x) &= H(f) - H^{pD}(f|x) - p_{|x=1} \cdot (H(f_{|x=1}) - H(f_{|x=0} \oplus f_{|x=1})),
\end{aligned}$$

and Equation 12.14 and Equation 12.12 is true. Likewise, the theorem for the nD expansion (Equation 12.13) can be proven. ∎

Example 12.8 *Given a switching function of three variables, calculate the entropy of Shannon and Davio expansions with respect to all variables. The results are summarized in Table 12.4. We observe that the minimal value of the information theoretical measure corresponds to Shannon expansion with respect to the variable x_2.*

TABLE 12.4
Choosing the type of expansion for switching function of three variables (Example 12.8).

	$H^S(f_1\|x)$	$H^{pD}(f_1\|x)$	$H^{nD}(f_1\|x)$
x_1	0.88	0.95	0.88
x_2	**0.67**	0.88	0.75
x_3	0.98	0.98	0.95

12.4.3 Optimization of variable ordering in a decision tree

The entropy based optimization of decision tree design can be described as the optimal (with respect to the information criterion) node selection process. A path in the decision tree starts from a node and finishes in a terminal node. Each path corresponds to a term in the final expression for f.

The criterion for choosing the decomposition variable x and the expansion type $\omega \in \{S, pD, nD\}$ is that the conditional entropy of the function with respect to this variable has to be minimum:

$$H^{\omega}(f|x) \rightarrow MIN.$$

The entropy based algorithm for minimization of AND/EXOR expressions is introduced in the example below. In this algorithm, the ordering restriction is relaxed. This means that

- ▶ Each variable appears once in each path, and
- ▶ The orderings of variables along the paths may be different.

Example 12.9 *The design of an AND/EXOR decision tree for the hidden weighted bit function is given in Figure 12.11. The order of variables in tree is evaluated based on measure of entropy of switching function f with respect to variable x_1 x_2, and x_3. According to the criterion of minimum entropy, x_1 is assigned to the root, and the other assignments are shown in Figure 12.11. The quasi-optimal Reed-Muller expression corresponding to this tree is:*

$$f = x_2x_3 \oplus x_1\overline{x}_3$$

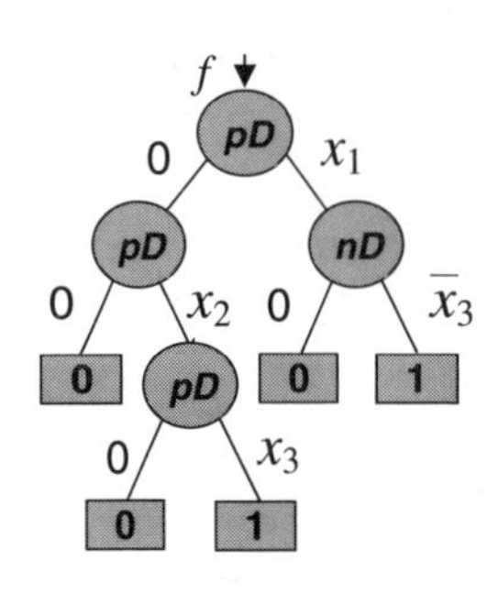

x_1	x_2	x_3	f
0	0	0	0
0	0	1	0
0	1	0	0
0	1	1	1
1	0	0	1
1	0	1	0
1	1	0	1
1	1	1	1

Step 1. *Choose the variable x_1 and pD expansion for the root node*

$H^{pD}(f|x_1) = 0.91$ *bit*
Decision: select the $f_0 = f_{|x_1=0}$

Step 2. *Choose the variable x_2 and pD expansion for the next node*

$H^{pD}(f|x_2) = 0.5$ *bit*
$f_0 = f_{|x_2=0} = 0$
Decision: select
$f_1 = f_{|x_2=0} \oplus f_{|x_2=1}$

Step 3. *Select pD expansion for the variable x_3*

$f_0 = f_{|x_3=0} = 0$ *and*
$f_1 = f_{|x_3=0} \oplus f_{|x_3=1} = 1$
Decision: select
$f_0 = f_{|x_1=0} \oplus f_{|x_1=1}$

Step 4. *Choose the variable x_3 and select nD expansion*

$f_0 = f_{|x_3=1} = 0$ *and*
$f_1 = f_{|x_3=0} \oplus f_{|x_3=1} = 1$

FIGURE 12.11
AND/EXOR decision tree design (Example 12.9).

Example 12.10 *Design of the Shannon tree based on sum-of-products expression given the hidden weighted bit function. The Shannon tree is shown in Figure 12.12.*

12.5 Information measures in the $\mathcal{N}$-hypercube

It has been shown that information-theoretical measures for logic networks can be evaluated by decision trees. In this section we focus on the details of information measures in $\mathcal{N}$-hypercube based on information measures in decision trees.

A useful property of an $\mathcal{N}$-hypercube is that compared with decision trees and diagrams it is possible to obtain information measure without recalculation after changing the order of variables. The example below illustrates this property.

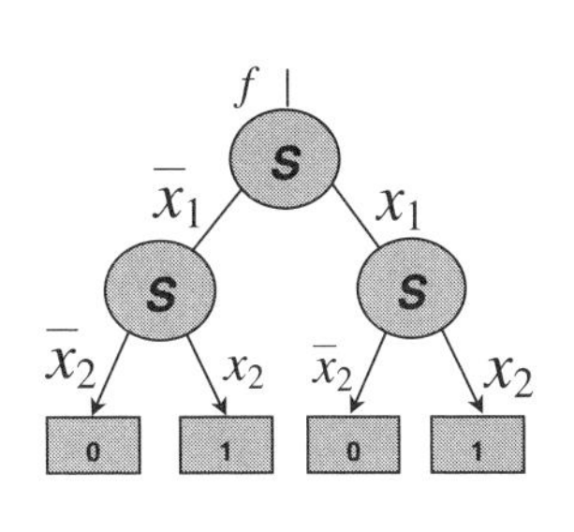

x_1	x_2	x_3	f
0	0	0	0
0	0	1	0
0	1	0	0
0	1	1	1
1	0	0	1
1	0	1	0
1	1	0	1
1	1	1	1

Entropy of the function

$$H(f) = -\ {}^1/_2 \cdot \log {}^1/_2 - {}^1/_2 \cdot \log {}^1/_2 = 1\ bit/pattern$$

Conditional entropy with respect to the variable x_1

$$H(f|x_1) = -\ {}^3/_8 \cdot \log {}^3/_8 - {}^1/_8 \cdot \log {}^1/_8 - {}^1/_8 \cdot \log {}^1/_8 - {}^3/_8 \cdot \log {}^3/_8 = 0.81\ bit/pattern$$

Sum-of-products expression
$f = \overline{x}_3 \cdot x_1 \vee x_3 \cdot x_2$

FIGURE 12.12
Shannon decision tree design (Example 12.10).

Example 12.11 *Figure 12.13 illustrates the calculation of entropy on $\mathcal{N}$-hypercube. Starting with the root, where entropy is maximal, we approach variables in sequence. Approaching x_1 reveals information about this variable, etc. Approaching terminal nodes means that the entropy becomes 0.*

Example 12.12 *Let the order of variables of a two variables switching function be*

$$\{x_1, x_2\}.$$

In Figure 12.14a, the corresponding decision tree and $\mathcal{N}$-hypercube are depicted. Let us change the order of variables: $\{x_2, x_1\}$. It follows from Figure 12.14b that it is necessary to recalculate information estimation in the decision tree, but we do not need to recalculate in the $\mathcal{N}$-hypercube.

Information computing of two-variable functions is given in Table 12.1. Here we suppose that input patterns are generated with equal probabilities. An alternative approach is based on calculation of input and output entropy assuming that input patterns are generated with different probabilities.

Example 12.13 *Calculation of information and entropy on the tree for a switching function given a truth table [0 1 0 1] is illustrated in Figure 12.15.*

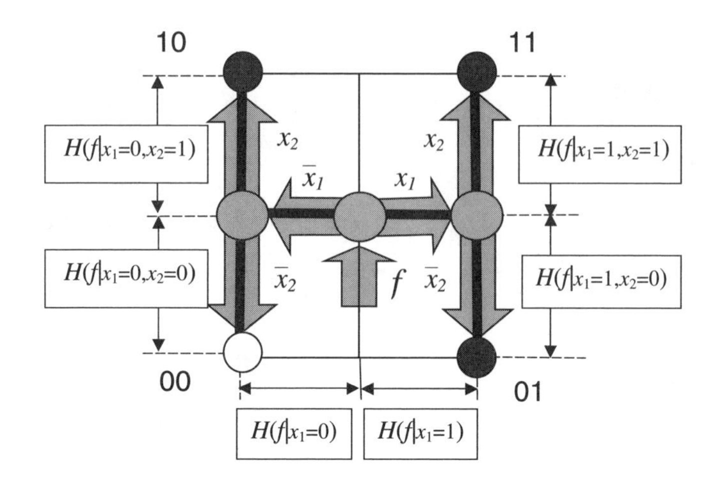

FIGURE 12.13
Measure of entropy on an $\mathcal{N}$-hypercube (Example 12.11).

12.6 Information-theoretical measures in multivalued functions

In this section, information-theoretical measures are applied to multiple-value functions. The focus is S, pD, and nD expansions and decision tree design in terms of entropy and information.

Information-theoretical measures can be applied to m-valued functions. For calculation, the logarithm base m is applied, e.g. $\log_3$ for ternary function, $\log_4$ for quaternary function, etc. The example below demonstrates the technique for computing computing information-theoretical characteristics for the function given by a truth table.

Example 12.14 *Computing entropy, conditional entropy and mutual information for a 4-valued function f given its truth column vector* [0000 0231 0213 0321]^T *are shown in Figure 12.16.*

12.6.1 Information notation of S expansion

The information measures of expansion in GF(4) are given in Table 12.6, where $J_i(x)$, $i = 0, \dots, k-1$, are the characteristic functions, denoted by $J_i(x) = 1$, if $x = i$ and $J_i(x) = 0$, otherwise. The average entropy is equal to

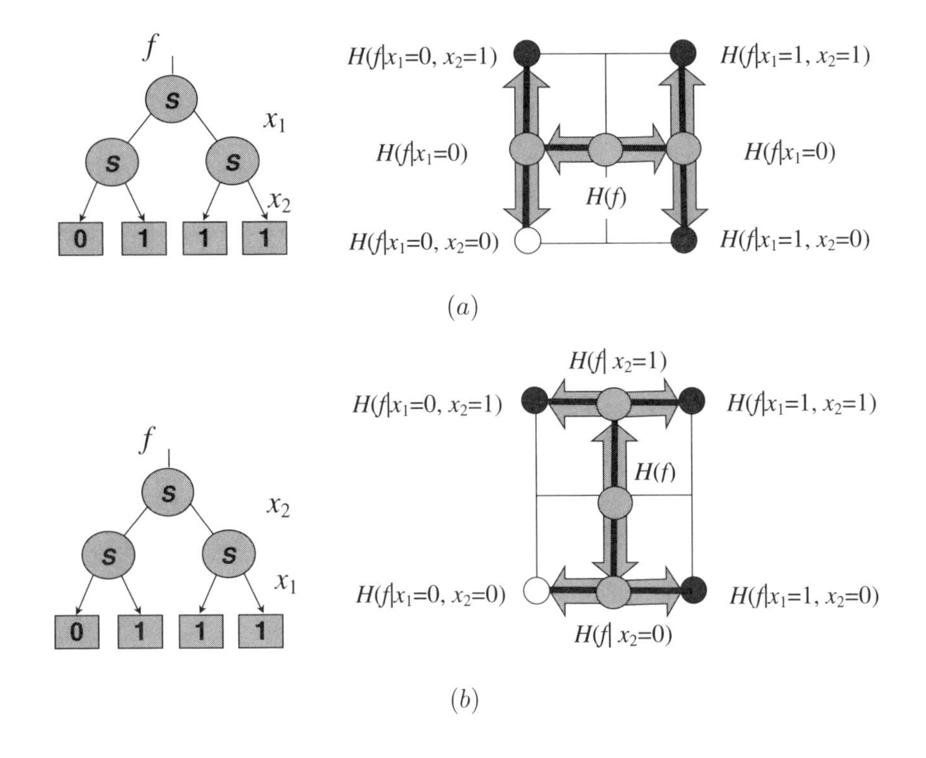

FIGURE 12.14
Order of variables in decision tree and rotation of an $\mathcal{N}$-hypercube (Example 12.12).

conditional entropy $H(f|x)$ of function f with respect to x:

$$H^S(f|x) = H(f|x). \tag{12.15}$$

12.6.2 Information notations of pD and nD expansion

The following theorem is the key to information measures.

Theorem 12.2 *(Information merit) For a completely specified 4-valued function f, information carried by couples (x, pD), (x, nD') differs from (x, S) by:*

$$\begin{aligned}\triangle I^{pD} = {}^1/_4 \cdot (H(f_{|x=1}) - H(f_1) + H(f_{|x=2}) \\ -H(f_2) + H(f_{|x=3}) - H(f_3)),\end{aligned} \tag{12.16}$$

$$\begin{aligned}\triangle I^{nD'} = {}^1/_4 \cdot (H(f_{|x=0}) - H(f_0) + H(f_{|x=2}) \\ -H(f_2) + H(f_{|x=3}) - H(f_3)).\end{aligned} \tag{12.17}$$

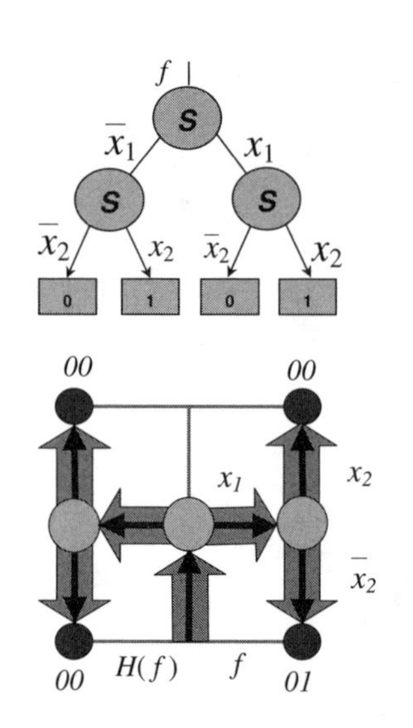

Input entropy

$$\begin{aligned} H_{in} =& -2(1-p)\times\log_2(1-p) \\ & -2p\times\log_2 p \\ =& -2(1-0.707)\times\log_2(1-0.707) \\ & -2\times 0.707\times\log_2 0.707 \\ =& \ 1.745\ bit \end{aligned}$$

Output entropy

$$\begin{aligned} H_{out} =& -0.707^2\times\log_2 0.707^2 \\ & -(1-0.707)^2\times\log_2(1-0.707)^2 \\ =& \ 0.804\ bit \end{aligned}$$

Loss of information

$$\begin{aligned} H_{loss} &= H_{out} - H_{in} \\ &= 1.745 - 0.804 = 0.941\ bit \end{aligned}$$

FIGURE 12.15
Information measures on an $\mathcal{N}$-hypercube (Example 12.13).

Equivalent relations can be obtained for the couples $(x, nD^{''})$, $(x, nD^{'''})$.

PROOF Since $H^S(f|x) = H(f|x)$ and $I^S(f;x) = I(f;x)$, then for S expansion, we can write $I(f;x) = H(f) - H(f|x)$. The information merit can be denoted by $I^{\omega}(f;x) = I^S(f;x) + \triangle I^{\omega}$. Taking into consideration that, for completely specified 4-valued function, $p_{|x\neq 0} = {}^3/_4$, we can evaluate $\triangle I^{pD}$ by the expression:

$$\begin{aligned} \triangle I^{pD} &= H(f|x) - {}^1/_4 \cdot (H(f_{|x=0}) + H(f_1) + H(f_2) + H(f_3)) \\ &= {}^1/_4 \cdot (H(f_{|x=1}) - H(f_1) + H(f_{|x=2}) - H(f_2) + H(f_{|x=3}) - H(f_3)), \end{aligned}$$

and then Equation 12.16 is true. Analogously, the same theorem can be proven for $nD^{'}$, $nD^{''}$ and $nD^{'''}$ expansions. ∎

12.6.3 Information criterion for decision tree design

The main properties of the information measure are

- The recursive character of S, pD and nD expansions and their generalization for the 4-valued case, and
- The possibility of choosing a decomposition variable and expansion type based on the information measure.

TABLE 12.5
Information measures of elementary switching functions of two variables.

Function		Information estimates
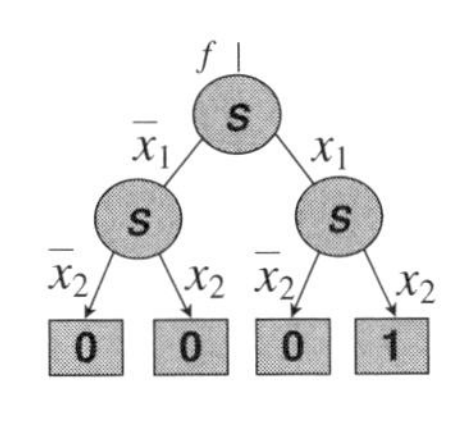	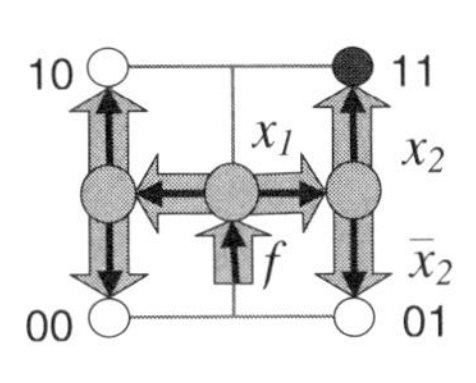	$f = x_1 x_2$ $H(f) = -\ p_{\|f=0} \cdot \log_2 p_{\|f=0}$ $-\ p_{\|f=1} \cdot \log_2 p_{\|f=1}$ $p_f = 0.5 \cdot 0.5$ $= 0.25$ $H(f) = -0.25 \cdot \log_2 0.25$ $-0.75 \cdot \log_2 0.75$ $= 0.8113\ bit$
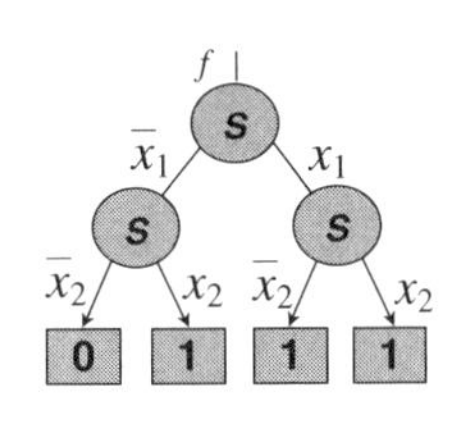	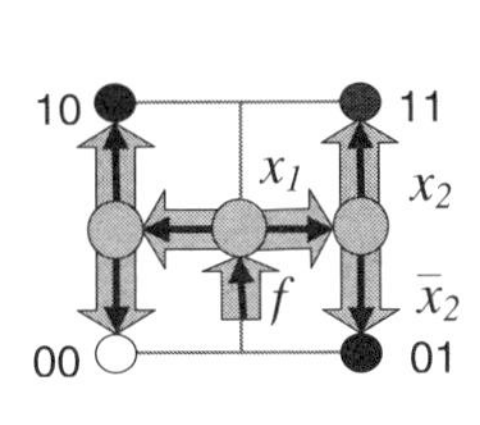	$f = x_1 \vee x_2$ $H(f) = -p_{\|f=0} \cdot \log_2 p_{\|f=0}$ $-p_{\|f=1} \cdot \log_2 p_{\|f=1}$ $p_f = 1 - (1 - 0.5) \cdot (1 - 0.5)$ $= 0.75$ $H(f) = -0.75 \cdot \log_2 0.75$ $-0.25 \cdot \log_2 0.25$ $= 0.8113\ bit$
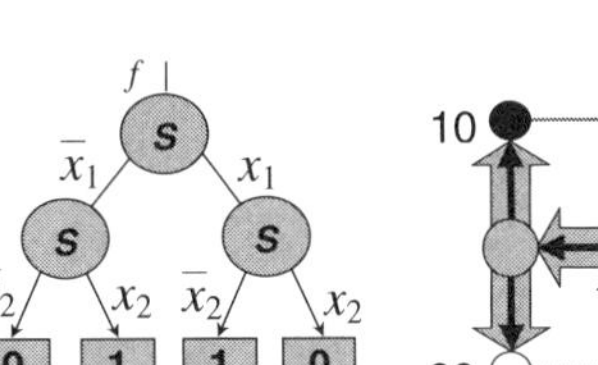	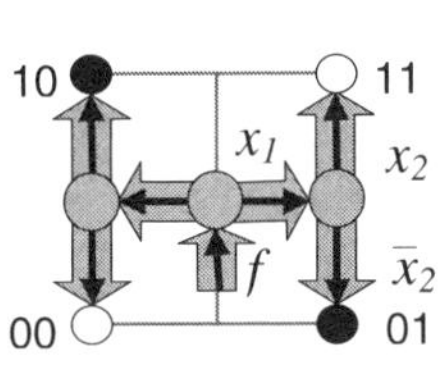	$f = x_1 \oplus x_2$ $H(f) = -p_{\|f=0} \cdot \log_2 p_{\|f=0}$ $-p_{\|f=1} \cdot \log_2 p_{\|f=1}$ $p_f = 0.5 \cdot 0.5 + 0.5 \cdot 0.5$ $= 0.5$ $H(f) = -0.5 \cdot \log_2 0.5$ $-0.5 \cdot \log_2 0.5$ $= 1\ bit$
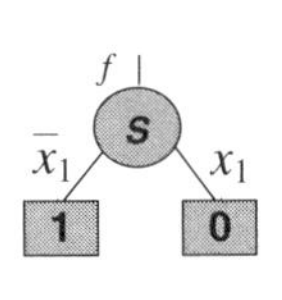	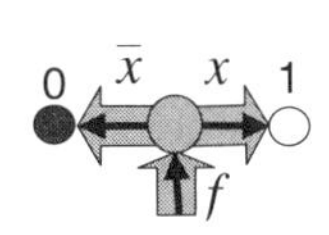	$f = \bar{x}$ $H(f) = p_{\|f=0} \cdot \log_2 p_{\|f=0}$ $-p_{\|f=1} \cdot \log_2 p_{\|f=1}$ $p_f = 1 - 0.5$ $= 0.5$ $H(f) = -0.5 \cdot \log_2 0.5$ $-0.5 \cdot \log_2 0.5$ $= 1\ bit$

x_1	x_2	f
0	0	0
0	1	0
0	2	0
0	3	0
1	0	0
1	1	3
1	2	1
1	3	0
2	0	2
2	1	0
2	2	1
2	3	3
3	0	0
3	1	3
3	2	2
3	3	1

The probabilities of the logic function values are

$$p_{|f=0} = {}^7/_{16}, \quad p_{|f=1} = p_{|f=2} = p_{|f=3} = {}^3/_{16}$$

The entropy of the logic function f is

$$H(f) = -{}^7/_{16} \cdot \log_2 {}^7/_{16} - 3 \cdot {}^3/_{16} \cdot \log_2 {}^3/_{16} = 1.88 \; bit$$

The conditional entropy of the logic function f with respect to the variable x_1 is

$$H(f|x_1) = -{}^4/_{16} \cdot \log_2 1 - 12 \cdot {}^1/_{16} \cdot \log_2 {}^1/_4 = 1.5 \; bit.$$

The conditional entropy with respect to variable x_2 is

$$H(f|x_2) = 1.25 \; bit$$

The mutual information for the logic function f and the variables x_1 and x_2 is

$$I(f;x_1) = 0.38 \; bit$$
$$I(f;x_2) = 0.63 \; bit$$

FIGURE 12.16
Information-theoretical measures of a 4-valued function (Example 12.14).

Decision tree design can be interpreted as an optimized (with respect to information criterion) node selection process.

The criterion for choosing decomposition variable x and expansion type $\omega \in \{S, pD, nD\}$ is that the conditional entropy of the logic function given variable has to be minimal

$$H^{\omega}(f|x) = MIN(H^{\omega_j}(f|x_i) \mid \forall \; pairs \; (x_i, \omega_j)) \tag{12.18}$$

In the algorithm, the ordering restriction is relaxed. This means that (i) each variable appears once in each path and (ii) the order of variables along with each path may be different.

Example 12.15 *Consider the design of a decision tree for the logic function f from Example 12.14.*

Step 1. Choose variable x_2 and 4–pD expansion for root node, because the minimal entropy is $H^{pD}(f|x_2) = 0.75$ bit. Functions $f_0 = f_{|x_2=0}$ and $f_3 = f_{|x_2=0} + f_{|x_2=1} + f_{|x_2=2} + f_{|x_2=3}$ both take logic value 0. Select the function $f_1 = f_{|x_2=1} + 3f_{|x_2=2} + 2f_{|x_2=3}$.

Step 2. Choose variable x_1 and pD expansion for the next node. The successors are constant: $f_0 = 0$, $f_1 = 0$, $f_2 = 3$ and $f_3 = 1$. Select the function $f_2 = f_{|x_2=1} + 2f_{|x_2=2} + 3f_{|x_2=3}$.

TABLE 12.6
Information measures of Shannon and Davio expansions in GF(4).

Type	Information theoretical measures
S	$H^S(f\|x) = \overbrace{p_{\|x=0} \cdot H(f_{\|x=0})}^{Leaf\ 1} + \overbrace{p_{\|x=1} \cdot H(f_{\|x=1})}^{Leaf\ 2} + \underbrace{p_{\|x=2} \cdot H(f_{\|x=2})}_{Leaf\ 3} + \underbrace{p_{\|x=3} \cdot H(f_{\|x=3})}_{Leaf\ 4}$
pD	$H^{pD}(f\|x) = p_{\|x\neq 0} \cdot (H(f_1) + H(f_2) + H(f_3))/3 + p_{\|x=0} \cdot H(f_{\|x=0})$
nD′	$H^{nD'}(f\|x) = p_{\|x\neq 1} \cdot (H(f_0) + H(f_2) + H(f_3))/3 + p_{\|x=1} \cdot H(f_{\|x=1})$
nD″	$H^{nD''}(f\|x) = p_{\|x\neq 2} \cdot (H(f_0) + H(f_1) + H(f_3))/3 + p_{\|x=2} \cdot H(f_{\|x=2})$
nD‴	$H^{nD'''}(f\|x) = p_{\|x\neq 3} \cdot (H(f_0) + H(f_1) + H(f_2))/3 + p_{\|x=3} \cdot H(f_{\|x=3})$

Step 3. Select $H^{nD'}$ expansion for variable x_1. The successors are constant: $f_0 = 0$, $f_1 = 1$, $f_2 = 0$ and $f_3 = 1$.

The decision tree obtained is shown in Figure 12.17(a). The corresponding Reed-Muller expression is

$$f = 3 \cdot x_2 \cdot x_1^2 + x_2 \cdot x_1^3 + x_2^2 \cdot {}^{1-}x_1 + x_2^2 \cdot {}^{3-}x_1.$$

By analogy, the logic expression corresponding to decision tree showed in Figure 12.17(b), is

$$f = 2 \cdot x_2 \cdot J_1(x_1) + 3 \cdot x_2 \cdot J_2(x_1) + 2 \cdot x_2^2 \cdot J_2(x_1) + 3 \cdot x_2^2 \cdot J_3(x_1).$$

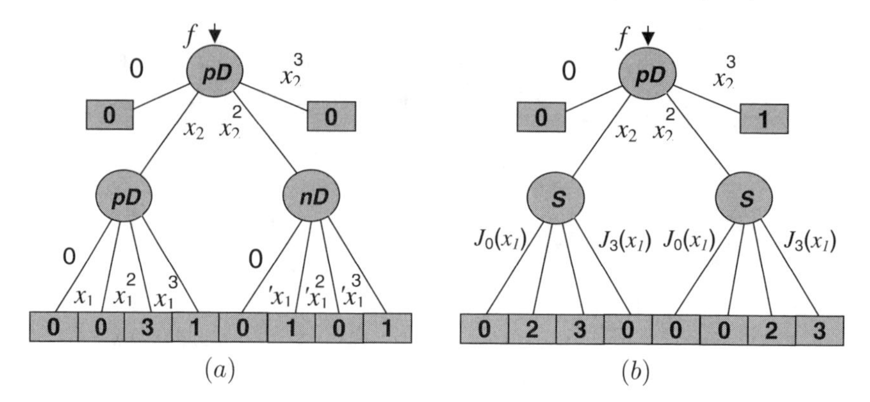

FIGURE 12.17
Decision tree design (Example 12.15).

12.7 Summary

1. In 3-D structures on the molecular/atomic level, information carriers become compatible with partially distributed sources/receivers/transmitters, and so measures that are inherent to the nature of information processing on a nanoscale level are required.

2. Information theoretical measures may combine both static and dynamic attributes. The information content of a logic function is a natural attribute of the function and it is technology-independent. The information content defines the complexity of function implementation, and thus can be used to estimate a lower bound on some physical (topological) parameters with respect to various implementations. Thus, it captures the fundamental characteristic of logic function behavior. Entropy, as spatial measurement in $\mathcal{N}$-hypercube space, can be viewed as a contribution to the information content, with respect to all nodes of the embedded decision diagram.

3. The technique of decision trees and diagrams is revised from an information-theoretical point of view. Shannon, positive Davio and negative Davio expansion is formulated in terms of entropy. Based on the information-theoretic approach, an arbitrary decision tree or decision diagram can be designed. In each step of the decision making process, the variable and type of expansion is chosen based on the information estimations.

4. Information-theoretical measures can be applied to multivalued functions. Instead of a binary signal that takes values 0 and 1, the multilevel signal carries information by m levels.

12.8 Problems

Problem 12.1 For the elementary switching functions AND, OR, EXOR, NAND, NOR of two variables, calculate

(*a*) The information carried by the variables and function (follow the Example 12.1)
(*b*) The relative information of the variables and function (follow the Example 12.2)
(*c*) Conditional entropy
(*d*) Shannon entropy of the variables and function (follow the Example 12.3)
(*e*) Mutual information between the variables and function (follow the Example 12.4)
(*f*) Joint entropy

Problem 12.2 Calculate the entropy for the completely specified function given in Figure 12.18a.

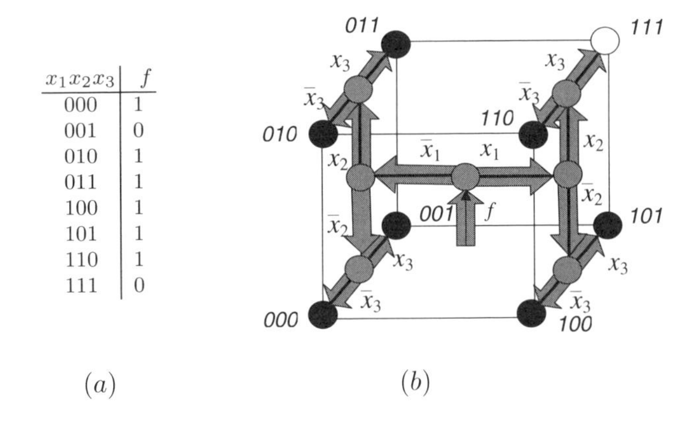

$x_1x_2x_3$	f
000	1
001	0
010	1
011	1
100	1
101	1
110	1
111	0

FIGURE 12.18
Representations of a switching function of three variables (Problems 12.2 and 12.4).

Problem 12.3 Calculate entropies of the switching functions and the function with respect to variables x_1, x_2 and x_3:

(*a*) $f = \overline{x}_1x_2 \vee x_1x_2\overline{x}_3 \vee x_3$
(*b*) $f = x_1 \oplus x_1x_2x_3 \oplus x_1x_3$
(*c*) $f = (x_1 \vee x_2)(\overline{x}_2 \vee x_3)$

Problem 12.4 In Figure 12.18b, a $\mathcal{N}$-hypercube of a 3-input AND function is given. Use the criterion of minimal entropy to evaluate the optimal way of calculation on the hypercube.

Problem 12.5 A 2-output switching function of five variables is given by a truth table in Table 12.7.

(a) The entropy of functions f_1 and f_2
(b) The conditional entropy f_1 with respect to variables x_1, x_2, x_3, x_4, and x_5
(c) Derive AND/OR decision tree using Shannon expansion and variable order by entropy criterion
(d) Derive AND/EXOR decision tree using Davio expansion and variable order by entropy criterion

Hint: the probabilities of output values are $p_{|f_1=0} = {}^{14}/_{32}$ and $p_{|f_1=1} = {}^{18}/_{32}$.

TABLE 12.7
Truth table of the two-output switching function of five variables (Problem 12.5).

x_1	x_2	x_3	x_4	x_5	f_1	f_2	x_1	x_2	x_3	x_4	x_5	f_1	f_2
0	0	0	0	0	0	0	1	0	0	0	0	0	0
0	0	0	0	1	0	1	1	0	0	0	1	0	1
0	0	0	1	0	0	0	1	0	0	1	0	0	0
0	0	0	1	1	0	1	1	0	0	1	1	0	1
0	0	1	0	0	0	0	1	0	1	0	0	1	0
0	0	1	0	1	0	1	1	0	1	0	1	1	1
0	0	1	1	0	0	0	1	0	1	1	0	1	0
0	0	1	1	1	0	0	1	0	1	1	1	1	0
0	1	0	0	0	1	1	1	1	0	0	0	1	1
0	1	0	0	1	1	1	1	1	0	0	1	1	1
0	1	0	1	0	1	1	1	1	0	1	0	1	1
0	1	0	1	1	1	1	1	1	0	1	1	1	1
0	1	1	0	0	1	1	1	1	1	0	0	1	1
0	1	1	0	1	1	1	1	1	1	0	1	1	1
0	1	1	1	0	0	0	1	1	1	1	0	1	0
0	1	1	1	1	0	0	1	1	1	1	1	1	0

Problem 12.6 In Figure 12.19, the logic network and four decision trees are given.

(a) Justify that these trees correspond to the same logic network
(b) Show that application of Shannon expansion in the root node, and choosing the variable x_2 at the first step of decision tree design is more preferable according to information theoretical criteria

Hint: Calculate $H^S(f_1|x)$, $H^{pD}(f_1|x)$, and $H^{nD}(f_1|x)$, $x = x_{1,2}, x_3, x_4, x_5$ and choose the expansion with minimal entropy; calculate also the information merit of pD and nD expansions.

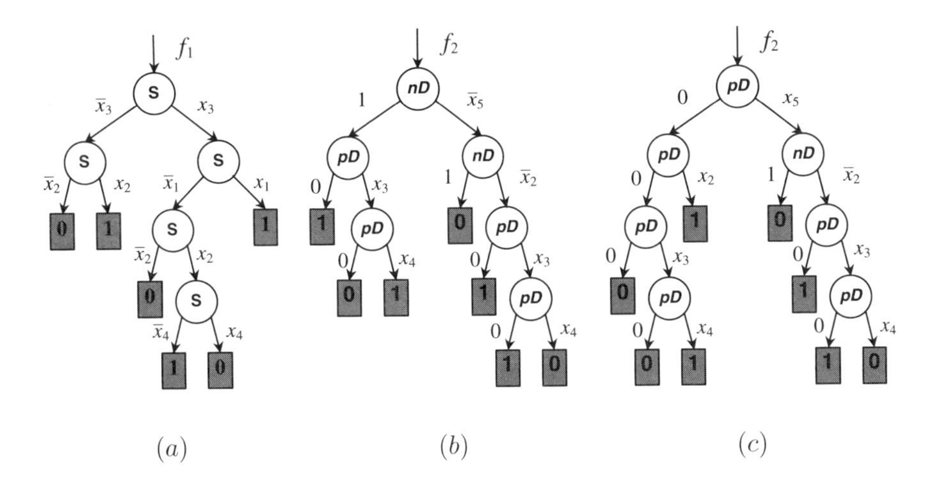

FIGURE 12.19
Two-output five-input logic circuit ordered decision tree based on Shannon expansion (a), free decision tree based on Shannon expansion (b), decision tree based on Davio expansion (c) (Problem 12.6).

Problem 12.7 Based on information theoretical measures, justify that decision tree in Figure 12.20 corresponds to logic circuit.

Problem 12.8 Information is a measurable quantity which is independent of the physical medium by which it is conveyed. The most appropriate measure of information is similar to the measure of entropy. In logic design, Shannon entropy is used. Physicists have emphasized thermodynamics entropy. There is an identity of meaning as well as a form. Much of the published work on information theory discusses the relation between *information thermodynamic* entropies. For example, one can start to study the problem with the paper by Gershenfeld [13]. Entropy is inherently associated with energy and temperature. Entropy in classical thermodynamics is measured as a difference from some arbitrary origin, and this parallels the feature that information also is measured as a *difference* between the state of knowledge of the recipient before and after the communication of information.

Remarks on information measures in decision diagrams

The first approach. For a completely specified switching function $f:\ p(x = 0) = p(x = 1) = 1/2$, and since each node of BDD is an instance of Shannon expansion, a probability assignment algorithm in a down-top fashion works as follows: $p(f) = 1/2p(f|x = 0) + 1/2p(f|x = 1)$.

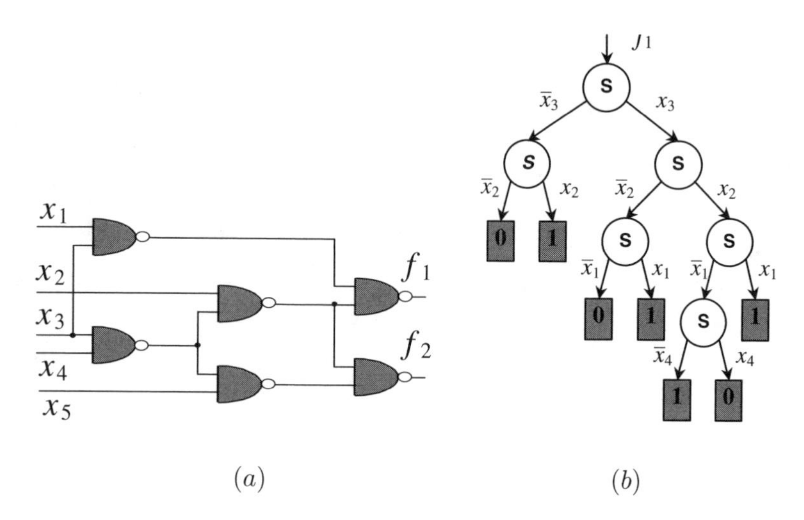

FIGURE 12.20
Two-output five-input logic circuit (a) and ordered decision tree based on Shannon expansion (b) (Problem 12.7).

The second approach. Consider $p(leaf|f=1)=1$ and $p(leaf|f=0)=0$, and output probability $p(f=1)=p(root)$ and $p(f=0)=1-p(root)$. Apply the recursive strategy:

$$p(node) = p(x=0)p(edgel) + p(x=1)p(edger).$$

In this way, calculate conditional and joint probabilities for computing conditional entropy. Thus, for joint probability $p(f=1, x=1)$ it is necessary to set $p(x=1)=1$ and $p(x=0)=0$ before BDD traversal. This allows to calculate the whole range of probabilities using only one BDD traversal.

For example, for the switching function $f = x_3 \vee x_2 \vee x_1$ with truth vector **F**=[10001111]:

$$H(f) = -5/8 \cdot log_2(5/8) - 3/8 \cdot log_2(3/8) = 0.96 \; bit,$$
$$H(f|x1) = -1/8 \cdot log_2(1/4) - 3/8 \cdot log_2(3/4) - 4/8 \cdot log_2(4/4) - 0 = 0.41 \; bit.$$

By the same computations we have $H(f|x_2) = 0.91$ bit, $H(f|x_3) = 0.91$ bit.

A down-top approach with assigning $p(leaf|f=1)=1$ gives us $p(f=1) = p(root) = 0.625$ (BDD with three nodes). The result of setting $p(x_2=0)=1$ is a conditional probability $p(f|x_2=0) = 0.75$.

12.9 Further reading

The problem of decision making in the presence of uncertainty is recognized as being of great importance within the field of logic nanoIC design. Many methods rely on the use of numerical information to handle imperfections.

Historical remarks. The entropy principle of conquering uncertainty has a long history. In 1850, Rudolf Clausius, a German physicist, introduced entropy as a quantity that remains constant in the absence of heat dissipation. Entropy has since been interpreted as the amount of disorder in the system. Indeed, in thermodynamics, entropy is defined as the thermodynamic probability of the internal particles of a system while holding the external properties constant. A hundred years later, in 1948, Shannon suggested a measure to represent the information by a numerical value, nowadays known as *Shannon entropy*. Since then, the term "uncertainty" is interchangable with the term "entropy."

Shannon decomposition. In 1938, Shannon introduced a method for the decomposition of switching functions [32] known as *Shannon expansion*. In state-of-the-art of decision diagram technique, Shannon expansion of a switching function f with respect to a variable x_i is used in the form

$$f = \overline{x}_i f_0 \vee x_i f_1,$$

where $f_0 = f|_{x_i=0}$ and $f_1 = f|_{x_i=1}$. Here $f = f|_{x_i=a}$ denotes the cofactor of f after assigning the constant a to the variable x_i.

Shannon entropy. In 1948, Shannon suggested a measure to represent the information in numerical values, denoted as the *Shannon entropy* [33].

The Shannon information theory has been developed for many applications in circuit design. The latest characterization of a computing system as a communication system is consistent with von Neumann concept of a computer. The bit strings of information are understood as messages to be communicated from a messenger to a receiver. Each message i is an event that has a certain probability of occurrence p_i with respect to its inputs. The measure of information produced when one event is chosen from the set of N events is the entropy of message i: $-\sum_{i \in N} p_i \log p_i$.

In state-of-the-art decision diagram technique, the information theoretical notation of a Shannon expansion of a switching function f with respect to a variable x is used in the form

$$H^S(f|x) = p_{|x=0} \cdot H(f_{|x=0}) + p_{|x=1} \cdot H(f_{|x=1}),$$

where $H(f_{|x=0})$ and $H(f_{|x=1})$ is entropy of function f given $x = 0$ and $x = 1$ respectively.

Fundamentals of information theory. The amount of randomness in a probability distribution is measured by its entropy (or information). In a fundamental sense, the concept of information proposed by Shannon captures only the case when unlimited computing power is available. However, computational cost may play a central role.

This and other aspects of information theory can be found in [3, 5, 19, 24].

Thermodynamics entropy. In communication and logic design, Shannon entropy is used. Physicists have emphasized thermodynamics entropy. Relationship of these different measures has been considered in many papers, for instance, Gershenfeld's [13].

Applications in logic design. The most important results in this field can be found in the book *Artificial Intelligence in Logic Design* edited by S.N. Yanushkevich, Kluwer Academic Publishers, 2004, that includes nine papers on the fundamentals of logic functions manipulation based on artificial intelligence paradigm, evolutionary circuit design, information measures in circuit design, and logic design of nanodevices.

Testing. The analysis is based upon a model where all signals are assumed to have certain statistical properties. The dynamic flavor of entropy has been studied in many papers to express testability (observability and controllability) measures for gate level circuits. For example, Agraval has shown that the probability of the fault detection can be maximized by choosing test patterns that maximize the information at the output [1]. The problem of the construction of sequential fault location for permanent faults has been considered by Varshney et al. [38].

Decision trees and diagrams. Information theoretical measures have been used in [10, 14, 15, 18, 34, 35] in decision trees and diagrams design. Entropy based strategies for minimization of logic functions have been studied in [10, 18, 41]. These results are related to the earlier work by Ganapathy and Rajaraman [12] and Hartmann et al. [15] on conversion of decision tables (truth tables of logic functions) into decision trees. Methods of information theory were used in Popel's study of continuous data representation and multivalued decision diagrams [26].

Power dissipation. Existing techniques for power estimation at gate and circuit levels can be divided in dynamic and static. These techniques rely on probabilistic information the input stream. The average switching activity per node (gate) is the main parameter that needs to be correctly determined. These and related problems are the focus many researchers. For example, in [22, 23, 31], it is demonstrated that the average switching activity in circuit can be calculated using either entropy or information energy averages.

Finite state machines. Most of algorithms for minimization of state assignments in finite state machines target reduced average switching per transition, i.e., average Hamming distance between states. Several papers have used entropy based models to solve the above problem. In particular, Tyagi's paper [37] provides theoretical lower bound on the average Hamming distance per transition for finite state machines based on information theoretical methods.

Search space reduction in optimization. Tomaszewska et al. introduces a two phase algorithm to detect symmetry in logic functions [36]. In the first phase, the search space is reduced by using information properties of symmetric functions. In the second phase, the exact method based on logic differences is applied to recognizing symmetries. The design of the algorithm consists of several formal steps, namely, formal definition of symmetry, deriving the necessary and sufficient conditions to detect the symmetry, measuring the necessary conditions in terms of information (the first phase of the algorithm), and measuring sufficient conditions in terms of logic derivatives (aims to find symmetry in the reduced search space in the second phase of the algorithm). It was demonstrated via experiments that a search space can be reduced by 82% at the first phase. Note that properties of symmetry play an important role in logic design [6, 7, 42]. Also, information measures in optimization technique are considered by Jozwiak [17].

Set of Pairs of Functions to be Distinguished (SPFDs) is a method to represent the flexibility of a node in a multilevel network [40]. An SPFD attached to a node of a network specifies which pairs of primary input minterms can be or have to be distinguished. This can be understood as the information content of the node, since it indicates what information the node contributes to the network.

Related works are [8, 9] where the problem of *synthesis flexibility* based on information theoretical measure is focused. Using an information theoretical approach, it is possible to verify not only that a network achieves the target functionality, but also that this network can be automatically *corrected* to achieve this. Let Net_f be a network implementing the *target function* f and Net_g be the *given network* with the set X of primary inputs. Denote by $V(Net_g)$ the set of outputs, internal outputs and primary inputs of the network Net_g and constants. Conditional entropy $\mathbf{H}(f|V(Net_g)$ is an information measure, reflecting an ambiguity of values of the target function f given network Net_g.

Given function f and network Net_g, there exists a pair $v_i, v_j \in V(Net_g)$ such that

$$\mathbf{H}(f|v_i, v_j) = 0,$$

then there exists a logic function ϕ such that $f = \phi(v_i, v_j)$ and Net_g is 1-neighbor of Net_f. During entropy computation the necessary data to form the truth table of the function ϕ is obtained.

Evolutionary circuit design. There have already been some approaches to evolutionary circuit design. The main ides is that evolutionary strategy would inevitably explore a much richer set of possibilities in the design space that are beyond the scope of traditional methods. In [2, 8, 9, 20, 21] evolutionary strategy and information theoretical measures were used in circuit design (see also references in Chapter 11).

Information engine, computational work, and complexity. A deep and comprehensive analysis of computing systems' information engine has been done by Watanabe [39].

The relationship between function complexity and entropy is conjectured by Cook and Flynn [11]. The complexity of a switching function is expressed by the cost of implementing the function as a combinational network.

Hellerman has proposed so-called *logic entropy* [16]. Computation is considered as a process that reduces the disorder (or entropy) in the space of solutions while finding a result. The number of decisions required to find one correct answer in the space of solutions has been defined as *entropy of computation*, or *logic entropy* calculated as $log\frac{S}{A}$, where s is the number of solutions, A is the number of answers. This definition is consistent with the Shannon entropy provided that the space of solutions is all possible messages (bit strings) of a given length. The answer is one of the messages, so the entropy is the numbers of bits required to specify the correct answer. The term *logical entropy* owes its name to the fact that it depends on the number of logic operations required to perform the computation. In the beginning of the computation, the entropy (disorder) is maximum, at the end of computation the entropy is reduced to zero.

The other form of entropy is *spatial entropy*, and it is relevant to mapping the computation onto a domain where data travels over a physical distance. The data communication process is a process of removal of spatial entropy, while performing logical operations is aimed at removal of logical entropy (disorder). The spatial entropy of a system is a measure of the effort needed to bring data from the input location to the output locations. The removal of the spatial entropy corresponds to reduction of the distance between the input and the output.

Other applications. The paper by Pavlidis et al. [25] is an excellent example of how to apply the information theoretical approach to topological structures. They formulated the problem encoding information on some medium using printed technology as a set of following conflicting requirements: the code to have a high density of information, to read the code reliable, to minimize the cost of the printing process, and to minimize the cost of the reading equipment.

The results on entropy-based analysis of natural language processing have been reported by Berger [4].

In machine learning, information theory has been recognized as a useful criterion [27]. To classify objects from knowledge of a training set of examples whose classes are previously known, a decision tree rule induction method known as the ID3 algorithm was introduced by Quinlan [28]. The method is based on recursive partitioning of the sample space and defines classes structurally by using decision trees. A number of improved algorithms exist such as C4.5, C5, CHAID and CART which use general to specific learning in order to build simple knowledge based systems by inducing decision trees from a set of examples [29, 30], and the method of quantitatively information of logical expressions developed by Zhong and Ohsuga [43].

12.10 References

[1] Agraval V. An information theoretic approach to digital fault testing. *IEEE Transactions on Computers*, 30(8):582–587, 1981.

[2] Aguirre AH, and Coello CA. Evolutionary synthesis of logic circuits using information theory. In: Yanushkevich SN., Ed., *Artificial Intelligence in Logic Design.* Kluwer, Dordrecht, pp. 285–311, 2004.

[3] Ash RB. *Information Theory.* John Wiley & Sons, New York, 1967.

[4] Berger A, Pietra SD, and Pietra VD. A maximum entropy approach to natural language processing. *Computational Linguistics*, 22(1):39–71, 1996.

[5] Brillouin L. *Science and Information Theory.* Academic Press, New York, 1962.

[6] Butler J, Dueck G, Shmerko V, and Yanushkevich S. On the number of generators of transeunt triangles. *Discrete Applied Mathematics*, 108:309–316, 2001.

[7] Butler J, Dueck G, Shmerko V, and Yanushkevich S. Comments on SYMPATHY: fast exact minimization of fixed polarity Reed-Muller expansion for symmetric functions. *IEEE Transactions on Computer-Aided Design of Integrated Circuits and Systems*, 19(11):1386–1388, 2000.

[8] Cheushev VA, Yanushkevich SN, Moraga C, and Shmerko VP. Flexibility in logic design. An approach based on information theory methods. *Research Report 741*, Forschungsbericht, University of Dortmund, Germany, 2000.

[9] Cheushev VA, Yanushkevich SN, Shmerko VP, Moraga C, and Kolodziejczyk J., Remarks on circuit verification through the evolutionary circuit design. In *Proceedings IEEE 31st International Symposium on Multiple-Valued Logic*, pp. 201–206, 2001.

[10] Cheushev VA, Shmerko VP, Simovici D., and Yanushkevich SN. Functional entropy and decision trees. In *Proceedings: IEEE 28th International Symposium on Multiple-Valued Logic*, pp. 357–362, Japan, 1998.

[11] Cook RW, and Flynn MJ. Logical network cost and entropy. *IEEE Transactions on Computers*, 22(9):823-826, 1973.

[12] Ganapathy S, and Rajaraman V. Information theory applied to the conversion of decision tables to computer programs. *Communications of the ACM*, 16:532–539, 1973.

[13] Gershenfeld N. Signal entropy and the thermodynamics of computation. *IBM Systems Journal*, 35:577–586, 1996.

[14] Goodman RM, and Smyth P. Decision tree design from a communication theory standpoint. *IEEE Transactions on Information Theory*, 34(5):979–994, 1988.

[15] Hartmann CRP, Varshney PK, Mehrotra KG, and Gerberich CL. Application of information theory to the construction of efficient decision trees. *IEEE Transactions on Information Theory*, 28(5):565–577, 1982.

[16] Hellerman L. A measure of computation work. *IEEE Transactions on Computers*, 21(5):439–446, 1972.

[17] Jozwiak L. Information relationships and measures in application to logic design. In *Proceeding IEEE 29th International Symposium on Multiple-Valued Logic*, pp. 228–235, Freiburg, Germany, 1999.

[18] Kabakcioglu AM, Varshney PK, and Hartman CRP. Application of information theory to switching function minimization. *IEE Proceedings*, Pt E, 137(5):389–393, 1990.

[19] Lo H, Spiller T, and Popescu S. *Introduction to Quantum Computation and Information*. World Scientific, Hackensack, NJ, 1998.

[20] Łuba T, Moraga C, Yanushkevich SN, Shmerko VP, and Kolodziejczyk J. Application of design style in evolutionary multi-level networks synthesis. In *Proceedings IEEE Symposium on Digital System Design*, pp. 156–163, Maastricht, Netherlands, 2000.

[21] Luba T, Moraga C, Yanushkevich SN, Opoka M, and Shmerko VP. Evolutionary multi-level network synthesis in given design style. In *Proceeding IEEE 30th International Symposium on Multiple-Valued Logic*, pp. 253–258, 2000.

[22] Marculescu D, Marculesku R, and Pedram M. Information theoretic measures for power analysis. *IEEE Transactions on Computer Aided Design of Integrated Circuits and Systems*, 15(6):599–610, 1996.

[23] Marculescu R, Marculesku D, and Pedram M. Sequence compaction for power estimation: theory and practice. *IEEE Transactions on Computer Aided Design of Integrated Circuits and Systems*, 18(7):973–993, 1999.

[24] Martin NFG, and England JW, *Mathematical Theory of Entropy*. Addison-Wesley, Reading, MA, 1981.

[25] Pavlidis T, Szwartz J, and Wang Y. Information encoding with two-dimensional bar codes. *Computer*, 18–28, June, 1992.

[26] Popel DV. Conquering uncertaity in multiple-valued logic design. Evolutionary synthesis of logic circuits using information theory. *Artificial Intelligence Review, the International Journal*, Kluwer Academic Publishers, 20(3-4):419–433, 2003.

[27] Principe JC, Fisher III JW, and Xu D. Information theoretic learning. In Simon Haykin, Ed., *Unsupervised Adaptive Filtering*, John Wiley & Sons, New York, 2000.

[28] Quinlan JR. Induction of decision trees. In *Machine Learning*, Vol. 1, pp. 81-106, Kluwer Academic Press, Dordrecht, 1986.

[29] Quinlan JR. Probabilistic Decision Trees. In Kockatoft Y, and Michalshi R. Eds.*Machine Learning, Vol. 3: An AI Approach*, Kluwer Academic Press, Dordrecht, 1990, pp. 140-152.

[30] Quinlan JR. Improved use of continuos attributes in C4.5. *Journal of Artificial Intelligence Research*, 4:77–90, 1996.

[31] Ramprasad S, Shanbhag NR, and Hajj IN. Information-theoretic bounds on average signal transition activity. *IEEE Transactions on Very Large Scale Integration (VLSI) Systems*, 7(3):359–368, 1999.

[32] Shannon C. A Symbolic analysis of relay and switching circuits. *Transactions AIEE*, 57:713-723, 1938.

[33] Shannon C. A Mathematical theory of communication. *Bell Systems Technical Journal*, 27:379–423, 623–656, 1948.

[34] Shmerko VP, Popel DV, Stanković RS, Cheushev VA, and Yanushkevich SN. Entropy based algorithm for 4-valued functions minimization. In *Proceedings IEEE 30th International Symposium on Multiple-Valued Logic*, pp. 265–270, 2000.

[35] Shmerko VP, Popel DV, Stanković RS, Cheushev VA, and Yanushkevich SN. Information theoretical approach to minimization of AND/EXOR

expressions of switching functions. In *Proceedings IEEE International Conference on Telecommunications*, pp. 444–451, Yugoslavia, 1999.

[36] Tomaszewska A, Dziurzanski P, Yanushkevich SN, and Shmerko VP. Two-phase exact detection of symmetries, In *Proceeding IEEE 31st International Symposium on Multiple-Valued Logic*, pp. 213–219, 2001.

[37] Tyagi A. Entropic bounds of FSM switching. *IEEE Transactions on Very Large Scale Integration (VLSI) Systems*, 5(4):456–464, 1997.

[38] Varshney P, Hartmann C, and De Faria J. Application of information theory to sequential fault diagnosis. *IEEE Transactions on Computers*, 31:164–170, 1982.

[39] Watanabe H. A basic theory of information network. *IEICE Transactions Fundamentals*, E76-A(3):265-276, 1993.

[40] Yamashita S, Sawada H, and Nagoya A. SPFD: a method to express functional flexibility. *IEEE Transactions on Computer-Aided Design of Integrated Circuits and Systems*, 19(8):840–849, 2000.

[41] Yanushkevich SN, Shmerko VP, Dziurzanski P, Stanković RS, and Popel DV. Experimental verification of the entropy based method for minimization of switching functions on pseudo-ternary decision trees. In *Proceedings IEEE International Conference on Telecommunications in Modern Satellite, Cable and Broadcasting Services*, pp. 452–459, Yugoslavia, 1999.

[42] Yanushkevich S, Butler J, Dueck G, and Shmerko V. Experiments on FPRM expressions for partially symmetric logic functions. In *Proceedings IEEE 30th International Symposium on Multiple-Valued Logic*, pp. 141–146, 2000.

[43] Zhong N, and Ohsuga S. On information of logical expression and knowledge refinement. *Transactions of Information Processing Society of Japan*, 38(4):687–697, 1997.

Index

D

E

F

G

H

I

T

U

V

W

Z